PERGAMON MATERIALS SERIES
VOLUME 1

CALPHAD (Calculation of Phase Diagrams): A Comprehensive Guide

PERGAMON MATERIALS SERIES

Series Editor: Robert W. Cahn FRS
Department of Materials Science and Metallurgy, University of Cambridge, UK

Vol. 1 **CALPHAD (Calculation of Phase Diagrams): A Comprehensive Guide**
by N. Saunders and A. P. Miodownik

A selection of further titles:

Non-equilibrium Processing of Materials edited by C. Suryanarayana
Phase Transformations in Titanium- and Zirconium-based Alloys
by S. Banerjee and P. Mukhopadhyay
Wettability at High Temperatures by N. Eustathopoulos, M. G. Nicholas and B. Drevet
Ostwald Ripening by S. Marsh
Nucleation by A. L. Greer and K. F. Kelton
Underneath the Bragg Peaks: Structural Analysis of Complex Materials
by T. Egami and S. J. L. Billinge
The Coming of Materials Science by R. W. Cahn

PERGAMON MATERIALS SERIES

CALPHAD

Calculation of Phase Diagrams

A Comprehensive Guide

by

N. Saunders

Thermotech Ltd., Guildford, UK

and

A. P. Miodownik

Professor Emeritus,
School of Mechanical and Materials Engineering,
University of Surrey, Guildford, UK

PERGAMON

UK Elsevier Science Ltd, The Boulevard, Langford Lane, Kidlington, Oxford OX5 1GB
USA Elsevier Science Inc., 655 Avenue of the Americas, New York, NY 10010, USA
JAPAN Elsevier Science Japan, 9–15 Higashi-Azabu I-chome, Minato-ku, Tokyo 106, Japan

---

Library of Congress Cataloging in Publication Data

Saunders, N. (Nigel)
CALPHAD (calculation of phase diagrams) : a comprehensive guide / by N. Saunders and A. P. Miodownik.
p. cm. — (Pergamon materials series : v. 1)
Includes bibliographical references.
ISBN 0-08-042129-6 (alk. paper)
1. Phase diagrams—Data processing. 2. Thermochemistry—Data processing. I. Miodownik, A. P. (A. Peter) II. Title.
III. Series.
QD503.S265 1998
530.4'74--DC21 98–15693
CIP

British Library Cataloguing in Publication Data

A catalogue record for this book is available from the British Library

ISBN 0-08-0421296

Transferred to digital printing 2005

Contents

CHAPTER 5
THERMODYNAMIC MODELS FOR SOLUTION AND COMPOUND PHASES 91

CHAPTER 6
PHASE STABILITIES 129

Series preface

My editorial objective in this new series is to present to the scientific public a collection of texts that satisfies one of two criteria: the systematic presentation of a specialised but important topic within materials science or engineering that has not previously (or recently) been the subject of full-length treatment and is in rapid development; or the systematic account of a broad theme in materials science or engineering. The books are not, in general, designed as undergraduate texts, but rather are intended for use at graduate level and by established research workers. However, teaching methods are in such rapid evolution that some of the books may well find use at an earlier stage in university education.

I have long editorial experience both in covering the whole of a huge field — physical metallurgy or materials science and technology — and in arranging for specialised subsidiary topics to be presented in monographs. My intention is to apply the lessons learnt in more than 35 years of editing to the objectives stated above. Authors have been invited for their up-to-date expertise and also for their ability to see their subjects in a wider perspective.

I am grateful to Elsevier Science Ltd., who own the Pergamon Press imprint, and equally to my authors, for their confidence.

The first book in the series, presented herewith, is on a theme of major practical importance that is developing fast and has not been treated in depth for over 25 years. I commend it confidently to all materials scientists and engineers.

ROBERT W. CAHN, FRS
(Cambridge University, UK)

Preface

This book could not have been written without the help and encouragement of many individuals within the CALPHAD community who have given valuable viewpoints through technical conversations, supplying references and responding to questionnaires. First and foremost we would like to acknowledge Larry Kaufman without whose pioneering efforts this community would never have come into existence, and thank Robert Cahn for his support and efforts in ensuring that this book was finally written. Special thanks go to Imo Ansara (Grenoble) for his input during the early stages of the book, and Catherine Colinet (Grenoble) and Ursula Kattner (NIST) for their kind efforts in reading and commenting on the chapters concerning ordering and computational methods. One of us (NS) would also like to thank Bo Sundman for providing valuable insight to the chapter on Thermodynamic Models and arranging for a visiting position at The Royal Institute of Technology, Stockholm, Sweden, where it was possible to begin work, in earnest, on the book. We would also like to acknowledge the long-standing collaboration we have had with Colin Small (Rolls-Royce plc) who has been a tireless supporter for the use of CALPHAD in real industrial practice.

We are also indebted to the University of Surrey (APM/NS) and the University of Birmingham (NS) for allowing us the academic freedom to explore many of the strands that have been woven into the fabric of this book and for providing invaluable library facilities.

Last, and by no means least, we would like to thank our families for the forbearance they have shown during the many hours we have spent working on this book.

NIGEL SAUNDERS
PETER MIODOWNIK

Foreword

The comprehensive guide to the development and application of the CALculation of PHAse Diagram (CALPHAD) techniques presented by Saunders and Miodownik is a unique and up-to-date account of this rapidly expanding field, since it was combined with computer methods in the early 1970s. The most distinctive character of the methodology is its aim to couple the phase diagrams and thermochemical properties in an attempt to explicitly characterise all of the possible phases in a system. This includes phases that are stable, metastable and unstable over the widest possible range of temperature, pressure and composition. In contrast to treatments which attempt to locate phase boundaries by the application of factors such as a critical electron–atom ratio, electron density or electron vacancy number, the thermochemical basis of the CALPHAD approach rests explicitly on the notion that the phase boundary is the result of competition between two or more competing phases. This important distinction has provided a much wider horizon and a much more fertile field for exploitation and testing than narrower contemporary theories.

Right from the beginning, the acronym CALPHAD was chosen to signal the vision of the early 'Calphadians', who chose this very challenging problem as a focus for their energies. The history, examples and references that the authors have interwoven in this book provide a clear insight into the remarkable progress achieved during the past 25 years. They have captured the flavor of the early years, showing how the practitioners of the CALPHAD method developed the computer techniques, databases and case studies required to create a new field of intellectual and industrial activities, which is now applied vigorously all over the world. The authors guide the interested reader through this extensive field by using a skilful mixture of theory and practice. Their work provides insight into the development of this field and provides a reference tool for both the beginner and any accomplished worker who desires to assess or extend the capabilities that the CALPHAD methodology brings to a host of new multi-component systems.

LARRY KAUFMAN
(*Cambridge, Massachusetts, USA*)

Chapter 1

Introduction

This book is intended to be a comprehensive guide to what has become known as CALPHAD. This is an acronym for the CALculation of PHAse Diagrams but it is also well defined by the sub-title of the CALPHAD journal, *The Computer Coupling of Phase Diagrams and Thermochemistry*. It is this coupling which, more than any other factor, defines the heart of this subject area.

Phase diagrams have mainly been the preserve of a limited number of practitioners. This has been partly due to the difficulties many scientists and engineers have in interpreting them, especially at the ternary and higher-order level. Their use has also been seen as rather academic, because almost all real materials are multi-component in nature and phase diagrams are generally used to represent only binary and ternary alloys. The CALPHAD method has altered this viewpoint because it is now possible to predict the phase behaviour of highly complex, multi-component materials based on the extrapolation of higher-order properties from their lower-order binary and ternary systems. Furthermore, the method can be coupled with kinetic formalisms to help understand and predict how materials behave in conditions well away from equilibrium, thus considerably enhancing its value.

One of the objectives of this book is to act as a benchmark for current achievements, but it also has a number of other important general objectives. Despite the undoubted success of the CALPHAD approach, any methodology based on thermodynamic concepts is often erroneously perceived as being difficult to follow, and even considered as unlikely to have a direct practical application because of its association with equilibrium situations. The authors have therefore set themselves the goals of removing such misconceptions and making the book readable by any scientifically competent beginner who wishes to apply or extend the concepts of CALPHAD to new areas.

The book begins with a chapter describing the history and growth of CALPHAD. This provides a useful point of departure for a more detailed account of the various strands which make up the CALPHAD approach. Chapters 3 and 4 then deal with the basic thermodynamics of phase diagrams and the principles of various experimental techniques. This is because one of basic pillars of the CALPHAD approach is the concept of coupling phase diagram information with all other available thermodynamic properties. It is a key factor in the assessment and characterisation of the lower-order systems on which the properties of the higher-order systems are based. In order to optimise such coupling, it is not only necessary to understand the assumptions which are made in the thermodynamic functions being used, but also to grasp the inherent level of errors associated with the various experimental techniques that are used to determine both phase diagrams and associated thermodynamic properties.

Chapter 3 defines thermodynamic concepts, such as ideal/non-ideal behaviour, partial/integral quantities and simple regular solution theory. This allows the relationship between the general topological features of phase diagrams and their underlying thermodynamic properties to be established and acts as a stepping stone for the discussion of more realistic models in Chapter 5. Chapter 4 deals with the advantages and disadvantages of various methods for the determination of critical points and transus lines as well as enthalpies of formation, activities, heat capacities and associated properties. The complementary nature of many measurements is stressed, together with the need to combine the various measurements into one overall assessment. This chapter is not intended to be a treatise on experimental methods but provides a necessary background for the intelligent assessment of experimental data, including some appreciation of why it may be necessary to include weighting procedures when combining data from different sources.

Chapter 5 examines various models used to describe solution and compound phases, including those based on random substitution, the sub-lattice model, stoichiometric and non-stoichiometric compounds and models applicable to ionic liquids and aqueous solutions. Thermodynamic models are a central issue to CALPHAD, but it should be emphasised that their success depends on the input of suitable coefficients which are usually derived empirically. An important question is, therefore, how far it is possible to eliminate the empirical element of phase diagram calculations by substituting a treatment based on first principles, using only wave-mechanics and atomic properties. This becomes especially important when there is an absence of experimental data, which is frequently the case for the metastable phases that have also to be considered within the framework of CALPHAD methods.

Chapter 6 therefore deals in detail with this issue, including the latest attempts to obtain a resolution for a long-standing controversy between the values obtained by thermochemical and first-principle routes for so-called 'lattice stabilities'. This chapter also examines (i) the role of the pressure variable on lattice stability, (ii) the prediction of the values of interaction coefficients for solid phases, (iii) the relative stability of compounds of the same stoichiometry but different crystal structures and (iv) the relative merits of empirical and first-principles routes.

Another area where empirical, semi-empirical and more fundamental physical approaches overlap is in the case of ordering processes. Chapter 7 carefully analyses the advantages and disadvantages of many competing methods that have all been used in determining the relationship between the degree of structural order, composition and temperature, including the important question of short-range order. The chapter includes a description of mainstream ordering models such as the Monte Carlo method (MC), the Bragg–Williams–Gorsky (BWG) model and the Cluster Variation Method (CVM) as well as a number of hybrid and empirical routes. There is also reference to the role of lattice vibrations and the effect of kinetic stabilisation of ordered states that are not always considered under this heading.

Magnetic ordering is dealt with separately in Chapter 8. This is because it is necessary to have a very accurate description of the magnetic component of the Gibbs energy in order to have any chance of accounting for the phase transformations in ferrous materials, which are still one of the most widely used material types in existence. Furthermore, it is necessary to handle situations where the end-members of the system exhibit different kinds of magnetism. The various effects caused by both internal and external magnetic fields are enumerated, and the reasons for using different approaches to structural and magnetic ordering are also discussed.

Although the previous three chapters have been concerned with placing the CALPHAD methodology on a sound physical basis, the over-riding objective of such a method is to generate a reliable and user-friendly output that reflects various properties of multicomponent industrial materials. The last three chapters therefore return to the more practical theme of how this can be achieved.

Chapter 9 deals with the general principles of computational thermodynamics, which includes a discussion of how Gibbs energy minimisation can be practically achieved and various ways of presenting the output. Optimisation and, in particular, 'optimiser' codes, such as the Lukas programme and PARROT, are discussed. The essential aim of these codes is to reduce the statistical error between calculated phase equilibria, thermodynamic properties and the equivalent experimentally measured quantities.

Chapter 10 provides an exhaustive description of how these techniques can be applied to a large number of industrial alloys and other materials. This includes a discussion of solution and substance databases and step-by-step examples of multi-component calculations. Validation of calculated equilibria in multi-component alloys is given by a detailed comparison with experimental results for a variety of steels, titanium- and nickel-base alloys. Further selected examples include the formation of deleterious phases, complex precipitation sequences, sensitivity factor analysis, intermetallic alloys, alloy design, slag, slag-metal and other complex chemical equilibria and nuclear applications.

Although Chapter 10 clearly validates CALPHAD methodology for the calculation of phase equilibria in complex industrial alloys, there are many processes that depart significantly from equilibrium. There may be a systematic but recognisable deviation, as in a casting operation, or quite marked changes may occur as in the metastable structures formed during rapid solidification, mechanical alloying or vapour deposition. The limitations of a 'traditional' CALPHAD approach have long been understood and a combination of thermodynamics and kinetics is clearly a logical and desirable extension of the CALPHAD methodology, especially if these are designed to use the same data bases that have already been validated for equilibrium calculations. The penultimate chapter therefore describes a number of successful applications that have been made in treating deviations from equilibrium, for both solid- and liquid-phase transformations, including the Scheil-Gulliver solidification model and its various modifications.

The final chapter presents a view of where CALPHAD may go in the future. Such a process involves, by its very nature, some rather personal views. For this we do not apologise, but only hope that much of what is suggested will be achieved in practice when, at some point in time, other authors attempt a review of CALPHAD.

Chapter 2

History of CALPHAD

Chapter 2
History of CALPHAD

2.1. INTRODUCTION

The history of CALPHAD is a chronology of what can be achieved in the field of phase equilibria by combining basic thermodynamic principles with mathematical formulations to describe the various thermodynamic properties of phases. The models and formulations have gone through a series of continuous improvements and, what has become known as the CALPHAD approach, is a good example of what can be seen as a somewhat difficult and academic subject area put into real practice. It is indeed the art of the possible in action and its applications are wide and numerous.

The roots of the CALPHAD approach lie with van Laar (1908), who applied Gibbs energy concepts to phase equilibria at the turn of the century. However, he did not have the necessary numerical input to convert his algebraic expressions into phase diagrams that referred to real systems. This situation basically remained unchanged for the next 50 years, especially as an alternative more physical approach based on band-structure calculations appeared likely to rationalise many hitherto puzzling features of phase diagrams (Hume-Rothery *et al.* 1940).

However, it became evident in the post-war period that, valuable as they were, these band-structure concepts could not be applied even qualitatively to key systems of industrial interest; notably steels, nickel-base alloys, and other emerging materials such as titanium and uranium alloys. This led to a resurgence of interest in a more general thermodynamic approach both in Europe (Meijering 1948, Hillert 1953, Lumsden 1952, Andrews 1956, Svechnikov and Lesnik 1956, Meijering 1957) and in the USA (Kaufman and Cohen 1956, Weiss and Tauer 1956, Kaufman and Cohen 1958, Betterton 1958). Initially much of the work related only to relatively simple binary or ternary systems and calculations were performed largely by individuals, each with their own methodology, and there was no attempt to produce a co-ordinated framework.

2.2. THE EARLY YEARS

Meijering was probably the first person to attempt the calculation of a complete ternary phase diagram for a real system such as Ni–Cr–Cu (Meijering 1957). This example was particularly important because mere interpolation of the geometric features from the edge binary systems would have yielded an erroneous diagram. It was also a pioneering effort in relation to the concept of lattice stabilities; Meijering realised that to make such a calculation possible, it was necessary to deduce a value for the stability of f.c.c. Cr, a crystal structure that was not directly accessible by experiment.

His attempt to obtain this value by extrapolation from activity measurements was an important milestone in the accumulation of such lattice-stability values. That these early results have been only marginally improved over the years (Kaufman 1980) is quite remarkable, considering the lack of computing power available at that time. Apart from correctly reproducing the major features of the phase diagram concerned, it was now possible to give concrete examples of some of the thermodynamic consequences of phase-diagram calculations. Band theory was stressing the electronic origin of solubility limits, but the thermodynamic approach clearly showed that the solubility limit is not merely a property of the solution concerned but that it also depends on the properties of the coexisting phases and, of course, also on temperature. It was also shown that a retrograde solidus has a perfectly sound explanation, and "did not fly in the face of natural law" (Hansen 1936).

The capacity to give a *quantitative* description of all the topological features of phase diagrams was to be the crucial issue which ultimately convinced the scientific community of the necessity for a global thermodynamic approach. Developments in the USA were complementary to the work in Europe in several crucial respects. Firstly, the activity was initially linked to practical problems associated with steels (Kaufman and Cohen 1956, 1958). Secondly, there was, from the outset, a vision of producing an extensive database from which phase diagrams could be calculated on a permutative basis. The single-minded determination to combine these two aspects, and to gather together all the major workers in the field on a world-wide basis, must be considered to be the basic driving force for the eventual emergence of CALPHAD. It was, however to take some 15 years before the concept was officially realised.

Such a lengthy incubation period between vision and fruition seems very long in retrospect, but is on par with similar developments in other areas of science and technology. It reflects the time taken for individuals to meet each other and agree to work together and also the time taken for the scientific and technological community to devote adequate funds to any new activity. Thus a conference on the physical chemistry of solutions and intermetallic compounds, convened in 1958 at the National Physical Laboratory, did not lead to any noticeable increase in communal activities, although it did marginally increase a number of bilateral contacts.

In parallel to Kaufman and co-workers in the USA, work was taking place in Sweden stemming from the appointment of Mats Hillert to the Royal Institute of Technology in Stockholm in 1961. Kaufman and Hillert studied together at MIT and Hillert was very keen to make the use of thermodynamic calculations a key

References are listed on pp. 26–29.

feature of the department there. There was also considerable encouragement from both the government and the steel industry for such a programme of work. Long-term funding was made available and in the mid-1960s a group from Tohuku University at Sendai, Japan, were amongst the early workers that took part in this work. They brought with them experimental techniques that were to prove invaluable in validating the early calculations of Hillert *et al.* (1967). On their return to Japan, these workers, who included Professor Nishizawa, formed another centre which furthered such thermodynamic activities in their home country (Hasebe and Nishizawa 1972, Ishida *et al.* 1989, Nishizawa 1992).

Work on the characterisation of steels started with quite rudimentary polynomial descriptions of the thermodynamic properties of iron. A more fundamental approach which could include magnetism was developed by (Weiss and Tauer 1956, Tauer and Weiss 1958) and eventually led to a quite sophisticated model (Kaufman *et al.* 1963). However, magnetic alloys were clearly not the best test vehicle with which to convince a sceptical scientific community of the value of a general thermodynamic approach, although a knowledge of the so-called lattice stabilities was undoubtedly the vital ingredient required to create a viable database. Alloys based on titanium therefore seemed a more likely starting point for the development of a database and a project therefore began with characterisation of Ti and Zr pressure/temperature diagrams (Kaufman 1959a).

The projected database also required lattice stabilities for elements in crystal structures that were sufficiently unstable so as not to appear under attainable combinations of temperature and pressure. The next few years saw the discovery and application of methods for deducing working values of these key parameters (see Chapter 6). In addition, the calculation of phase diagrams also required a complementary set of interaction parameters between the various elements or species in the system. Although these were available for specific systems, the production of a whole set of these parameters would be a time-consuming step. Moreover the situation in iron-base alloys was complicated by the presence of magnetic effects (see Chapter 8). Kaufman therefore decided to test his general methodology on a series of non-magnetic transition metal binary alloys using an ideal solution model, in which the interaction parameters are set to zero (Kaufman 1967). Despite this rather drastic assumption, the work presented at this Battelle meeting constituted the first demonstration of the potential power of the CALPHAD approach.

The use of an ideal-solution model meant that there were a number of instances where calculated and experimental results were quantitatively at variance. However, the approach very successfully predicted the general form of most of the phase diagrams, for example whether they were peritectic or eutectic, and accounted for the appearance of intermediate phases in systems such as Cr–Rh. That the approach could do this using such simple and internally self-consistent models is a demonstration of the inherent power of CALPHAD methods. The importance of this first step therefore cannot be overestimated, although its significance was not appreciated by many members of the scientific community at the time. In fact these initial results were criticised by both experimentalists and theoreticians. The former considered the calculations too inaccurate and were in general not prepared to consider the potential improvements of using a more elaborate system of interaction coefficients. The latter were mostly concerned with potential errors in the selected values of the lattice stabilities and the supposed superiority of a more fundamental approach based on band structure calculations. These initial criticisms required very persistent refutations and the criticisms from some of the theoreticians were somewhat galling as the band-structure approach at this time had no possibility of handling phase equilibria at anything other than absolute zero.

Not surprisingly, Kaufman found the Battelle meeting discouraging. The values of lattice stabilities proposed by Brewer (1967) were substantially different to those proposed on thermodynamic grounds, but there was at least the possibility of a dialogue on this issue, and so Kaufman (1966a) wrote, in a letter to Brewer, that

> ...*"There have been the same tired old band theory discussions and little seems to have happened over the last thirty years; the attendees do not seem to be aware of the most important directions for attacking the problem."*

Although this attempt to launch the CALPHAD approach in the scientific community at large did not seem to have much success, Kaufman had by this time considerably extended and improved his modelling. He had incorporated regular solutions together with a semi-empirical method for deriving the necessary interaction coefficients from other atomic parameters (see Chapter 6). Details of this procedure were published as an Air Force Contract Report (Kaufman 1959b), but it was clearly desirable to make it available in the open literature. Providentially, letters were being exchanged at this time between Kaufman and Hume-Rothery (Kaufman 1966b) concerning the thermodynamic representation of liquidus and solidus lines in iron-base alloys (Buckley and Hume-Rothery 1963), which resulted in a modified approach by these authors (Sinha and Hume-Rothery 1967). This correspondence evolved into a lengthy discussion on the mode of presentation of thermodynamic equilibria in general, and the difficulties of changing prevailing views in particular, which ended in the commissioning of a review.

Hume-Rothery was to prove a fair, if demanding, editor, and the result was an important review on the stability of metallic phases as seen from the CALPHAD viewpoint (Kaufman 1969). The relevant correspondence provides a fascinating insight into his reservations concerning the emerging framework that Kaufman had in mind. Hume-Rothery had spent most of his life on the accurate determination of experimental phase diagrams and was, in his words (Hume-Rothery 1968), "...not unsympathetic to any theory which promises reasonably accurate calculations of phase boundaries, and saves the immense amount of work which their experimental determination involves".

But while Hume-Rothery agreed that increasing accuracy had to be obtained by stages, he held on to his opinion that the calculations being presented were inadequate in many specific cases (e.g Mo–Rh). His view was that while it was immensely valuable to make the calculations and see what happened it would be difficult to persuade him that such calculations represented real progress and that the methodology should be included in the proposed review. Interestingly he concluded (Hume-Rothery 1968) "... however I would like to be convinced of the contrary!".

Kaufman took immense pains to restate his position and counter Hume-Rothery's objections, sensing that this was indeed a crucial watershed in gaining acceptance for his point of view. He agreed that the task of computing phase equilibria accurately was a very difficult one, and that it might even be impossible. Nonetheless it seemed clear that the preconditions for tackling this challenge involved certain axiomatic requirements. Firstly, an emphasis on the competition between all principal phases; secondly, the provision of an explicit description which could be applied to many systems; and thirdly, the availability of a numerical input which could provide a quantitative description applicable over a wide range of temperature, composition and pressure. It was clear that his current description of phase equilibria did not satisfy these requirements perfectly, but it did offer an idealised description of the desired ultimate framework. No previous attempt had simultaneously included lattice stability descriptions for an extensive number of the elements, a generalised treatment of interaction coefficients in terms of group number, size factor and internal pressure contributions, combined together with a suitable Gibbs energy minimisation routine (Kaufman 1968a).

Kaufman (1968b) also made it clear that the use of more realistic descriptions, such as sub-regular solution models, would necessitate the determination of many more parameters and thought that: "Until such time as our knowledge of solution theory and the physical factors which control the properties of solutions might permit these parameters to be determined, it is better to continue with a simpler model." This conclusion was of course also conditioned by the limited computer memory available at the time, which prevented the use of more complex models with the subsequent increase in number of parameters which they entailed.

Throughout the editorial stages of the emerging review it was continually necessary to spell out the differences between (a) the use of an ideal solution model, (b) the use of a regular solution model with parameters derived solely from atomic properties and finally (c) the use of interaction parameters derived by feedback from experiment. A proper understanding of the differences between these three approaches lay at the heart of any realistic assessment of the value of calculations in relation to experimentally determined diagrams.

The discussion between Hume-Rothery and Kaufman also dealt with the apparent conflict between different sets of competing lattice stabilities. It was particularly worrying to him that Brewer (1967) and Engel (1964) proposed values based on spectroscopic data that could differ by almost an order of magnitude in some instances. To an outsider, these were simply alternatives which had equal validity, because the effect of changing such values on phase diagram calculations was not immediately obvious. To Kaufman, with much computing experience behind him, some of the consequences of adopting the Brewer values were totally unacceptable. In particular, Kaufman found difficulty in accepting some of the resulting low melting points for metastable phases and the subsequent requirement that there should be highly composition-dependent interaction parameters in the vicinity of the pure metals (Brewer 1967). All this was aggravated by the fact that the methods used by Brewer and Engel were not transparent enough to allow other workers to make their own calculations. Kaufman was very concerned at the uphill struggle required to establish an acceptable set of lattice stabilities and was disappointed that there had not been a widespread appreciation of this concept (Kaufman 1968c).

Lattice-stability values obtained by electron energy calculations also differed from those obtained by thermochemical routes, but at that time such calculations were still at a relatively rudimentary stage and it was assumed that the two sets of values would eventually be related. However, there is no doubt that lack of agreement in such a fundamental area played a part in delaying a more general acceptance of the CALPHAD methodology.

1970 saw the publication of the first textbook devoted to the quantitative computation of phase diagrams (Kaufman and Bernstein 1970) whose appeal was shown by its translation into several languages, including Russian in 1972. Some reviewers, however, evidently still thought that the approach was much too empirical and had no real physical basis (Walbaum 1972). With hindsight, it is easy to see that the slow acceptance of the emerging CALPHAD technique was due to its attempt to occupy a semi-empirical middle ground, sandwiched between a rigorously physical basis and experimentally determined equilibria. Many scientists were certainly not yet willing to accept that such pragmatism was necessary in order to allow the method to be extended to multi-component systems.

The other major hurdle in the early 1970s was to find a way of tackling the industrially important case of steels. All around the world researchers were being confronted with the need to use markedly non-linear free-energy expressions, which were associated with the magnetism of iron, and there were substantial problems in incorporating such terms into the formalism for non-magnetic alloys used at the time. Although a breakthrough appeared to have been made for the non-magnetic transition elements (Pettifor 1969a, 1969b) by 1971, it was still thought that real progress would not be achieved in this area until more sophisticated band-structure calculations became available (Ducastelle 1971).

In order to progress with steels, it was necessary to pursue semi-empirical methods (Weiss and Tauer 1956, Tauer and Weiss 1958) which culminated in the seminal paper describing the thermodynamics of iron (Kaufman *et al.* 1963). This included the concept of two different magnetic states for f.c.c. iron, which had been applied to rationalise the Invar properties of iron-nickel alloys (Weiss 1963). However, this concept was viewed with considerable scepticism. Some support

References are listed on pp. 26–29.

eventually emerged from band-structure calculations (Roy and Pettifor 1977) and the two gamma states hypothesis was shown to have considerable potential in other iron-base alloys (Miodownik, 1977, 1978b). However, at the time it was considered to be too complex to merit inclusion in the general thermodynamic framework.

Although the basic foundations of CALPHAD were being laid in these formative years, the necessary ideas were still essentially being exchanged by correspondence, with isolated cases of personal contact. The number of papers in this field was however increasing slowly but surely, and three closely spaced meetings, at Brunel and Sheffield in 1971 and Munich in 1972, eventually provided an opportunity for the correspondents from different groups to meet each other personally. These meetings confirmed and crystallised the need for a dedicated forum that could deal solely with the emerging topic of computer-aided calculation of phase diagrams. In order to facilitate funding, the international nature of the proposed meetings and the idea of rotating meetings were stressed from the outset. An initial bilateral agreement to generate official meetings between French and American experts in this field was drafted (Kaufman 1973a) and invitations simultaneously extended to representatives from the UK, Sweden and Germany. It is interesting to note both the clarity with which the objectives were defined (Kaufman and Ansara 1973) and the fact that these objectives still form part of current CALPHAD activities today:

> *"We are at present all using different computer methods to obtain tie-line solutions. We are also using slightly different formulations for the excess Gibbs energy of solution. In some cases the differences may be more semantic than real, but in any case, if we can all employ equivalent computer programmes, we could concentrate on the problem of defining system parameters in order to achieve universal interchangeability of results. We believe that substantial progress could be made in a short period of time if we could arrange to work together for one week at one of our facilities to define the problems, disband, carry out some individual activities, and meet again for a week at a second facility to compare results and chart future activities."*

The first meeting was scheduled to take place in Boston in 1973 with the second meeting following at Grenoble in 1974. These meetings, held in a workshop format, were to provide opportunities for intimately exploring the details of phase boundary calculations in addition to assessing the broader problems of data acquisition and representation. The potential benefits to the scientific community of developing a uniform description of phases and generating interchangeable data, as well as the ongoing nature of the suggested meetings, gained financial support from the French Ministry of Foreign Affairs (Ansara 1973), the National Science Foundation of the USA (Reynik 1973) and the National Physical Laboratory (NPL) (Spencer 1973). This set a pattern for future support and established the principle that funds should be available to defray the travel costs of individuals with special expertise and to co-ordinate the efforts of various institutions. Significantly, the original proposal also contained the suggestion that the report covering the first two CALPHAD conferences could be used as the basis for a larger workshop on phase-diagram calculations which might follow these meetings. Such a meeting was duly organised by the National Bureau of Standards (NBS) at Gaithersburg in August 1975. Although not organised by the CALPHAD group, this was labelled CALPHAD IV, since it took place shortly after CALPHAD III was held in the UK in April of that year.

The private correspondence associated with setting up the first CALPHAD meeting indicates that the arrival of a new generation of younger personnel, who were more familiar with computing, played an important part in the embryonic CALPHAD. For instance, although specifically invited to the first meeting, Kubaschewski (1973) had by then moved to Aachen and was unsure whether he could make a meaningful contribution: "Although I am in favour of your CALPHAD Project, I beg you to note that my interests end where those of CALPHAD begin. I make myself largely responsible for the provision of critical and consistent thermochemical data of binary alloy systems.... However if funds could be made available for my new colleague Dr. I. Barin, who is looking after our computer programming, he could attend in my place".

The extra computing expertise was of course valuable, but Kaufman was at pains to reply that the objectives set for the CALPHAD group were to *couple* computational methods with all available thermochemical data and that experience in assessing such data would be invaluable (Kaufman 1973b). Kubaschewski was to join Barin for the second meeting in Grenoble and played a major role in promoting the concept of self-consistent thermodynamic information. The importance of the self-consistency engendered by such coupling is still underestimated by newcomers to the CALPHAD procedure even today. The selection of interaction parameters is always a delicate balance between a purely mathematical treatment and a weighting process which inevitably requires an element of subjective experience.

As the selected values allow predictions to be made on a variety of experimental quantities, for instance heats of solution, vapour pressures, and specific heats as well as the position of phase boundaries, it is important to find a method of optimising the interaction parameters so as to give the minimum deviation with respect to all these properties. A least squares method of optimising data was being developed by Mager *et al.* (1972) at Stuttgart and with great generosity was subsequently made freely available to all interested parties prior to its eventual publication (Lukas *et al.* 1977). This accelerated the characterisation of binary systems substantially and contributed greatly to the goal of integrating a wide range of thermodynamic information. Equally, it also made it clearer to the non-specialist that a fully characterised phase diagram could provide a much wider range of properties than was generally thought possible.

2.3. THE INTERMEDIATE YEARS

The 1970s and 1980s proved to be the time when CALPHAD established itself as an

References are listed on pp. 26–29.

accepted tool in the general armoury of materials modelling. While, by necessity, it retained a level of empiricism, there was a continuous attempt throughout this period to provide a more physical basis for the modelling process. The number of people involved in CALPHAD methods was also increasing appreciably and new researchers were to profoundly influence the subject area as a whole.

As a further means of providing long-term financial support, and expand the potential of the CALPHAD methodology, it was also decided to form a non-profit-making organisation (Kaufman and Nesor 1975) that would: "Undertake to conduct meetings of interested persons, organise and operate seminars and other educational experiences and publish results of research and descriptions of educational programmes." More specifically, the company was empowered to: "Engage in the activities of research, fund-raising, education, publishing and consulting, insofar as these lead to the further understanding of thermodynamic and thermochemical research and process...."

CALPHAD Inc. was duly incorporated in Massachusetts in 1975. One of the first fund-raising options to be pursued was the possibility of publishing a journal which would act as a focus for papers in this newly emerging field. This eventually came to fruition with the publication of the CALPHAD journal by Pergamon Press in 1977. The growth of CALPHAD activities was now clearly accelerating and an Alloy Data Centre formed at the NBS included CALPHAD calculations as one of its inputs.

A number of important workers who, up to that time, had been working in relative isolation, now became more involved in CALPHAD conferences and publications. The Canadian group, led by Pelton, had evolved is own representation of thermodynamic quantities in a specific set of non-metallic systems (Pelton and Flengas 1969, Pelton and Schmalzried 1973, Bale and Pelton 1974) but found that many of the concepts being outlined for alloys could also be applied in their area. This was to eventually lead to a Facility for Analysis of Chemical Thermodynamics (F*A*C*T) (Bale and Pelton 1979) and the foundation of the Centre of Research for Thermodynamic Calculations at Montreal. In due course there would be further collaboration (Thompson *et al.* 1983) between F*A*C*T and the SOLGASMIX programme of Eriksson (1975).

By 1976 it was becoming clear that the increasing volume of results in the field of electronic structure calculations should be brought to the attention of the CALPHAD community. Providentially, a workshop on Electronic Structure and Phase Stability in Metals and Alloys was held later that year at Liège. This gave an opportunity for CALPHAD members to ascertain which physicists could best verbally bridge the gap between a physics-based approach and the more chemically orientated background of most of the CALPHAD practitioners. CALPHAD VI thus became the first meeting to which physicists were invited in order to assess the feasibility of combining data obtained by first-principle calculations with CALPHAD techniques (Pettifor 1977). In a number of ways this meeting started the process of collaborative discussion between CALPHAD practitioners and the physics community, which continues to the present day. At the same time, speakers such as Miedema and Machlin, who had developed semi-empirical techniques for deriving missing interaction parameters (Miedema *et al.* 1973, 1975, Machlin 1974, 1977), were also invited to explain the background to their methods.

1977 was to provide two significant milestones. The first of these was a major workshop on Applications of Phase Diagrams in Metallurgy and Ceramics jointly organised by a consortium of defence establishments at the NBS. Representatives from leading research and development organisations were invited to address the current need for phase diagram information on a world-wide scale and, although the calculation of phase diagrams formed only a small part of the programme, the importance of the meeting lay in the scale of the projected requirements for multi-component systems. It was clear that such a task could not be tackled by conventional methods on a realistic time-scale and, consequently, suggested priorities included a survey of the status and merits of various predictive methods. Amongst the competitors were the PHACOMP method and various pseudo-potential techniques. PHACOMP is essentially based on the concept that certain types of phases, such as σ, occur at a critical values of electron-to-atom ratios or the equivalent number of holes in the d-band. However, PHACOMP requires the input of additional arbitrary parameters which means that, while it has some limited usefulness for the purposes of retrospective systematisation, it has no real predictive power. The various pseudo-potential methods were more soundly based but, at the time, were not concerned with anything beyond simple binary systems. By comparison, the CALPHAD technique appeared much more promising.

Although the NBS meeting was very effective in stimulating funding for the production of more relevant and usable phase diagrams, the same measure of support did not materialise for CALPHAD techniques. The self-consistency arising from coupling thermochemical data with phase equilibria was still not appreciated. The potential of computer calculations was acknowledged, but there remained a general reluctance to accept results for multi-component calculations without a major associated experimental programme. However, an increasing audience was being made aware of a viable alternative to the older, more labour-intensive methods of handling phase equilibria.

The second event of 1977 was the publication of the first volume of the CALPHAD journal, which both acted as a cumulative record of progress in making calculations and as an invaluable depository of validated parameters. The aim of the journal was not primarily to be a vehicle for publicity, but it was very useful to have the first issue of the CALPHAD journal available for distribution at the NBS meeting. There was no mandatory requirement to publish the papers and discussions presented at the CALPHAD conferences in the journal. Nevertheless, subsequent issues in the first year of publication of the CALPHAD journal included several key papers given at previous meetings, together with summaries of these

References are listed on pp. 26–29.

meetings. These included the seminal paper on first-principle calculations by Pettifor (1977) and the publication of the Lukas Programme for the statistical coupling of input from various thermodynamic and phase diagram sources (Lukas *et al.* 1977).

In the mid-1970s, the cost of sophisticated computers made it difficult for academic departments to requisition the necessary equipment for dealing with thermodynamic problems. The Royal Institute of Technology was one of the few materials science institutions that acquired a powerful minicomputer as early as 1976. This meant that there was more computing power, and a greater capacity for interactive computing, available in Stockholm than in most other materials departments at that time. This clearly had a major impact on software development. Even so, the limitations of the available memory (128 kilobytes!) made it essential to compress data in a very economical way. There was initially a problem with the language facilities on their NORD 10 computer, but eventually a specially structured architecture was devised which formed the basis for the early versions of the programme known as POLY (Jansson 1984a). At this particular point the Stockholm group also decided to develop their own module for the assessment of experimental data, which eventually led to the creation of the PARROT module (Jansson 1984b).

The CALPHAD conference of 1979 in Stockholm provided a good opportunity to absorb major developments which were taking place in both computing hardware and software. In the late 1970s the extension by Sundman and Ågren of the two-sub-lattice model of Hillert and Staffansson (1970) into a general model which could account for multiple sub-lattices and multi-component alloys was being undertaken. This was reported to the conference and details of the model published two years later (Sundman and Ågren 1981). As one of the key papers on general modelling, this laid the foundation for a substantial proportion of the CALPHAD assessments that have been made to date. The suite of programmes later to become known as *Thermo-Calc* was also initiated at this time, and still remains one of the most widely used software packages for thermodynamic computation (Sundman 1991).

In England, a corresponding suite of programmes had evolved at the NPL via early mathematical steps in the 1960s (Kubaschewski and Chart 1965) and led to the development of a commercial module in 1974. At the same time the MANLABS suite of programmes was being developed and marketed in the USA (Kaufman 1974). By 1978 the NPL programmes were marketed as ALLOYDATA and, together with an extensive substance database, were being used for many interesting non-metallic applications (Chart *et al.* 1980). The programmes were included as part of a more general Metallurgical and Thermo-chemical Data Service (MTDS) at the NPL and after substantial redevelopment in the 1980s they became more generally known as MTDATA (Davies *et al.* 1991). Further details of these and other programmes can be found in Bale and Eriksson (1990).

One of the driving forces for the formation of CALPHAD was the concept of providing a unified database for use by the various calculation programmes

(Kaufman and Ansara 1973). SOLGASMIX (Eriksson 1975) and F*A*C*T, which largely dealt with non-metallic systems (Thompson *et al.* 1983), had interchangeable file structures and data formats. But this was not the case for most of the programmes, which still tended to have their own specific data format structure. Tomiska (1986a, 1986b) showed that the various polynomial formalisms in use at the time could be transformed mathematically into each other. This meant that transfer of data was possible, providing the data for the elements was consistent, even though different models for excess terms were being used by various centres. However, this was often done on a rather *ad hoc* personal basis. There was clearly a need to systematise the interchange of data and this led to the formation of a consortium of the major groups in France, Germany, Sweden and the United Kingdom known as the Scientific Group Thermodata Europe (SGTE) (Ansara 1983) and eventually to the publication of a unified database for the elements (Dinsdale 1991). This allowed anyone with a general interest in phase-diagram calculations to use much more consistent elemental data, and interchange data with other people.

1979 marked an important milestone, as these developments had now generated a series of parallel symposia devoted to the application of thermodynamic calculations, held under the umbrella of major national organisations. These included the AIME symposium on the Calculation of Phase Diagrams and Thermochemistry of Alloy Phases at Milwaukee in the USA and the Symposium on the Industrial Use of Thermochemical Data organised by the NPL and the Chemical Society at the University of Surrey in the UK. These meetings served to make a larger audience aware of the potential of the CALPHAD technique, as keynote papers based on proven software developments were given at both meetings.

The primary objective of the Milwaukee symposium was to establish cross-linkages amongst metallurgists and material scientists working in the calculation of phase diagrams and to demonstrate that advances in this field could be readily applied in many areas of physical, chemical and process metallurgy. In this context, the paper by Hillert (1979), on the availability of new methods of presenting data and utilising data, was a significant departure from presenting new models or applications. Although an apparently innocuous subject, this embedded some revolutionary ideas. Up to that time, potential users usually had great difficulty in interpreting conventional ternary diagrams and there was usually a total mental block to considering anything with a larger number of components. It was therefore of great importance to break through this barrier and show the potential user that the required information from a multi-component system could also be extracted and digested in an easily assimilated form. Hillert (1980) used a telling analogy with chess as a means to overcome the reluctance of total newcomers to use the CALPHAD approach:

''When considering what fundamentals are required in order to use thermodynamic data for practical purposes, it may be helpful to consider thermodynamics as a game

References are listed on pp. 26–29.

and to consider the question how one can learn to play that game well. For inspiration, one may first look at another game, the game of chess. The rules are very simple to learn, but it was always very difficult to become a good player The situation has now changed due to the application of computers. Today there are programmes for playing chess which can beat almost any expert. It seems reasonable to expect that it should also be possible to write programmes for 'playing thermodynamics', programmes which should be almost as good as the very best thermodynamics expert."

This approach was very timely, as by now such meetings were attracting industrial interest; of the total attendees at the Surrey meeting, 29% came from academic institutions, 26% from government research institutions and 45% from industry, forming an excellent juxtaposition of application and theory. The meeting was summarised as providing an illustration of the power of reliable databases for technology, databases whose credibility had been established by the professional evaluation of experimental data. It was pointed out that such major codification, condensation and integration of expensively won knowledge not only benefits fundamental science but has also a capability for quantitative interpolation, extrapolation and new directions of understanding (Westrum 1980). This meeting was effectively to be the forerunner of a separate series of international meetings devoted to user applications of phase diagrams, sponsored internationally by various groupings of metallurgical and materials societies in later years.

The next few years saw an increased interest in molten salts and silicates (Pelton and Blander 1984) as well as geological contributions, which were typically concerned with aluminates and spinels (Navrotsky 1983). Geological applications seemed to be an ideal growth area, which was accordingly earmarked for further attention at a future CALPHAD meeting. However, although attention has steadily increased on basic oxide systems, geological applications have remained in abeyance until much more recent times (Shi *et al.* 1994, 1996). Another notable application in the early eighties was by Eriksson (1983), who demonstrated the scope of the SolGasMix programme by applying it to the planetary condensation of solar gases containing more than 30 elements.

During this period, various aspects of Miedema's methods for predicting the heat of formation of binary compounds were assembled and eventually published in book form (de Boer *et al.* 1988). This included the application of the technique to predict the thermodynamic behaviour of some ternary compounds. Whilst only applicable to a restricted set of crystallographic structures, this was nevertheless a significant development, as a common objection to the CALPHAD approach was that the existence of ternary compounds could never be predicted solely from binary data.

Another area which saw increasing attention was semiconductor materials. Ishida *et al.* (1989) constructed a database which combined the calculation of phase equilibria in III–V compounds with the calculation of band gaps in the same systems. This was to be the forerunner of future attempts to expand the database beyond the basic thermodynamic information needed for phase diagram calculations, while (Mohri *et al.* 1989) introduced the effect of an elastic component into such calculations in the same subject area.

The 1980s also saw systematic attempts to couple thermodynamics and kinetics in concentrated alloys. In contrast to geological applications, this field was to blossom. At first the main applications were in rapid solidification processing where a number of papers showed how the underlying thermodynamics of a system could control glass-forming ability (Saunders and Miodownik 1985) as well as the metastable crystalline structures formed during vapour deposition (Saunders and Miodownik 1987). Collaborative work begun at this time between the Royal Institute of Technology (KTH) in Sweden and the Max Planck Institute (MPI) for Iron Research at Düsseldorf led to the development of a substantial programme called DICTRA, which simultaneously solved both the relevant diffusion and thermodynamic equations which control phase transformations in both the liquid and solid states (Ågren 1992).

Despite all these advances there remained a surprising resistance to the general acceptance of CALPHAD techniques in many quarters. As there was no shortage of validated applications, it was thought that the reason might be connected with deficiencies in the conventional methods by which phase diagrams had been taught to the past generation of students. Teaching each class of diagram by rote, as well as being exceedingly time consuming and topologically complicated if extended beyond binary systems, rarely produced any real appreciation of phase equilibria. The use of computational techniques as a teaching aid was therefore placed on the agenda, and it was shown that suitable software could certainly make the conventional methods much more interesting and efficient (Hayes 1984). However, a more radical approach was made by Hillert (1984), who suggested that the conventional approach should be abandoned altogether and replaced by the direct calculation of any desired critical temperatures and volume fractions. This raised considerable controversy at the time but in fact was probably a correct diagnosis of future trends.

CALPHAD XIV, which was organised on the Campus of MIT at Massachusetts in 1985, emphasised the growing dichotomy between phase stabilities calculated by electron energy methods and the values that were currently used by the CALPHAD community. On the one hand the latter could point to the fact that their values were self-consistent and also produced excellent agreement with experimental results even in multi-component systems. On the other hand the physics community was now extremely confident of the theoretical background for their values of the phase stabilities at 0 K. The debate was continued a few months later, before a wider audience, at a symposium on computer modelling of phase diagrams organised by AIME in Toronto.

A paper by Miodownik (1986) suggested that significant changes in some of the thermochemical parameters could be obtained by adopting different entropies of

References are listed on pp. 26–29.

fusion for the metastable states of certain elements such as tungsten. This approach was more fully explored by Saunders *et al.* (1988) who attempted a self-consistent reappraisal of Kaufman's original values and showed how recent experimental information on entropies of fusion could alter Kaufman's values for lattice stabilities to magnitudes significantly closer to *ab initio* values. This process used metastable melting points for different crystal structures of the elements which were still very close to those used by Kaufman, and a flurry of papers followed (see Chapter 6) which placed limits on the values for phase stability that could still lead to acceptable phase diagrams. While clearly bringing the two sides closer together, significant discrepancies still remained between a thermochemical and *ab initio* approach to the lattice stabilities of specific elements. The desirable goal of a single set of lattice stabilities acceptable to both communities was therefore still not in place.

It now looks as though all the outstanding discrepancies between thermochemical (TC) and first-principle (FP) lattice stabilities are associated with cases where the postulated metastable allotrope is mechanically unstable to shear and would spontaneously collapse at 0 K (Craeivich *et al.* 1994); a possibility previously suggested by Pettifor (1988). This means that it is no longer meaningful to compare the TC and FP values at 0 K unless it is also possible to derive an expression for the anomalous entropy of such systems (Fernandez Guillermet *et al.* 1995). The use of extrapolated TC lattice stabilities for phase-diagram calculations has remained justifiable, with tighter definitions. The main challenge now lies in properly formulating the temperature dependence for the lattice stabilities obtained by FP calculations. Until this problem has been solved it is clear that phase-diagram calculations from 'first principles' will not, in general, be able to match the accuracy obtained by more conventional CALPHAD techniques. In the meantime a simple but entirely empirical treatment (Chang *et al.* 1995) has already shown that the two treatments are consistent with each other once a suitable expression for the entropy has been formulated.

2.4. THE LAST DECADE

The most obvious sign that CALPHAD had finally been accepted, was the number of papers included in a series of internationally organised meetings devoted to emphasising the applications of phase diagrams. The first such meeting was held in 1986 at Orlando, USA, followed by one at Petten, Holland, in 1991, with a third meeting at Rosemount, USA, in 1994. Proceedings of these conferences were published and provide good examples of theory and practice being combined (Kaufman 1987, Hayes 1991, Nash and Sundman 1995). All such meetings exposed the subject to a much wider audience , while new techniques and new fields of potential application were being explored inside the continuing series of annual CALPHAD meetings. Practical questions of measuring thermodynamic properties were gradually being left to a third set of biannual international meetings on the thermodynamics of alloys. The CALPHAD approach was also beginning to feature in physics-based meetings such as the NATO conference on the statics and dynamics of phase transformations, which was held in Rhodes almost simultaneously with the Jerusalem meeting. This was probably the first time that an audience of physicists had been given a thorough overview of the CALPHAD approach, especially the insistence on self-consistency obtained by coupling different forms of thermodynamic and experimental information (Miodownik 1994, Inden 1994).

As far as the mainstream CALPHAD meetings were concerned, 1987 and 1988 saw separate sessions given to semiconductor systems and surface equilibria, with a greater emphasis being placed on oxides and ceramics, including the now popular ceramic super-conducting systems. One of the advantages of a first-principles approach is that one can study the effect of small distortions, so that it also becomes possible to approach the energetics of shear transformations and non-equilibrium effects such as the martensite transformation. For the first time a CALPHAD conference contained a whole session devoted to this topic and this included an application in the field of the new ceramic superconductors based on the YBCO (yttrium–barium–copper–oxygen system), showing how rapidly phase equilibrium calculations were adapting to current systems of topical interest.

In 1990 CALPHAD XIX took place immediately before the international conference on user aspects of phase diagrams at Petten. This was the first time a CALPHAD meeting was located in Holland, although a strong group had been attending from that country ever since CALPHAD III. This meeting was organised jointly by the University of Utrecht and Philips Research Laboratories. The background of these organisers ensured that, for the first time, there was a substantial number of presentations devoted to organic systems (e.g., Oonk *et al.* 1986) and there was also a greater emphasis on associated physical properties (Grimwall *et al.* 1987).

The next two CALPHAD meetings were both affected to some extent by world events outside the control of the organisers. CALPHAD XX took place in India at Jamshedpur in 1991 at a time which coincided with the onset of the Gulf War, and this severely restricted attendance. The 1992 CALPHAD meeting took place in Jerusalem, and while numbers had increased again, they were still affected by residual tensions in the Middle East. However, despite the smaller numbers, these were still very useful and successful meetings. Meetings in Barcelona (1993), Madison (1994), Kyoto (1995) and Erice (1996) saw numbers revert towards the underlying trend of previous years. The growth of the mainstream CALPHAD meetings in terms of both the number of participants and participating countries is shown in (Fig. 2.1).

The period has been marked by growing interest in CALPHAD techniques outside Europe and America, which was evidenced by the choice of "Progress in CALPHAD" as the title for the 1992 Honda memorial lecture to the Japanese

References are listed on pp. 26–29.

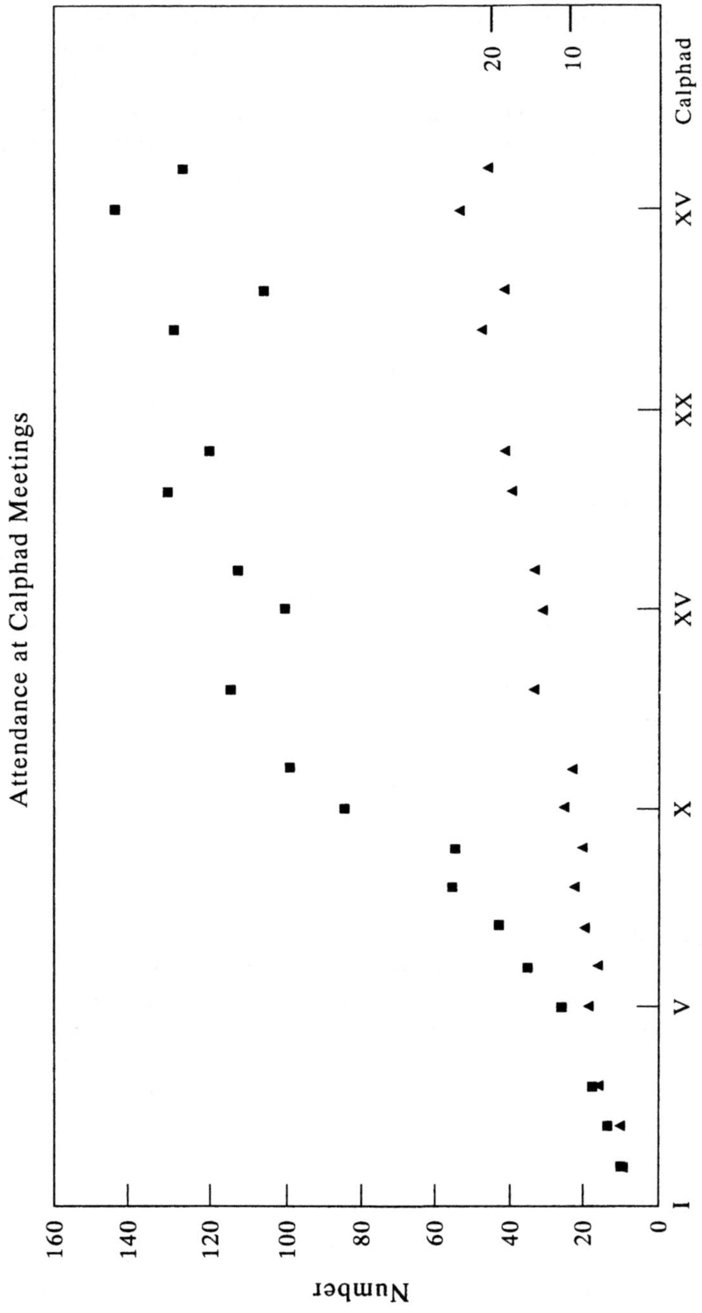

Figure 2.1 ■, Total attendance, left-hand scale. ▲, Number of participating countries, right-hand scale.

Institute of Metals (Nishizawa 1992). Several papers dealing with the CALPHAD approach were presented in the same year at the CAMSE (Computer Applications in Materials) meeting in Yokahama, and this resulted in a joint meeting of the two groups at Madison, USA, in 1994, with the next CALPHAD meeting being held in Japan for the first time. In 1995 the institution of the Ringberg workshop on unary systems marked the realisation that a more sophisticated computational framework was needed for the next decade. With this in view a further two such meetings were scheduled for 1996 and 1997.

2.5. THE CURRENT STATUS OF CALPHAD

Clearly CALPHAD has now come of age and is at a watershed where complex phase equilibria calculations can now be performed as a routine operation and yet have also been placed on a sound physical basis. Computer programmes exist which can perform complex calculations on a PC and which can therefore be operated at any location without extensive prior expertise. Furthermore, it is possible in many cases to predict phase equilibria in multi-component alloys to a degree which is close to that expected from experiment (see Chapter 10). It is therefore a branch of science that is mature and, indeed, has already entered the next stage of development, which emphasises the need to concentrate on extending its range of applicability.

Solution databases now exist for a number of the major metallic alloy systems such as steels, Ni- based superalloys and other alloy systems, and highly accurate calculation have been made which even a few years ago would have been considered impossible. The number of substance databases are increasing and the numbers of substances they include is reaching well into the thousands. Substance and solution databases are increasingly being combined to predict complex reactions such as in gas evolution in cast-irons and for oxidation reactions, and it is already possible to consider calculations of extreme complexity such as the reactions which may occur in the burning of coal in a industrial power generator or the distribution of elements in meteorites.

In parallel with such endeavours there is a continuing effort to replace empirical treatments by more physically based approaches. There are certainly two areas in which significant advances are likely. Firstly, while it is has for some time been possible to consider a slag system, it remains a goal to fully combine this with models for concentrated metallic solutions which will properly predict complex solid-state reactions. A second area is the modelling of ordered systems. While it is possible to use sub-lattice modelling for systems such as Ni–Al, these techniques still remain for the most part quite empirical. The *Cluster Variation Method* (CVM) has already achieved wide recognition as one of the methods for the modelling of ordering systems and has the advantages that short-range ordering is properly incorporated. Increasingly available computer power now makes it possible to

References are listed on pp. 26–29.

explore more sophisticated avenues. Although such techniques as the CVM (de Fontaine 1979) and Monte Carlo method (Binder 1986) have a much earlier origin in the physics literature, a significant milestone is rapidly approaching when fundamental treatments of ordering are going to be embedded into phase-diagram calculations instead of just being treated as ordering phenomena *per se*.

Combined inputs from first principles and statistical mechanics can now yield diagrams which show most of the topological features of the experimental phase diagram (Sanchez 1984, Sigli and Sanchez 1984, Colinet *et al.* 1985), which is an impressive achievement. However, the time needed to make a full multi-component calculation is still prohibitive. A solution to this problem will also have the benefit that associated properties such as antiphase domain boundary energies, stacking faults, etc., and also the energetics of various metastable ordering states can then also be predicted more accurately.

The main computing advances are likely to come from application-orientated research programmes where CALPHAD methods will become increasingly used as part of larger, more comprehensive modelling programmes (Mahin 1997, Bannerjee *et al.* 1997). One of the most significant features of current CALPHAD methods has been the combining of kinetics and thermodynamics. Thermo-Calc and ChemSage have a common programming interface which can be used by external programmes to provide detailed thermodynamic information. DICTRA is a software package to simulate diffusional transformations which utilises the subroutines of Thermo-Calc in a more closely integrated fashion and is a vital step in the progression of this combined field. There will also be a need for increased computing efficiency, robustness and speed of calculation. As CALPHAD methods become more popular, there will also be a need for an increased degree of user-friendliness and built-in help modules that guide non-specialists to cope with some of the pitfalls that can be associated with more complex equilibria.

The present state of art of CALPHAD is therefore at a high level and it is a very varied and powerful approach to solving problems associated with thermochemistry and phase equilibria. It has been of interest to the present authors to document how a new scientific approach has come to fruition and it echoes a prescient lecture given by Professor Köster on the development of new ideas and concepts in science (Köster 1956). Some excerpts are given below:

> *"It is fascinating to trace the devious paths a research worker may have followed before he arrived at his final knowledge and discoveries. Sometimes it may be the gradual perfection of experimental apparatus or the systematic pursuit of an idea; on other occasions success may arise from the deliberate solution of a problem only he believes possible at all, or the recognition of the importance of some apparently insignificant factor; it may even be the appreciation of a phenomenon already known, but completely neglected; or again success may come simply by sudden inspiration, or as the outcome of a long, patient and plodding series of experiments."...*
>
> *"But one thing is common to all true research. Its results are won in the course of*

unceasing, stubborn wrestling with problems in the service of knowledge.'"... "For each deeper insight, each original discovery, arises not only from cool calculation but is also the outcome of creative force and imaginative perception."

The year this lecture was delivered coincided exactly with the point in time when the idea of the CALPHAD was conceived. In retrospect, the excerpts above show that the evolution of the CALPHAD technique is an excellent example of the many factors that enter into the realisation of any new idea.

REFERENCES

Ågren, J. (1992) *ISIJ International*, **32,** 291.
Andrews, K. W. (1956) *J. Iron & Steel Inst.*, **184,** 414.
Ansara, I. (1973) Private Communication to Kaufman, L. April 13.
Ansara, I. (1983) Private Communication: Registration of (S.G.T.E) under Ordinance 9-821, September 23, Domaine Universitaire, St. Martin d'Heres Cedex, France.
Bale, C. W. and Eriksson, G. (1990) *Can. Met. Quart.*, **29,** 105.
Bale, C. W. and Pelton, A. D. (1974) *Met. Trans.*, **5,** 2323.
Bale, C. W, Pelton, A. D. (1979) *F*A*C*T User Guide* (1st edition). McGill University, Ecole Polytechnique, Montreal.
Bannerjee, D. K., Kattner, U. and Boettinger, W. (1997) *Solidification Processing 1997*, eds Beech, J. and Jones, H. (University of Sheffield, UK), p. 354.
Betterton, W. (1958) *Acta Met.*, **6,** 205.
Binder, A. R. (1986) *Monte-Carlo Methods in Statistical Physics*, ed. Binder, A. R. (Springer Verlag, Berlin).
Brewer, L. (1967) *Acta Met.*, **15,** 553.
Buckley, R. A. and Hume-Rothery, W. (1963) *J. Iron & Steel. Inst.*, **201,** 227.
Chang, Y. A, Colinet, C., Hillert, M., Moser, Z., Sanchez, J. M., Saunders, N., Watson, R. E. and Kussmaul, A. (1995) *CALPHAD*, **19,** 481.
Chart, T. G., Dinsdale, A. T. and Putland, F. H. (1980) *Industrial Use of Thermochemical Data*, ed. Barry, T. I. (Chemical Society, London), p. 235.
Colinet, C., Pasturel, A. and Hicter, P. (1985) *CALPHAD*, **9,** 71.
Craievich, P. J., Weinert, M., Sanchez, J. and Watson, R. E. (1994) *Phys. Rev. Lett.*, **72,** 3076.
Davies, R. H., Dinsdale, A. T., Hodson, S. M., Gisby, J. A., Pugh, N. J., Barry, T. I. and Chart, T. G. (1991) in *User Aspects of Phase Diagrams*, ed. Hayes, F. H. (Inst. Metals, London), p. 140.
de Boer, F. R., Boom, R., Mattens, W. C. M., Miedema, A. R., and Niessen, A. K. (1988) in *Cohesion in metals: Cohesion and Structure Vol. 1.* (Elsevier Science, Amsterdam), p. 644.
de Fontaine, D. (1979) *Solid State Physics*, **34,** 73.
Dinsdale, A. T. (1991) *CALPHAD*, **15,** 317.
Ducastelle, F. (1971) Private Communication to Kaufman, L. September 17.
Engel, N. (1964) *ASM Trans. Quarterly*, **57,** 619.
Eriksson, G. (1975) *Chem. Scripta*, **8,** 100.

Eriksson, G. (1983) presented at Calphad XII (Liege).
Fernandez Guillermet, A., Ozolins, V., Grimwall, G. and Korling, M. (1995) *Phys. Rev. B*, **51,** 10364.
Grimwall, G., Thiessen, M. and Fernandez Guillermet, A. (1987) *Phys. Rev. B*, **36,** 7816.
Hansen, M. (1936) *Bibliography of Alloys*. (Springer Verlag, Berlin).
Hasebe, M. and Nishizawa, T. (1972) *Bull. Jap. Inst. Metals*, **11,** 879.
Hayes, F. (1984) Contribution to Calphad XIII, Villard les Lans, France.
Hayes, F. (1985) *J. Less Common Metals*, **114,** 89.
Hayes, F. (1991) *User Aspect of Phase Diagrams*, ed. Hayes, F. (Inst. Metals, London).
Hillert, M. (1953) *Acta Met.*, **1,** 764.
Hillert, M. (1979) in *Calculation of Phase Diagrams and Thermochemistry of Alloy Phases*, eds Chang, Y. A. and Smith, J. F. (Met. Soc. AIME, Warrendale, PA), p. 1.
Hillert, M. (1980) in *Conference on the Industrial Use of Thermochemical Data*, ed. Barry, T. (Chemical Society, London), p. 1.
Hillert, M. and Staffansson, L.-I. (1970) *Acta Chem. Scand.*, **24,** 3618.
Hillert, M., Wada, T. and Wada, H. (1967) *J. Iron & Steel. Inst.*, **205,** 539.
Hume-Rothery, W. (1968) Private Communication to Kaufman, L. June 20.
Hume-Rothery, W., Reynolds, P. W. and Raynor, J. V. (1940) *J. Inst. Metals*, **66,** 191.
Inden, G. (1994) in *Statics and Dynamics of Alloy Phase Transformations*, eds Turchi, P. and Gonis, A. (NATO ASI Series B: Physics Vol. 319). (Plenum Press, New York).
Ishida, K., Tokunaga, H., Ohtani, H. and Nishizawa, T. (1989) *J. Crystal Growth*, **98,** 140.
Jansson, B. (1984a) TRITA-MAC 0233, Royal Institute of Technology, S-100 44 Stockholm, Sweden.
Jansson, B. (1984b) TRITA-MAC 0234, Royal Institute of Technology, S-100 44 Stockholm, Sweden.
Kaufman, L. (1959a) *Acta Met.*, **7,** 575.
Kaufman, L. (1959b) ManLabs Research Report No. 2.
Kaufman, L. (1966a) Private Communication to Brewer, L. March 25.
Kaufman, L. (1966b) Private Communication to Hume-Rothery, W. May 31.
Kaufman, L. (1967) in *Phase Stability in Metals and Alloys*, eds. Rudman, P. S. *et al.* (McGraw-Hill, New York), p. 125.
Kaufman, L (1968a) Private Communication to Hume-Rothery, W. June 25.
Kaufman, L (1968b) Private Communication to Hume-Rothery, W. July 31.
Kaufman, L. (1968c) Private Communication to Hillert, M. November 1.
Kaufman, L. (1969) *Prog. Mater. Sci.*, **14,** 55.
Kaufman, L. (1973a) Private Communication to Reynik, R. (NSF). March 29.
Kaufman, L. (1973b) Private Communication to Kubaschewski, O. August 23.
Kaufman, L. (1974) ManLabs-Brochure on ManLabs-NPL Materials Data Bank.
Kaufman, L. (1980) in *Conference on the Industrial Use of Thermochemical Data*, ed. Barry, T. (Chemical Society, London), p. 215.
Kaufman, L. (1987) *User Applications of Phase Diagrams*, ed. Kaufman, L. (ASM, Metals Park, OH).
Kaufman, L. and Ansara, I. (1973) Private Communication to Spencer, P. J and Kubaschewski, O. (NPL) and Hillert, M. and Kirchner (KTH), April 30.
Kaufman, L. and Bernstein, H. (1970) *Computer Calculations of Phase Diagrams*. (Academic Press, New York).

Kaufman, L. and Cohen, M. (1956) *J. Metals*, **10,** 1393.
Kaufman, L and Cohen, M. (1958) *Prog. Metal. Phys.*, **7,** 165.
Kaufman, L. and Nesor, H. (1975) Articles of Organisation for Calphad Inc., Mass. November.
Kaufman, L., Clougherty, E. V. and Weiss, R. J. (1963) *Acta Met.*, **11,** 323.
Köster, W. (1956) *J. Inst. Metals*, **85,** 113.
Kubaschewski, O. and Chart, T. (1965) *J. Inst. Metals*, **93,** 329.
Kubaschewski, O. (1973) Private Communication to Kaufman, L, August 29.
Lukas, H. L., Henig, E. Th. and Zimmermann, B. (1977) *CALPHAD*, **1,** 225.
Machlin, E. S. (1974) *Acta Met.*, **22,** 95.
Machlin, E. S. (1977) *CALPHAD*, **1,** 361.
Mager, T., Lukas, H. L. and Petzow, G. (1972) *Z. Metallkde.*, **63,** 638.
Mahin, K. W., (1997) *Advanced Materials and Processes*, **151**, 20.
Meijering, J. L. (1948) *Philips Res. Rept.*, **3,** 281.
Meijering, J. L. (1957) *Acta Met.*, **5,** 257.
Miedema, A. R., Boom, R. and de Boer, F. R. (1975) *J. Less Common Metals*, **41,** 283.
Miedema, A. R., de Boer, F. R. and de Chatel, P. F. (1973) *J. Phys. F*, **3,** 1558.
Miodownik, A. P. (1977) in *Applications of Phase Diagrams in Metallurgy and Ceramics.* NBS special publication 496, Gaithesburg, p. 1479.
Miodownik, A. P. (1978) in *Physics and Application of Invar Alloys, Vol. 3*, ed. Saito, H. Maruzen, p. 288.
Miodownik, A. P. (1986) in *Computer Modelling of Phase Diagrams*, ed. Bennett, L. (Met. Soc. AIME, Warrendale), p. 253.
Miodownik, A. P. (1994) in *Statics and Dynamics of Alloy Phase Transformations*, ed. Turchi, P. and Gonis, A. (NATO ASI Series B: Physics Vol. 319). (Plenum Press, New York), p. 45.
Mohri, T. Koyanagi, K, Ito, T and Watanabe, K. (1989) *Jap. J. Appl. Phys.*, **28,** 1312.
Nash, P. and Sundman, B. (1995) *Applications of Thermodynamics in the Synthesis and Processing of Materials*, ed. Nash, P. and Sundman, B. (TMS, Warrendale, PA).
Navrotski, A. (1983) presented at Calphad XII (Liège).
Nishizawa, T. (1992) Materials Trans. *Jap. Inst. Metals*, **33,** 713.
Oonk, H. A., Eisinga, P. J. and Rouwer, N. B. (1986) *CALPHAD*, **10,** 1.
Pelton, A. D. and Blander, M. (1984) in *2nd Int. Symp. on Metallurgical Slags & Fluxes*, eds Fine, A. H and Gaskell, D. R. (Met. Soc. AIME, Warrendale, PA), p. 281.
Pelton, A. D. and Flengas, S. N. (1969) *Can. J. Chem.*, **47,** 2283.
Pelton, A. D. and Schmalzried, H. (1973) *Met. Trans.*, **4,** 1395.
Pettifor, D. G. (1969a) Private Communication to Kaufman, L. September 8.
Pettifor, D. G. (1969b) *J. Phys. C*, **2,** 1051.
Pettifor, D. G. (1977) *CALPHAD*, **1,** 305.
Pettifor, D. G. (1988) Private Communication to Saunders *et al.* (1988).
Reynik, R. (1973) Private Communication May 13.
Roy, D. M. and Pettifor, D. G. (1977) *J. Phys. F*, **7,** 183.
Sanchez, F. J (1984) in *High Temperature Alloys: Theory and Design*, ed. Stiegler, O. (Met. Soc. AIME, Warrendale, PA), p. 83.
Saunders, N. and Miodownik A. P. (1985) *CALPHAD*, **9,** 283.
Saunders, N. and Miodownik A. P. (1987) *J. Mat. Sci.*, **22,** 629.

Saunders, N., Miodownik, A. P. and Dinsdale, A. T. (1988) *CALPHAD*, **12,** 351.
Shi, P., Saxena, S. K., Zhang, Z. and Sundman, B. (1994) *CALPHAD*, **18,** 47.
Shi, P., Saxena, S. K., Zhang, Z. and Sundman, B. (1996) *CALPHAD*, **20,** 93.
Sigli, C. and Sanchez, J. M. (1984) *CALPHAD*, **8,** 221.
Sinha, A. K. and Hume-Rothery, W. (1967) *J. Iron & Steel Inst.*, **206,** 1145.
Spencer, P. J. (1973) Private Communication to Kaufman, L. August 9.
Sundman, B. (1991) *User Aspects of Phase Diagrams*, ed. Hayes, F. H. (Inst. Metals, London), p. 130.
Sundman, B. and Ågren, J. (1981) *J. Phys. Chem.*, **42,** 297.
Svechnikov, V. N. and Lesnik, A. G. (1956) *Fisika Metallov. Metallovedenie*, **3**(1), 87.
Tauer, K. J. and Weiss, R. J. (1958) *J. Phys. Chem. Solids*, **4,** 135.
Thompson, W. T., Pelton, A. D. and Bale,C. W. (1983) *CALPHAD*, **7,** 137.
Tomiska, J. (1986a) *CALPHAD*, **10,** 91.
Tomiska, J. (1986b) *CALPHAD*, **10,** 239.
van Laar, J. J. (1908) *Z. Physik. Chem.*, **63,** 216; **64,** 257.
Walbaum, D. R. (1972) Review of Kaufman and Bernstein (1970) in *American Mineralogist* (May/June), p. 1009.
Weiss, R. J. (1963) *Proc. Phys. Soc.*, **82,** 281.
Weiss, R. J. and Tauer, K. J. (1956) *Phys. Rev.*, **102,** 1490.
Westrum, E. F. (1980) in *Conference on the Industrial Use of Thermochemical Data*, ed. Barry, T. (Chemical Soc., London), p. 416.

Chapter 3
Basic Thermodynamics

Chapter 3
Basic Thermodynamics

3.1. INTRODUCTION

Thermodynamics is the science which underpins the CALPHAD method. Although the practice of calculating phase diagrams is computer intensive and relies heavily on mathematical manipulation, it would not be possible to begin without model descriptions of the underlying thermodynamics of the system of interest. A further strength of thermodynamics is that the underlying principles and laws are applicable to any type of system and hence the CALPHAD approach can be used equally well in many disciplines, for example geology, chemical and process engineering, materials science and nuclear technology; the list is very long.

The aim of this chapter is not to go into great depth on this subject but to introduce the reader to the key concepts which will be discussed later in the book. The chapter begins by describing the first and second laws of thermodynamics and introducing the reader to the ideas of enthalpy, entropy and Gibbs energy. There will be examples of how Gibbs energy can be used to indicate the direction of a chemical reaction and how heat evolution can be predicted. The concept of Gibbs energy will then be applied to solution phases, for example liquids, and the concept of partial quantities introduced. The reader will then be introduced to the calculation of phase diagrams using a series of quite simple examples.

3.2. THE FIRST LAW OF THERMODYNAMICS

Thermodynamics is concerned with energy and the way energy is transferred. It is a science of the macroscopic world but its effects are applied even at the microscopic scale. The first law introduces the basic thermodynamic concepts of work, heat and energy and can be defined as follows: *Energy can neither be created or destroyed in a system of constant mass, although it may be converted from one form to another.*

This law is also called the *Law of Conservation of Energy*. The diagram below (Fig. 3.1), after Mackowiak (1966), represents a system of constant mass and energy which is enclosed in the circle. The arrows represent the interconvertability of the various types of energy. For example, heat energy can be converted to mechanical work, which is the basis of internal combustion engines.

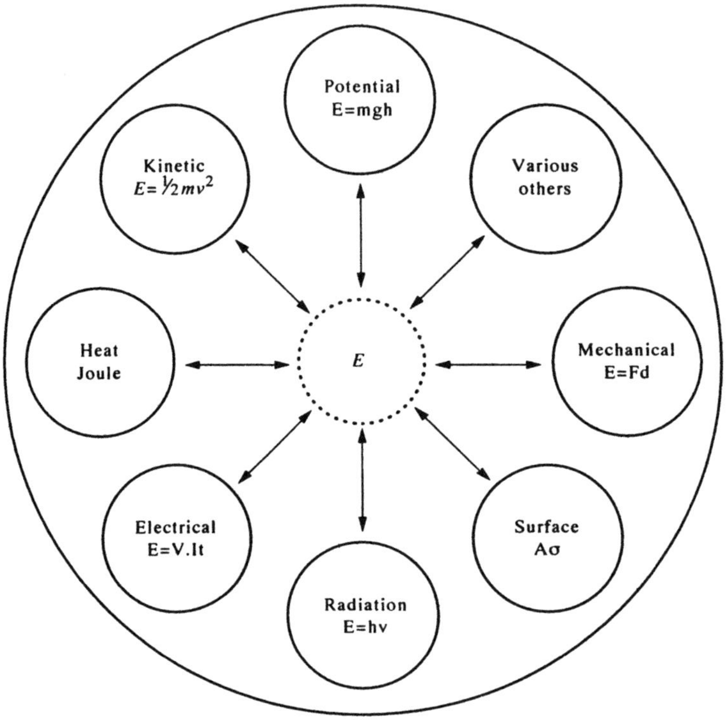

Figure 3.1. A closed system of constant mass and energy showing the interconvertability of various forms of energy.

If external energy is supplied to the system it must absorb the energy either by increasing its own *internal energy* and/or by doing *work*. As energy can neither be created or destroyed, the following relationship is obtained:

$$\Delta U = q - w \tag{3.1}$$

where ΔU is the change in internal energy (defined as the sum of all the kinetic energies and energies of interaction of the particles in the system), q is the heat (or energy) supplied and w is the work done by the system. Some of the fundamental equations of thermodynamics will now be derived.

3.2.1 The definition of enthalpy and heat capacity

If q_p joules of energy are introduced to a gas confined in a cylinder at pressure p (Fig. 3.2(a)), the internal energy of the gas increases and to maintain the pressure at a constant p the gas expands (Fig. 3.2(b)).

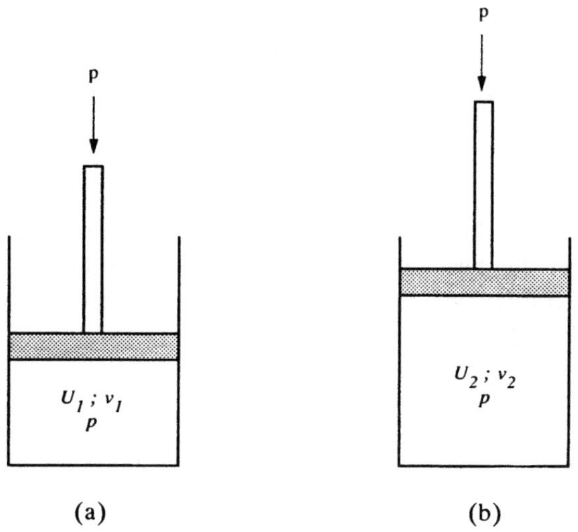

Figure 3.2.

To keep the pressure constant, the gas has done work by pushing the piston during expansion and this work is given by

$$w = p\Delta v \tag{3.2}$$

where Δv is the change in volume of the gas. Substituting into Eq. (3.1) gives

$$\Delta U = q_p - p\Delta v \tag{3.3}$$

or, alternatively

$$\Delta(U + pv) = q_p. \tag{3.4}$$

The term on the left-hand side $(U + pv)$ is very important in thermodynamics and is called the *enthalpy*. It is given the symbol H. Thus by definition

$$\Delta H = q_p. \tag{3.5}$$

An increase in enthalpy therefore denotes the heat absorbed during a process at constant pressure. For the case of constant volume the process is simpler as the system does no work on its surroundings, i.e., $p\Delta v = 0$. Therefore,

$$\Delta U = q_v \tag{3.6}$$

where q_v is the heat absorbed by the system with constant volume. The heat capacity of a system, C, can now be defined as the quantity of heat necessary to raise a system by a temperature, δT, and can be expressed as a differential

$$C = \frac{dq}{dT}. \tag{3.7}$$

C is temperature dependent and is often described using an empirical formula such as

$$C = a + bT + cT^{-1} + dT^{-2}. \tag{3.8}$$

Equation (3.7) is true for heat capacities at either constant pressure or constant volume. However, the heat capacity will be different if the system is at constant pressure (C_p) or constant volume (C_v). For the case of constant pressure, combining Eqs (3.5) and (3.7) gives

$$C_p = \frac{dH}{dT} \tag{3.9}$$

and for the case of constant volume, combining Eqs (3.6) and (3.7) gives

$$C_v = \frac{dU}{dT}. \tag{3.10}$$

3.2.2 Enthalpy of formation

The enthalpy of formation is defined as the heat evolved or absorbed when an alloy or compound is formed from its elements. For example, in the formation of CO_2

$$C_{(s)} + O_{2(g)} = CO_{2(g)} \tag{3.11}$$

there is an evolution of heat and such a reaction is termed *exothermic*. In this circumstance the enthalpy of formation is then a *negative* quantity. Conversely, during the reaction

$$C_{(s)} + CO_{2(g)} = 2CO_{(g)} \tag{3.12}$$

heat is absorbed and in this circumstance the reaction is *endothermic*. The enthalpy of formation is then a *positive* quantity.

3.2.3 Hess's law

The enthalpy of formation or reaction can be either measured directly or calculated using Hess's law. This states that the overall change in the enthalpy of a chemical reaction is the same whether it takes place in one step or in several steps. Therefore

$$\Delta H = \Delta H_1 + \Delta H_2 + \Delta H_3 + \cdots + \Delta H_n \tag{3.13}$$

where $\Delta H_{1,2,3...n}$ are the individual enthalpy changes which take place at each of the steps. This can be used to calculate a heat of reaction in the following way. Suppose that a moles of reactant A and b moles of B react as follows:

$$aA + bB + \Delta H^{o} \rightarrow cC + dD. \tag{3.14}$$

The heat absorbed/evolved, ΔH^{o}, during the reaction is given by

$$a\Delta H_A + b\Delta H_B + \Delta H^{o} = c\Delta H_C + d\Delta H_D \tag{3.15}$$

ΔH^{o} is then found if $\Delta H_{A,B,C,D}$ are known. ΔH is not temperature independent and Kirchhoff's law can be used to define how ΔH varies with temperature.

3.2.4 Kirchhoff's law

Figure 3.3 shows how to determine the enthalpy change ΔH_2 at T_2 starting with a knowledge of a chemical reaction at temperature T_1 whose enthalpy change is ΔH_1.

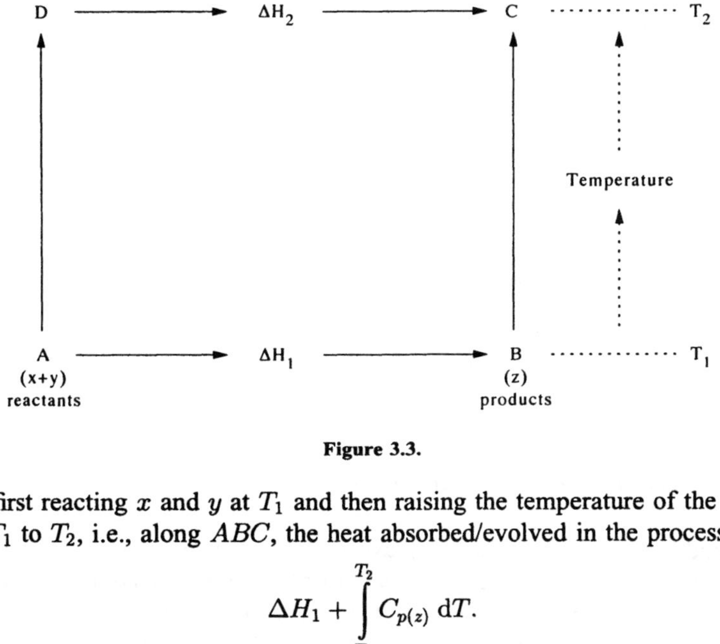

Figure 3.3.

By first reacting x and y at T_1 and then raising the temperature of the products from T_1 to T_2, i.e., along ABC, the heat absorbed/evolved in the process will be

$$\Delta H_1 + \int_{T_1}^{T_2} C_{p(z)}\, \mathrm{d}T. \tag{3.16}$$

The second way of obtaining the same result is to raise the temperature from T_1 to T_2 and then reacting x and y at T_2, i.e., along DC. This gives

$$\int_{T_1}^{T_2} \left(C_{p(x)} + C_{p(y)}\right) \mathrm{d}T + \Delta H_2. \tag{3.17}$$

From Hess's law Eq. (3.16) = Eq. (3.17) and, therefore,

$$\int_{T_1}^{T_2} \left(C_{p(x)} + C_{p(y)}\right) \mathrm{d}T + \Delta H_2 = \Delta H_1 + \int_{T_1}^{T_2} C_{p(z)} \, \mathrm{d}T \tag{3.18}$$

which gives

$$\Delta H_2 - \Delta H_1 = \int_{T_1}^{T_2} \left[C_{p(z)} - \left(C_{p(x)} + C_{p(y)}\right)\right] \mathrm{d}T = \int_{T_1}^{T_2} \Delta C_p \, \mathrm{d}T \tag{3.19}$$

with

$$\Delta C_p = C_{p(z)} - \left(C_{p(x)} + C_{p(y)}\right) \tag{3.20}$$

where ΔC_p is the difference between heat capacity of product and reactants. Eq. (3.18) can then be written as

$$\Delta H_2 = \Delta H_1 + \int_{T_1}^{T_2} \Delta C_p \, \mathrm{d}T \tag{3.21}$$

and often tabulated as

$$\Delta H_T = \Delta H_{298.15} + \int_{298.15}^{T} \Delta C_p \, \mathrm{d}T \tag{3.21a}$$

where ΔH_T is the enthalpy of formation at temperature T and $\Delta H_{298.15}$ is the enthalpy of formation at 298.15 K.

3.3. THE SECOND LAW OF THERMODYNAMICS

The second law of thermodynamics plays a vital part in any reaction, whether this is a simple combustion process or a complex phase transformation in a steel. The first law of thermodynamics considered the heat/work/energy involved in reactions, but this is not sufficient to decide whether a reaction will proceed in any given direction; it is the so-called 'free energy' of a reaction whose sign is crucial.

Take a reversible cycle in a system where p, v, T, etc., change so slowly that the system passes through a series of small, equilibrium, isothermal steps before arriving back to its initial position. Let δq_a be the heat absorbed during one of these steps when the temperature is T_a. It is found experimentally that the sum of $\delta q_a/T_a$ for all of the various steps taken during the cycle is equal to zero, i.e.,

$$\sum \delta q/T = 0 \tag{3.22}$$

References are listed on p. 57.

thus $\delta q/T$ is a *state function*. This means that the value of $\delta q/T$ depends on the state of the system rather than its history, i.e., its value is independent of the path taken. The name entropy was given to the quantity $\delta q/T$ which was given the symbol δS. It was found that *the entropy of a closed system can only increase (in an irreversible reaction) or stay constant (in equilibrium)*. This is the *second law of thermodynamics*.

The function which decides whether a process will occur is the *free energy change*, ΔG, where

$$\Delta G = \Delta U + p\Delta v - T\Delta S \tag{3.23}$$

and for a reaction to proceed ΔG must be *negative*. For a process at constant pressure Eq. (3.23) is written as

$$\Delta G = \Delta H - T\Delta S \tag{3.24a}$$

and ΔG is known as the *Gibbs energy* change of the process. At constant volume ($\Delta v = 0$) Eq. (3.23) becomes

$$\Delta F = \Delta U - T\Delta S \tag{3.24b}$$

where ΔF is called the *Helmholtz energy*.

3.3.1 The Gibbs–Helmholtz equation

The Gibbs–Helmholtz equation (Eq. (3.25) below) can be conveniently used to calculate the enthalpy if the rate of change of Gibbs energy with temperature is known. ΔS is obtained from Eq. (3.24a) by differentiating it with respect to temperature, so $\mathrm{d}\Delta G/\mathrm{d}T = \Delta S$. Substituting back into Eq. (3.24a) gives the relationship

$$\Delta H = \Delta G - T\frac{\mathrm{d}\Delta G}{\mathrm{d}T}. \tag{3.25}$$

3.3.2 Calculation of entropy and Gibbs energy change from heat capacities

From Eq. (3.7) and for the limiting case of an infinitely small step in temperature

$$\mathrm{d}q/\mathrm{d}T = C_p. \tag{3.26}$$

The entropy change is

$$\mathrm{d}S = \mathrm{d}q/T. \tag{3.27}$$

Combining Eqs (3.26) and (3.27) gives

$$\mathrm{d}S = \frac{C_p}{T}\,\mathrm{d}T. \tag{3.28}$$

If it is assumed that S_0 is the entropy at $T = 0$ K, then the entropy change ΔS in going to temperature T can be calculated from

$$\Delta S = S_T - S_0 = \int_0^T \frac{C_p}{T}\,\mathrm{d}T \tag{3.29}$$

or the absolute entropy at T is given by

$$S_T = S_0 + \int_0^T \frac{C_p}{T}\,\mathrm{d}T. \tag{3.30}$$

In standard tables this is commonly converted to

$$S_T = S_{298} + \int_{298}^T \frac{C_p}{T}\,\mathrm{d}T \tag{3.31}$$

where S_{298} is the entropy of the substance at 298 K. It then follows that the Gibbs energy of a substance is given by

$$G_T = H_{298} + \int_{298}^T C_p\,\mathrm{d}T - T\left(S_{298} + \int_{298}^T \frac{C_p}{T}\mathrm{d}T\right). \tag{3.32}$$

3.3.3 The physical nature of entropy

Traditional thermodynamics gives a clear definition of entropy but unfortunately does not tell us what it is. An idea of the physical nature of entropy can be gained from statistical thermodynamics. Kelvin and Boltzmann recognised that there was a relationship between entropy and probability (cf., disorder) of a system with the entropy given by

$$S = k \log_e W \tag{3.33}$$

where k is Boltzmann's constant and W is a measure of the disorder of a system. This means the maximum entropy is associated with the greatest disorder. The consequences of this can be understood by considering a condensed and gaseous phase. The gas has a very high degree of disorder compared to a condensed phase, hence (1) it will absorb more heat and (2) it will become stabilised as the temperature is increased. This concept has profound implications in the field of phase transformations.

References are listed on p. 57.

3.4. THE THIRD LAW OF THERMODYNAMICS

It is more problematical to define the third law of thermodynamics compared to the first and second laws. Experimental work by Richards (1902) and Nernst (1906) led Nernst to postulate that, as the temperature approached absolute zero, the entropy of the system would also approach zero. This led to a definition for the third law of thermodynamics that at a temperature of absolute zero the entropy of a condensed system would also be zero. This was further refined by Planck (1911) who suggested this be reworded as *the entropy of a pure element or substance in a perfect crystalline form is zero at absolute zero.*

This law is particularly useful as it allows the total entropy of a substance to be obtained if sufficiently low-temperature enthalpy or heat-capacity measurements are available. However, questions remain as to its validity when considering metastable crystalline forms and the law, as stated, would not apply to defect-stabilised structures and amorphous phases.

3.5. THERMODYNAMICS AND CHEMICAL EQUILIBRIUM

In a reversible reaction where A, B, C and D are *species* consisting of the atoms, molecules, ions, etc., involved in the reaction, the chemical reaction can be expressed by the equation $A + B \rightleftharpoons C + D$. In this case the forward arrow represents the reaction proceeding from left to right and backward arrow the reverse reaction. It is possible for the reaction to occur in either direction and the extent to which this occurs depends on the temperature.

For instance, at room temperature when two moles of hydrogen gas (H_2) react with one mole of graphite (C), there is a complete conversion of the reactants into one mole of methane gas (CH_4). However, if the reaction is carried out at high temperatures and constant pressure, it is found that the reaction does not proceed to completion and even after a prolonged time at that temperature and pressure, some hydrogen gas and graphite remain. The reaction thus reaches a state of *chemical equilibrium* where the rates of forward and reverse reactions are equal and a *dynamic equilibrium* is reached.

3.5.1 The law of mass action and the equilibrium constant

The law of mass action states that *the velocity of a reaction at a given temperature is proportional to the product of the active masses of the reacting substances.* To illustrate this law consider the reaction

$$A + B \rightleftharpoons C + D.$$

According to the law of mass action the rate of forward reaction (v_1) is proportional to the product of the active masses of A and B denoted $[A]$ and $[B]$

$$v_1 = k_1.[A][B]. \qquad (3.34)$$

Similarly the rate of reverse reaction (v_2) is

$$v_2 = k_2.[C][D] \qquad (3.35)$$

where k_1 and k_2 are proportionality constants. Since the equilibrium is a dynamic one, i.e., $v_1 = v_2$, then we have

$$k_1.[A][B] = k_2.[C][D] \qquad (3.36)$$

and thus

$$\frac{k_1}{k_2} = \frac{[C][D]}{[A][B]} = K \qquad (3.37)$$

where K is the equilibrium constant of the reaction at constant temperature. In general for any reversible reaction $aA + bB \rightleftharpoons cC + dD$

$$K_c = \frac{[C]^c[D]^d}{[A]^a[B]^b}. \qquad (3.38)$$

These concentration terms refer to the values at equilibrium and K_c is denoted the *concentration equilibrium constant*. An example of this expression can be appreciated by considering the following chemical reaction which occurs in an aqueous solution

$$\underset{aA}{Na_2CO_3} + \underset{bB}{CaCl_2} \rightleftharpoons \underset{cC}{CaCO_3} + \underset{dD}{2NaCl}.$$

Here the values of a, b and c are equal to 1 and d is equal to 2. K_c therefore is given by

$$K_c = \frac{[CaCo_3]^1[NaCl]^2}{[Na_2CO_3]^1[CaCl_2]^1}. \qquad (3.39)$$

The equilibrium constant then gives information on how far the reaction has proceeded. If the reaction has proceeded almost to completion the equilibrium constant is large, while if little reaction has occurred it is small. In general three equilibrium constants may be used in reactions according to the ways in which the active masses are expressed:

(a) K_c = the concentration equilibrium constant where $[A]$, $[B]\ldots$, etc., are given in terms of the molal concentration of the species as shown above. This is usually used for aqueous solutions.
(b) K_p = the pressure equilibrium constant where $[A]$, $[B]\ldots$, etc., are given in

References are listed on p. 57.

terms of the partial pressures of the species as shown above. This is used for reactions involving gaseous species.
(c) K_a = the equilibrium constant expressed in terms of activities of the species and can be used for gases, solids or liquids.

The above example considered a *homogeneous* reaction where all of the reactants and products are in the same state, i.e., in an aqueous solution. However, many reactions involve mixtures of gases, liquids, solids, etc. In this case they are defined as *heterogeneous* reactions.

3.5.2 The Van't Hoff isotherm

Consider a reaction taking place at constant temperature and pressure as defined by

$$aA + bB + \cdots \rightleftharpoons cC + dD + \cdots$$

where A, B, C, etc., are species in a reaction and a, b, c, etc., are the moles of the species in question. The general Gibbs energy change of the reaction is given by the difference in the partial Gibbs energies of the reactants and the products and defined below as

$$G = (c\bar{G}_C + d\bar{G}_D + \cdots) - (a\bar{G}_A + b\bar{G}_B + \cdots). \tag{3.40}$$

It is usual to give the Gibbs energies with respect to the standard states of the species ($G^{\circ}_{A,B,C,D...}$). The standard state of the species is simply its stable form at the temperature of interest and G° is simply the Gibbs energy of the species in its stable state. For example, water is solid (ice) below 0°C and $G^{\circ}_{H_2O}$ would be the Gibbs energy of ice, while above 100°C $G^{\circ}_{H_2O}$ refers to the Gibbs energy of water. By definition a gas is in its standard state when its pressure is 1 atm. Equation (3.40) can then be rewritten as

$$G - G^{\circ} = (c\bar{G}_C + d\bar{G}_D + \cdots) - (a\bar{G}_A + b\bar{G}_B + \cdots) \tag{3.41}$$

where G° is the standard Gibbs energy change associated with the reaction and $\bar{G}_{A,B,C,D...}$ is the partial Gibbs energies of the species $A, B, C, D\ldots$ with respect to their respective standard states. Substituting from Eq. (3.57) for $\Delta\bar{G}$ (see section 3.6.2) gives

$$\begin{aligned} G - G^{\circ} =& (cRT\log_e a_C + dRT\log_e a_D + \cdots) \\ & - (aRT\log_e a_A + bRT\log_e a_B + \cdots) \end{aligned} \tag{3.42}$$

using the identity $x\log_e y = \log_e y^x$ and $\log_e x - \log_e y = \log_e(x/y)$, Eq. (3.42) can be rewritten as

$$G - G^o = RT \log_e \left[\frac{a_C^c . a_D^d \dots}{a_B^b . a_C^c \dots} \right]. \tag{3.43}$$

In the equilibrium state the reactants and products are in equilibrium with each other (i.e. $G = 0$) and the activity product given in the brackets is the equilibrium constant K. The standard Gibbs energy of the reaction (G^o) is therefore

$$G^o = -RT \log_e \mathrm{K}_a. \tag{3.44}$$

A practical example of the use of this relationship is in the oxidation of metals. Taking the reaction

$$\mathrm{Me_{(s)}} + \mathrm{O_{2(g)}} \rightleftharpoons \mathrm{MeO_{2(s)}}$$

the standard Gibbs energy change of the reaction at temperature T is then given by

$$G^o = -RT \log_e \left[\frac{a_{\mathrm{MeO_2}}}{a_{\mathrm{Me}} a_{\mathrm{O_2}}} \right]. \tag{3.45}$$

As Me and MeO_2 are pure species in their standard states, by definition their activities are equal to 1 and as the activity of $O_2 = p_{\mathrm{O_2}}/p^o_{\mathrm{O_2}}$ Eq. (3.45) becomes

$$G^o = -RT \log_e \left[\frac{p^o_{\mathrm{O_2}}}{p_{\mathrm{O_2}}} \right]. \tag{3.46}$$

As the standard state of O_2 is 1 atm this further reduces to

$$G^o = -RT \log_e \left[\frac{1}{p_{\mathrm{O_2}}} \right].$$

Values for G^o are known for various oxidation reactions and can be plotted on diagrams called *Ellingham diagrams* as shown in Fig. (3.4). Such diagrams are very useful to predict reactions associated with oxidation and reduction.

3.6. SOLUTION PHASE THERMODYNAMICS

In many cases the CALPHAD method is applied to systems where there is solubility between the various components which make up the system, whether it is in the solid, liquid or gaseous state. Such a system is called a solution, and the separate elements (i.e., Al, Fe...) and/or molecules (i.e., NaCl, CuS...) which make up the solution are defined as the components. The model description of solutions (or solution phases) is absolutely fundamental to the CALPHAD process and is dealt with in more detail in chapter 5. The present chapter will discuss concepts such as ideal mixing energies, excess Gibbs energies, activities, etc.

References are listed on p. 57.

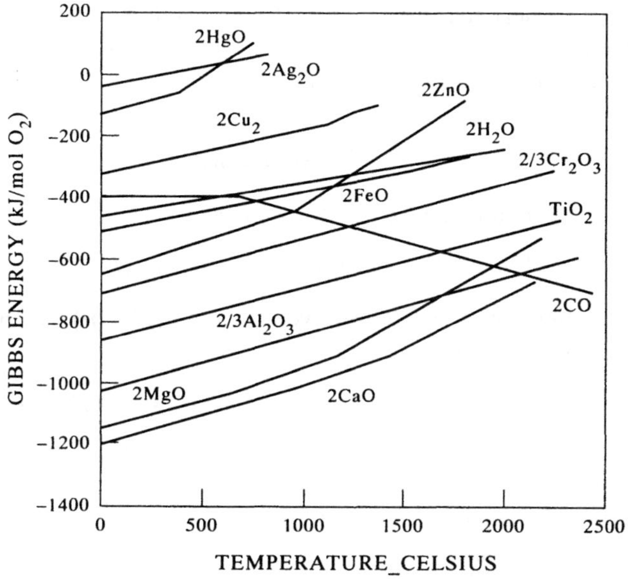

Figure 3.4. Ellingham diagram showing standard Gibbs energy of formation of various oxides as a function of temperature.

3.6.1 Gibbs energy of binary solutions

***3.6.1.1 Ideal mixing*.** First take the example of a crystal with a total of N sites available for the occupation of atoms or molecules, n of which are occupied by A atoms/molecules and $(N - n)$ are occupied by B atoms/molecules. In this case it can be shown that the total number of ways of distributing them (W) is given by

$$W = \frac{N!}{n!(N-n)!} \tag{3.47}$$

some of which are shown diagrammatically below (Fig. 3.5) where 1 atom/molecule of B mixes with 8 atoms/molecules of A.

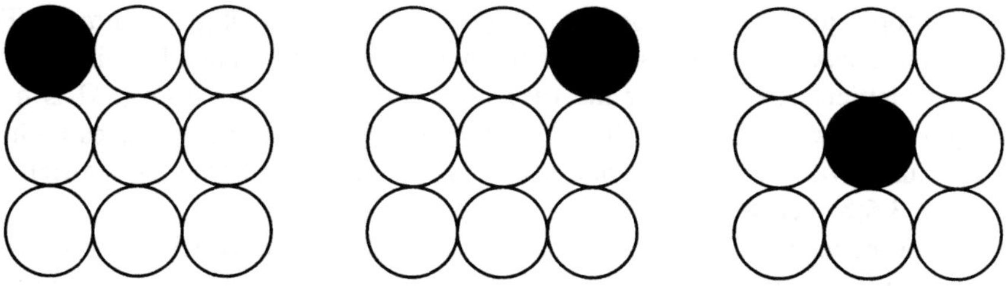

Figure 3.5. Schematic representation of random mixing of atom/molecule of B in A.

In this case it is clear that the number of configurations the system can adopt depends on the number of possible positions in which the B atom or molecule can place itself. Replacing W in Eq. (3.47) gives

$$S = k.\log_e\left[\frac{N!}{n!(N-n)!}\right]. \tag{3.48}$$

Using Stirling's approximation Eq. (3.48) becomes

$$S = k\big[N\log_e N - n\log_e n - (N-n)\log_e(N-n)\big]. \tag{3.49}$$

As the mole fractions of A and B are given by $x_a = (N-n)/N$ and $x_b = n/N$ respectively, Eq. (3.49) reduces to

$$S = -Nk(x_a \log_e x_b + x_b \log_e x_b) = -R(x_a \log_e x_b + x_b \log_e x_b) \tag{3.50}$$

as $R = Nk$. This then defines the *ideal entropy change on mixing*. If there are no repulsive or attractive interactions between atoms A and B the solution is called *ideal* and the Gibbs energy of mixing is given by

$$G_{\text{mix}}^{\text{ideal}} = RT(x_a \log_e x_b + x_b \log_e x_b). \tag{3.51}$$

3.6.1.2 Non-ideal mixing. In reality there are mixing energies associated with attractive and repulsive interactions between A and B atoms and a further *excess mixing energy*, $G_{\text{mix}}^{\text{xs}}$, must be considered. The simplest way to consider these interactions is via the *regular solution model*. In this case

$$G_{\text{mix}}^{\text{xs}} = x_a\, x_b.\Omega \tag{3.52}$$

where Ω is the regular solution interaction energy parameter and is related to the energies of the bonds between A and B atoms. When Ω is positive this is equivalent to *repulsive* interactions and negative values of Ω are associated with *attractive* interactions. Equations (3.51) and (3.52) can be combined to give

$$G_{\text{mix}} = x_a\, x_b\, \Omega + RT(x_a \log_e x_b + x_b \log_e x_b). \tag{3.53}$$

In the case of a system such as Fe–Ni, Ω is *negative* and this, combined with the ideal entropy, produces a smoothly changing curve with a single minimum (Fig. 6(a)). In this case a continuous solid solution is formed, i.e., Ni and Fe mix freely in the f.c.c. lattice. However, for the case of a system such as Cu–Ag, Ω is quite strongly *positive*. In this case the addition of the ideal entropy produces a Gibbs energy curve with two minima, one at the Cu-rich end the other at the Ag-rich end (see Fig. 3.6(b) below).

The consequence of the positive interactions is that alloys which lie between x_1 and x_2 can lower their Gibbs energy by forming *two-phase* structures. One phase which is A-rich with composition x_1 and the other which is B-rich with composition

References are listed on p. 57.

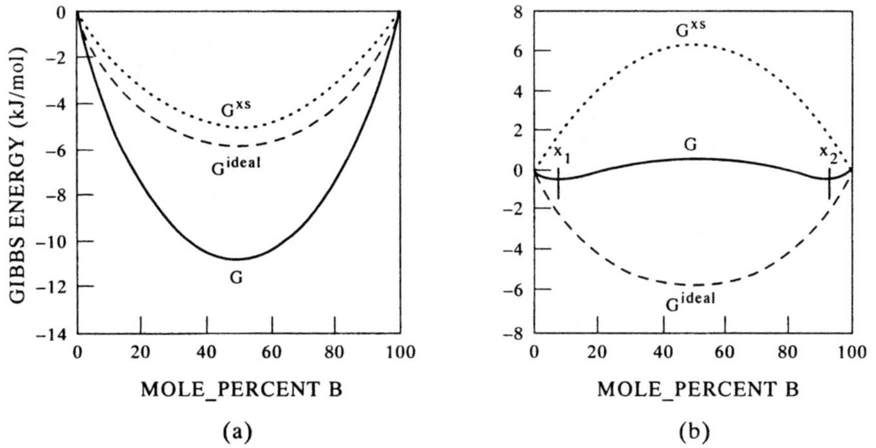

Figure 3.6. Gibbs energy vs composition diagrams of *A–B* systems showing the effect of (a) attractive and (b) repulsive interactions.

x_2. The lowering of Gibbs energy, by forming multi-phase structures rather than a series of continuous solutions, is the reason for some of the fundamental features of alloy phase diagrams and will be discussed later in section 3.7.

3.6.2 Partial Gibbs energy and activity in binary solutions

When the components *A* and *B* mix as a continuous solution, the Gibbs energy of the alloy at any composition is given by Eq. (3.53). However, another fundamental quantity which arises from the mixing of *A* and *B* is the *partial Gibbs energy* of mixing. G_{mix} at any composition is related to the partial Gibbs energy of *A* and *B* in the solution by the relationship

$$G_{\mathrm{mix}} = x_a \bar{G}_a + x_b \bar{G}_b \tag{3.54}$$

where $\bar{G}_a$ and $\bar{G}_b$ are the partial Gibbs energies of *A* and *B* in the solution given with respect to the Gibbs energy of pure *A* or *B* respectively. Because of this G_{mix} is often called the integral Gibbs energy of mixing. The relationship between the quantities is shown graphically below (Fig. 3.7).

The partial Gibbs energy is then related to the derivative of the Gibbs energy vs composition curve, i.e., the change in Gibbs energy as *A* or *B* is added to the alloy in question. If $\bar{G}_{\mathrm{a,b}}$ is negative the *A* or *B* atoms are readily bound in the solution. However, if $\Delta\bar{G}_a$ or $\Delta\bar{G}_b$ is positive there is a tendency for *A* or *B* atoms to disassociate themselves from the solution. The effect of this is shown in the *activity* of *A* or *B* in the solution which can be defined as

$$a_i = \frac{p_i}{p_i^{\mathrm{o}}} \tag{3.55}$$

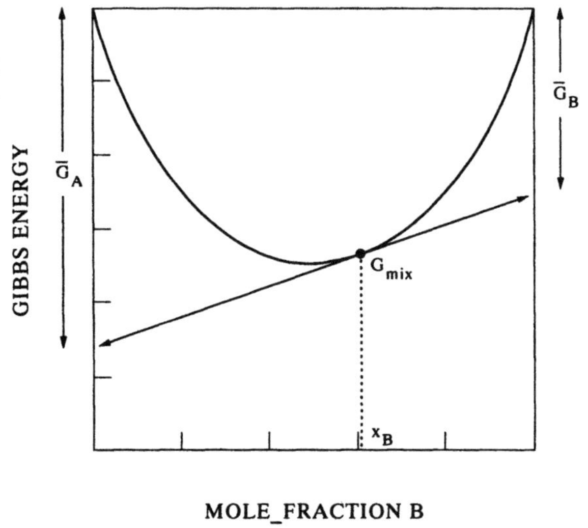

Figure 3.7. Gibbs energy vs composition diagrams of an *A–B* system showing relationship between partial and integral Gibbs energies.

where a_i is the activity of the component i in the solution, p_i is the measured vapour pressure of i above 1 mole of the alloy and p_i^o is the standard vapour pressure of 1 mole of the component i. It therefore follows that the activity of the pure component i is equal to 1. In an 'ideal' solution where no interactions exist between A and B

$$p_i^{\text{ideal}} = x_i\, p_i^o \tag{3.56}$$

and the activity is given simply by

$$a_i^{\text{ideal}} = x_i. \tag{3.57}$$

The plot of activity vs composition is then a straight line as shown in Fig. 3.8 for the activity of B in a solution of A and B.

However, if the B atom is bound into the solution because of negative interactions (i.e., $\Omega - ve$) the vapour pressure of B above the alloy is less than if the mixing was ideal. In this circumstance there is a negative deviation from ideality and the plot of activity vs composition is as shown in Fig. 3.9(a). In the case where there are positive interactions ($\Omega + ve$) there is a positive deviation from ideality as shown in Fig. 3.9(b), and the vapour pressure of B is greater than if mixing is ideal.

The activity is related to the partial Gibbs energy by the relationship

$$\bar{G}_i = RT \log_e(a_i) \tag{3.58}$$

and in an ideal solution where $a_i = x_i$, Eq. (3.58) becomes

References are listed on p. 57.

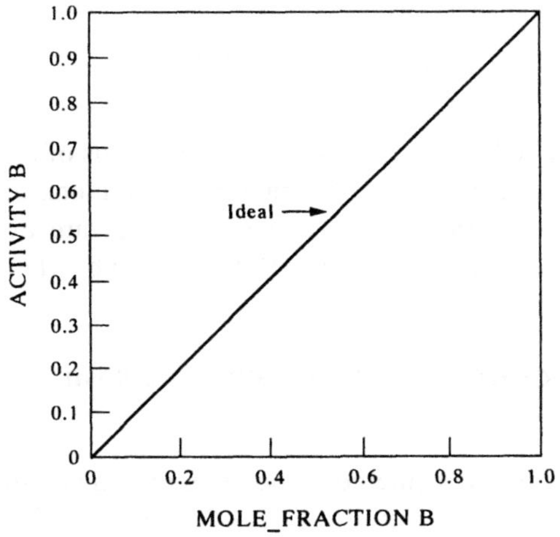

Figure 3.8. Activity vs composition diagrams of an *A–B* system showing ideal behaviour of *B* in *A*.

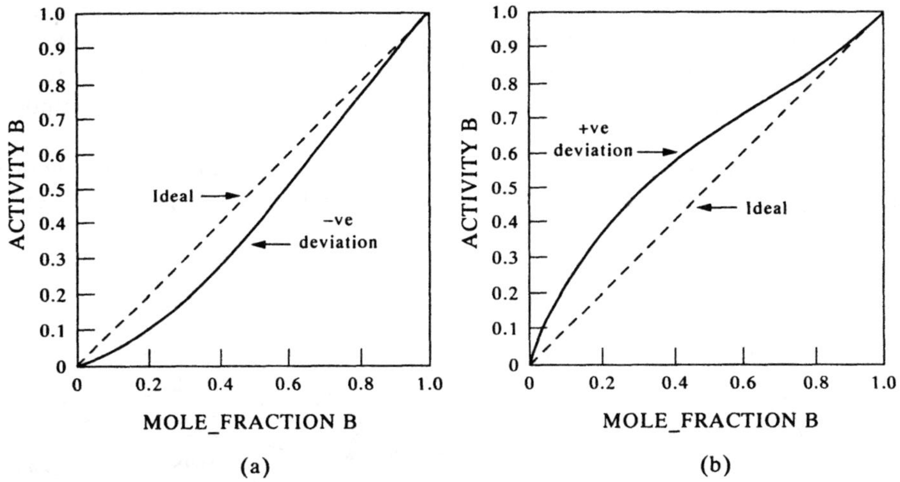

Figure 3.9. Activity vs composition diagrams of an *A–B* system showing (a) negative and (b) positive deviation from ideality.

$$\bar{G}_{\text{ideal}} = RT \log_e(x_i) \qquad (3.59)$$

The relationship between Eqs (3.58) and (3.59) then allows calculation of the partial *excess* energy of mixing of *i* arising from non-ideal interactions as

$$\bar{G}_i^{xs} = RT \log_e(a_i) - RT \log_e(x_i) \qquad (3.60)$$

or

$$\bar{G}_i^{xs} = RT \log_e \left(\frac{a_i}{x_i} \right) . \tag{3.61}$$

The term (a_i/x_i) is called the *activity coefficient*. Methods of measuring the activity and activity coefficient are given in Chapter 5. Tabulations of Gibbs energies of mixing of numerous systems defined in various ways are available in standard reference books, such as Hultgren *et al.* (1973) and Kubaschewski *et al.* (1993).

3.7. THERMODYNAMICS OF PHASE EQUILIBRIA AND SOME SIMPLE CALCULATED PHASE DIAGRAMS

A phase diagram is often considered as something which can only be measured directly. For example, if the solubility limit of a phase needs to be known, some physical method such as microscopy would be used to observe the formation of the second phase. However, it can also be argued that if the thermodynamic properties of a system could be properly measured this would also define the solubility limit of the phase. The previous sections have discussed in detail unary, single-phase systems and the quantities which are inherent in that system, such as enthalpy, activity, entropy, etc. This section will deal with what happens when there are various equilibria between different phases and includes a preliminary description of phase-diagram calculations.

Figure 3.10(a) shows one of the simplest forms of phase diagram, a system with a miscibility gap. It is characterised by a high-temperature, single-phase field of α

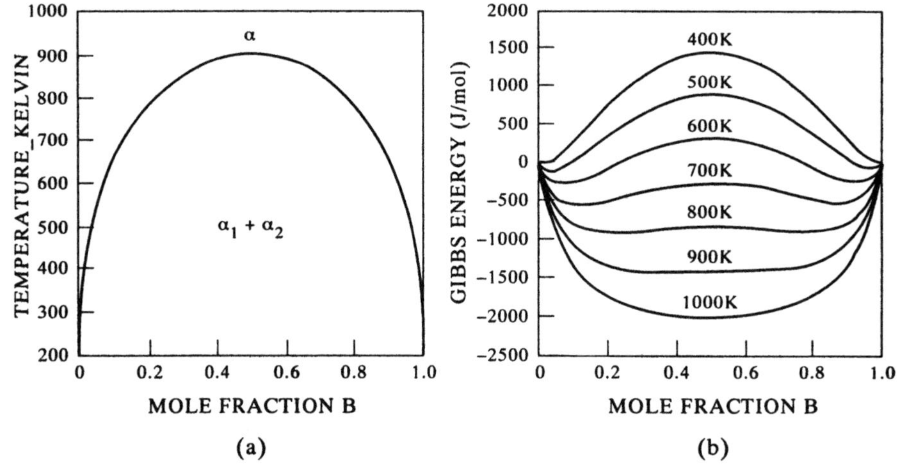

Figure 3.10. (a) Phase diagram for an *A–B* system showing a miscibility gap and (b) respective *G/x* curves at various temperatures.

References are listed on p. 57.

which separates into a two-phase field between α_1 and α_2 below a critical temperature of 900 K. This occurs because of the repulsive interactions between the components *A* and *B*. The Gibbs energy diagrams at different temperatures (Fig. 3.10(b)) show that at higher temperatures the *G/x* diagram is smooth, exhibiting a single minimum, while at lower temperatures it exhibits a central 'hump' and the formation of two distinct minima. While a single minimum exists, the Gibbs energy of the alloy is always at its lowest as a single phase. However, below 900 K the system has further possibilities to lower its Gibbs energy.

Figure 3.11 shows the *G/x* diagram at 600 K. If an alloy of composition x_0 were single phase it would have a Gibbs energy G_0. However, if it could form a mixture of two phases, one with composition x'_{α_1} and the other with composition x'_{α_2}, it could lower its total Gibbs energy to G', where G' is defined by the equation

$$G' = \frac{x_0 - x'_{\alpha_1}}{x'_{\alpha_2} - x'_{\alpha_1}} G'_{\alpha_2} + \frac{x'_{\alpha_2} - x_0}{x'_{\alpha_2} - x'_{\alpha_1}} G'_{\alpha_1} \tag{3.62}$$

and G'_{α_1} and G'_{α_2} are respectively the Gibbs energies of α at composition x'_{α_1} and x'_{α_2}.

A further separation to compositions x''_{α_1} and x''_{α_2} sees a further reduction of the Gibbs energy of the two-phase mixture to G''. This process can continue but is limited to a critical point where the compositions correspond to $x^E_{\alpha_1}$ and $x^E_{\alpha_2}$ where any further fluctuation in the compositions of the two phases causes the Gibbs energy of the mixture to rise. This point is then a critical position and the phases α_1

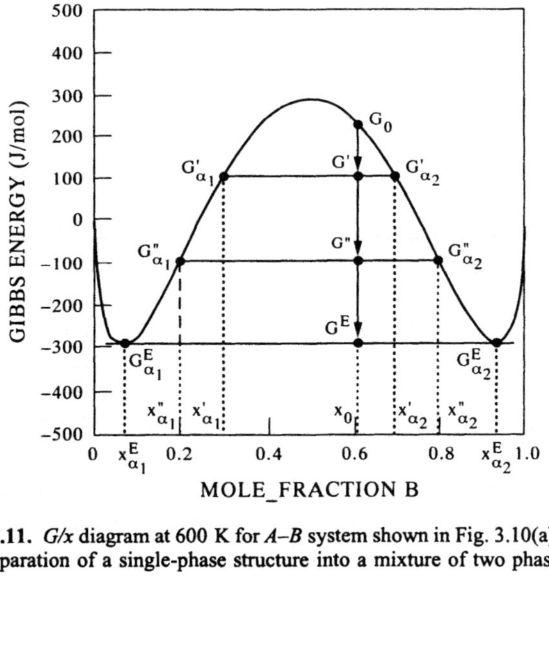

Figure 3.11. *G/x* diagram at 600 K for *A–B* system shown in Fig. 3.10(a) showing separation of a single-phase structure into a mixture of two phases.

and α_2 with compositions $x^E_{\alpha_1}$ and $x^E_{\alpha_2}$ respectively are defined as being in equilibrium with each other. At this point it is convenient to define the fraction of each phase using the equation

$$N_{\alpha_1} = \frac{x^E_{\alpha_2} - x_0}{x^E_{\alpha_2} - x^E_{\alpha_1}} \tag{3.63a}$$

and

$$N_{\alpha_2} = \frac{x_0 - x^E_{\alpha_1}}{x^E_{\alpha_2} - x^E_{\alpha_1}} \tag{3.63b}$$

where $N^P_{\alpha_1}$ and $N^P_{\alpha_2}$ are the number of moles of α_1 and α_2 respectively.

This critical position of equilibrium can be defined in two fundamentally different ways. The first is that the system *A–B* with composition x_0 has reached an equilibrium where its Gibbs energy is at a minimum. The second definition is that phases α_1 and α_2 with compositions $x^E_{\alpha_1}$ and $x^E_{\alpha_2}$ are in equilibrium because the chemical potentials of *A* and *B* are equal in both phases. The importance of these two definitions becomes clearer if we consider how it would be possible to write a computer programme to find $x^E_{\alpha_1}$ and $x^E_{\alpha_2}$.

One method would be to use Eq. (3.62) and utilise a Newton–Raphson technique to perform a Gibbs energy minimisation with respect to the composition of either *A* or *B*. This has an advantage in that only the integral function need be calculated and it is therefore mathematically simpler. The other is to minimise the difference in potential of *A* and *B* in the two phases using the relationships

$$\bar{G}^\alpha_A = G^\alpha_A + \left(1 - x^\alpha_A\right) \mathrm{d}\frac{G^\alpha_A}{\mathrm{d}x^\alpha_A} \tag{3.64}$$

and

$$\bar{G}^\alpha_B = G^\alpha_B + \left(1 - x^\alpha_B\right) \mathrm{d}\frac{G^\alpha_B}{\mathrm{d}x^\alpha_B}. \tag{3.65}$$

This has added mathematical complexity but it can be a very sensitive parameter for convergence which can be achieved quite rapidly. The calculation of the phase diagram is then achieved by calculating phase equilibria at various temperatures below 900 K and plotting the phase boundaries for each temperature.

When the points in Fig. 3.12 are joined by a line-fitting routine this yields the phase diagram shown in Fig. 3.10(a). In the more general case there is usually equilibria between crystallographically distinct phases. In this case it is necessary to define some reference states. So far only the case of Gibbs energy of mixing has been considered for solution phases. However, when two or more crystallographically distinct phases are considered, the Gibbs energy at the 'end-points' of the mixing curve must be considered. Perhaps the easiest way to consider this is to

References are listed on p. 57.

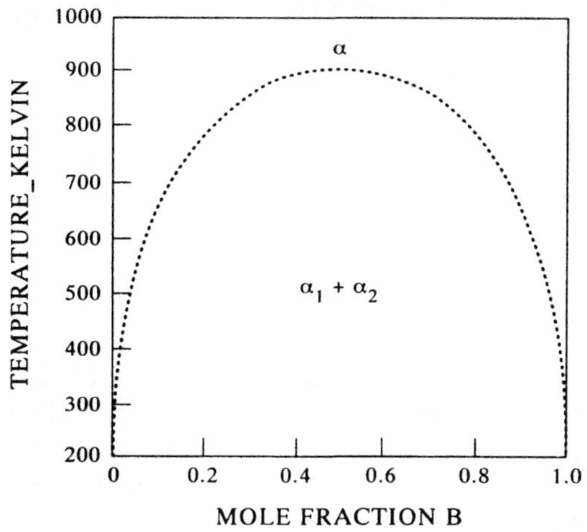

Figure 3.12. Calculated phase diagram of an *A–B* system with a miscibility gap.

take a calculation of phase equilibria for Ni–Cu between 1000–1500 K where there are two phases, f.c.c. and liquid, which form continuous solutions between Cu and Ni.

To describe the Gibbs energy of this system, the Gibbs energy of Cu and Ni must be described in both the liquid and solid states. This means we must add the reference states for the liquid and f.c.c. phases in pure Cu and Ni, to the mixing energies of the liquid and f.c.c. phases. The Gibbs energy is then described by the general formula

$$G = G_{\text{ref}} + G_{\text{mix}}^{\text{xs}} + G_{\text{mix}}^{\text{ideal}} \tag{3.66}$$

where $G_{\text{mix}}^{\text{ideal}}$ is given by Eq. (3.51). G_{ref} is then a simple linear extrapolation of the Gibbs energies of the liquid and f.c.c. phases using the equations

$$G_{\text{ref}}^{\text{f.c.c.}} = x_{\text{Cu}}\, G_{\text{Cu}}^{\text{f.c.c.}} + x_{\text{Ni}}\, G_{\text{Ni}}^{\text{f.c.c.}} \tag{3.67}$$

and

$$G_{\text{ref}}^{\text{liq}} = x_{\text{Cu}}\, G_{\text{Cu}}^{\text{liq}} + x_{\text{Ni}}\, G_{\text{Ni}}^{\text{liq}} \tag{3.68}$$

where $G_{\text{Cu,Ni}}^{\text{f.c.c.}}$ and $G_{\text{Cu,Ni}}^{\text{liq}}$ are the Gibbs energies of the f.c.c. and liquid phases in pure Cu and Ni respectively. More generally the reference and ideal mixing terms may be written as

$$G_{\text{ref}} = \sum_i x_i\, G_i^{\text{o}} \tag{3.69}$$

where x_i is the mole fraction of component i and G_i^{o} is the Gibbs energy of the phase in question at pure i. The ideal Gibbs energy of mixing can be written as

$$G_{\text{mix}}^{\text{ideal}} = RT \sum_{i} x_i \log_e x_i. \tag{3.70}$$

The $G_{\text{mix}}^{\text{xs}}$ term is usually more complicated than given in Eq. (3.52) and will take into account non-regular terms such that

$$G_{\text{mix}}^{\text{xs}} = \sum_{i=1}^{n} \sum_{j=i+1}^{n+1} x_i\, x_j \left[\Omega_{ij}^0 + \Omega_{ij}^1 (x_i - x_j) + \Omega_{ij}^2 (x_i - x_j)^2 \cdots \right]. \tag{3.71}$$

Figure 3.13(a) shows the *G/x* diagram for Ni–Cu at 1523 K with the f.c.c. Ni and Cu taken as the reference states. It can be seen that Ni-rich alloys are more stable in the f.c.c. phase while Cu-rich alloys are more stable in the liquid phase. At some point near the centre of the composition axis, x_0, the Gibbs energy of f.c.c. and liquid phases are the same. If this intersection is taken as a start point it can be seen that the Gibbs energy of an alloy of composition x_0 can lower its Gibbs energy by forming a two-phase mixture of f.c.c. and liquid with the compositions $x_{\text{f.c.c.}}^1$ and x_{liq}^1 respectively (Figure 3.13(b)). Similarly to the case of a miscibility gap, the Gibbs energy of alloy x_0 can be further minimised until the compositions of the f.c.c. and liquid phases reach $x_{\text{f.c.c.}}^{\text{E}}$ and $x_{\text{liq}}^{\text{E}}$ respectively, at which point the phases are in equilibrium. This process is then performed over a range of temperatures and the phase diagram plotted as shown in Fig. 3.14.

It can now be seen that if the thermodynamic properties of the liquid and f.c.c. phases could be determined it would be possible to calculate a phase diagram without the need for experimental determination of the phase diagram itself. Indeed, this was done by Pascoe and Mackowiak (1970, 1971) more than 25 years ago and the diagram was in good agreement with that observed experimentally. Ni–Cu presents

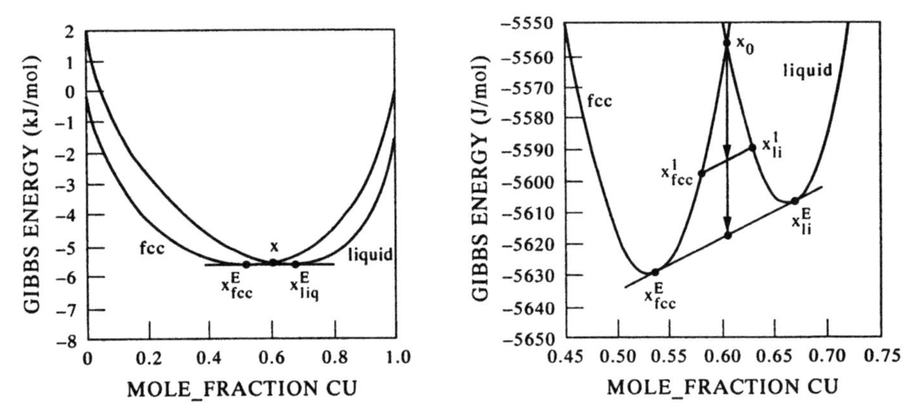

Figure 3.13. Calculated *G/x* diagrams for Ni–Cu at 1523 K.

References are listed on p. 57.

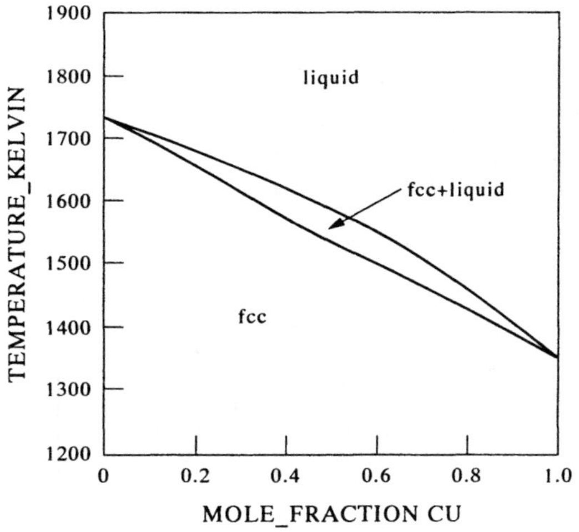

Figure 3.14. Calculated phase diagram for Ni–Cu.

an ideal case-study and in practice it is not usual for such a complete description of the thermodynamic properties of a system to exist. In such circumstances a variety of measurement techniques are used to help develop the thermodynamic models, and how this is done will be discussed in more detail in Chapter 5.

3.7.1 Topological features of phase diagrams calculated using regular solution theory

It is instructive to look at how changes in phase diagrams can be produced by systematically varying the regular solution parameter of the Gibbs excess energy of mixing as defined by Eq. (3.52). Such a systematic approach has been previously demonstrated by Pelton (1983) and it is worthwhile discussing the changes and types of diagram in more detail. Figure 3.15 shows a series of calculated phase diagrams based on an *A–B* system involving two phases, one a solid, the other liquid.

The first row shows the case where large repulsive interactions exist in the solid phase, producing a large miscibility gap, and the liquid regular solution parameter (Ω^{liq}) has been varied between −20 and +30 kJ mol^{-1}. The first diagram with Ω^{liq} = −20 kJ mol^{-1} is characterised by a deep eutectic trough and would be typical of a system such as Au–Si. Such systems are often useful for brazing or soldering applications and can also be good glass formers. As Ω becomes less negative the general topology of the diagram remains similar, even for the case where Ω^{liq} = +10 kJ mol^{-1} the system is characterised by a eutectic, although it is quite shallow. Such behaviour is typical of systems such as Ag–Bi and Ag–Pb or Cd–Zn and Cd–

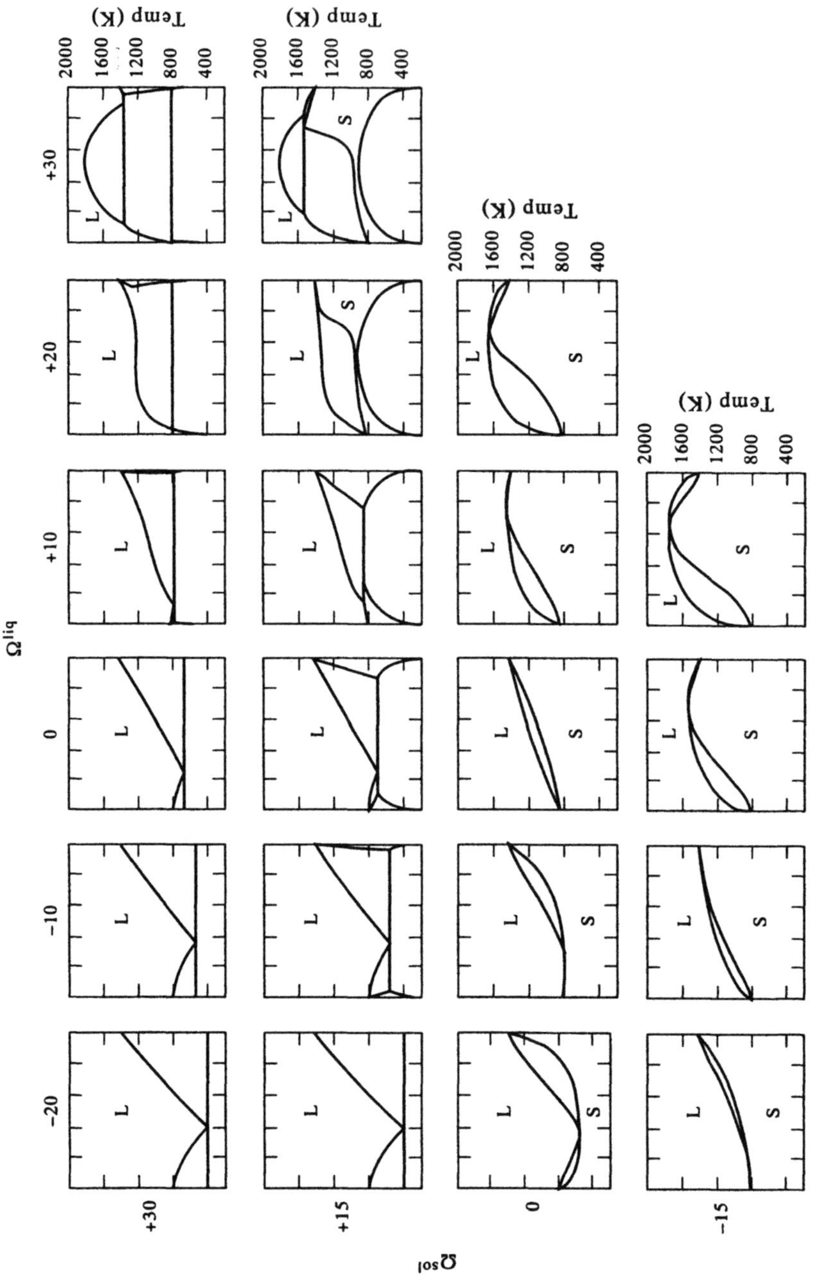

Figure 3.15. Topological features of phase diagrams calculated using regular solution theory.

References are listed on p. 57.

Pb, where retrograde solubility for the solid in equilibrium with the liquid can also occur. As a critical value of Ω^{liq} is approached the liquid forms its own miscibility gap and the diagram then exhibits two forms of liquid invariant reaction, the lower temperature reaction being either eutectic or peritectic, while the higher temperature reaction becomes monotectic. Examples of such systems are Cu–Pb and Cu–Tl. When Ω^{liq} becomes even larger, the top of the liquid miscibility gap rises above scale of the graph and there is little solubility of either element in the liquid. Such a diagram is typical of Mg systems such as Mg–Fe or Mg–Mn.

The second longitudinal series shows what happens when the regular solution parameter of the solid is made less positive (Ω^{sol} = +15 kJ mol^{-1}). Initially, when Ω^{liq} is negative, there is little apparent difference from the first series. There is a greater extent of solid solubility, but otherwise the type of diagram is very similar. However, as Ω^{liq} becomes positive the reaction type changes to become peritectic in nature and is typical of such systems as Ag–Pt and Co–Cu. When Ω^{liq} reaches +20 kJ mol^{-1} the solid miscibility gap becomes apparent and the shape of the liquidus and solidus looks rather strange, with almost a rectangular appearance. Such a diagram is typical of systems such as Au–Pt, Cu–Rh or Ti–W.

The third and fourth longitudinal series takes the solid interactions to be either ideal, i.e., $\Omega^{sol} = 0$, or negative, Ω^{sol} = –15 kJ mol^{-1}. The form of the diagrams is now completely changed, and they are characterised by complete solubility in the solid phase. The main features to note are (i) minima in the liquid when $\Omega^{liq} > \Omega^{sol}$, (ii) maxima when $\Omega^{liq} < \Omega^{sol}$ and (iii) a smooth liquidus and solidus when $\Omega^{liq} \approx \Omega^{sol}$.

The above series of calculations helps demonstrate that all types of topology of phase diagrams involving simple liquid and solid solutions can be calculated within the same simple framework, and diagrams with increasing complexity, i.e., increasing number of phases, compounds, allotropic changes in the elements, etc., can also be routinely handled.

REFERENCES

Nernst, W. H. (1906) *Nachr. Ges. Wiss. Göttingen* (Math.-Phys. K1), p. 1.

Pelton, A. D. (1983) in *Physical Metallurgy: 3rd Edition*. eds Cahn, R. W. and Haasen, P. (Elsevier Science, Amsterdam), p. 328.

Pascoe, G. and Mackowiak, J. (1970) *J. Inst. Metals*, **98,** 253.

Pascoe, G. and Mackowiak, J. (1971) *J. Inst. Metals*, **99,** 103.

Planck, M. (1911) *Thermodynamik, 3rd Edition*. (Veit & Comp., Leipzig), p. 279.

Richards, T. W. (1902) *Z. Physikal. Chem.*, **42,** 129.

Mackowiak, J. (1966) *Physical Chemistry for Metallurgists*. (George Allen and Unwin), London.

Chapter 4

Experimental Determination of Thermodynamic Quantities and Phase Diagrams

Chapter 4
Experimental Determination of Thermodynamic Quantities and Phase Diagrams

4.1. INTRODUCTION

The intention of this chapter is to introduce the reader to the types of techniques which are available for determining phase diagrams and thermodynamic quantities such as enthalpies of formation, activities, heat capacities, etc. It is not intended to be comprehensive in its coverage of methods but requisite references are given in the chapter if the reader wishes to go more deeply into this area.

The first section concerning thermodynamic measurements will be divided into three parts after Kubaschewski *et al.* (1993). The first of these will deal with calorimetric techniques which offer probably the most versatile method for determining thermodynamic quantities such as enthalpy, enthalpies of formation/mixing, heat capacities, standard entropies etc. While calorimetric techniques are very powerful for establishing 'integral' and partial enthalpies, they are limited when dealing with quantities such as the partial Gibbs energy and its associated quantities, such as the activity and activity coefficient. The second and third subsections will therefore deal with methods for determining these quantities by gas phase equilibria and electromotive forces (EMF) respectively.

The second section will concentrate on methods of determining phase diagrams. The first part will examine non-isothermal methods, such as differential thermal analysis and cooling curve determinations, while the second will concentrate on isothermal methods, such as metallography, X-ray measurements, etc. The various limitations of both methods will be discussed and some novel techniques introduced.

4.2. EXPERIMENTAL DETERMINATION OF THERMODYNAMIC QUANTITIES

4.2.1 Calorimetric methods

In principle, all of the thermodynamic quantities associated with pure substances can be measured using calorimeters which measure heat evolution and absorption arising either during heating and cooling of a substance, or from a reaction. The reaction can be a straightforward chemical one between two components or a phase transformation, for example, the austenite → ferrite transformation in a steel or the crystallisation of a glass. The reliability of a calorimeter is governed by a number of important factors, such as heat conduction, its own heat capacity, the efficiency by which heat is transferred between sample and calorimeter, and it cannot be sufficiently emphasised that the sound design and correct use of a calorimeter is critical when considering the reliability of the final result.

Broadly speaking calorimeters can be classified into four types (Kubaschewski *et al.* 1993):

(i) *Isothermal calorimeters.* In this calorimeter the temperature of the surrounding (T_s) is equalised to the temperature of the calorimeter (T_c); a good example is the Bunsen ice calorimeter.

(ii) *Adiabatic calorimeters.* In this case $T_s = T_c$ but the temperature is not constant. These are mainly used for the determination of heat capacities.

(iii) *Heat-flow calorimeters.* These are adiabatic in nature but $T_s - T_c$ is maintained constant instead of keeping $T_s = T_c$.

(iv) *Isoperibol calorimeters.* In this case the enclosure of the calorimeter is held at a constant temperature and T_c is measured at various stages during the reaction.

The isothermal and isoperibol calorimeters are well suited to measuring heat contents from which heat capacities may be subsequently derived, while the adiabatic and heat-flow calorimeters are best suited to the direct measurement of heat capacities and enthalpies of transformation.

4.2.1.1 Measurement of enthalpy and heat capacity

Heat contents can be measured accurately by a number of techniques based on a 'drop' method. This involves heating the sample to a high temperature and dropping it directly into a calorimeter held at a lower temperature. The calorimeter then measures the heat evolved while the sample cools to the temperature of the calorimeter. The temperature at which the sample is initially heated is varied and a plot of $H_T - H_{298.15}$ vs temperature is drawn (Fig. 4.1). Heat capacities can then be calculated using Eq. (3.9). A popular calorimeter for this is the diphenyl ether calorimeter (Hultgren *et al.* 1958, Davies and Pritchard 1972) but its temperature range is limited below about 1050 K.

More recently, calorimeters have been designed to extend the range of this method and a notable example has been the development of levitation calorimetry. This has led to the measurement of the heat contents of a number of high-melting-point elements such as W (Arpachi and Frohberg 1984), Mo (Treverton and Margave 1970, Berezin *et al.* 1971, Betz and Frohberg 1980a) and Nb (Betz and Frohberg 1980b) where the latter is shown in Fig. 4.2. The advantage of an electromagnetic levitation method is that the sample can be held at temperature without contact with an external holding device and pyrometry is utilised as the technique for measuring temperature. The above 'drop' techniques suffer from the disadvantage that C_p values are derived from the heat content curves and are not directly measured. They

References are listed on pp. 85–87.

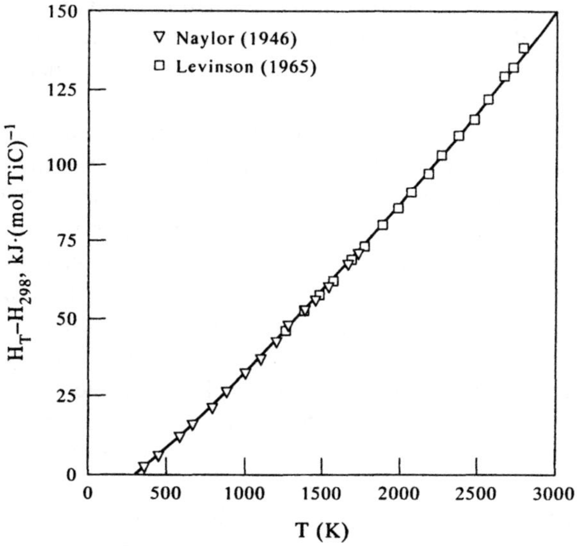

Figure 4.1. Heat content of stoichiometric TiC after Jonsson (1996) with experimental data of Naylor (1946) and Levinson (1965) superimposed.

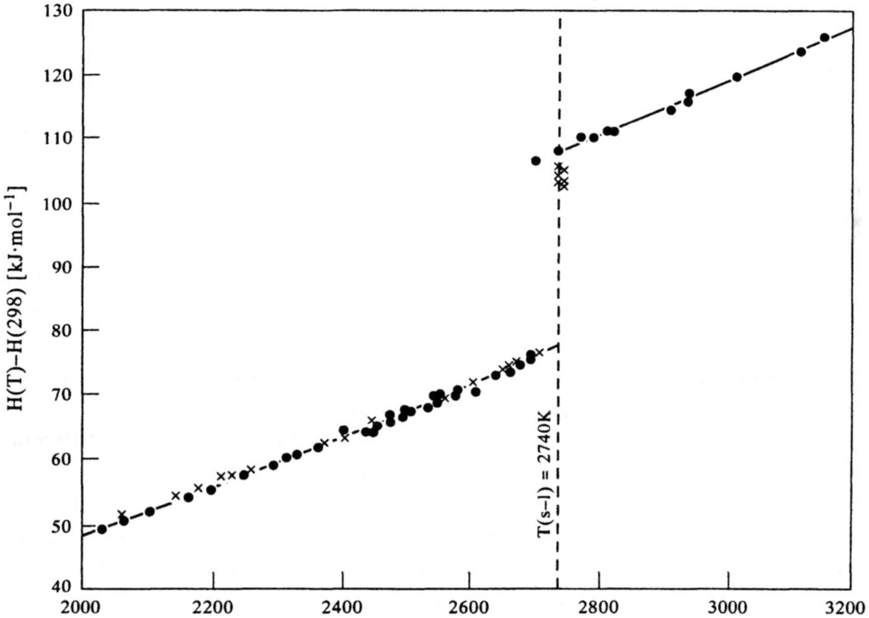

Figure 4.2. Heat content of Nb measured (•) by Betz and Frohberg (1980b) with results (×) of Sheindlin *et al.* (1972) also shown.

are not sensitive enough to accurately measure small changes in energy associated with, for example, order/disorder transformations or magnetic transitions. In these cases a direct method of measuring heat capacity is preferable, such as an adiabatic calorimeter.

The principle behind an adiabatic calorimeter is the measurement of the heat absorbed or evolved by the sample during a continuous heating process. This requires that the calorimeter is stringently calibrated so that its heat capacity is known accurately. Any deviation from this heat capacity is then due to the heat capacity of the sample. If the amount of heat required to raise the temperature of the calorimeter is high in comparison to that for the sample, then problems with accuracy will occur. Correct design minimises this problem and, for example, in a high-temperature adiabatic calorimeter (Sale 1970) approximately 90% of the energy supplied to the specimen heater was used to heat the specimen and only 10% in heating the heater assembly and thermocouples. This method can yield very reliable results but transformations must occur fast enough so that equilibrium is continuously achieved during heating, otherwise reliability of measurement will be degraded. Recent advances have been made in high-temperature regimes using a 'pulse' heating method. This can be done by heating the sample using a well-defined electrical 'pulse' of energy (Cezairliyan 1969) and monitoring the specimen temperature closely to calculate the heat capacity.

4.2.1.2 Measurement of enthalpies of transformation. While enthalpies of transformation can be measured using calorimeters (see for example Fig. 4.2) other techniques can also be applied such as differential scanning calorimetry (DSC) and differential thermal analysis (DTA). DSC offers an immediate quantification of the heat absorbed/evolved during a transformation and, as such, is much more widely used for measuring thermodynamic properties during phase transformations. In DSC the sample and reference material are heated separately and conditions are maintained such that the temperature of the sample, T_s, is kept equal to that of the reference material, T_r. The heat that is required to maintain this balance is then a direct measurement of the heat evolved/absorbed during the transformation and is plotted as a function of temperature. Integration of the area within the DSC peak is then a direct measurement of the enthalpy of transformation.

The principle behind DTA is measurement of the temperature difference between the sample of interest and a reference sample which is inert during a heating and/or cooling cycle and whose thermal behaviour is accurately known. A typical configuration would be that the sample and reference material are placed in separate chambers within a heating block which is contained within the furnace. A thermocouple is then placed in both chambers directly in contact with the sample and connected in series. The furnace and sample block are then heated and the temperature of the sample and reference material monitored. There is a difference in temperature between the sample and reference material which is a function of (1) their different C_p values and (2) the heat absorbed or evolved due to a transformation. This temperature difference is recorded continuously and the area of the DTA peak is then proportional to the heat evolved or absorbed during the transformation. It is not possible to directly obtain the enthalpy of transformation using DTA; this must be done by careful calibration at each sample to obtain reliable thermodynamic information. In practice DTA is not as powerful as DSC for obtaining enthalpies of transformation. However, because of its sensitivity to temperature changes in the sample, it is very useful for determining the temperatures for phase changes, i.e., liquidus temperatures and melting points.

The techniques above are very useful for looking at pure substances or alloys of a known composition. However, to obtain enthalpies of formation it is necessary to examine some form of reaction which involves two or more starting materials. Some form of reaction between these materials occurs and there is a subsequent heat evolution/absorption associated with the reaction.

The problems associated with direct reaction calorimetry are mainly associated with (1) the temperature at which reaction can occur; (2) reaction of the sample with its surroundings; and (3) the rate of reaction which usually takes place in an uncontrolled manner. For low melting elements such as Zn, Pb, etc., reaction may take place quite readily below 500°C. Therefore, the materials used to construct the calorimeter are not subjected to particularly high temperatures and it is easy to select a suitably non-reactive metal to encase the sample. However, for materials such as carbides, borides and many intermetallic compounds these temperatures are insufficient to instigate reaction between the components of the compound and the materials of construction must be able to withstand high temperatures. It seems simple to construct the calorimeter from some refractory material. However, problems may arise if its thermal conductivity is very low. It is then difficult to control the heat flow within the calorimeter if some form of adiabatic or isothermal condition needs to be maintained, which is further exacerbated if the reaction rates are fast.

One of the simplest calorimetric methods is 'combustion bomb calorimetry'. In essence this involves the direct reaction of a sample material and a gas, such as O or F, within a sealed container and the measurement of the heat which is produced by the reaction. As the heat involved can be very large, and the rate of reaction very fast, the reaction may be explosive, hence the term 'combustion bomb'. The calorimeter must be calibrated so that heat absorbed by the calorimeter is well characterised and the heat necessary to initiate reaction taken into account. The technique has no constraints concerning adiabatic or isothermal conditions but is severely limited if the amount of reactants are small and/or the heat evolved is small. It is also not particularly suitable for intermetallic compounds where combustion is not part of the process during its formation. Its main use is in materials thermochemistry where it has been used in the determination of enthalpies of formation of carbides, borides, nitrides, etc.

Kubaschewski and Dench (1955) designed a direct-reaction calorimeter which allowed them to systematically measure the enthalpies of formation of a variety of transition metal-aluminides and it is somewhat similar in concept to the 'combustion bomb' calorimeter. The method involves electrically heating the unreacted metal powders of the transition metal and aluminium to a temperature where reaction takes place rapidly. The amount of electrical energy needed to do this is then recorded. The maximum temperature of reaction is measured and the calorimeter allowed to cool to room temperature. The calorimeter, with the reacted sample, is then re-heated to the maximum temperature recorded during the actual reaction and the amount of electrical energy to do this recorded. The difference between the electrical energy input during the reaction run and the subsequent calibration run is the energy given off by the reaction. This technique was extensively applied and subsequent work has shown that results obtained by this method were, in the main, quite reliable.

Adiabatic calorimeters have also been used for direct-reaction calorimetry. Kubaschewski and Walter (1939) designed a calorimeter to study intermetallic compounds up to 700°C. The procedure involved dropping compressed powders of two metals into the calorimeter and maintaining an equal temperature between the main calorimetric block and a surrounding jacket of refractory alloy. Any rise in temperature due to the reaction of the metal powders in the calorimeter was compensated by electrically heating the surrounding jacket so that its temperature remained the same as the calorimeter. The heat of reaction was then directly a function of the electrical energy needed to maintain the jacket at the same temperature as the calorimeter. One of the main problems with this calorimeter was the low thermal conductivity of the refractory alloy which meant that it was very difficult to maintain true adiabatic conditions.

Kleppa (1955) overcame this problem by using aluminium as the material for the calorimeter and surrounding jacket. This substantially improved its ability to maintain adiabatic conditions and it was successfully used for more than 10 years. However, the main limitation was that its temperature capability was governed by the low melting point of aluminium, which meant that its main use was for reactions which took place below ~500°C.

Recently, Gachon and Hertz (1983) devised a calorimeter for measuring the enthalpy of formation of a variety of intermetallic compounds in the temperature range 800–1800 K (Gachon *et al.* 1985). A calorimetric cell, consisting of 17 Pt/Rh thermocouples in series, surrounds a working crucible where a sample of compacted elemental powders is held. The assembly is heated to a temperature where reaction takes place and the signal from the calorimetric cell is recorded. The integral of the signal from the calorimetric cell as a function of time is proportional to the heat of reaction. With calibration it is then possible to obtain the heat formation of the compound or alloy of interest. A similar device was also constructed by Kleppa and Topor (1989). The main drawbacks to this method are

References are listed on pp. 85–87.

reaction of the sample with the crucible and possible incomplete reaction. While Gachon *et al.* (1985) had significant problems with reactivity of the crucible and sample, they appeared to achieve complete reaction in alloys based on transition metals with Ti, Zr and Hf. However, Kleppa (1994) found substantial problems if the operating temperature was too low and the reaction enthalpy small. It was still possible, however, to obtain reliable results for an extensive range of borides, germanides and aluminides and many of these have been reported by Kleppa (1994).

Another method to obtain enthalpies of formation of compounds is by solute–solvent drop calorimetry. This method was pioneered by Tickner and Bever (1952) where the heat formation of a compound could be measured by dissolving it in liquid Sn. The principle of the method is as follows. If the heat evolved in the dissolution of compound *AB* is measured, and the equivalent heat evolved in the dissolution of the equivalent amount of pure *A* and *B* is known or measured, the difference provides the enthalpy of formation of the compound *AB*. Kleppa (1962) used this method for determining enthalpies of formation of a number of Cu-, Ag- and Au-based binaries and further extended the use of the method to high-melting-point materials with a more generalised method.

This can be demonstrated for the determination of the enthalpy of formation of LaB_6 by reaction with Pt. In this case, although the melting points of Pt and LaB_6 are 1769°C and 2715°C respectively, an alloy of the two will produce a low-melting-point liquid. Knowledge of the phase diagram is then essential, so that the likely temperature and composition range of the liquid region formed on alloying Pt and LaB_6 is known. In this case a mixture of 0.72 moles of Pt and 0.04 moles of LaB_6 will become fully liquid at 1400 K. Therefore, if the heats involved in the following reactions are known

(1) $0.72\mathrm{Pt}\,(s, 298) + 0.04\mathrm{LaB_6}\,(s, 298) \Rightarrow \mathrm{Pt_{0.72}\,La_{0.04}\,B_{0.24}}\,(l, 1400)$

(2) $0.72\mathrm{Pt}(s, 298) + 0.04\mathrm{La}(s, 298) + 0.24\mathrm{B}(s, 298) \Rightarrow \mathrm{Pt_{0.72}La_{0.04}B_{0.24}}(l, 1400)$

the difference can be used to define the enthalpy of formation of LaB_6.

4.2.2 Gas phase equilibria techniques

The general field of measuring thermodynamic quantities using gas-phase methods is based simply on Eq. (3.55) where the activity of component *i* (a_i) is equal to the vapour pressure of *i* in the alloy divided by the equilibrium vapour pressure of pure *i*.

$$a_i = \frac{p_i}{p_i^o} \tag{4.1}$$

Hence all techniques rely in some way on measuring the vapour pressure of the component of interest. There are a number of methods which can be applied and these can be broadly broken down into three categories.

4.2.2.1 Static methods for measurement of vapour pressures. An experimental rig would typically contain the sample material of interest held in a container at the required temperature. At least two possibilities then exist for measuring the vapour pressure. The first is to seal the system and in some way measure the vapour pressure. Typically a container holding the sample is evacuated and equilibrium is reached with the vapour given off by the sample. The technique for measuring the vapour pressure can be a direct one, for example by a mercury manometer, or by a more indirect method such as a radiation absorption method. The absorption method utilises the fact that atoms or molecules of the various vapour species have characteristic absorption wavelengths. If a source such as a light emitter can then provide a wavelength corresponding to this characteristic wavelength the intensity of the light transmitted through the gas can be used as a measure of the amount vapour species with that characteristic wavelength. Reviews of on static and quasistatic methods have been made by Clopper *et al.* (1967) and Norman and Winchell (1970).

4.2.2.2 The dew-point and non-isothermal isopiestic methods. These are methods where a temperature gradient is maintained between a sample held at a high temperature and a pure metal which is held at a lower temperature. In the dew-point method the vapour pressure of the volatile component is determined by direct observation of the temperature at which it condenses.

4.2.2.3 The Knudsen effusion and Langmuir free-evaporation methods. In both of these cases the rate of vaporisation of a substance in vacuum is measured and the following formula is utilised

$$p = \frac{m}{tA}\sqrt{\frac{2\pi RT}{M}} \tag{4.2}$$

where p is the vapour pressure, m is the mass of the vapour species with a molecular weight M which evaporates from an area, A, in a time, t.

In the Knudsen effusion method a substance is enclosed in a sealed container into which a very small hole is drilled. This hole must be knife-edged and the mean free path of the vapour must be 10 times the diameter of the hole. In its simplest form an experiment proceeds as follows. The Knudsen cell, with sample in it, is carefully weighed and then heated in a vacuum at the requisite temperature for a set time. The cell is then re-weighed and the weight loss is measured. However, it is now more usual to continuously measure the weight of the cell. If the molecular weight and surface area of the sample is known the vapour pressure can be found.

References are listed on pp. 85–87.

Unfortunately, the method is not straightforward. It is necessary to consider a number of other factors, such as preferential depletion of a species from the surface of the sample, reaction of the gas with the cell itself, and the effusion of the gas which does not always follow ideal behaviour. Recently much work has been done to improve accuracy through better understanding of the method itself, and much early work pre-1960 is now considered to be unreliable (Komarek 1972). The method also relies on knowledge of the molecular weight of the vapour species and there may be more than one species involved. To this end the method is now often linked with mass spectroscopy (Komarek 1972, Kubaschewski *et al.* 1993) so that composition of the gas is better understood.

In the Langmuir free-evaporation method, the sample is suspended freely in a vacuum system with no container surrounding it. As very low levels of vapour pressure can be measured it has advantages over the Knudsen method where the lower limit is about 10^{-3} atm. (Kubaschewski *et al.* 1993). It is therefore more useful in materials with high sublimation energies and therefore inherently low vapour pressures. It has a further advantages in that there is no container with which to react, but there are more significant problems associated with temperature measurement.

4.2.3 Electromotive force measurements

The electromotive force (EMF) generated by electrochemical cells can be used to measure partial Gibbs energies which, like vapour pressure measurements, distinguishes these methods from other techniques that measure integral thermodynamic quantities. Following Moser (1979), a typical cell used to obtain results on Zn–In–Pb is represented in the following way:

Zn (Liq) | $ZnCl_2$ in (LiCl--KCl) Eut | Zn--In--Pb (Liq).

In this case a fused salt electrolyte, incorporating $ZnCl_2$ dissolved in a eutectic mixture of LiCl and KCl, is used to determine the activity gradient between pure liquid Zn and a ternary Zn–In–Pb alloy, also in the liquid state. The EMF arises from the potential between pure Zn at unit activity to Zn in the alloy where the activity has a value, a_1. The ruling equation then becomes

$$\bar{G}_1 - \bar{G}_0 = -nFE = RT \log_e a_1 \tag{4.3}$$

where n is the charge, F the Faraday constant and the other symbols have their usual meaning. For many alloys the relation between EMF and temperature is linear and, in the simplest case, changes in partial enthalpy and entropy can be readily calculated by the following equations:

$$\Delta\bar{H}_1 = nF\left(\frac{\mathrm{d}E}{\mathrm{d}T} - E\right) \tag{4.4}$$

and

$$\Delta \bar{S}_1 = nF\frac{\mathrm{d}E}{\mathrm{d}T}. \tag{4.5}$$

These partial quantities may either be used directly or transformed by means of the Gibbs–Duhem equation in the overall assessment of the system in question (Moser 1979).

The principle of the method is simple, but many factors have to be considered before a given cell will yield reproducible and accurate results (Kubaschewski *et al.* 1993, Komarek 1973, Komarek and Ipser 1984). These include the choice and preparation of the electrolyte, the materials of construction of the cell, as well as scrupulous attention to experimental arrangements. First and foremost the basic equations hold only for reversible cells and assume that the only mechanism of conduction in the electrolyte is via a single ionic species. The best test for the reversibility and proper functioning of the working cell is to check that the same EMF values are obtained both for the increase and decrease in several repeated temperature runs and/or by placing the same materials in both half-cells to obtain a null result. However, at high temperatures most electrolytes generally also feature an element of electronic conduction, so that it is necessary to introduce a transport number (t_{ion}) into Eq. (4.3), otherwise the measured EMF will overestimate the ionic component. To allow for reference electrodes whose activities are not unity, this equation should be expanded to

$$\bar{G}_2 - \bar{G}_1 = -nEF = RTt_{\mathrm{ion}} \log_{\mathrm{e}} \frac{a_2}{a_1}. \tag{4.6}$$

Some general principles allow the best range of operation to be determined for solid electrolytes. n-type conduction is associated with low oxygen pressures, followed at higher pressures by a regime of ionic conduction, with near ideal transport numbers, while at still higher pressures the overall conduction pattern is increasingly associated with hole conduction. Also, the region of ionic conduction is most dominant when the diffusion coefficient of oxygen is high. The optimum choice of electrolyte can therefore be matched to the conditions of the study. Some cells employ a series of electrolytes in contact with each other in order to cope with wide ranges of oxygen activity and to minimise reactions at the relevant electrodes, but further terms need then to be added to take into account junction potentials (Kubaschewski *et al.* 1993).

Both liquid and solid electrolytes can be used, ranging from molten halides, such as a eutectic mixture of LiCl and KCl, to very sophisticated solid-state electrolytes such as calcia or yttria stabilised zirconia, CSZ, YSZ, which are conductors of oxygen ions.

$$\mathrm{Pt<Ni, NiO> \mid CaO{-}ZrO_2 \mid <Fe, FeO>Pt}.$$

Cells of this type were first introduced by Kiukkola and Wagner (1957). β-alumina has been increasingly used, largely in relation to the diffusion of sodium ions, but it can also be used to study the diffusion of Ag and Cu. Calcium floride, saturated with calcium sulphide, and other combinations, including yttrium sulphide, have been suggested for the measurement of sulphur potentials in oxidising environments above 500°C (Kleykamp 1983). A new class of electrolyte, which conducts S or C and operate above 1000°C, has been developed by Alcock and Li (1990) utilising a dispersed phase, in which the fluorine gradient can be measured indirectly, following previous work on three-phase equilibria (Jacob and Iwase 1982).

The overall accuracy of EMF methods requires that the voltage measurement should be carried out without the passage of current through the cell. Therefore, either the measuring circuit should have as high a resistance as possible, or alternatively measurements can be carried out under different conditions and extrapolated to zero current (Rose *et al.* 1948). It is also very important to avoid any parasitic voltages such as those which may occur from thermoelectric effects arising from temperature gradients (Kubaschewski *et al.* 1993). Other factors that enter into the reproducibility of the results is the necessity to ensure that the electrolyte remains uncontaminated from reaction with the materials used for construction of the cell, including the electrodes. To avoid displacement reactions the metal, which acts as the reference electrode, must be less noble than the other components of the investigated alloy. Other problems to be avoided are reaction with the atmosphere or water vapour and any changes in concentration due to high vapour pressure in the temperature range of measurements.

Assuming all these factors have been taken into account, it should be possible to make measurements to within 1 mV (Charette and Flengas 1968). The accuracy decreases at very low partial pressures of oxygen, when electronic contributions play a more significant role, and problems in design are greatly increased. Dench and Kubaschewski (1969) were successful in refining their experimental system and obtaining a precision of 0.2 mV using the solid state cell

$$\mathrm{Pt} \mid \mathrm{Co, CoO} \mid \mathrm{ZrO_2{-}{-}CaO} \mid \mathrm{Ni, NiO} \mid \mathrm{Pt.}$$

Bergman and Ågren (1984) used a similar cell to study the properties of MnO–NiO and also conducted a detailed analysis of the standard deviation of measured EMFs as a function of composition, showing that this could vary substantially across the system.

If results are required at very high temperatures, as in experiments related to steel making, even short-term survival makes severe demands on the construction of the cell (Komarek and Ipser 1984). However, oxygen concentration cells have been employed with molten ionic slags to determine the thermodynamics of oxide formation in iron between 1500–1600°C (Kay 1979). Other applications include the use of YSZ for studies of semiconducting systems (Sears and Anderson 1989, Lee *et al.* 1992).

In conclusion, the EMF method can provide good accuracy for activity and activity coefficients and, therefore, offers a useful alternative to vapour pressure measurements (Jacob *et al.* 1973). However, enthalpy and entropy values calculated from slopes dE/dT can be associated with much higher errors (Moser 1979). It is therefore highly desirable to combine information from EMF measurements with other data, such as calorimetric information, in order to obtain reliable entropies. Such a combination of EMF and calorimetric results may completely change results for certain systems. For instance, investigations on the liquid Mg–In system gave positive values of $\Delta \bar{S}^{xs}_{Mg}$ when calculated using only EMF data but, in combination with calorimetric studies, resulted in negative values of $\Delta \bar{S}^{xs}_{Mg}$.

4.3. EXPERIMENTAL DETERMINATION OF PHASE DIAGRAMS

It is an interesting consequence of CALPHAD methods that, while it is possible to make very accurate predictions for the phase equilibria of multi-component systems if accurate thermodynamic characterisations of the requisite binary and ternary sub-systems are available, it is in reality impossible to make predictions for the phase diagrams of binary systems to any degree of accuracy which would be acceptable for practical purposes. This means that experimental determination of phase diagrams, particularly binary and ternary systems, is a critical factor in the development of CALPHAD methods. The aim of this section is to give a brief introduction into some methods of measuring phase diagrams giving a few specific examples of practical work and, as such, is not intended to provide a comprehensive coverage of phase diagram determination. Critical reviews of binary systems by Hansen and Anderko (1958), Elliott (1965) and Shunk (1969) and now the *Journal of Phase Equilibria*, and its predecessor publication *The Bulletin of Alloy Phase Diagrams* published by ASM International, provide the reader with extensive publications concerning phase diagrams with numerous references to original work.

4.3.1 Non-isothermal techniques

Non-isothermal techniques are basically those where a sample is heated or cooled through a transformation and some property of the alloy changes as a consequence of the transformation. Most usually it is the heat effect associated with the transformation which is measured, but other effects such as magnetic or dilatometric changes can be measured.

It should be noted that these techniques are inherently non-equilibrium in nature in that it is often extremely difficult for a complete equilibrium transformation to take place during the heating or cooling cycle. However, having made this observation, techniques such as DTA have been very successful in defining positions of

References are listed on pp. 85–87.

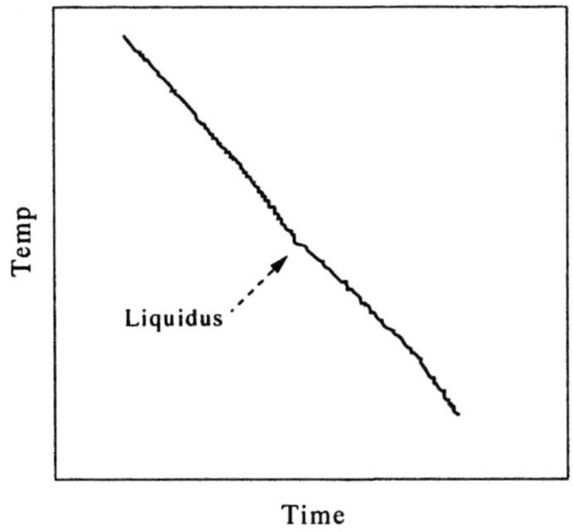

Figure 4.3. Cooling curve to determine liquidus point.

certain phase boundaries such as liquidus temperatures and invariant reactions such as eutectics. The reason for this is that it is often the start of transformation which is measured rather than the complete transformation itself. This can be seen for a liquidus point where an inflection in a cooling curve defines the onset of the solidification process (Fig. 4.3). This is more nucleation controlled rather than growth controlled and if the efficacy of potential nucleation agents in the system is high and the cooling rate is slow, of the order of 1°C min^{-1}, then very accurate results may be obtained. The method can be enhanced by plotting $\partial T/\partial t$ instead of T.

However, the situation is different if one considers the total transformation, including the solidus and peritectic type reactions where substantial solid state diffusion is needed to obtain complete equilibrium. Unless very slow cooling rates are used, or some further control mechanism utilised in the experiment, it is quite common to observe significant undercooling below the equilibrium temperature of transformation. The following sections will briefly describe determinations of phase diagrams where non-isothermal techniques have been successfully used, and possible problems associated with non-equilibrium effects will be discussed.

4.3.1.1 Thermal analysis techniques. The Cu–Sn system is an excellent example of what can be achieved with simple cooling curve measurements and it all also is one of the first quantitative determinations of a liquidus of a phase diagram. Heycock and Neville (1890, 1897) took liquid alloys and cooled them through the liquidus and measured the temperature using simple thermocouples. The time at each temperature measurement was also noted and cooling curves of temperature vs time graphically drawn. An inflection point determined the onset of solidification (Fig. 4.3) and this defined the liquidus. Figure 4.4 shows the series of liquidus temperatures measured for the whole of the Cu–Sn system. It is remarkable to note

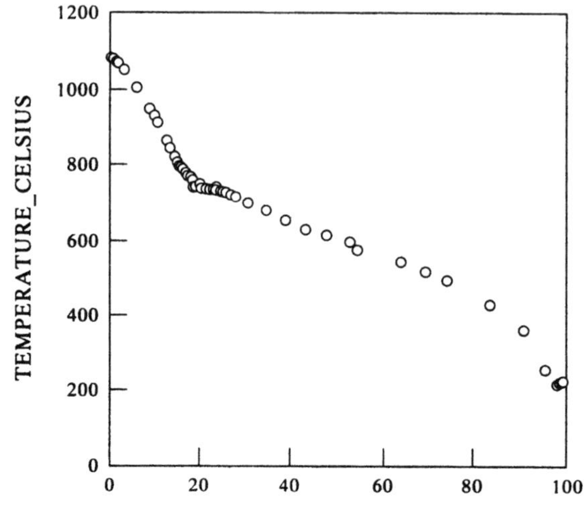

Figure 4.4. Liquidus points for Cu–Sn determined by Heycock and Neville (1890, 1897).

that the quality of their results has more than stood the test of time and a century later they are still considered as one of the defining sets of experiments for this system (Hansen and Anderko 1958, Saunders and Miodownik 1990).

More advanced techniques are now available and section 4.2.1.2 described differential scanning calorimetry (DSC) and differential thermal analysis (DTA). DTA, in particular, is widely used for determination of liquidus and solidus points and an excellent case of its application is in the In–Pb system studied by Evans and Prince (1978) who used a DTA technique after Smith (1940). In this method the rate of heat transfer between specimen and furnace is maintained at a constant value and cooling curves determined during solidification. During the solidification process itself cooling rates of the order of 1.25°C min^{-1} were used. This particular paper is of great interest in that it shows a very precise determination of the liquidus, but clearly demonstrates the problems associated with determining solidus temperatures.

Figure 4.5 shows the liquidus and solidus for In–Pb determined during cooling from the liquid; the solidus, in this case, is a non-equilibrium boundary. This was demonstrated by Evans and Prince (1978) when they annealed as solidified alloys for several hours just below their measured solidus. On re-heating melting occurred at a higher temperature than had been measured by DTA. The samples were then annealed for several hours just below the new measured melting temperature before being heated until melting was just observed again. The cycle was repeated until the solidus temperature became reproducible and the temperature then plotted as the true solidus. This is compared with the 'apparent solidus' on the initial cooling experiment in Fig. 4.5. This paper shows the problems that can occur with

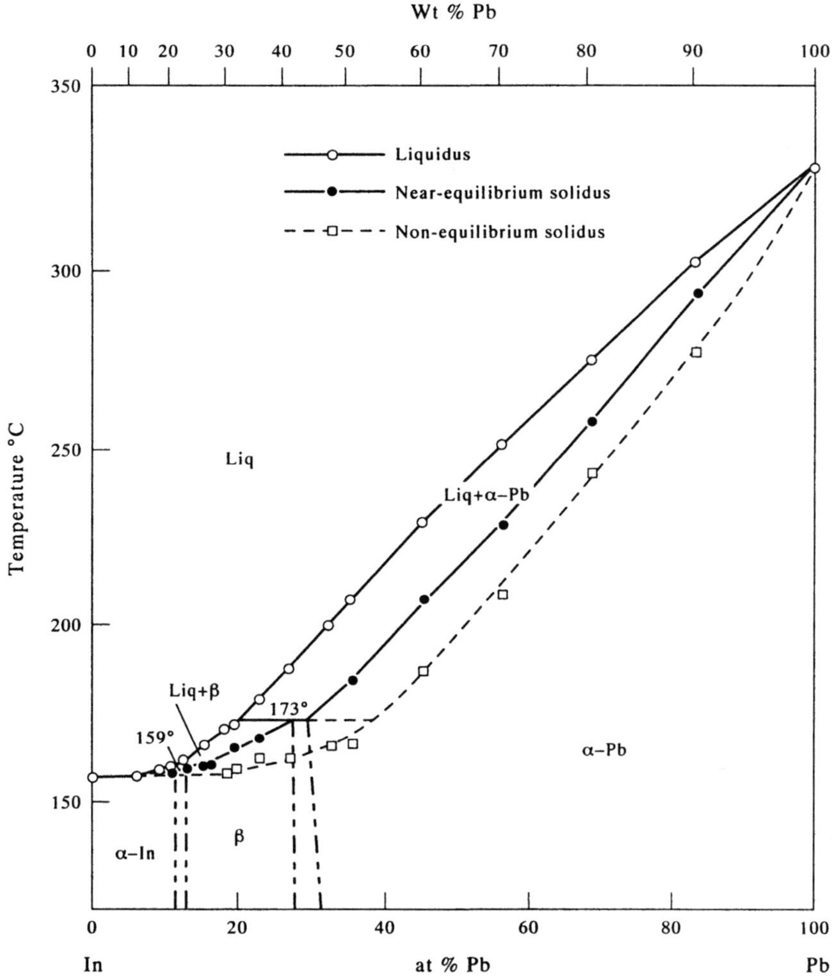

Figure 4.5. Experimentally measured liquidus (○) and solidus (□) points measured by using DTA by Evans and Prince (1978) for In–Pb. (•) refers to the 'near-equilibrium' solidus found after employing re-heating/cycling method.

measurements using non-isothermal methods but also shows how these can be overcome to provide true equilibrium values.

4.3.1.2 Chemical potential techniques. Phase diagrams can be determined by 'chemical potential' methods such as EMF. In this case the activity of one or both of the components is measured during a cooling or heating cycle and a series of characteristic breaks define phase boundaries. Figure 4.6 shows a series of Al–Sn

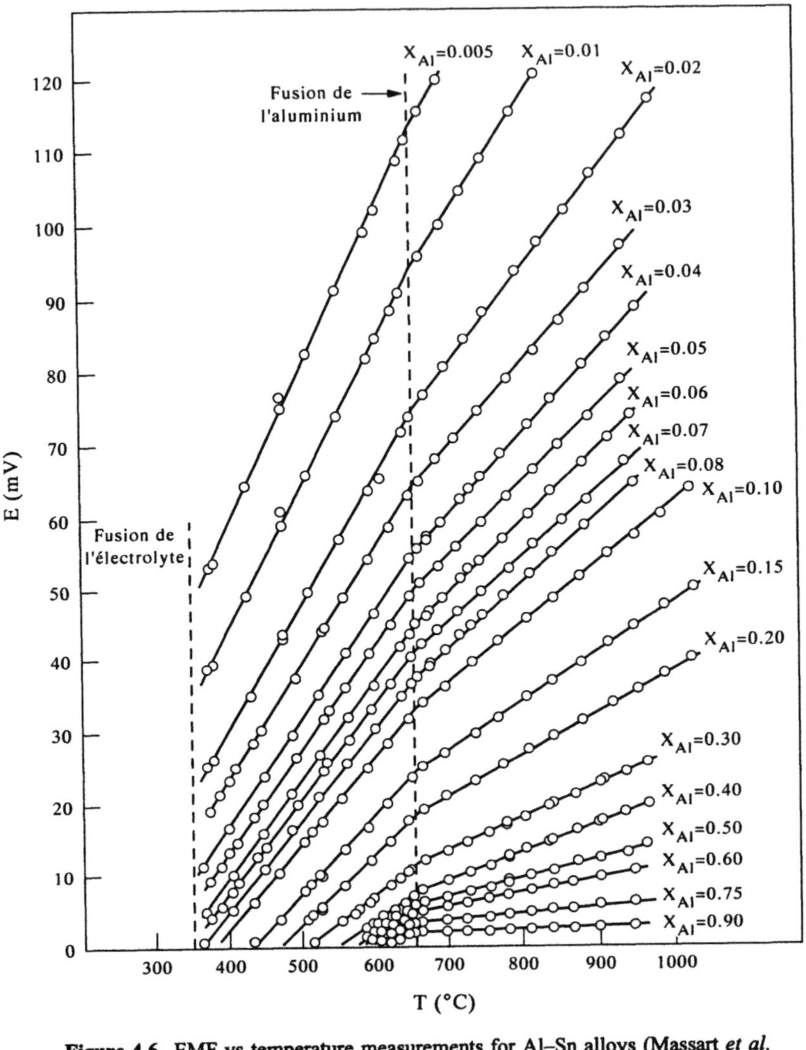

Figure 4.6. EMF vs temperature measurements for Al–Sn alloys (Massart *et al.* 1965).

alloys (Massart *et al.* 1965) where the EMF has been measured over a temperature range of 350–950°C. They characteristically show a break at the melting point of Al where the Al reference electrode solidifies. On further cooling the electrode voltage falls to zero when the Al-liquidus is reached and this defines the equilibrium liquidus. Work done at the same laboratory by Massart *et al.* (1966) on Al–In and by Martin-Garin *et al.* (1966) further demonstrated the applicability of the method to Al alloys. This technique is potentially powerful as it is possible to

References are listed on pp. 85–87.

define *both* the thermodynamic properties of the system as well as the phase diagram. Further, the point at which the phase boundary is reached is unambiguously identified, whereas in some thermal analysis techniques this point is not always clearly defined.

4.3.1.3 Magnetic susceptibility measurements. Magnetic susceptibility measurement is an interesting technique for determining phase boundaries in magnetic systems as there is a distinct change in magnetic properties during a phase transition. In this technique the alloy is suspended on a pendulum and a magnetic field is applied. The sample is then heated or cooled through the phase transition and the sample is deflected from its position due to the change in its inherent magnetism. The magnetic field is then altered by changing the applied current and the pendulum brought back to its original position. The magnetic susceptibility (χ) is then defined as

$$\chi = (K/m).I^2 \tag{4.7}$$

where K is a constant, m is the mass of the sample and I is the compensating current. The magnetic susceptibility is then plotted as a function of temperature and sharp changes can be seen. Figure 4.7 shows a plot of χ vs temperature for a Fe-0.68at%Nb alloy (Ferrier *et al.* (1964)) where the phase boundaries are delineated very clearly. These results were then combined with DTA measurements for the

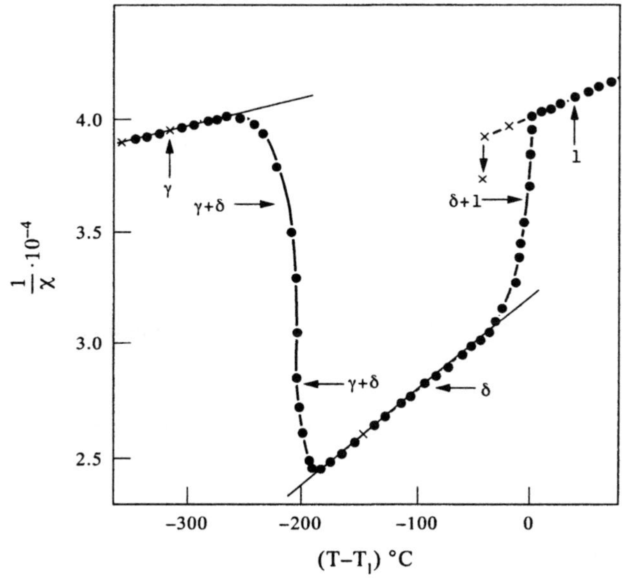

Figure 4.7. Magnetic susceptibility (χ) vs temperature measurements for a Fe-0.68at%Nb alloy (Ferrier *et al.* 1964). • = heating, × = cooling.

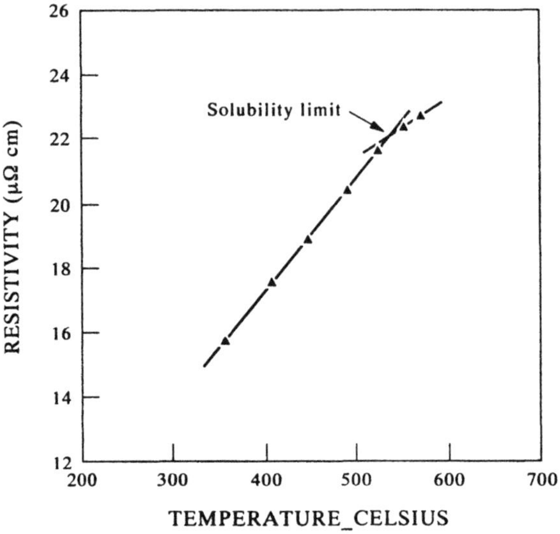

Figure 4.8. Plot of resistivity vs temperature for a Al-12.6at%Li alloy (Costas and Marshall 1962).

liquidus to define the Fe–Nb phase diagram between 1200 and 1550°C (Ferrier *et al.* (1964)). Other examples of the use of the magnetic susceptibility technique can be found in Fe–P (Wachtel *et al.* 1963) and Fe–Si (Übelacker 1965).

4.3.1.4 Resistivity methods. This is a simple technique by which the resistivity of an alloy is measured as a function of temperature. As for the case of magnetic susceptibility, quite distinct changes in resistivity occur during phase changes, and it is a relatively simple matter to measure these changes during a heating or cooling cycle. Figure 4.8 shows a plot of resistivity vs temperature for a Al-12.6at%Li alloy from Costas and Marshall (1962) and a clear transition at the solubility limit of the alloy is observed.

4.3.1.5 Dilatometric methods. This can be a sensitive method and relies on the different phases taking part in the phase transformation having different coefficients of thermal expansion. The expansion/contraction of a sample is then measured by a dilatometer. Cahn *et al.* (1987) used dilatometry to examine the order–disorder transformation in a number of alloys in the Ni–Al–Fe system. Figure 4.9 shows an expansion vs temperature plot for a $(Ni_{79.9}Al_{20.1})_{0.87}Fe_{0.13}$ alloy where a transition from an ordered $L1_2$ compound (γ') to a two-phase mixture of γ' and a Ni-rich f.c.c._Al phase (γ) occurs. The method was then used to determine the $\gamma'/(\gamma+\gamma')$ phase boundary as a function of Fe content, at a constant Ni/Al ratio, and the results are shown in Fig. 4.10. The technique has been used on numerous other occasions,

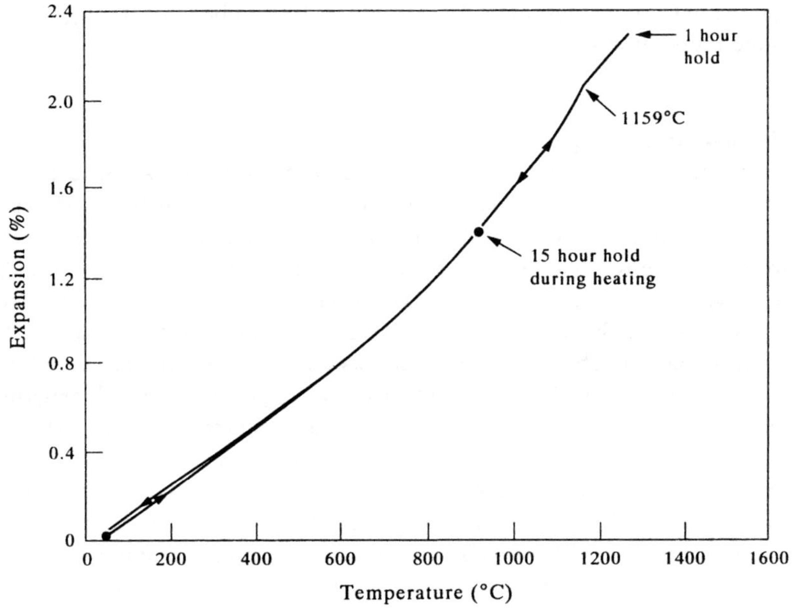

Figure 4.9. Expansion vs temperature plot for a $(Ni_{79.9}Al_{20.1})_{0.87}Fe_{0.13}$ alloy showing $\gamma'/\gamma' + \gamma$-phase boundary at 1159°C from Cahn *et al.* (1987).

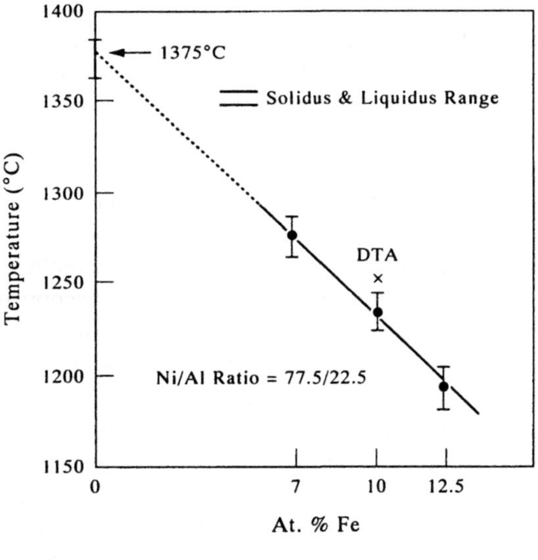

Figure 4.10. $\gamma'/\gamma' + \gamma$-phase boundary as a function of Fe content, at a constant Ni/Al ratio = 77.5/22.5 (from Cahn *et al.* 1987).

for example in Co–Fe (Masumoto 1926) and Cd–Li (Grube *et al.* 1932), usually for solid-state transformations.

4.3.2 Isothermal techniques

Isothermal methods are inherently closer to equilibrium as the sample can be left at temperature for substantial periods to allow equilibrium to be reached. However, in practice it can still be difficult to know the point at which this occurs. In some cases equilibrium can be reached rapidly, as can be seen for the $\gamma'/(\gamma + \gamma')$ boundary in Ni–Al above 1000°C which has been well established by many workers (see for example Fig. 4.1 in Saunders (1996)). However, in other systems such as Ni–Nb some considerable doubt existed as to the exact placement of the Ni solvus line with Ni_3Nb even though there had been a large number of experimental determinations (see for example Nash (1986) and Okamoto (1992)). There is no easy answer as to how long it is necessary to hold a sample at any given temperature to ensure equilibrium is reached; however, some simple type of diffusion analysis can be used as a first approximation. For example, a minimum holding time can be defined by the diffusion distance at temperature being greater than the grain size of the sample. In principle some of non-isothermal methods can be used in an isothermal way. In this case the composition is varied, rather than temperature, and a property such as EMF measured as a function of concentration.

The accuracy of some isothermal techniques, particularly those that rely on observation of phases, is limited by the number of different compositions that are prepared. For example, if two samples are separated by a composition of 2at%, and one is single-phase while the other two-phase, then formally the phase boundary can only be defined to within an accuracy of 2at%. This makes isothermal techniques more labour intensive than some of the non-isothermal methods. However, because it is now possible to directly determine compositions of phases by techniques such as electron microprobe analysis (EPMA), a substantially more quantitative exposition of the phase equilibria is possible.

4.3.2.1 Metallography. This is one of the best established methods for determining phase diagrams. In principle some form of optical microscopy is used to characterise the microstructure of the alloy and define whether it is single-phase or two-phase, etc. Because of the simplicity of preparing an optical sample, usually some form of grinding and polishing technique, it is very economic. There can be complications associated with the preparation of the polished surface, for example the use of the correct chemical etching medium. However, handbooks exist which detail etches and procedures for a wide range of materials types. Metallographic techniques do limit the accuracy of phase boundary and determination but they can lead to a quite rapid determination of the general form of the phase diagram. To demonstrate this Fig. 4.11 shows metallographic results from an experimental

References are listed on pp. 85–87.

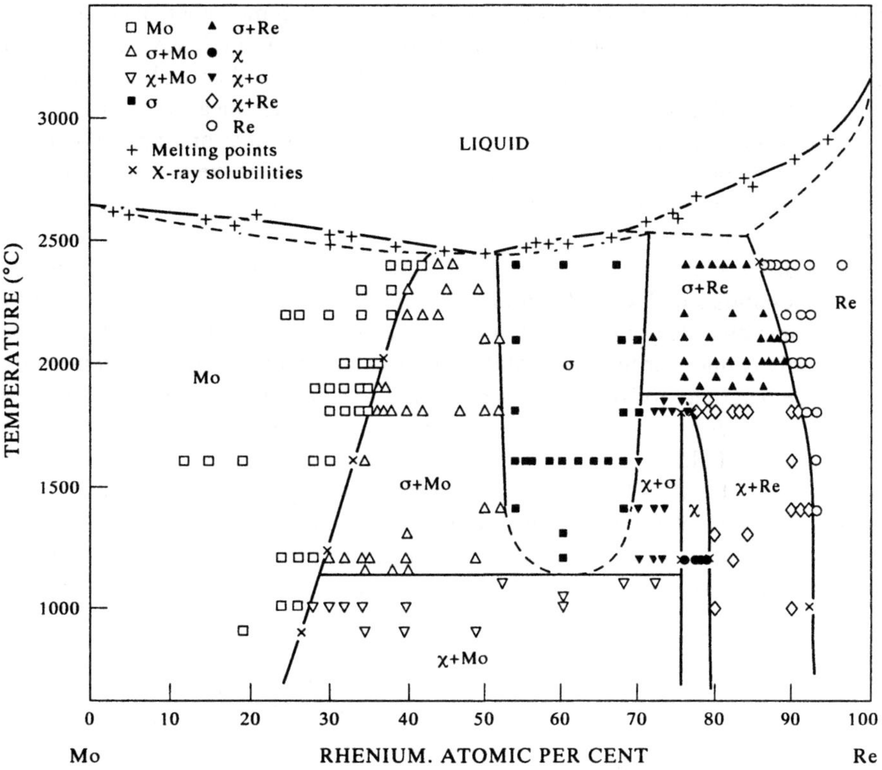

Figure 4.11. Equilibrium phase diagram for the Mo–Re system after Knapton (1958–59).

determination of the solid-state region of the Mo–Re system (Knapton 1958–59).

The previous paragraph has considered only optical methods, mainly for historical reasons, and the fact that the majority of phase-diagram work has probably been done using optical techniques. However, electron microscopy can be used instead, which has the considerable advantage that resolution is greatly increased. Further, the use of backscattered electron images can provide an immediate qualitative delineation of phases. As the image is controlled by the composition of the various phases, phases which contain higher levels of heavy elements appear light and conversely phases with lighter elements appear dark.

4.3.2.2 X-rays. In some ways this is an even simpler method of determining a phase diagram as sample preparation can be easy. The sample is placed in some form of X-ray diffractometer and its characteristic X-ray pattern obtained. If the crystal structure of the phases is known, the identification of phase fields can be rapid and reliable. X-ray methods do not tend to be a primary method for determination of phase diagrams and are more often used to support some other technique such metallography. In this context there are two further uses for X-ray methods. The first is to do with identification of crystal structure, the other with a more exact determination of phase boundaries using lattice parameter measurement.

Identification of unknown crystal structures and determination of phase fields by X-rays can be problematical if the characteristic patterns of the various phases are quite similar, for example in some b.c.c._A2-based ordered phases in noble-metal-based alloys. However, in many cases the characteristic patterns of the phases can be quite different and, even if the exact structure is not known, phase fields can still be well established. Exact determination of phase boundaries is possible using lattice-parameter determination and this is a well-established method for identifying solvus lines for terminal solid solutions. The technique simply requires that the lattice parameter of the phase is measured as a function of composition across the phase boundary. The lattice parameter varies across the single-phase field but in the two-phase field becomes constant. Figure 4.12 shows such a phase-boundary determination for the $HfC_{(1-x)}$ phase where results at various temperatures were used to define the phase boundary as a function of temperature (Rudy 1969). As can be seen, the position of is defined exactly and the method can be used to identify phase fields across the whole composition range.

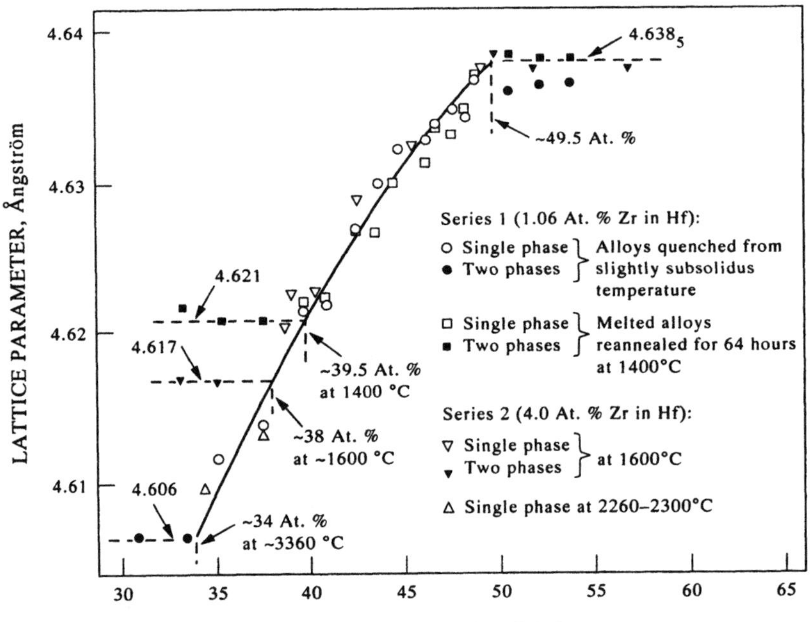

Figure 4.12. Lattice parameter vs composition measurements for $Hf_{(1-c)}C$ (Rudy 1969).

4.3.2.3 Quantitative determination of phase compositions in multi-phase fields. This topic can cover a broad range of techniques but, in practice, usually means measurement by electron microscopy in some form. The measurement of composition using characteristic X-rays given off when a sample is bombarded by high-energy electrons was pioneered by Castaing (1952). His initial instrumentation utilised optical methods to position the focused electrons but, later, as imaging using secondary and backscattered electrons became possible, resolution levels became sufficient to measure compositions of phases in actual heat-treated samples. Some of the earliest such work dates from the early 1960s, for example in the study of solid-state-phase equilibria in the Nb–Sn system (Wyman *et al.* 1962, Levinstein and Buehler 1964, Schadler *et al.* 1964).

The most significant advantage of these more quantitative methods is that, in a binary system, only one sample is needed to determine the position of both phase boundaries in a two-phase field. Further, if the alloy lies in the two-phase field over a wide range of temperatures, it is feasible that only one alloy need be used to fix the phase boundaries over this range of temperature. In a ternary system the analogous position is found with three-phase fields and, as these also define the limiting tie-lines of the three sets of two-phase fields, substantial information can be gained from the accurate determination of only one alloy. More recently transmission electron microscopy (TEM) has been used which is particularly valuable when microstructures are very fine as, for example, found in γTiAl alloys (Chen *et al.* 1994).

Quantitative analysis can now be performed at an even finer level using atom probe and field ion microscopy techniques. In these methods samples are prepared with sharp tips (radius 10–50 nm) which are subjected to high positive voltage, giving rise to high electrical fields at the tip surface. Pulses of electrical energy are then superimposed which causes the surface atoms to be vapourised and ionised. The atoms are then collected and identified using sophisticated measuring devices. As atoms are removed layer by layer this means that some spatial resolution is possible and distinct changes can be seen as phase boundaries are crossed. Such methods have been used with particular success in Ni-based superalloys (Blavette *et al.* 1988, Duval *et al.* 1994). More recently these techniques have been expanded so that a 3D reconstruction of the tip can be made showing the position of various phases (Blavette *et al.* 1993).

4.3.2.4 Sampling/equilibriation methods. The sampling method considers mainly equilibria involving the liquid phase where an alloy is held in a liquid + solid region until equilibrium is considered to be reached. Some of the liquid is then removed and its composition measured, thus defining the liquidus composition. The method relies on ensuring that the liquid and solid are sufficiently separated so that only the liquid is removed. This can usually be done straightforwardly as the density of the liquid and solid are different and gravity is then sufficient to provide the separation. Such methods have proved successful in determining the liquidus compositions of a number of Al-alloys where solubility levels are low due to the formation of high melting Al-rich intermetallics. Such systems cause problems in DTA experiments as liquidus slopes can be steep and the amount of solid formed is very small. Consequently the heat effect associated with the transformation is small and the exact point of the liquidus is difficult to determine.

A novel technique for the accurate determination of tie-lines in liquid/solid two-phase fields was first used by Willemin *et al.* (1986) in the Ni–Al–Ta system. The technique relies on holding an alloy just below its liquidus and quenching. The centre of the dendrite is then the part that was in equilibrium with the liquid at the temperature of holding. Microprobe analysis is subsequently made across the primary dendrite arms and composition profiles determined. The composition at the centre of the dendrite is clearly located in the concentration profiles by either maxima, when $x^{\mathrm{sol}} > x^{\mathrm{liq}}$, or minima, when $x^{\mathrm{sol}} < x^{\mathrm{liq}}$. This technique has also been used with good success in Ni–Al–Ti (Willemin and Durand-Charre 1990).

4.3.2.5 Diffusion couples. These methods have become increasingly popular with the general, increased use of EPMA. The technique relies on the ability to quantitatively determine the diffusion paths as two samples of different metals are joined together and allowed to react at high temperatures. Essentially the method is non-equilibrium in that equilibrium would only be reached when complete homogenisation of diffusion couple took place. However, during the diffusion process the diffusion path follows equilibrium tie-lines within the various two-phase fields of a binary or ternary alloy because the alloy equalises the potential of the various components at the phase interface. Figure 4.13 shows such a path of a diffusion couple between alloys, *A* and *B*, in the Ni–Al–Fe system at 1000°C (Cheng and

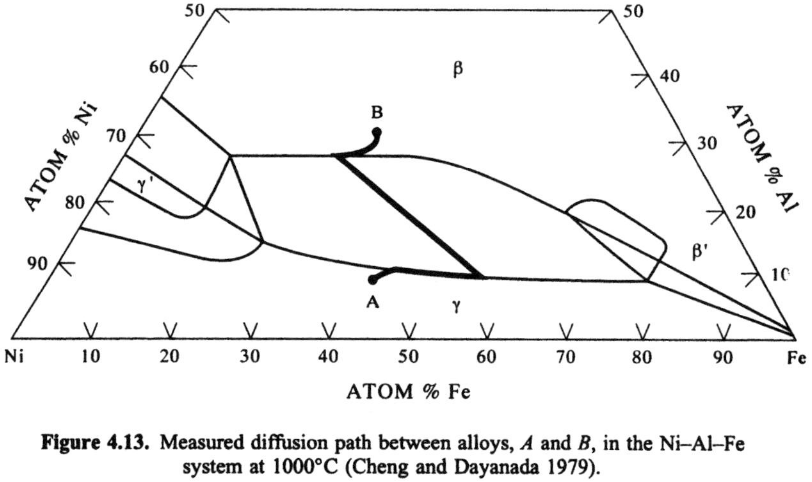

Figure 4.13. Measured diffusion path between alloys, *A* and *B*, in the Ni–Al–Fe system at 1000°C (Cheng and Dayanada 1979).

References are listed on pp. 85–87.

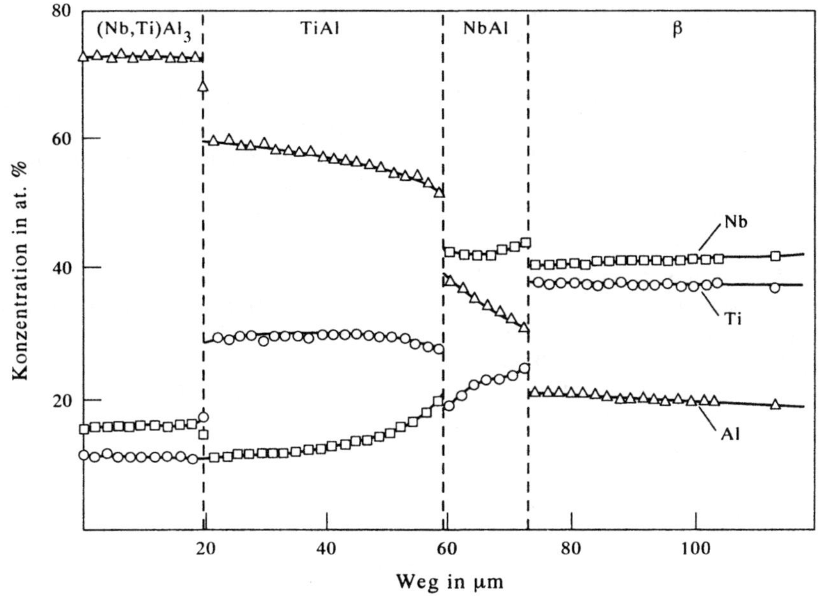

Figure 4.14. Concentration profile in a diffusion couple from the Al–Nb–Ti system at 1200°C (Hellwig 1990).

Dayanada 1979) and Fig. 4.14 shows the measured composition profiles of the various elements in a diffusion couple from Al–Nb–Ti at 1200°C (Hellwig 1990). In the latter case the diffusion path crosses three two-phase fields, (Nb, Ti) Al_3 + TiAl, TiAl + Nb_2Al, and β + Nb_2Al. Other good examples of this technique in practice can be found in the work of van Loo and co-workers (1978, 1980, 1981).

REFERENCES

Alcock, B. and Li, B. (1990) *Solid State Ionics*, **39**, 245.

Arpachi, E. and Frohberg, M. G. (1984) *Z. Metallkde.*, **75**, 614.

Beretzin, B. Ya., Chekhovskoi, V. Ya. and Sheindlin, A. E. (1971) *High Temp.–High Press.*, **3**, 169.

Bergman, W. and Ågren, J. (1984) TRITA-MAC-0232, Div. Physical Metallurgy, Royal Institute of Technology, Stockholm, Sweden.

Betz, G. and Frohberg, M. G (1980a) *High Temp.–High Press.*, **12**, 169.

Betz, G. and Frohberg, M. G (1980b) *Z. Metallkde.*, **71**, 451.

Blavette, D., Caron, P. and Khan, T. (1988) in *Superalloys 1988*, eds Reichman, S. *et al.* (The Metallurgical Society, Warrendale, PA), p. 305.

Blavette, D., Bostel, A., Sarrau, J. M., Deconihout, B. and Menand, A. (1993) *Nature*, **363**, 432.

Cahn, R. W., Siemers, P. A., Geiger, J. E. and Bardhan (1987) *Acta Metall.*, **35**, 2737.

Charette, G. G. and Flengas, S. N. (1968) *J. Electrochem. Soc.*, **115**, 796.
Castaing, R. (1952) Thèse, ONERA Publ. 55.
Cezairliyan, A. (1969) *High Temp.–High Press.*, **1**, 517.
Chen, Z., Jones, I. P., Saunders, N. and Small, C. (1994) *Scripta Met. Mater.*, **13**, 1403.
Cheng, G. H. and Dayanada, M. A. (1979) *Met. Trans. A*, **10A**, 1407.
Clopper, P. R., Altman, R. L. and Margrave, J. L. (1967) in *The Characterisation of High Temperature Vapours*, ed. Margave, J. L. (J. Wiley & Sons, New York), p. 48.
Costas, L. P. and Marshall, R. P. (1962) *Trans. AIME*, **224**, 970.
Davies, D. V. and Pritchard, J. (1972) *J. Chem. Thermodynamics*, **4**, 9.
Dench, W. A. and Kubaschewski, O. (1969) *High-Temp.–High Press.*, **1**, 357.
Duval, S., Chambreland, S., Caron, P. and Blavette, D. (1994) *Acta Met. Mater.*, **42**, 185.
Elliott, R. P. (1965) *Constitution of Binary Alloys, 1st Supplement*. (McGraw-Hill, New York).
Evans, D. S. and Prince, A. (1978) *Metal Science*, **12**, 600.
Ferrier, A., Uebelacker, E. and Wachtel, E. (1964) *Compt. Rend.*, **258**, 5424.
Gachon, J. C. and Hertz, J. (1983) *CALPHAD*, **7**, 1.
Gachon, J. C., Charles, J. and Hertz, J. (1985) *CALPHAD*, **9**, 29.
Grube, G., Vosskühler, H and Vogt, H. (1932) *Z. Elektrochem.*, **38**, 869.
Hansen, M. and Anderko, K. (1958) *Constitution of Binary Alloys, 2nd edition*. (McGraw-Hill, New York).
Hellwig, A. (1990) Dr. Ing. Thesis, Univ. Dortmund, Germany.
Heycock, C. T. and Neville, F. H. (1890) *J. Chem. Soc.*, **57**, 379.
Heycock, C. T. and Neville, F. H. (1897) *Phil. Trans. Roy. Soc. (London)*, **A189**, 47–51; 62–66.
Hultgren R., Newcombe, P, Orr., R. L. and Warner, L. (1958) *Met. Chem. NPL Symposium No. 9* Ed. Kubaschewski, O. (HM Stationery Office, London).
Jacob, K. T. and Iwase, M. (1982) *Z. Metallkde.*, **73**, 316.
Jacob, K. T., Alcock, C. B. and Palamutcu, T. (1973) *Acta Met.*, **21**, 1003.
Jonsson, S. (1993) *Phase relationships in quaternary hard materials*, Ph.D. Thesis, Royal Institute of Technology, Stockholm, Sweden.
Kay, D. A. R. (1979) *Rev. Int. Hautes Temp. Refract.*, **16**, 21.
Kiukkola, K. and Wagner, C. (1957) *J. Electrochem. Soc.*, **104**, 379.
Kleppa, O. (1955) *J. Am. Chem. Soc.*, **59**, 175.
Kleppa, O. (1962) *J. J. Phys. Radium*, **23**, 763.
Kleppa, O. (1993) *J. Phase Equilibria*, **15**, 240.
Kleppa, O. and Topor, L. (1989) *Thermochim. Acta*, **139**, 291.
Kleykamp, H. (1983) *Ber. Bunsenges. Phys. Chem.*, **87**, 777.
Knapton, A. G. (1958–59) *J. Inst. Metals*, **87**, 62.
Komarek, K. L. (1972) in *Metallurgical Chemistry*, ed. Kubaschewski, O. (H.M. Stationery Office, London), p. 1.6-1.
Komarek, K. L. (1973) *Z. Metallkde.*, **64**, 325.
Komarek, K. L. and Ipser, H. (1984) *Pure & Appl. Chem.*, **56**, 1511.
Kubaschewski, O. and Dench, W. A. (1955) *Acta. Met.*, **3**, 339.
Kubaschewski, O. and Walter, A. (1939) *Z. Elektrochem.*, **45**, 631.
Kubaschewski, O., Alcock, C. B. and Spencer, P. J. (1993) *Materials Thermochemistry: 6th Edition* (Pergamon Press, Oxford).
Lee, H. D., Misra, S. and Anderson, T. J. (1992) *EMF Studies and Assessment of the* Ga–As *System*, presented at CALPHAD XXI, Jerusalem, 14–19 June.

Levinson, L. S. (1965) *J. Chem. Phys.*, **42,** 2891.
Levinstein, H. J. and Buehler, E. (1964) *Trans. AIME*, **230,** 1314.
Martin-Garin, R., Massart, G., Desre, P. and Bonnier, E. (1966) *C. R. Acad. Sci. Paris Serie C*, **262,** 335.
Massart, G., Durand, F. and Bonnier, E. (1965) *Bull. Soc. Chim. France*, p. 87.
Massart, G., Durand, F. and Bonnier, E. (1966) *C. R. Acad. Sci. Paris Serie C*, **262,** 000.
Masumoto, H. (1926) *Sci. Rep. Tohoku Univ.*, **15,** 469.
Moser, Z. (1979) in *Calculation of Phase Diagrams and Thermochemistry of Alloy Phases*, eds Chang, Y. A. and Smith, J. F. (Met. Soc. AIME, Warrendale), p. 242.
Nash, P. and Nash, A. (1986) *Bull. Alloy Phase Diagrams*, **7,** 124.
Naylor, B. F. (1946) *J. Amer. Chem. Soc.*, **68,** 370.
Norman, J. H. and Winchell, P. (1970) in *Physico-Chemical Measurements in Metal Research*, ed. Rapp, R. A. (Interscience, New York), p. 131.
Okamoto, H. (1992) *J. Phase Equilibria*, **13,** 444.
Rose, B. A., Davis, G. J. and Ellingham, H. J. T. (1948) *Disc. Faraday. Soc.*, **(4),** 154.
Rudy, E. (1969) *Compendium of Phase Diagram Data*, Tech. Rep. AFML-TR-65-2 Part V, Wright-Patterson A.F.B., Ohio.
Sale, F. R. (1970) *J. Sci. Instr.*, **3,** 653.
Saunders, N. (1996) in *Superalloys 1996*, eds Kissinger, R. D. *et al.* (TMS, Warrendale), p. 101.
Saunders, N. and Miodownik, A. P. (1990) *Bull. Alloy Phase Diagrams*, **11,** 278.
Schadler, H. W., Osika, L. M., Salvo, G. P. and DeCarlo, V. J. (1964) *Trans. AIME*, **230,** 1074.
Sears, B. and Anderson, T. J. (1989) *Solid State Electrochemical Study of Liquid* In–Bi *Alloys*, presented at CALPHAD XVIII, Stockholm, 28 May–2 June.
Sheindlin, A. E., Berezin, B. Ya. and Chekhovskoi, V. Ya. (1972) *High Temp–High Press.*, **4,** 611.
Shunk, F. A. (1969) *Constitution of Binary Alloys, 2nd Supplement* (McGraw-Hill, New York).
Smith, C. S. (1940) *Trans. AIME*, **137,** 236.
Tickner, L. N. and Bever, M. B. (1952) *J. Met.*, **4,** 941.
Treverton, J. A. and Margave, J. L. (1970) in *Proceedings 5th Symposium on Thermophysical Properties*, ed. Bonilla, C. F. (ASM, Ohio), p. 484.
Übelacker, E. (1965) *C. R. Acad. Sci. Paris Serie C*, **261,** 976.
van Loo, F. J. J., Bastin, G. F. and Leenen, A. J. H. (1978) *J. Less Common Metals*, **57,** 111.
van Loo, F. J. J., Vrolijk, J. W. G. A. and Bastin, G .F. (1981) *J. Less Common Metals*, **77,** 121.
van Loo, F. J. J., Bastin, G. F., Vrolijk, J. W. G. A. and Hendricks, J. J. M. (1980) *J. Less Common Metals*, **72,** 225.
Wachtel, E., Urbain, G. and Übelacker, E. (1963) *Compt. Rend.*, **257,** 2470.
Willemin, P., and Durand-Charre, M. (1990) *J. Mater. Sci.*, **25,** 168.
Willemin, P., Dugue, O., Durand-Charre, M. and Davidson, J. H. (1986) *Mat. Sci. Tech.*, **2,** 344.
Wyman, L. L., Cuthill, J. R., Moore, G. A, Park, J. J. and Yakowitz, H. (1962) *J. Res. Natl. Bur. Std.*, **A66,** 351.

Chapter 5

Thermodynamic Models for Solution and Compound Phases

Chapter 5

Thermodynamic Models for Solution and Compound Phases

5.1. INTRODUCTION

Thermodynamic modelling of solution phases lies at the core of the CALPHAD method. Only rarely do calculations involve purely stoichiometric compounds. The calculation of a complex system which may have literally 100 different stoichiometric substances usually has a phase such as the gas which is a mixture of many components, and in a complex metallic system with 10 or 11 alloying elements it is not unusual for all of the phases to involve solubility of the various elements. *Solution phases* will be defined here as any phase in which there is solubility of more than one component and within this chapter are broken down to four types: (1) random substitutional, (2) sublattice, (3) ionic and (4) aqueous. Others types of solution phase, such as exist in polymers or complex organic systems, can also be modelled, but these four represent the major types which are currently available in CALPHAD software programmes.

For all solution phases the Gibbs energy is given by the general formula

$$G = G^{\text{o}} + G^{\text{ideal}}_{\text{mix}} + G^{xs}_{\text{mix}} \tag{5.1}$$

where G^{o} is the contribution of the pure components of the phase to the Gibbs energy, $G^{\text{ideal}}_{\text{mix}}$ is the ideal mixing contribution and $G^{\text{xs}}_{\text{mix}}$ is the contribution due to non-ideal interactions between the components, also known as the Gibbs excess energy of mixing.

The random substitutional and sublattice models are intimately related, as a phase with random occupation of atoms on all sites can technically be considered as a phase containing a 'single sublattice' and mathematically the same general equations used in both cases. However, it is useful to separate the two types as the sublattice models implicitly define some internal, spatial substructure and give rise to site occupations which define stoichiometric compounds. Also, and very importantly, $G^{\text{ideal}}_{\text{mix}}$ and $G^{\text{xs}}_{\text{mix}}$ are governed by site occupation of the components in the various sublattices rather than the global concentration of the components themselves.

This chapter will begin, very briefly, with the thermodynamic representation of Gibbs energy for stoichiometric compounds before concentrating on the situation when mixing occurs in a phase.

5.2. STOICHIOMETRIC COMPOUNDS

The integral Gibbs energy, $G_{[T,P]}$, of a pure species or stoichiometric compound is given simply by the equation

$$G_{[T,P]} = H_{[T,P]} - TS_{[T,P]} \tag{5.2}$$

where $H_{[T,P]}$ and $S_{[T,P]}$ are the enthalpy and entropy as a function of temperature and pressure. Thermodynamic information is usually held in databases using some polynomial function for the Gibbs energy which, for the case of the Scientific Group Thermodata Europe (Ansara and Sundman 1987), is of the form (Dinsdale 1991)

$$G_{m[T]} - H_m^{\mathrm{SER}} = a + bT + cT\ln(T) + \sum_2^n d_n T^n. \tag{5.3}$$

The left-hand-side of the equation is defined as the Gibbs energy relative to a standard element reference state (SER) where $H_{\mathrm{m}}^{\mathrm{SER}}$ is the enthalpy of the element or substance in its defined reference state at 298.15 K, a, b, c and d_n are coefficients and n represents a set of integers, typically taking the values of 2, 3 and −1. From Eq. (5.3), further thermodynamic properties can be obtained as discussed in Chapter 6.

$$\begin{aligned} S &= -b - c - c\ln(T) - \sum n d_n T^{n-1} \\ H &= a - cT - \sum (n-1)\, d_n T^n \\ C_p &= -c - \sum n(n-1)\, d_n T^{n-1}. \end{aligned}$$

Equation (5.3) is given with respect to temperature but other terms associated with pressure and magnetism can be added. The representation of such effects is described more completely in Chapters 6 and 8.

5.3. RANDOM SUBSTITUTIONAL MODELS

Random substitutional models are used for phases such as the gas phase or simple metallic liquid and solid solutions where components can mix on any spatial position which is available to the phase. For example, in a simple body-centred cubic phase any of the components could occupy any of the atomic sites which define the cubic structure as shown below (Fig. 5.1).

In a gas or liquid phase the crystallographic structure is lost, but otherwise positional occupation of the various components relies on random substitution rather than any preferential occupation of site by any particular component.

References are listed on pp. 124–126.

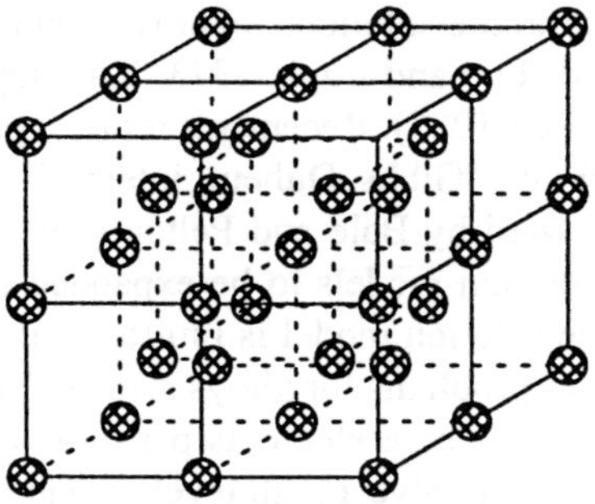

Figure 5.1. Simple body-centred cubic structure with random occupation of atoms on all sites.

5.3.1 Simple mixtures

5.3.1.1 Dilute solutions. There are a number of areas in materials processing where low levels of alloying are important, for example in refining and some age-hardening processes. In such cases it is possible to deal with solution phases by dilute solution models (Wagner 1951, Lupis and Elliott 1966). These have the advantage that there is a substantial experimental literature which deals with the thermodynamics of impurity additions, particularly for established materials such as ferrous and copper-based alloys. However, because of their fundamental limitations in handling concentrated solutions they will only be discussed briefly.

In a highly dilute solution, the solute activity (a_i) is found to closely match a linear function of its concentration (x_i). In its simplest form the activity is written as

$$a_i = \gamma_i^{\circ} x_i \tag{5.4}$$

where γ_i° is the value of the activity coefficient of i at infinite dilution. This is known as Henry's law. Equation (5.5) may be rewritten in terms of partial free energies as

$$\bar{G}_i = \bar{G}_i^{xs} + RT \log_e x_i \tag{5.5}$$

where $\bar{G}_i^{xs}$ has a constant value in the Henrian concentration range and is obtained directly from γ_i°. The expression can also be modified to take into account interactions between the solute elements and Eq. (5.5) becomes

$$\bar{G}_i = \bar{G}_i^{xs} + RT \log_e x_i + RT \sum_j \varepsilon_i^j x_j \tag{5.6}$$

where x_i is the concentration of solute i and ε_i^j is an interaction parameter taking into account the effect of mixing of component i and j in the solvent. There are various other treatments of dilute solutions, mainly associated with how secondary interactions are considered (Foo and Lupis 1973, Kirkaldy *et al.* 1978, Bhadeshia 1981, Enomoto and Aaronson 1985). Recently a more sophisticated dilute solution treatment consistent with the Gibbs–Duhem integration in more concentrated solutions has been put forward by Bale and Pelton (1990). This allows the solute range handled by dilute solution models to be expanded significantly.

Unfortunately, the dilute solution model is limited in its applicability to concentrated solutions. This causes problems for alloys such as Ni-based superalloys, high alloy steels, etc., and systems where elements partition strongly to the liquid and where solidification processes involve a high level of segregation. It is also not possible to combine dilute solution databases which have been assessed for different solvents. The solution to this problem is to use models which are applicable over the whole concentration range, some of which are described below.

5.3.1.2 Ideal solutions. An ideal substitutional solution is characterised by the random distribution of components on a lattice with an interchange energy equal to zero. The entropy due to configuration is easily calculated and is related to the proability of interchange of the components. The configurational entropy $S^{\rm conf}$ is given by

$$S^{\rm conf} = k \log_e W_P \tag{5.7}$$

where k is Boltzmann's constant and W_P is the number of configurations in which the components can be arranged for a given state. For a multicomponent system W_P is equal to the number of permutations given by

$$W_P = \frac{N!}{\prod_i n_i!} \tag{5.8}$$

where

$$N = \sum_i n_i \tag{5.9}$$

n_i is the number of components of i and N is the total number of components in the system. For one mole of components, N is equal to Avogadro's number. From Stirling's formula, $S^{\rm conf}$ is now equal to

$$S^{\rm conf} = -k \sum_i n_i \log_e \frac{n_i}{N}. \tag{5.10}$$

The ideal molar entropy of mixing is then given by

$$S^{\rm ideal}_{\rm mix} = -Nk \sum_i x_i \log_e x_i \tag{5.11}$$

References are listed on pp. 124–126.

where x_i is the mole fraction of component i. With the assumption that the interchange energy is zero it is possible to substitute the atoms without changing the energy of that state and $G_{\text{mix}}^{\text{ideal}}$ is then given by

$$G_{\text{mix}}^{\text{ideal}} = -TS_{\text{mix}}^{\text{ideal}} = RT\sum_i x_i \log_e x_i \tag{5.12}$$

where R is the gas constant. The Gibbs energy of an ideal solution phase will then be

$$G_m = \sum_i x_i G_i^o + RT\sum_i x_i \log_e x_i \tag{5.13}$$

with G_i^o defining the Gibbs energy of the phase containing the pure component i. For the case of gases, ideal mixing is often assumed and this assumption can often be quite reasonable. However, in condensed phases there is always some interaction between components.

5.3.1.3 Non-ideal solutions: regular and non-regular solution models. The regular solution model is the simplest of the non-ideal models and basically considers that the magnitude and sign of interactions between the components in a phase are independent of composition. Assuming the total energy of the solution (E_0) arises from only nearest-neighbours bond energies in a system $A - B$ then

$$E_0 = \omega_{AA} E_{AA} + \omega_{BB} E_{BB} + \omega_{AB} E_{AB} \tag{5.14}$$

where ω_{AA}, ω_{BB}, ω_{AB}, E_{AA}, E_{BB} and E_{AB} are the number of bonds and energies associated with the formation of different bond types *AA*, *BB* and *AB*. If there are N atoms in solution and the co-ordination number for nearest neighbours of the crystal structure is z, the number of bond types being formed in a random solution is

$$\omega_{AA} = \frac{1}{2} N z x_A^2 \tag{5.15a}$$

$$\omega_{BB} = \frac{1}{2} N z x_B^2 \tag{5.15b}$$

$$\omega_{AB} = N z x_A x_B \tag{5.15c}$$

where x_A and x_B are the mole fractions of A and B. Substituting Eqs (5.15 (a–c)) into Eq. (5.14) gives

$$E_0 = \frac{Nz}{2}\left(x_A E_{AA} + x_B E_{BB} + x_A x_B (2E_{AB} - E_{AA} - E_{BB})\right). \tag{5.16}$$

If the reference states are taken as pure A and B then Eq. (5.16) reverts to

$$H_{\text{mix}} = \frac{Nz}{2} x_A\, x_B\, (2E_{AB} - E_{AA} - E_{BB}) \tag{5.17}$$

where H_{mix} is the enthalpy of mixing. If the bond energies are temperature dependent there will also be an excess entropy of mixing leading to the well-known regular solution model for the Gibbs excess energy of mixing

$$G^{xs}_{\text{mix}} = x_A\, x_B\, \Omega \tag{5.18}$$

where Ω is now a temperature-dependent interaction parameter. When generalised and added to Eq. (5.13) this gives

$$G_m = \sum_i x_i\, G_i^{\circ} + RT \sum_i x_i \log_e x_i + \sum_i \sum_{j>i} x_i\, x_j\, \Omega_{ij}. \tag{5.19}$$

However, it has been realised for a long time that the assumption of composition-independent interactions was too simplistic. This led to the development of the sub-regular solution model, where interaction energies are considered to change linearly with composition. The following expression for G^{xs}_{mix} is then obtained as (Kaufman and Bernstein 1970)

$$G^{xs}_{\text{mix}} = x_i\, x_j\, (\Omega^i_{ij} x_i + \Omega^j_{ij} x_j). \tag{5.20}$$

Taking this process further, more complex composition dependencies to Ω can be considered and it is straightforward to show that a general formula in terms of a power series should provide the capability to account for most types of composition dependence (Tomiska 1980). The most common method is based on the Redlich–Kister equation and Eq. (5.19) is expanded to become

$$G_m = \sum_i x_i\, G_i^{\circ} + RT \sum_i x_i \log_e x_i + \sum_i \sum_{j>i} x_i\, x_j \sum_v \Omega^v_{ij}\, (x_i - x_j)^v \tag{5.21}$$

where Ω^v_{ij} is a binary interaction parameter dependent on the value of v. The above equation for G^{xs}_{mix} becomes regular when $v = 0$ and sub-regular when $v = 1$. In practice the value for v does not usually rise above 2. If it is found necessary to do so, it is probable that an incorrect model has been chosen to represent the phase.

Equation (5.21) assumes ternary interactions are small in comparison to those which arise from the binary terms. This may not always be the case and where evidence for higher-order interactions is evident these can be taken into account by a further term of the type $G_{ijk} = x_i\, x_j\, x_k\, L_{ijk}$, where L_{ijk} is an excess ternary interaction parameter. There is little evidence for the need for interaction terms of any higher order than this and prediction of the thermodynamic properties of substitutional solution phases in multi-component alloys is usually based on an assessment of binary and ternary terms. Various other polynomial expressions for the excess term have been considered, see for example the reviews by Ansara (1979) and Hillert (1980). All are, however, based on predicting the properties of the higher-order system from the properties of the lower-component systems.

Equation (5.21) is normally used in metallic systems for substitutional phases such as liquid, b.c.c., f.c.c., etc. It can also be used to a limited extent for ceramic systems and useful predictions can be found in the case of quasi-binary and quasi-ternary oxide systems (Kaufman and Nesor 1978). However, for phases such as interstitial solutions, ordered intermetallics, ceramic compounds, slags, ionic liquids and aqueous solutions, simple substitutional models are generally not adequate and more appropriate models will be discussed in Sections 5.4 and 5.5.

5.3.1.4 The extrapolation of the Gibbs excess energy to multi-component systems. As mentioned in the previous section, most methods of extrapolating the thermodynamic properties of alloys into multi-component systems are based on the summation of the binary and ternary excess parameters. The formulae for doing this are based on various geometrical weightings of the mole fractions (Hillert 1980). To demonstrate this principle, emphasis will be given to three of the main methods which have been proposed. For simplicity the binary systems will be represented as sub-regular solutions.

Muggianu's equation

The predominant method at the present time uses the equation developed by Muggianu *et al.* (1975). In this circumstance the excess energy in a multi-component system, as given by the last term of Eq. (5.21), expands for a sub-regular solution to

$$G^{xs}_{\text{mix}} = x_A x_B\{L^0_{AB} + L^1_{AB}(x_A - x_B)\} + x_B x_C\{L^0_{BC} + L^1_{BC}(x_B - x_C)\} \\ + x_A x_C\{L^0_{AC} + L^1_{AC}(x_A - x_C)\}. \qquad (5.22)$$

Kohler's equation

In the Kohler equation (Kohler 1960) G^{xs}_{mix} is described by

$$G^{xs}_{\text{mix}} = (x_A + x_B)^2 \frac{x_A}{x_A + x_B} \cdot \frac{x_B}{x_A + x_B} \left\{ L^0_{AB} + L^1_{AB}\left(\frac{x_A - x_B}{x_A + x_B}\right) \right\} \\ + (x_B + x_C)^2 \frac{x_B}{x_B + x_C} \cdot \frac{x_C}{x_B + x_C} \left\{ L^0_{BC} + L^1_{BC}\left(\frac{x_B - x_C}{x_B + x_C}\right) \right\} \\ + (x_A + x_C)^2 \frac{x_A}{x_A + x_C} \cdot \frac{x_C}{x_A + x_C} \left\{ L^0_{AC} + L^1_{AC}\left(\frac{x_A - x_C}{x_A + x_C}\right) \right\}. \qquad (5.23)$$

Although different in form to the Muggianu equation, it becomes identical in a binary system as the various terms $(x_i + x_j)$ terms become equal to 1.

Toop's equation

Both the Muggianu and Kohler equations can be considered symmetrical as they treat the components in the same way and do not differentiate between them. This is true for another method suggested by Colinet (1967) which is derived differently than either the Kohler or Muggianu equations but which can be reduced to the Muggianu equation (Hillert 1980). Toop's equation (1965) is essentially different in that it considers one of the binary systems does not behave in the same way as the others and the extrapolation is based essentially on two of the binaries having identical mole-fraction products with the mole-fraction products of the third being different. For a sub-regular solution Toop's equation breaks down to (Hillert 1980)

$$G^{xs}_{\text{mix}} = x_A x_B\{L^0_{AB} + L^1_{AB}(x_A - x_B - x_C)\} + x_A x_C\{L^0_{AC} + L^1_{AC}(x_A - x_C - x_B)\} + x_B x_C\left\{L^0_{BC} + L^1_{BC}\left(x_B - x_C + \frac{x_B - x_C}{x_B + x_C} x_A\right)\right\}. \quad (5.24)$$

Again this reduces to the normal sub-regular form in the binaries.

The geometrical representation of the Muggianu, Kohler and Toop equations is best summarised using the figures below.

In the Muggianu extrapolation it can be seen that the line from the ternary alloy composition to the edge binaries forms a right-angle to the binary. This has the consequence that, when the alloy composition is dilute in two of the components, the interaction parameter of these two components will approach regular behaviour because the term $(x_i - x_j)$ becomes small. This is the case even if the ratio $x_i : x_j$ differs substantially from unity. This is not important for systems where the interaction terms higher than sub-regular are small in comparison to the regular term. However, for systems such as Cr–Ni where the sub-regular part predominates, any composition dependence of Cr–Ni interactions in dilute solutions with other elements will be ignored. This could have significant consequences in the prediction of activity coefficients in low-alloy steels for example. Kohler's equation allows for the composition dependence of the binaries and in these circumstances

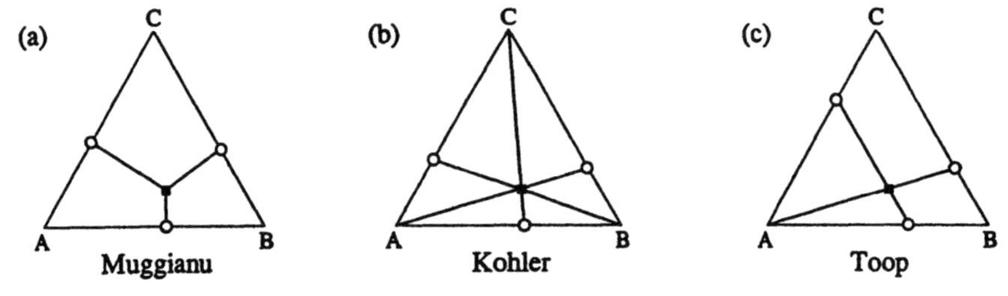

Figure 5.2. Geometrical constructions of (a) Muggianu, (b) Kohler and (c) Toop models.

may be preferable. In practice, the phase boundaries calculated by either the Muggianu and Kohler extrapolations seem to provide comparable results (Ansara *et al.* 1978), but it was noted that the choice of extrapolation method should receive more attention when exact knowledge of partial quantities such as activity coefficients is more critical. The Toop equation is not suitable for metallic systems but may be appropriate for some ionic liquid systems. However, it should be used with care in all cases as the extrapolation is dependent on which binary is chosen to behave differently, and it is possible to obtain three different answers depending on this choice.

5.4. SUBLATTICE MODELS

5.4.1 Introduction

A sublattice phase can be envisaged as being composed of interlocking sublattices (Fig. 5.3) on which the various components can mix. It is usually crystalline in nature but the model can also be extended to consider ionic liquids where mixing on particular 'ionic sublattices' is considered. The model is phenomenological in nature and does not define any crystal structure within its general mathematical formulation. It is possible to define internal parameter relationships which reflect structure with respect to different crystal types, but such conditions must be externally formulated and imposed on the model. Equally special relationships apply if the model is to be used to simulate order–disorder transformations.

Sublattice modelling is now one of the most predominant methods used to describe solution and compound phases. It is flexible and can account for a variety of different phase types ranging from interstitial phases such as austenite and ferrite in steels to intermetallic phases, such as sigma and Laves, which have wide homogeneity ranges. Furthermore, it can be extended to simulate Bragg–Williams–Gorsky ordering in metallic systems and be adapted to account for

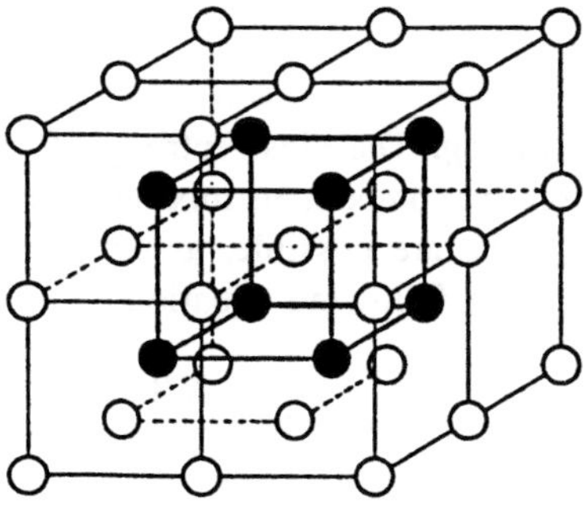

Figure 5.3. Simple body-centred cubic structure with preferential occupation of atoms in the body-centre and corner positions.

ionic liquids. This section will begin by giving a detailed description of the multiple sublattice model of Sundman and Ågren (1981) and give some specific examples of usage.

5.4.2 The generalised multiple sublattice model (Sundman and Ågren 1981)

5.4.2.1 Definition of site fractions. The multiple sublattice model is an extension of earlier treatments of the two-sublattice models of Hillert and Steffansson (1970), Harvig (1971) and Hillert and Waldenstrom (1977). It allows for the use of many sublattices and concentration dependent interaction terms on these sublattices. To work with sublattice models it is first necessary to define what are known as site fractions, y_i^s. These are basically the fractional site occupation of each of the components on the various sublattices where

$$y_i^s = \frac{n_i^s}{N^s} \tag{5.25}$$

n_i^s is the number of atoms of component i on sublattice s and N^s is total number of sites on the sublattice. This can be generalised to include vacancies, which are important to consider in interstitial phases, such that Eq. (5.25) becomes

$$y_i^s = \frac{n_i^s}{n_{\mathrm{Va}}^s + \sum_i n_i^s} \tag{5.26}$$

and n_{Va}^s is the number of vacancies on sublattice s. Mole fractions are directly related to site fractions by the following relationship

$$x_i = \frac{\sum_s N^s y_i^s}{\sum_s N^s (1 - y_{\mathrm{Va}}^s)}. \tag{5.27}$$

5.4.2.2 Ideal entropy of mixing. The ideal entropy of mixing is made up of the configurational contributions by components mixing on each of the sublattices. The number of permutations which are possible, assuming ideal interchanges within each sublattice, is given by the following equation

$$W_P = \prod_s \frac{N^s!}{\prod_i n_i^s!} \tag{5.28}$$

and the molar Gibbs ideal mixing energy is

References are listed on pp. 124–126.

$$G_{\mathrm{mix}}^{\mathrm{ideal}} = -TS_{\mathrm{mix}}^{\mathrm{ideal}} = RT\sum_{s} N^s \sum_{i} y_i^s \log_e y_i^s \tag{5.29}$$

where y_i^s includes contribution from vacancies.

5.4.2.3 Gibbs energy reference state. The Gibbs energy reference state is effectively defined by the 'end members' generated when only the pure components exist on the sublattice. Envisage a sublattice phase with the following formula $(A, B)_1(C, D)_1$. It is possible for four points of 'complete occupation' to exist where pure A exists on sublattice 1 and either pure B or C on sublattice 2 or conversely pure B exists on sublattice 1 with either pure B or C on sublattice 2. The composition space of the phase can then be considered in Fig. 5.4 below as consisting of four compounds, the so-called 'end members', at the corner of each square. The composition of the phase is then encompassed in the space between the four 'end member' compounds (Fig. 5.4(a)) and the reference energy surface will look like Fig. 5.4(b).

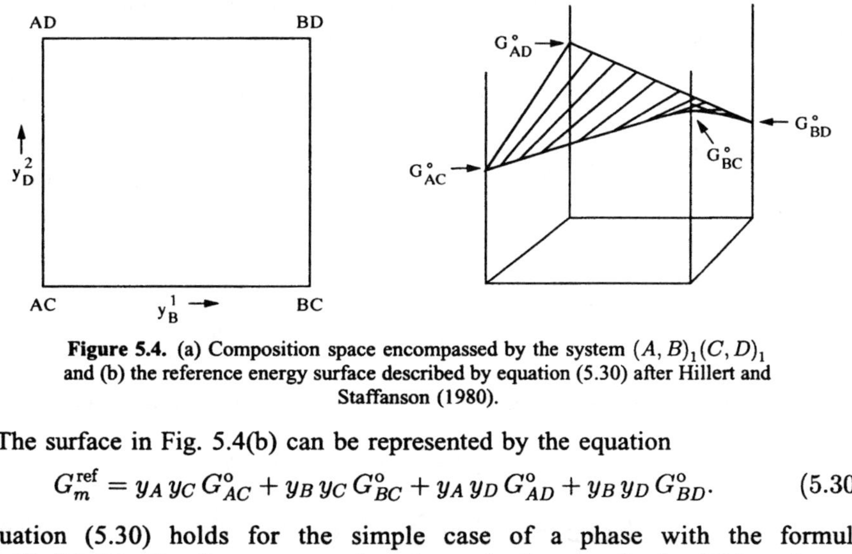

Figure 5.4. (a) Composition space encompassed by the system $(A, B)_1(C, D)_1$ and (b) the reference energy surface described by equation (5.30) after Hillert and Staffanson (1980).

The surface in Fig. 5.4(b) can be represented by the equation

$$G_m^{\mathrm{ref}} = y_A\, y_C\, G_{AC}^{\mathrm{o}} + y_B\, y_C\, G_{BC}^{\mathrm{o}} + y_A\, y_D\, G_{AD}^{\mathrm{o}} + y_B\, y_D\, G_{BD}^{\mathrm{o}}. \tag{5.30}$$

Equation (5.30) holds for the simple case of a phase with the formula $(A, B)_1(C, D)_1$. But for more complex phases the function for the Gibbs reference energy surface may be generalised by arranging the site fractions in a $(l + c)$ matrix if there are l sublattices and c components.

$$y = \begin{array}{cccccc} y_1^1 & y_2^1 & y_3^1 & \bullet & \bullet & y_c^1 \\ y_1^2 & y_2^2 & y_3^2 & & & y_c^2 \\ y_1^3 & y_2^3 & y_3^3 & & & \bullet \\ \bullet & & & & & \bullet \\ \bullet & & & & & \bullet \\ y_1^l & \bullet & \bullet & \bullet & \bullet & y_c^l \end{array}$$

Each row represents a sublattice and each column a component. The information contained in the subscript and superscript can be represented by a component array I, which defines one component for each sublattice. Using this notation a more generalised formula can be obtained (Sundman and Ågren 1981)

$$G_m^{\text{ref}} = \sum_I P_I(Y) \cdot G_I^{\text{o}} \tag{5.31}$$

where G_I^{o} represent the Gibbs energy of the compound defined by I and $P_I(Y)$ represents the corresponding product of site fractions from the Y matrix.

5.4.2.4 Gibbs excess energy of mixing. The method for describing the Gibbs excess energy can, again, be best shown by using a two-sublattice system $(A, B)_1(C, D)_1$ before generalising to a multi-component system. In this alloy A–C, A–D, B–C and B–D interactions are controlled by the Gibbs energy of the compounds AC, BC, AD and BD. Mixing on the sublattices controls A–B and C–D interactions and the simplest form of interaction is a regular solution format such that

$$G_{\text{mis}}^{xs} = y_A^1\, y_B^1\, L_{A,B:*}^0 + y_C^1\, y_D^1\, L_{*:C,D}^0 \tag{5.32}$$

where $L_{A,B:*}^0$ and $L_{*:C,D}^0$ denote regular solution parameters for mixing on the sublattices irrespective of site occupation of the other sublattice. A sub-regular model can be introduced by making the interactions compositionally dependent on the site occupation in the other sublattice.

$$\begin{aligned} G_{\text{mis}}^{xs} &= y_A^1\, y_B^1\, y_C^2\, L_{A,B:C}^0 + y_A^1\, y_B^1\, y_D^2\, L_{A,B:D}^0 \\ &\quad + y_C^1\, y_D^1\, y_A^2\, L_{A:C,D}^0 + y_C^1\, y_D^1\, y_B^2\, L_{B:C,D}^0 \end{aligned} \tag{5.33}$$

Finally some site fraction dependence to these parameters can be added such that

$$L_{A,B:C}^0 = y_A^1\, y_B^1\, y_C^2 \sum_v L_{A,B:C}^v \left(y_A^1 - y_B^1\right)^v \tag{5.34a}$$

$$L_{A,B:D}^0 = y_A^1\, y_B^1\, y_D^2 \sum_v L_{A,B:D}^v \left(y_A^1 - y_B^1\right)^v \tag{5.34b}$$

$$L_{A:C,D}^0 = y_A^1\, y_C^2\, y_D^2 \sum_v L_{A:C,D}^v \left(y_C^2 - y_D^2\right)^v \tag{5.34c}$$

$$L_{B:C,D}^0 = y_B^1\, y_C^2\, y_D^2 \sum_v L_{B:C,D}^v \left(y_C^2 - y_D^2\right)^v \tag{5.34d}$$

It is clear that this can be expanded to any number of sublattices and components and Eq. (5.33) can be generalised using the notation of Eq. (5.31)

$$G_{\text{mix}}^{xs} = \sum_{I1} P_{I1}(Y) \cdot L_{I1}. \tag{5.35}$$

$I1$ refers to a component array of the first order, where one sublattice contains two components, but the remaining sublattices are occupied by only one component with the summation being taken over all different $I1$. The type of component array that was introduced in Eq. (5.31) can then be denoted $I0$ and referred to as a component of the zeroth order. Equation (5.35) is general for the case of regular solutions but can be further extended to include interactions of a higher order, as in Eq. (5.34), by introducing the appropriate arrays IZ but with a further restriction that component array must not contain any component more than once in each sublattice. In this way the excess Gibbs energy can be written as

$$G^{xs}_{\text{mix}} = \sum_{Z>0} \sum_{IZ} P_{IZ}(Y) \cdot L_{IZ} \tag{5.36}$$

and the total energy of the phase including the reference energy, the ideal entropy and excess terms becomes

$$G^{\text{ref}}_m = \sum_{I0} P_{I0}(Y) \cdot G^{\text{o}}_{I0} + RT \sum_s N^s \sum_i y^s_i \log_e y^s_i + \sum_{Z>0} \sum_{IZ} P_{IZ}(Y) \cdot L_{IZ}. \tag{5.37}$$

5.4.3 Applications of the sublattice model

5.4.3.1 Line compounds. These are phases where sublattice occupation is restricted by particular combinations of atomic size, electronegativity, etc., and there is a well-defined stoichiometry with respect to the components. Many examples occur in transition metal borides and silicides, *III–V* compounds and a number of carbides. Although such phases are considered to be stoichiometric in the relevant binary systems, they can have partial or complete solubility of other components with preferential substitution for one of the binary elements. This can be demonstrated for the case of a compound such as the orthorhombic Cr_2B-type boride which exists in a number or refractory metal-boride phase diagrams. Mixing then occurs by substitution on the metal sublattice.

For example in the Fe–Cr–B system (Fig. 5.5) Cr_2B extends across the diagram at a constant stoichiometry of boron as $(Cr,Fe)_2B$ (Villars and Calvert 1985). Also shown is the corresponding extension of the tetragonal Fe_2B compound. Such phases are commonly called 'line compounds'. Generally, sublattice occupation can be described as

$$(A, B, C\ldots)_u(Z)_v$$

where A, B and C are the elements which mix on the first sublattice and Z is the element with fixed stoichiometry. The Gibbs energy of such a phase can be written

$$G_m = \sum_i y^1_i G^{\text{o}}_{i:z} + RTu \sum_i y^1_i \log_e y^1_i + \sum_i \sum_{j>i} y^1_i y^1_j \sum_v L^v_{ij} (y^1_i - y^1_j)^v \tag{5.38}$$

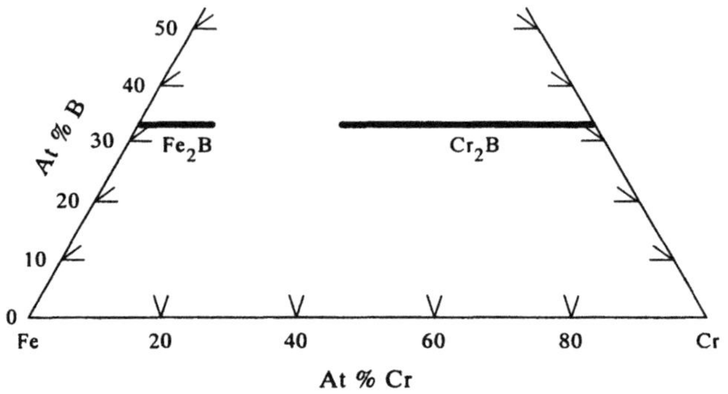

Figure 5.5. Partial isothermal section for Fe–Cr–B showing stoichiometric behaviour of the $(Cr,Fe)_2B$ compound.

where i and j denote the elements mixing on sublattice 1 while Z is the stoichiometric element. u is the molar number of sites in sublattice 1 and $G^o_{i:z}$ denotes the Gibbs energies of the various compounds when sublattice 1 is completely occupied by one of the components mixing on sublattice 1. The excess energy term is associated only with interactions between the components on the first sublattice. Because of the fixed stoichiometry, mole fractions can straightforwardly be expressed as $x_i = (u/(u+v))\, y^1_i$ and Eq. (5.38) becomes equivalent to the line compound models used by Kaufman and Bernstein (1970).

5.4.3.2 Interstitial phases. These are predominant in steels and ferrous-based alloys, where elements such as C and N occupy the interstitial sites of the ferrite and austenite lattices. In this case the structure of the phase can be considered as consisting of two sublattices, one occupied by substitutional elements, such as Fe, Cr, Ni, Mn, etc., and the other occupied by the interstitial elements, such as C or N, and interstitial vacancies (Va). As the concentration of C, N,..., etc., increases the interstitial vacancies are filled until there is complete occupation. The occupation of the sublattices is shown below as

$$(\mathrm{Fe, Cr, Ni, Mn}\ldots)_u\,(\mathrm{Va, C, N}\ldots)_v.$$

In the case of an f.c.c._A1 structure, with $u = v = 1$, the state of complete occupation of interstitial carbon corresponds to a MC carbide with the NaCl lattice. For the c.p.h._A3 structure, with $u = 1$ and $v = 0.5$, complete occupation of the interstitial sites by carbon gives the M_2C carbide with a hexagonal Fe_2N-type structure.

For the case of the f.c.c._A1, austenite phase in Cr–Fe–C, the Gibbs energy of the phase is represented by the formula $(Cr, Fe)_1\,(C, Va)_1$ and its Gibbs energy given by the following equation

$$G_m = y^1_{\mathrm{Cr}} y^2_{\mathrm{Va}} G^{\mathrm{o}}_{\mathrm{Cr:Va}} + y^1_{\mathrm{Fe}} y^2_{\mathrm{Va}} G^{\mathrm{o}}_{\mathrm{Fe:Va}} + y^1_{\mathrm{Cr}} y^2_{\mathrm{C}} G^{\mathrm{o}}_{\mathrm{Cr:C}}$$
$$+ y^1_{\mathrm{Fe}} y^2_{\mathrm{C}} G^{\mathrm{o}}_{\mathrm{Fe:C}} + RT\left(1 \cdot \left(y^1_{\mathrm{Cr}} \log_e y^1_{\mathrm{Cr}} + y^1_{\mathrm{Fe}} \log_e y^1_{\mathrm{Fe}}\right)\right.$$
$$\left. + 1 \cdot \left(y^2_{\mathrm{C}} \log_e y^2_{\mathrm{C}} + y^2_{\mathrm{Va}} \log_e y^2_{\mathrm{Va}}\right)\right)$$
$$+ y^1_{\mathrm{Cr}} y^1_{\mathrm{Fe}} y^2_{\mathrm{Va}} \left(\sum_v L^v_{\mathrm{Cr,Fe:Va}} \left(y^1_{\mathrm{Cr}} - y^1_{\mathrm{Fe}}\right)^v\right)$$
$$+ y^1_{\mathrm{Cr}} y^1_{\mathrm{Fe}} y^2_{\mathrm{C}} \left(\sum_v L^v_{\mathrm{Cr,Fe:C}} \left(y^1_{\mathrm{Cr}} - y^1_{\mathrm{Fe}}\right)^v\right)$$
$$+ y^1_{\mathrm{Cr}} y^2_{\mathrm{C}} y^2_{\mathrm{Va}} \left(\sum_v L^v_{\mathrm{Cr:C,Va}} \left(y^2_{\mathrm{C}} - y^2_{\mathrm{Va}}\right)^v\right)$$
$$+ y^1_{\mathrm{Fe}} y^2_{\mathrm{C}} y^2_{\mathrm{Va}} \left(\sum_v L^v_{\mathrm{Fe:C,Va}} \left(y^1_{\mathrm{C}} - y^1_{\mathrm{Va}}\right)^v\right). \quad (5.39)$$

The first two terms represent the Gibbs energy of f.c.c._A1 Cr and Fe because the second sublattice consists entirely of unfilled interstitial sites. The next two terms represent the Gibbs energy of the CrC and FeC compounds which are formed when all of the interstitial sites of the f.c.c._A1 lattice are filled with C. The second line is the ideal entropy term while the last four lines correspond to the excess term. It can be seen that, when $y^2_{\mathrm{C}} = 0$, the model reduces to Eq. (5.21) for the simple substitutional mixing of Cr and Fe. This is particularly useful as thermodynamic assessments of metallic binary systems can be extended to include interstitial elements without the need for changing any of the existing parameters.

5.4.3.3 Complex intermetallic compounds with significant variation in stoichiometry. In many binary systems intermetallic compounds may exhibit significant solubility and there may be more than two crystallographic sublattices to consider. This is the case for compounds such as the σ, μ, χ and Laves phases, and in such circumstances the sublattice model should, as closely as possible, follow the internal sublattice structure of the compound itself. Taking the case of σ, this consists of five crystallographically distinct sublattices occupied by atoms in the ratio 2:4:8:8:8 (Villars and Calvert 1991). The total number of atoms per unit cell is then 30. Full mixing of all components on all sublattices yields a prohibitively high number of variables, making a CALPHAD calculation in a multi-component system almost impossible. Even modelling the phase in a binary would be extremely difficult and some simplification is necessary.

This can be achieved for σ by considering two types of components differentiated almost exclusively by size considerations. The sublattice occupation can then be simplified. One possibility is to consider σ as either $(A, B)_{16}(A, B)_{10}(B)_4$ where the first and third sublattices and fourth and fifth sublattices are 'joined', i.e., site occupation of A and B atoms are considered to be equal on the sites. Alternatively $(A, B)_{16}(A)_{10}(B)_4$ may be used where mixing is considered only to occur on the fourth and fifth sublattices (Ansara *et al.* 1997). The latter occupation arises when solubility of B components is small in the second sublattice, as appears to be the case for most σ phases. The consequence of this simplification is that the sublattice formula for the Co–Mo σ phase is given by $(Co, Mo)_{16}(Co)_{10}(Mo)_4$ and the Gibbs energy is then

$$G_m = G^o_{\mathrm{Co:Co:Mo}} + G^o_{\mathrm{Mo:Co:Mo}} + RT\left(16 \cdot (y^1_{\mathrm{Co}} \log_e y^1_{\mathrm{Co}} + y^1_{\mathrm{Mo}} \log_e y^1_{\mathrm{Mo}})\right) + y^1_{\mathrm{Co}} y^1_{\mathrm{Mo}} y^2_{\mathrm{Co}} y^3_{\mathrm{Mo}} \left(\sum_v L^v_{\mathrm{Co,Mo:Co:Mo}} \left(y^1_{\mathrm{Co}} - y^1_{\mathrm{Mo}}\right)^v\right). \quad (5.40)$$

As $y^2_{\mathrm{Co}} = y^3_{\mathrm{Mo}} = 1$, Eq. (5.40) becomes rather like Eq. (5.21) in that mixing is simplified to two components on a single sublattice but the points of complete occupation do not reach pure Co or Mo. The expressions for Gibbs energy become more complex when, with the addition of further elements, mixing can occur on all sublattices as in the case of Ni–Co–Mo–W. Here the sublattice formula would be $(Ni, Co, Mo, W)_{16}(Ni, Co)_{10}(Mo, W)_4$, but this degree of complexity can be handled straightforwardly by most current software.

5.4.3.4 Order–disorder transformations. The previous examples considered strict site preference for the components in sublattice phases. For example, in the $(Cr, Fe)_2B$ compound, B is not considered to mix on the metal sublattice, nor are Cr and Fe considered to mix on the B sublattice. This strict limitation on occupancy does not always occur. Some phases, which have preferential site occupation of elements on different sublattices at low temperatures, can disorder at higher temperatures with all elements mixing randomly on all sublattices.

It was demonstrated by Sundman (1985) and later by Ansara *et al.* (1988) that an order–disorder transformation could be modelled by setting specific restrictions on the parameters of a two-sublattice phase. One of the first phases to be considered was an A_3B-ordered compound. In such circumstances the sublattice formula $(A, B)_3(A, B)$ can be applied and the possible relationships between site fractions and mole fractions are given in Figure 5.6. The dashed lines denoted $x_B = 0.25$, 0.5 and 0.75 show variations in order of the phase while the composition is maintained constant. When these lines cross the diagonal joining A_3A and B_3B the phase has disordered completely as $y^1_B = y^2_B = x_B$. As the lines go toward the boundary edge the phase orders and, at the side and corners of the composition square, there is complete ordering of A and B on the sublattices.

The *two-sublattice order–disorder model* (2SLOD) requires first that the Gibbs energy should always have an extremum along the diagonal representing the disordered state with respect to fluctuations in site fractions at constant composition. Further, when the disordered phase is stable this extremum must be a minimum. By assuming that $G^o_{A:A}$ and $G^o_{B:B}$ are zero and applying the above conditions it is possible to define interconnected parameters for the various values

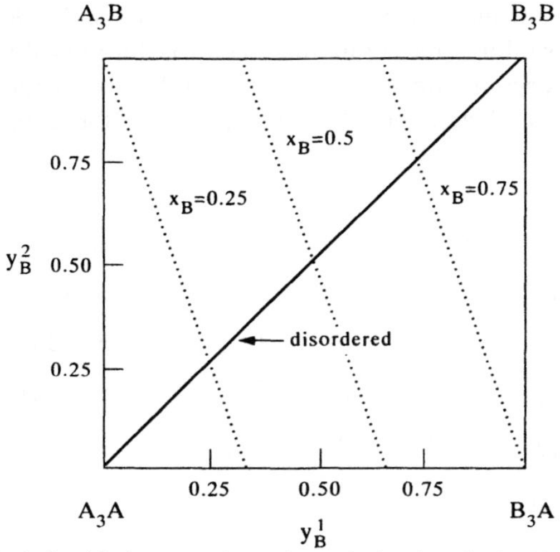

Figure 5.6. Relationship between site and atomic fractions in the 2SLOD model for Ni_3Al.

of $G^o_{A:B}$ and $G^o_{B:A}$ and the excess mixing terms, such that the A_3B phase is ordered at low temperature but disorders as the temperature is increased. For the case of an A_3B compound one solution gives (Ansara *et al.* 1988)

$$\begin{aligned}
G^o_{\mathrm{A:B}} &= u_1\\
G^o_{\mathrm{B:A}} &= u_2\\
L^0_{\mathrm{A,B:A}} &= 3u_1 + u_2/2 + 3u_3\\
L^0_{\mathrm{A,B:B}} &= u_1/2 + 3u_2 + 3u_3\\
L^0_{\mathrm{A:A,B}} &= u_2/2 + u_3\\
L^0_{\mathrm{B:A,B}} &= u_1/2 + u_3\\
L^1_{\mathrm{A,B:A}} &= 3u_4\\
L^1_{\mathrm{A,B:B}} &= 3u_5\\
L^1_{\mathrm{A:A,B}} &= u_4\\
L^1_{\mathrm{B:A,B}} &= u_5\\
L^0_{\mathrm{A,B:A,B}} &= 4u_4 - 4u_5.
\end{aligned} \tag{5.41}$$

The above terms give the ordering contribution to the total energy, and to provide the necessary disordered energy it was necessary to add further terms. This is done by using the relationships

$$x_A = u \cdot y^1_A + v \cdot y^2_A \quad \text{and} \quad x_B = u \cdot y^1_B + v \cdot y^2_B \tag{5.42}$$

where u and v are the number of sites on sublattices 1 and 2. Replacing x_A and x_B in Eq. (5.21), expanding and comparing with the formula for the two-sub-lattice model, an equivalence in Gibbs energy between the disordered substitutional solid solution and a two-sublattice model can be obtained if the following parameters are used (Saunders 1989):

$$
\begin{aligned}
G^{\mathrm{o}}_{\mathrm{A:B}} &= uv \cdot L^0_{\mathrm{AB}} + uv(u-v) \cdot L^1_{\mathrm{AB}} + uv(u-v)^2 \cdot L^2_{\mathrm{AB}} \\
G^{\mathrm{o}}_{\mathrm{B:A}} &= uv \cdot L^0_{\mathrm{AB}} - uv(u-v) \cdot L^1_{\mathrm{AB}} + uv(u-v)^2 \cdot L^2_{\mathrm{AB}} \\
L^0_{\mathrm{A,B:A}} &= u^2 \cdot L^0_{\mathrm{AB}} + 3u^2v \cdot L^1_{\mathrm{AB}} + u^2v(5v-2u) \cdot L^2_{\mathrm{AB}} \\
L^0_{\mathrm{A,B:B}} &= u^2 \cdot L^0_{\mathrm{AB}} - 3u^2v \cdot L^1_{\mathrm{AB}} + u^2v(5v-2u) \cdot L^2_{\mathrm{AB}} \\
L^1_{\mathrm{A,B:A}} &= u^3 \cdot L^1_{\mathrm{AB}} + 4u^3v \cdot L^2_{\mathrm{AB}} \\
L^1_{\mathrm{A,B:B}} &= u^3 \cdot L^1_{\mathrm{AB}} - 4u^3v \cdot L^2_{\mathrm{AB}} \\
L^2_{\mathrm{A,B:A}} &= u^4 \cdot L^2_{\mathrm{AB}} \\
L^2_{\mathrm{A,B:B}} &= u^4 \cdot L^2_{\mathrm{AB}} \\
L^0_{\mathrm{A:A,B}} &= v^2 \cdot L^0_{\mathrm{AB}} + 3v^2u \cdot L^1_{\mathrm{AB}} + v^2u(5u-2v) \cdot L^2_{\mathrm{AB}} \\
L^0_{\mathrm{B:A,B}} &= v^2 \cdot L^0_{\mathrm{AB}} - 3v^2u \cdot L^1_{\mathrm{AB}} + v^2u(5u-2v) \cdot L^2_{\mathrm{AB}} \\
L^1_{\mathrm{A:A,B}} &= v^3 \cdot L^1_{\mathrm{AB}} + 4v^3u \cdot L^2_{\mathrm{AB}} \\
L^1_{\mathrm{B:A,B}} &= v^3 \cdot L^1_{\mathrm{AB}} - 4v^3u \cdot L^2_{\mathrm{AB}} \\
L^2_{\mathrm{A:A,B}} &= v^4 \cdot L^2_{\mathrm{AB}} \\
L^2_{\mathrm{B:A,B}} &= v^4 \cdot L^2_{\mathrm{AB}} \\
L^0_{\mathrm{A,B:A,B}} &= -24u^2v^2 \cdot L^2_{\mathrm{AB}}
\end{aligned}
\qquad (5.43)
$$

where $L^{0,1,2}_{AB}$ are the respective excess parameters from Eq. (5.21) (note L replaces Ω in the above notation). Adding the ordered and disordered part together provides the total Gibbs energy of the phase both in the ordered and disordered state. This method was used by Gros (1987), Ansara *et al.* (1988) and Saunders (1989) and, in general, calculations give quite reasonable results. However, the model proved to have some flaws (Saunders 1996). Firstly, when asymmetrical terms for the ordering energies are used (i.e., $u_1 \neq u_2$) they give rise to a residual, extraneous excess Gibbs energy when the phase disorders and there is no longer an equivalence in Gibbs energy between the 2SLOD model and the original disordered phase. Secondly, when the disordered part is extended to higher order systems there is an incompatibility with the substitutional model when sub-regular or higher terms are used for the various $L^{0,1,2}$ parameters in Eq. (5.43).

Some of the problems with the model were addressed in later work by Dupin

(1995) who used a more complex formulation for the ordering parameters. Firstly, the ordering contribution was separated from disordered contribution and added straightforwardly to the Redlich–Kister energy polynomial. This also made it simpler to combine with a Redlich–Kister model and removed the need for re-formatting existing phases already modelled using this format. Secondly, the extraneous excess energies from the ordering parameters were empirically removed such that the excess Gibbs energy due to the ordered parameters became zero on disordering. The Gibbs energy is then expressed as a sum of three terms

$$G_m = G_m^{\mathrm{dis}}(x_i) + G_m^{\mathrm{ord}}(y_i^1 y_i^2) - G_m^{\mathrm{ord}}(y_i^1 = x_i;\ y_i^2 = x_i) \tag{5.44}$$

where $G_m^{\mathrm{dis}}(x_i)$ is the Gibbs energy contribution of the disordered state, $G_m^{\mathrm{ord}}(y_i^1 y_i^2)$ is the Gibbs energy contribution due to ordering and $G_m^{\mathrm{ord}}(y_i^1 = x_i;\ y_i^2 = x_i)$ is a term which represents the extraneous excess energy contribution from the ordered parameters when the phase disorders. A more accurate representation of the Ni–Al diagram was achieved using this model (Dupin 1995) and the work was extended to Ni–Al–Ti–Cr and Ni–Al–Ta–Cr.

However, the empirical removal of the residual, extraneous energies, G_m^{ord} $(y_i^1 = x_i;\ y_i^2 = x_i)$, causes internal inconsistencies in the model. For example, in an ordered compound such as an *AB* or A_3B type, it is possible that the Gibbs energy which is *actually* calculated for the fully ordered state is quite different from that specified for G_{AB}^{ord} or $G_{A_3B}^{\mathrm{ord}}$ (Saunders 1996). It would therefore be better if the model was derived in such a way that these extraneous energies did not arise. This is actually the case when ordering energies are symmetrical in the form where $G^{\mathrm{o}}_{A:B} = G^{\mathrm{o}}_{B:A}$ and $L^0_{A,B:*} = L^0_{*:A,B} = -G^{\mathrm{o}}_{A:B}$ (Saunders 1989). This is equivalent to the conditions of the Bragg–Williams–Gorsky model of Inden (1975a, 1975b, 1977a, 1977b). However, this limits the model when it is applied to phases such as Ti_3Al and Ni_3Al where substantial asymmetries are apparent.

The empirical nature of the 2SLOD model is such that it cannot be considered a true ordering model in its own right and is therefore included in this chapter rather than the more fundamental chapter on ordering (Chapter 7). However, Sundman and Mohri (1990), using a hybrid sublattice model, showed it was possible to model the Cu–Au system such that it closely matched the phase diagram achieved by a Monte-Carlo method (see Chapter 7). This was done by combining a four-sublattice model, with composition-independent interaction energies, and a gas cluster model for the short-range ordering which was modified to account for a restriction in the degrees of freedom in the solid state. The sublattice model was equivalent to a classical Bragg–Williams (1934) treatment at low temperatures and therefore remains a basic ordering treatment. To model complex ordering systems such as Ti–Al it is almost certainly necessary to include at least third nearest-neighbour interchange energies (Pettifor 1995).

5.5. IONIC LIQUID MODELS

Phase diagrams containing ionic liquids, such as slag systems and molten salts, can be complex and show apparently contradictory behaviour. For example the SiO_2–CaO phase diagram shows liquid immiscibility as well as thermodynamically stable compounds. Immiscibility is usually associated with positive deviations from ideality and, at first sight, is not consistent with compound-forming systems which exhibit large negative deviations. Such features arise from a complex Gibbs energy change with composition and, although the Gibbs energy of mixing can be negative over the whole composition range, inflections in the mixing curve give rise to spinodal points with subsequent decomposition to two liquids (Taylor and Dinsdale 1990). The use of simple mixture models for ionic liquids has not been successful. They need large numbers of coefficients to mimic the sharp changes in enthalpy around critical compositions, and binary systems thus modelled tend to predict the behaviour of multi-component systems badly.

Many models have been proposed to account for the thermodynamic behaviour of ionic liquids and some important ones are listed below:

(1) Cellular models
(2) Modified quasichemical models
(3) Sublattice models
(4) Associated solution models

The above is not intended to be a definitive list but rather to indicate some of the more commonly used models at the present time. Other, more historical, models have been used extensively, for example the polymerisation models of Toop and Samis (1962) and Masson (1965), the models of Flood (1954), Richardson (1956) and Yokakawa and Niwa (1969). More recently the 'central atom' model by Satsri and Lahiri (1985, 1986) and the 'complex' model of Hoch and Arpshofen (1984) have been proposed. Each has been used with some success in lower-order systems, but the extension to multicomponent systems is not always straightforward.

5.5.1 The cellular model

Kapoor and Frohberg (1973) applied this model to the ternary system CaO–FeO–SiO_2. They envisaged that mixing occurred by formation of three 'asymmetric cells' obtained by a reaction between 'symmetric cells' of the metallic oxides and silica. For CaO–SiO_2, the formation energy for the asymmetric cell is denoted W_{IS}, where the subscripts I and S denote the combination of 'symmetric' cells from CaO and SiO_2. In addition interactions between the various symmetric and asymmetric cells were considered such that

$$\varepsilon_{II,SS} = 2\varepsilon_{IS,SS} \tag{5.45}$$

References are listed on pp. 124–126.

where $\varepsilon_{II,SS}$ denotes the interaction between the symmetric CaO and SiO_2 cells and $\varepsilon_{IS,SS}$ the interaction between the asymmetric cell formed between CaO and SiO_2 and the symmetric silica cell. This is expanded with the addition of FeO so that the two further asymmetric cells are formed with energies W_{JS} and W_{IJ}, where J denotes the FeO cell, and the additional interaction energies between the cells were related in the following way:

$$\varepsilon_{IJ,SS} = \varepsilon_{IS,SS} + \varepsilon_{JS,SS} \tag{5.45a}$$

where $\varepsilon_{JS,SS}$ is the interaction energy between the asymmetric cell formed between FeO and SiO_2 and the silica cell. It was further assumed that interactions terms were negligibly small in comparison to the cell formation energies and that $\varepsilon_{IS,SS}/kT$ and $\varepsilon_{JS,SS}/kT$ were small compared to unity. Based on these assumptions they were able to define the Gibbs energy of mixing in a system such as CaO–FeO–SiO_2 as

$$\begin{aligned}\frac{G_{\text{mix}}}{RT} = &-\frac{3N_S}{2}\log_e N_S - \frac{(1-N_S)}{2}\log_e(1-N_S) - N_I\log_e N_I - N_J\log_e N_J\\ &+ (N_I - R_{IS} - R_{IJ})\log_e(N_I - R_{IS} - R_{IJ})\\ &+ (N_J - R_{JS} - R_{IJ})\log_e(N_J - R_{JS} - R_{IJ})\\ &+ (N_S - R_{IS} - R_{JS})\log_e(N_S - R_{IS} - R_{JS})\\ &+ 2R_{IS}\log_e R_{IS} + 2R_{JS}\log_e R_{JS} + 2R_{IJ}\log_e R_{IJ}\\ &+ \frac{2W_{IS}R_{IS}}{RT} + \frac{2W_{JS}R_{JS}}{RT} + \frac{2W_{IJ}R_{IJ}}{RT}\\ &+ \frac{2\varepsilon(1-N_S)(N_S - R_{IS} - R_{JS})}{RT}\end{aligned} \tag{5.46}$$

with

$$N_i = \frac{n_i}{\sum_i n_i + 2n_S}$$

and

$$N_S = \frac{n_S}{\sum_i n_i + 2n_S}$$

where i denotes either CaO or FeO and n_i and n_S denote the number of moles of CaO, FeO and SiO_2. The interaction parameter ε is given by

$$\varepsilon = \frac{\sum_i n_i\,\varepsilon_{IS,SS}}{\sum_i n_i}$$

and the various values of R_{ij}s by

$$R_{ij} = \frac{r_{ij}}{\sum_i n_i + 2n_S} \tag{5.48a}$$

$$(N_I - R_{IS} - R_{IJ})\,(N_J - R_{JS} - R_{IJ}) = R_{IJ}^2 \exp\left(-\frac{2W_{IJ}}{RT}\right) \tag{5.48b}$$

$$(N_I - R_{IS} - R_{IJ})\,(N_S - R_{IS} - R_{JS}) = R_{IS}^2 \exp\left(-\frac{2W_{IS}}{RT}\right) \exp\left(-\frac{2\varepsilon(1 - N_S)}{RT}\right) \tag{5.48c}$$

$$(N_J - R_{JS} - R_{IJ})\,(N_S - R_{IS} - R_{JS}) = R_{IS}^2 \exp\left(-\frac{2W_{JS}}{RT}\right) \exp\left(-\frac{2\varepsilon(1 - N_S)}{RT}\right) \tag{5.48d}$$

where r_{ij} are the number of moles of the different asymmetric cell types. In Eq. (5.46) it can be seen that the first five lines correspond to the configurational entropy term. This is no longer ideal because of preferential formation of the various asymmetrical cells and the system effectively orders. This is a feature of ionic systems and gives rise to forms of quasichemical entropy which will also be discussed in the next section.

From the above equations, activities of metallic oxides and silica were calculated for various binary slag systems and CaO–FeO–SiO_2 and found to be in good agreement. The drawback of the model is that equivalents need to be made between various anionic and cationic types and problems can arise when these are polyvalent. Also, considering that only a ternary system was considered, the expressions for Gibbs energy are complex in comparison to the simple mixture and sublattice types. Gaye and Welfringer (1984) extended this model to multi-component systems but the treatment is too lengthy to consider here.

5.5.2 Modified quasichemical models

A modified form of the quasichemical model of Guggenheim (1935) and Fowler and Guggenheim (1939) has recently been developed by Pelton and Blander (1986a, 1986b, 1988) for application to ionic liquids. The model considers that the liquid has a strong tendency to order around specific compositions, associated with specific physical or chemical phenomena. This was considered to be a general feature of ionic liquids and that previous attempts at modelling of ordered salts and silicate slags were either too specific in nature to the type of system, i.e., oxide or

References are listed on pp. 124–126.

halide, or too complicated for general usage, particularly when considering extrapolation to higher-order systems. The model is therefore phenomenological in nature which has the advantage that it can be widely applied.

First, consider a binary liquid *A–B* in which *A* and *B* atoms mix substitutionally on a quasi-lattice with coordination number *Z*. There is the possibility that *A–B* pairs will be formed from *A–A* and *B–B* pairs by the following relation

$$(A-A)+(B-B)=2(A-B).$$

The molar enthalpy and entropy change of this reaction, denoted respectively ω and η by Pelton and Blander (1986a, 1986b, 1988), is given by $(\omega-\eta T)$. If this is zero the solution is ideal. However, if $(\omega-\eta T)$ is negative then there will be ordering of the mixture around the 50:50 composition and the enthalpy and entropy of mixing will show distinct minima at the *AB* composition. As ordering does not always occur at *AB* it is desirable to allow other compositions to be chosen for the position of the minima, which is done by replacing mole fractions with equivalent fractions y_A and y_B where

$$y_A=\frac{\beta_A x_A}{\beta_A x_A+\beta_B x_B} \quad \text{and} \quad y_B=\frac{\beta_B x_B}{\beta_A x_A+\beta_B x_B} \tag{5.49}$$

where β_A and β_B are numbers chosen so that $y_A=y_B=0.5$ at the composition of maximum ordering. Letting x_{AA}, x_{BB} and x_{AB} be the fractions of each type of pair in the liquid gives

$$H_{\text{mix}}=(\beta_A x_A+\beta_B x_B)\left(\frac{x_{AB}}{2}\right)\omega \tag{5.50}$$

$$\begin{aligned}S_{\text{mix}}^{\text{conf}}=&-\frac{RZ}{2}(\beta_A x_A+\beta_B x_B)\cdot\left(x_{AA}\log_e\frac{x_{AA}}{y_A^2}+x_{BB}\log_e\frac{x_{BB}}{y_B^2}+x_{AB}\log_e\frac{x_{AB}}{2y_A y_B}\right)\\&-R(x_A\log_e x_A+x_B\log_e x_B)\end{aligned} \tag{5.51}$$

and

$$S_{\text{mix}}^{xs}=(\beta_A x_A+\beta_B x_B)\left(\frac{x_{AB}}{2}\right)\eta \tag{5.52}$$

where *Z* is the average co-ordination number. Two mass balance equations can be written

$$2y_A=2x_{AA}+x_{AB} \tag{5.53a}$$

$$2y_B=2x_{BB}+x_{AB} \tag{5.53b}$$

and minimisation of the Gibbs energy then gives the ‘quasichemical equilibrium constant’

$$\frac{x_{AB}^2}{x_{AA}\,x_{BB}} = 4\exp\left(\frac{-2(\omega-\eta T)}{zRT}\right) \tag{5.54}$$

Eqs (5.53) and (5.54) can be solved for any given values of ω and η to provide x_{AA}, x_{BB} and x_{AB} which can then be used to define the enthalpy and entropy values given in Eqs (5.50)–(5.52). The choice of β_A and β_B can be found using the following equation

$$(1-r)\log_e(1-r) + r\log_e r = 2\beta_B\, r\log_e 0.5 \tag{5.55}$$

where $r = \beta_A/(\beta_A+\beta_B)$. Finally, some simple compositional dependence is defined for ω and η

$$\omega = \sum_{n=0} \omega^n\, y_B^n \tag{5.56a}$$

and

$$\eta = \sum_{n=0} \eta^n\, y_B^n. \tag{5.56b}$$

The approach has been used effectively for a variety of systems, for example oxide and silicate slags, salt systems, and is readily extendable to multi-component systems (Pelton and Blander 1984, 1986a, 1986b, Blander and Pelton 1984). The general applicability of the model is demonstrated by the recent work of Eriksson *et al.* (1993) who are in the process of creating a comprehensive database for the system SiO_2–Al_2O_3–CaO–MgO–MnO–FeO–Na_2O–K_2O–TiO_2–Ti_2O_3–ZrO_2–S.

5.5.3 Sublattice models

Essentially, sublattice models originate from the concepts of Temkin (1945) who proposed that two separate sublattices exist in a solid-state crystal for cations and anions. The configurational entropy is then governed by the site occupation of the various cations and anions on their respective sublattices. When the valence of the cations and anions on the sublattices are equal, and electroneutrality is maintained, the model parameters can be represented as described in Section 5.4.2. However, when the valence of the cations and anions varies, the situation becomes more complex and some additional restrictions need to be made. These can be expressed by considering equivalent fractions (f) which, for a sublattice phase with the formula $(I^{+i}, J^{+j}\ldots)(M^{-m}, N^{-n}\ldots)$, are given by

$$f_I = \frac{i y_I^1}{\sum_I i y_I^1} \tag{5.57a}$$

and

$$f_M = \frac{my_M^2}{\sum_M my_M^2} \tag{5.57b}$$

where i and m are the valences of I and M respectively. These replace site fractions in the general expressions for the Gibbs energy reference and excess mixing terms. The ideal mixing term is given as (Pelton 1988)

$$\frac{G_{\text{mix}}^{\text{ideal}}}{RT} = \left(\frac{1}{\sum_I iy_I^1}\right)\left(\sum_I y_I^1 \log_e y_I^1\right) + \left(\frac{1}{\sum_M my_M^2}\right)\left(\sum_M y_M^2 \log_e y_M^2\right). \tag{5.58}$$

This approach works well for systems where no neutral ions exist and a database for the system (Li, Na, K) (F, Cl, OH, CO_3, SO_4) has been developed (Pelton 1988) which gives good estimates for activities in multi-component liquids. However, the above approach is more limited in the presence of neutral ions which are necessary if one wishes to model ionic liquids which may contain neutrals such as S.

To overcome this problem an extension of the sublattice model was proposed by Hillert *et al.* (1985) which is now known as the ionic two-sublattice model for liquids. As in the previous case it uses constituent fractions as composition variables, but it also considers that vacancies, with a charge corresponding to the charge of the cations, can be introduced on the anion sublattice so that the composition can move away from the ideal stoichiometry and approach an element with an electropositive character. The necessary neutral species of an electronegative element are added to the anion sublattice in order to allow the composition to approach a pure element. The sublattice formula for the model can then be written as

$$\left(C_i^{+v_i}\right)_P \left(A_j^{-v_j},\, Va,\, B_k^0\right)_Q \tag{5.59}$$

where C represents cations, A anions, Va hypothetical vacancies and B neutrals. The charge of an ion is denoted v_i and the indices i, j and k are used to denote specific constituents. For the following description of the model, superscripts v_i, v_j and 0 will be excluded unless needed for clarity. The number of sites on the sublattices is varied so that electroneutrality is maintained and values of P and Q are calculated from the equation

$$P = \sum_j v_j\, y_{A_j} + Qy_{Va} \tag{5.60a}$$

and

$$Q = \sum_i v_i\, y_{C_i} \tag{5.60b}$$

Eqs (5.60a) and (5.60b) simply mean that P and Q are equal to the average charge on the opposite sublattice with the hypothetical vacancies having an induced charge equal to Q. Mole fractions for the components can be defined as follows:

Cations $$x_{C_i} = \frac{P y_{C_i}}{P + Q(1 - y_{Va})} \tag{5.61a}$$

Anions $$x_{D_i} = \frac{Q y_{D_i}}{P + Q(1 - y_{Va})}. \tag{5.61b}$$

The integral Gibbs energy for this model is then given by

$$\begin{aligned} G_m = {} & \sum_i \sum_j y_{C_i} y_{A_i} G^{\circ}_{C_i:A_j} + Q y_{Va} \sum_i y_{C_i} G^{\circ}_{C_i} + Q \sum_k y_{B_i} G^{\circ}_{B_i} \\ & + RT\left(P \sum_i y_{C_i} \log_e y_{C_i} + Q\left(\sum_j y_{A_j} \log_e y_{A_j} + y_{Va} \log_e y_{Va}\right.\right. \\ & \left.\left. + \sum_k y_{B_k} \log_e y_{B_k}\right)\right) \\ & + \sum_{i_1} \sum_{i_2} \sum_j y_{i_1} y_{i_2} y_j L_{i_1 i_2:j} + \sum_{i_1} \sum_{i_2} y_{i_1} y_{i_2} y^2_{Va} L_{i_1 i_2:Va} \\ & + \sum_i \sum_{j_1} \sum_{j_2} y_i y_{j_1} y_{j_2} L_{i:j_1 j_2} + \sum_i \sum_j y_i y_j y_{Va} L_{i:jVa} \\ & + \sum_i \sum_j \sum_k y_i y_j y_k L_{i:jk} + \sum_i \sum_k y_i y_k y_{Va} L_{i:Vak} \\ & + \sum_{k_1} \sum_{k_2} y_{k_1} y_{k_2} L_{k_1 k_2} \end{aligned} \tag{5.62}$$

where $G^{\circ}_{C_i:A_j}$ is the Gibbs energy of formation for $(v_1 + v_2)$ moles of atoms of liquid C_iA_j and $\Delta G^{\circ}_{C_i}$ and $\Delta G^{\circ}_{B_i}$ are the Gibbs energy of formation per mole of atoms of liquid C_1 and B_i respectively. The first line represents the Gibbs energy reference state, the second line the configurational mixing term and the last three lines the excess Gibbs energy of mixing. As the list of excess parameters is long and the subscript notation is complex, it is worth giving some specific examples for these parameters after Sundman (1996):

$L_{i_1 i_2:j}$ represents the interaction between two cations in the presence of a common anion; for example $L_{Ca^{+2},Mg^{+2}:O^{-2}}$ represents an interaction term from the system CaO–MgO.

$L_{i_1 i_2:Va}$ represents interactions between metallic elements; for example $L_{Ca,Mg}$ in Ca–Mg.

References are listed on pp. 124–126.

$L_{i:j_1j_2}$ represents interactions between two anions in the presence of a common cation; for example $L_{Ca^{+2}:OH^{-1},CO_3^{-2}}$ in $Ca(OH)_2$–$CaCO_3$.

$L_{i:jk}$ represents interactions between an anion and a neutral atom; for example $L_{Fe^{+2}:S^{-2},S}$ in the S-rich part of Fe–S.

$L_{i:Vak}$ represents interactions between a metal and a neutral; for example $L_{Fe^{+2}:Va,C}$ in Fe–C.

L_{k_1,k_2} represents interactions between two neutrals; for example $L_{Si_3N_4,SiO_2}$ in Si_3N_4–SiO_2. It should be noted that the cation is irrelevant here as the number of cation sites has become zero when only neutrals are taken into account.

The model is certainly complex, but perhaps no more so than previous ionic liquid models described in Sections 5.5.1 and 5.5.2. The initial experience (Selleby 1996) suggests that the number of terms needed to describe a ternary system such as Fe–Mn–S is quite similar for both ionic two-sublattice liquid and associate models (see next section). The modelling of ionic liquids is, in the main, complex and the advantages of the various techniques can only become apparent as they become more commonly used.

There are a number of immediate advantages to the two-sublattice ionic liquid model. The first is that it becomes identical to the more usual Redlich–Kister representation of metallic systems when the cation sublattice contains only vacancies. This immediately allows data from an assessed metallic system to be combined with data from an oxide system so that the full range of compositions is covered. In binary cases the model can be made equivalent to an associated model. For example in Cu–S the associated model would consider a substitutional solution of Cu, S and Cu_2S. If an ionic two-sublattice model with a formula $(Cu^{+1})_P(S^{-2}, Va, S)_Q$ is used, it is straightforward to derive parameters to give identical results to the associate model. However, it should be noted that the Gibbs energy of the two models does not remain equivalent if they are extended into ternary and higher-order systems (Hillert *et al.* 1985).

5.5.4 Associated solution models

There are a number of papers dealing with the associated models (Predel and Oehme 1974, Sommer 1977, Sharma and Chang 1979, Chuang *et al.* 1981, Björkman 1985) which are all very similar in principle. There are differences in the way the excess terms are handled, but all consider that some type of complex or associate is formed inside the liquid from a reaction between the components of the system in question. The thermodynamic properties of the liquid then depend predominantly on the Gibbs energy of formation of these complexes, or associates, rather than by interactions between the components. This gives rise to enthalpy of mixing diagrams which are characterised by sharp changes at critical compositions where the associate(s) exist and also by markedly non-ideal mixing entropies.

The derivation of Sommer (1977, 1980, 1982) can be used as an example. This considers the formation of a single associate with a formula A_iB_j within a binary system A–B. It is assumed the liquid contains n_{A_1} and n_{B_1} number of moles of 'free' A and B in equilibrium with $m_{A_iB_j}$ number of moles of the associate A_iB_j. The mole fractions of A, B and A_iB_j, in a binary alloy containing 1 mole of A and B atoms, are then given by the formula

$$x_A = n_{A_1} + i\,n_{A_iB_j}; \quad x_B = n_{B_1} + j\,n_{A_iB_j} \quad \text{and} \quad x_{A_iB_j} = n_{A_iB_j}. \tag{5.63}$$

The excess Gibbs energy of mixing is then given by the general formula

$$G^{xs}_{\text{mix}} = G^{\text{ass}} + G^{\text{reg}} \tag{5.64}$$

G^{ass} is the Gibbs energy due to the formation of the associate defined as

$$G^{\text{ass}} = n_{A_jB_j}\, G^{\text{o}}_{A_iB_j} \tag{5.65}$$

where $G^{\text{o}}_{A_iB_j}$ is the Gibbs energy of formation of one mole of the associate. G^{reg} considers the Gibbs energy due to the interactions between the components A and B themselves and with the associate A_iB_j such that

$$G^{\text{reg}} = G^{\text{reg}}_{A,B}\frac{n_{A_1}\,n_{B_1}}{n} + G^{\text{reg}}_{A,A_iB_j}\frac{n_{A_1}\,n_{A_iB_j}}{n} + G^{\text{reg}}_{B,A_iB_j}\frac{n_{B_1}\,n_{A_iB_j}}{n} \tag{5.66}$$

where $n = n_{A_1} + n_{B_2} + n_{A_iB_j}$. The configurational entropy is given simply by

$$S^{\text{conf}}_{\text{mix}} = -R\big(n_{A_1}\log_e x_A + n_{B_1}\log_e x_B + n_{A_iB_j}\log_e x_{A_iB_j}\big). \tag{5.67}$$

The model was applied to numerous ionic melts and good agreement was found with experimental results (Sommer 1982). One of the main criticisms of the associate formalism has been that, although the concept considers associates or complexes to be 'diffuse' in nature, the mathematical formalism implies that distinct molecules exist, which is harder to justify. However, if this stringent view is relaxed, it can be seen that the model merely implies some underlying structure to the liquid, which is quite reasonable, and it does provide functions which allow for the temperature dependence of the enthalpy of mixing. A more serious criticism is that some knowledge of the relevant associate is necessary before Eqs (5.64)–(5.67) can be applied. The most appropriate associate can be selected by fitting of experimental results for enthalpies of mixing which is sufficient in a large number of cases. However, in some systems there may be a number of different associates and it is not always obvious as to which types actually exist. The main advantages

References are listed on pp. 124–126.

of the associate models is that they allow a simple strategy to be adopted for optimisation. It is further easy to define ternary and higher-order associates and extend the model to multi-component systems. For example, Björkman (1985) showed this to good effect in the system Fe–O–SiO_2.

To complete this section it is interesting to show the equivalence between the ionic two-sublattice model and the associate model as demonstrated by Hillert *et al.* (1985). Equation (5.62) can be simplified for a system $(A^{+vA})_p(B^{+vB}, Va^{-vA}, B^o)_Q$ where $+va = -vb$ to become, for one mole of atoms,

$$G = \frac{y_B\, G^o_{A(-vB/+vA)B_1} + RT(y_B \log_e y_B + y_b \log_e y_b + y_{Va} \log_e y_{Va}) + G^{xs}_{\text{mix}}/vA}{\left(1 - \dfrac{y_B\, v_B}{vA}\right)} \tag{5.68}$$

where y_b is the site fraction occupation of the neutral B^o on sublattice Q. Equations (5.65)–(5.67) will yield for one mole of atoms

$$G = \frac{\begin{pmatrix} x_{A_iB_j} G^{\text{ass}} + RT\left(x_{A_iB_j} \log_e x_{A_iB_j} + x_{A_1} \log_e x_{A_1} + x_{B_1} \log_e x_{B_1}\right) \\ +x_{A_1}\, x_{B_1}\, G^{\text{reg}}_{A_1,B_1} + x_{A_1}\, x_{A_iB_j} G^{\text{reg}}_{A_1,A_iB_j} + x_{B_1}\, x_{A_iB_j} G^{\text{reg}}_{B_1,A_iB_j} \end{pmatrix}}{(1 + x_{A_iB_j}(i + j - 1))}. \tag{5.69}$$

The two models become identical for the case where $i = j = vA = -vB = 1$ and it is then possible to show the following identities: (1) $x_{A_iB_j} \equiv y_B$, (2) $x_{B_1} \equiv y_b$ and (3) $x_{A_1} \equiv y_{Va}$.

As $y_A = 1$, the C^{reg} term has the following equivalence

$$x_{A_iB_j}\, G^{\text{ass}} = y_A\, y_B\, G^o_{A^{+vA}:B^{-vB}} \tag{5.70}$$

and the excess terms of the associate show the following equivalences

$$\begin{aligned} x_{A_1}\, x_{B_1}\, G^{\text{reg}}_{A,B} &= y_A\, y_{Va}\, y_B\, L_{A^{+vA}:B^o,Va^{-vA}} \\ x_{A_1}\, x_{A_iB_j}\, G^{\text{reg}}_{A,A_iB_j} &= y_A\, y_{Va}\, y_B\, L_{A^{+vA}:B^{-vB},Va} \\ x_{B_1}\, x_{A_iB_j}\, G^{\text{reg}}_{B,A_iB_j} &= y_A\, y_b\, y_B\, L_{A^{+vA}:B^oB-vB}. \end{aligned} \tag{5.71}$$

They can also be made identical in the general case if the conditions $i = -vB/vA$ and $j = 1$. The equivalences break down in ternary and higher-order systems as there is the introduction of more compositional variables in the associate model than for the two-sublattice case. This was considered (Hillert *et al.* 1985) to demonstrate the advantages of the sub-lattice model, but as mentioned previously it turns out that the number of excess terms to describe Fe–Mn–S is very similar.

5.6 AQUEOUS SOLUTIONS

Aqueous solutions form a large area for thermodynamic modelling and hold great importance in their own right. However, they are, by definition, basically dilute solutions and almost all models deal exclusively with the thermodynamic properties of the aqueous solution, only considering some possible precipitation reactions of insoluble, mainly stoichiometric, compounds. The integration of these models into concentrated solution databases which are often used at high temperatures has therefore not been seriously undertaken by any group as far as the authors are aware. There has, however, been quite significant use of aqueous models in conjunction with substance databases for prediction of thermodynamic properties and solubility products in a number of important areas such as corrosion and geochemistry. To this end some discussion of aqueous modelling is worthwhile.

One of the main conceptual differences between the models discussed so far and aqueous solutions is that the units which are used to define thermodynamic functions are often different. This is because they apply to the properties which are actually measured for aqueous systems, and molarity (c_i) and molality (m_i) are far more common units than mole fraction. Molarity is defined as

$$c_i = \frac{10^{-3} n_i}{V} \tag{5.72}$$

and molality as

$$m_i = \frac{n_i}{n_w M_w} \tag{5.73}$$

where M_w is the molar mass of water and V the volume of the solution in m^3. The partial ideal mixing term for component i, ($\bar{G}_i^{\mathrm{ideal}}$) can be defined with respect to molality as

$$\bar{G}_i^{\mathrm{ideal}} = \bar{G}_i^{\mathrm{o}} + RT \log_e (m_i) \tag{5.74}$$

where $\bar{G}_i^{\mathrm{o}}$ is the chemical potential of component i in its standard state. This can be combined with some term for the partial excess Gibbs energy so that the chemical potential of i can be defined. The excess Gibbs energy term is defined using the ionic strength (I) where

$$I = 0.5 \sum_i m_i z_i^2 \tag{5.75}$$

where z_i is the charge in electron units of the component i. This can be used to define the limiting excess Gibbs energy of an electrolyte solution containing cations of charge z^+ and anions of charge z^- via the Debye–Hückel limiting law as

$$\frac{G^{xs}}{(n_w M_w RT)} = -\frac{4}{3} A I^{3/2}. \tag{5.76}$$

Differentiation of Eq. (5.76) with respect to the amount of solute leads to the expression

$$\log_e \gamma_i = -z_i^2 A\sqrt{I}. \tag{5.77}$$

The above equation assumes that the ions are point charges and interact in a continuous dielectric. It is essentially correct in the limit, but problems arise when considering finite concentrations of solute where an extended Debye–Hückel expression may be more appropriate

$$\log_e \gamma_i = \frac{-z_i^2 A\sqrt{I}}{(1 + Ba_i\sqrt{I})}. \tag{5.78}$$

Here A and B are temperature-dependent parameters which are properties of pure water only and a_i is the effective diameter of the hydrated ion. This expression can be used up to ionic strengths of 5×10^{-2} (Davies 1985) but for higher concentrations further terms are needed. One of the simplest expressions is to consider a linear term with respect to ionic strength such that

$$\log_e \gamma_i = \frac{-z_i^2 A\sqrt{I}}{(1 + Ba_i\sqrt{I})} + \beta I \tag{5.79}$$

or, more generally,

$$\log_e \gamma_i = \frac{-z_i^2 A\sqrt{I}}{(1 + Ba_i\sqrt{I})} + \sum_i \beta_i I^i. \tag{5.80}$$

By considering Ba_i as a single adjustable parameter, data can then be fitted using only Ba_i and β as variables. A further simplification is to make Ba_i equal to unity and add a further constant ionic strength term which gives the Davies equation (Davies 1962)

$$\log_e \gamma_i = -z_i^2 A\left(\frac{\sqrt{I}}{(1 + \sqrt{I})} - 0.3I\right). \tag{5.81}$$

Further expansions are possible following Bromley (1973) and Zemaitis (1980) such that

$$\log_{10} \gamma_i = \frac{-z_i^2 A\sqrt{I}}{(1 + \sqrt{I})} + \frac{(0.06 + 0.6\beta)z_i^2}{\left(1 + \frac{1.5I}{z_i^2}\right)^2} + \beta + \chi I^2 + \delta I^3. \tag{5.82}$$

A further approach is to consider a more specific form of ion interaction and replace Eq. (5.80) with the Bronsted–Guggenheim equation (Guggenheim and Turgeon 1955)

$$\log_e \gamma_i = \frac{-z_i^2 A\sqrt{I}}{(1 + Ba_i\sqrt{I})} + \sum_j \varepsilon_{(i,j,I_m)} m_j. \tag{5.83}$$

The last term represents ion interactions between i and j as a function of ionic strength. A simplification of Eq. (5.83) was suggested by Pitzer and Brewer (1961) such that

$$\log_{10} \gamma_i = \frac{-0.5107 z_i^2 \sqrt{I}}{(1 + \sqrt{I})} + \sum_j \varepsilon_{(i,j)} m_j \tag{5.84}$$

where the interaction coefficients become concentration independent. The Debye–Hückel term then has similarities with the Davies equation in that Ba_i is made equal to unity.

One of the main models which is available in CALPHAD calculation programmes is that based on Pitzer (1973, 1975), Pitzer and Mayorga (1973) and Pitzer and Kim (1974). The model is based on the development of an explicit function relating the ion interaction coefficient to the ionic strength and the addition of a third virial coefficient to Eq. (5.83). For the case of an electrolyte MX the excess Gibbs energy is given by

$$\frac{G_{\text{mix}}^{xs}}{RT} = f(I) + \sum_i \sum_j \lambda_{ij}(I) m_i m_j + \sum_i \sum_j \sum_k \mu_{ijk} m_i m_j m_k \tag{5.85}$$

with

$$f(I) = \frac{-4A^{\phi} I \log_e(1 + 1.2\sqrt{I})}{1.2} \tag{5.86}$$

where $\lambda_{ij}(I)$ and μ_{ijk} are second and third virial coefficients from the non-electrostatic part of G_{mix}^{xs} and A^{ϕ} is the Debye–Hückle limiting slope. μ_{ijk} is taken to be independent of ionic strength and set to zero if ions i, j and k have the same charge sign. Equation (5.86) is similar in form to the excess terms considered for simple mixture phases, except that some specific terms arising from the unique properties of aqueous systems are included. Differentiation of Eq. (5.85) with respect to concentration yield the following expression for activity coefficients of cation (M), anion (X) and neutral species (N) (Clegg and Brimblecombe 1989)

$$\begin{aligned}\log_e \gamma_M = {} & z_M^2 F + \sum_a m_a(2B_{Ma} + ZC_{Ma}) + \sum_c m_c\left(2\Phi_{Mc} + \sum_a m_a \Psi_{Mca}\right) \\ & + \sum_a \sum_{<a'} m_a m_{a'} \Psi_{Maa'} + |z_M| \sum_c \sum_a m_c m_a C_{ca} \\ & + 2\sum_n m_n \lambda_{Mn} + 3\sum_j \sum_k m_j m_k \mu_{Mjk}\end{aligned} \tag{5.87a}$$

References are listed on pp. 124–126.

$$\log_e \gamma_X = z_X^2 F + \sum_c m_c(2B_{cX} + ZC_{cX}) + \sum_a m_a\left(2\Phi_{aX} + \sum_c m_c \Psi_{cXa}\right)$$
$$+ \sum_c \sum_{<c'} m_c\, m_{c'} \Psi_{Xcc'} + |z_x| \sum_c \sum_a m_c\, m_a\, C_{ca}$$
$$+ 2\sum_n m_n\, \lambda_{Xn} + 3\sum_j \sum_k m_j\, m_k\, \mu_{Xjk} \tag{5.87b}$$

$$\log_e \gamma_N = 2\sum_n m_n\, \lambda_{Nn} + 2\sum_c m_c\, \lambda_{Nc} + 2\sum_a m_a\, \lambda_{Na} + 3\sum_i \sum_j m_i\, m_j\, \mu_{Nij} \tag{5.87c}$$

where the subscripts c, a, and n represent cations, anions and neutrals respectively. Summation indices $c < c'$ and $a < a'$ denote sums over all distinguishable pairs of cations and anions while summations with indices i, j and k cover all species and are unrestricted. $Z = \sum_i m_i |z_i|$ while F is defined as

$$F = -A^\phi\left(\frac{\sqrt{I}}{(1+b\sqrt{I})} + \frac{2\log_e(1+b\sqrt{1})}{b}\right) + \sum_c \sum_a m_c\, m_a\, B'_{ca}$$
$$+ \sum_c \sum_{<c'} m_c\, m_{c'} \Phi'_{cc'} + \sum_a \sum_{<a'} m_a\, m_{a'} \Phi'_{aa'}. \tag{5.88}$$

The ionic strength dependence of the coefficients, B_{MX}, is

$$B^\phi_{MX} = \beta^0_{MX} + \beta^1_{MX} \exp\left(-\alpha\sqrt{I}\right) \tag{5.89}$$

$$B_{MX} = \beta^0_{MX} + \beta^1_{MX}\, g\left(\alpha\sqrt{I}\right) \tag{5.90}$$

$$B'_{MX} = \frac{\beta^1_{MX}\, g'\left(\alpha\sqrt{I}\right)}{I} \tag{5.91}$$

with the function g and g' given by

$$g = \frac{2\left(1 - \left(1 + \alpha\sqrt{I}\right)\exp\left(-\alpha\sqrt{I}\right)\right)}{\alpha^2 I} \tag{5.92a}$$

$$g' = \frac{-2\left(1 - \left(1 + \alpha\sqrt{I} + \alpha^2 I/2\right)\exp\left(-\alpha\sqrt{I}\right)\right)}{\alpha^2 I} \tag{5.92b}$$

where α is usually assigned a value of 2 for electrolytes of low charge. The mixture terms Φ^φ, Φ and Φ' are defined as

$$\Phi^{\phi}_{ij} = \theta_{ij} + ^{E}\theta_{ij}(I) + I^{E}\theta'_{ij}(I) \quad (5.93)$$

$$\Phi_{ij} = \theta_{ij} + ^{E}\theta_{ij}(I) \quad (5.94)$$

$$\Phi'_{ij} = ^{E}\theta'_{ij}(I) \quad (5.95)$$

θ_{ij} is a constant for each pair of ions. $^{E}\theta_{ij}(I)$ and $^{E}\theta'_{ij}(I)$ are universal functions that are required when ions of differing charge types are mixed.

As can be seen from the above equations, modelling of concentrated aqueous solutions becomes complex and it is not possible within the scope of this book to do justice to what is a major field in its own right. There are a number of articles which can be recommended to give the reader some idea of what can be achieved with aqueous models. Harvie and Weare (1980) developed a model for calculating mineral solubilities in natural waters based on the Pitzer equations which achieved good accuracy up to strengths of 20 m. This work was extended by Harvie *et al.* (1982, 1984) so that a database exists for the prediction of mineral solubilities in water up to high ionic strengths at 25°C for the system Na, K, Mg, Ca, H, Cl, SO_4, OH, HCO_3, CO_3, CO_2, H_2O. More recently Clegg and Brimblescombe (1989) presented a database, again using the Pitzer model, for the solubility of ammonia in aqueous and multi-component solutions for the system Li, Na, Mg, NH_4, Mg, Ca, Sr, Ba, F, Cl, Br, I, OH, CNS, NO_3, NO_2, ClO_3, ClO_4, S, SO_3, CH_3COO, HCOO and $(COO)_2$, H_2O. Further models and applications can be found in articles by Zemaitis (1980) and Grenthe and Wanner (1988). The review of Davies (1985) also gives details of software programmes which are available for calculating equilibria in aqueous solutions.

REFERENCES

Ansara, I. (1979) *Int. Met. Reviews*, **22**, 20.
Ansara, I. and Sundman, B. (1987) in *Computer Handling and Dissemination of Data*, ed. Glaeser, P. S., p. 154. CODATA.
Ansara, I., Sundman, B. and Willemin, P. (1988) *Acta Metall.*, **36**, 977.
Ansara, I., Bernard, C., Kaufman, L. and Spencer, P. (1978) *CALPHAD*, **2**, 1.
Ansara, I., Chart, T. G., Hayes, F. H., Pettifor, D. G., Kattner, U., Saunders, N. and Zeng, K. (1997) *CALPHAD*, **21**, 171.
Bale, C. W. and Pelton, A. D. (1990) *Met. Trans. A*, **21A**, 1997.
Bhadeshia, H. K. D. H (1981) *Metal Science*, **15**, 175.
Björkman, B. (1985) *CALPHAD*, **9**, 271.
Blander, M. and Pelton, A. D. (1984) in *Second International Symposium on Metallurgical Slags & Fluxes*, eds Fine, A. H. and Gaskell, D. R. (Met. Soc. AIME, New York), p. 295.
Bragg, W. L. and Williams, E. J. (1934) *Proc. Roy. Soc.*, **A145**, 69.
Bromley, L. A. (1973) *J. Amer. Inst. Chem. Eng.*, **19**, 313.

Chuang, Y.-Y., Hsieh, K.-C. and Chang, Y. A. (1981) *CALPHAD*, **5**, 277.
Clegg, S. L. and Brimblecombe, P. (1989) *J. Phys. Chem.*, **93**, 7237.
Colinet, C. (1967) D.E.S. University of Grenoble, France.
Davies, C. W. (1962) *Ion Association* (Butterworths, Washington).
Davies, R. H (1985) in *Critical Thermodynamics in Industry: Models and Computation*, ed. Barry, T. I. (Blackwell Scientific, Oxford), p. 40.
Dinsdale, A. T. (1991) *CALPHAD*, **15**, 317.
Dupin, N. (1995) PhD. Thesis, l'Institut National Polytechnique de Grenoble, France.
Enomoto, M. and Aaronson, H. I. (1985) *CALPHAD*, **9**, 43.
Eriksson, G., Wu, P. and Pelton, A. D. (1993) *CALPHAD*, **17**, 189.
Flood, H. *et al.* (1954) *Z. Anorg. Allg. Chem.*, **276**, 289.
Foo, E.-H. and Lupis, C. H. P. (1973) *Acta Met.*, **21**, 1409.
Fowler, R. H. and Guggenheim, E. A. (1939) *Statistical Thermodynamics* (Cambridge University Press, Cambridge).
Gaye, H. and Welfringer, J. (1984) in *Proceedings of Second Symposium on Metallurgical Slags and Fluxes*, eds Fine, H. A. and Gaskell, D. R. (Met. Soc. AIME, New York), p. 357.
Grenthe, I. and Wanner, H. (1988) *Guidelines for the extrapolation to zero ionic strength*, Report TDB-2 to OECD Nuclear Energy Agency, Data Bank, F-91191 Gif-sur-Yvette, France (December 5, 1988).
Gros, J. P. (1987) Dr. Ing. Thesis, l'Institut National Polytechnique de Grenoble, France.
Guggenheim, E. A. (1935) *Proc. Roy. Soc.*, **A148**, 304.
Guggenheim, E. A. and Turgeon, J. C. (1955) *Trans. Farad. Soc.*, **51**, 747.
Harvie, C. E. and Weare, J. H. (1980) *Geochim. Cosmochim. Acta*, **44**, 981.
Harvie, C. E., Euguster, H. P. and Weare, J. H. (1982) *Geochim. Cosmochim. Acta*, **46**, 1603.
Harvie, C. E., Møller, N. and Weare, J. H. (1984) *Geochim. Cosmochim. Acta*, **48**, 723.
Harvig, H. (1971) *Acta Chem. Scand.*, **25**, 3199.
Hillert, M. (1980) *CALPHAD*, **4**, 1.
Hillert, M. and Steffansson, L.-I. (1970) *Acta Chem. Scand.*, **24**, 3618.
Hillert, M. and Waldenstrom, M. (1977) *CALPHAD*, **1**, 97.
Hillert, M., Jansson, B., Sundman, B. and Ågren, J. (1985) *Met. Trans. A*, **16A**, 261.
Hoch, M. and Arpshofen, M. (1984) *Z. Metallkde.*, **75**, 23 and 30.
Inden, G. (1975a) *Z. Metallkde.*, **66**, 577.
Inden, G. (1975b) *Z. Metallkde.*, **66**, 648.
Inden, G. (1977a) *Z. Metallkde.*, **68**, 529.
Inden, G. (1977b) *J. de Physique, Colloque*, **C7**, 373.
Kapoor, M. L. and Frohberg, G. M. (1973) in *Chemical Metallurgy of Iron and Steel* (Iron and Steel Institute, London), p. 17.
Kaufman, L. and Bernstein, H. (1970) *Computer Calculation of Phase Diagrams* (Academic Press, New York).
Kaufman, L. and Nesor, H. (1978) *CALPHAD*, **2**, 35.
Kirkaldy, J. S., Thomson, B. A. and Baganis, E. A. (1978) in *Hardenability Concepts with Applications to Steel*, eds Kirkaldy, J. S. and Doane, D. V. (AIME, Warrendale), p. 82.
Kohler, F. (1960) *Monatsh. Chem*, **91**, 738.
Lupis, C. H. P. and Elliott, C. H. P. (1966) *Acta Met.*, **4**, 529 and 1019.
Masson, C. R. (1965) *Proc. Roy. Soc.*, **A287**, 201.
Muggianu, Y. M., Gambino, M. and Bros, J. P. (1975) *J. Chim. Phys.*, **22**, 83.

Pelton, A. D. and Blander, M. (1984) in *Second International Symposium on Metallurgical Slags & Fluxes*, eds Fine, A. H. and Gaskell, D. R. (Met. Soc. AIME, New York), p. 281.
Pelton, A. D. (1988) *CALPHAD*, **12**, 127.
Pelton, A. D. and Blander, M. (1986a) *Met. Trans. B*, **17B**, 805.
Pelton, A. D. and Blander, M. (1986b) in *Computer Modelling of Phase Diagrams*, ed. Bennett, L. H. (TMS, Warrendale), p. 19.
Pelton, A. D. and Blander, M. (1988) *CALPHAD*, **12**, 97.
Pettifor, D. G. (1995) private communication.
Pitzer, K. S. (1973) *J. Phys. Chem.*, **77**, 268.
Pitzer, K. S. (1975) *J. Soln. Chem.*, **4**, 249.
Pitzer, K. S. and Brewer, L. (1961) *Thermodynamics; 2nd Edition* (McGraw-Hill, New York).
Pitzer, K. S. and Kim, J. J. (1974) *J. Am. Chem. Soc.*, **96**, 5701.
Pitzer, K. S. and Mayorga, G. (1973) *J. Phys. Chem.*, **77**, 2300.
Predel, B. and Oehme, G. (1974) *Z. Metallkde*, **65**, 509.
Richardson, F. D. (1956) *Trans. Farad. Soc.*, **52**, 1312.
Sastri, P. and Lahiri, A. K. (1985) *Met. Trans. B*, **16B**, 325.
Sastri, P. and Lahiri, A. K. (1986) *Met. Trans. B*, **17B**, 105.
Saunders, N. (1989) *Z. Metallkde.*, **80**, 894.
Saunders, N. (1996) *CALPHAD*, **4**, 491.
Selleby, M. (1996) unpublished research.
Sharma, R. C. and Chang, Y. A. (1979) *Met. Trans. B.*, **10B**, 103.
Sommer, F. (1977) *CALPHAD*, **2**, 319.
Sommer, F. (1980) *Z. Metallkde*, **71**, 120.
Sommer, F. (1982) *Z. Metallkde.*, **73**, 72 and 77.
Sundman, B. (1985) private communication.
Sundman, B. (1994) unpublished research.
Sundman, B. (1996) private communication.
Sundman, B. and Mohri, T. (1990) *Z. Metallkde*, **81**, 251.
Sundman, B. and Ågren, J. (1981) *J. Phys. Chem. Solids*, **42**, 297.
Taylor, J. R. and Dinsdale, A. T. (1990) *CALPHAD*, **14**, 71.
Temkin, M. (1945) *Acta Phys. Chim. USSR*, **20**, 411.
Tomiska, J. (1980) *CALPHAD*, **4**, 63.
Toop, G. W. (1965) *Trans. Met. Soc. Aime*, **233**, 855.
Toop, G. W. and Samis, C. S. (1962) *Can. Met. Quart.*, **1**, 129.
Villars, P. and Calvert, L. D. (1991) *Pearson's Handbook of Crystallographic Data for Intermetallic Phases; 2nd Edition* (ASM International, Materials Park, OH).
Wagner, C. (1951) *Thermodynamics of Alloys*, Addison-Wesley, Reading, Mass.
Yokokawa, T. and Niwa, T. (1969) *Trans. Japan Inst. Met.*, **10**, 1.
Zemaitis, J. F. (1980) in *Themodynamics of Aqueous Systems with Industrial Applications*, ed. Newman, S. A. (American Chemical Society), p. 227.

Chapter 6
Phase Stabilities

Chapter 6
Phase Stabilities

6.1. INTRODUCTION

The CALPHAD approach is based on the axiom that complete Gibbs energy versus composition curves can be constructed for all the structures exhibited by the elements right across the whole alloy system. This involves the extrapolation of G/x curves of many phases into regions where they are either metastable or unstable and, in particular, *the relative Gibbs energy for various crystal structures of the pure elements of the system must therefore be established.* By convention these are referred to as *lattice stabilities* and the Gibbs energy differences between all the various potential crystal structures in which an element can exist need to be characterised as a function of temperature, pressure and volume.

The basic question is how to perform extrapolations so as to obtain a consistent set of values, taking into account various complications such as the potential presence of mechanical instability. Additional complications arise for elements which have a magnetic component in their Gibbs energy, as this gives rise to a markedly non-linear contribution with temperature. This chapter will concern itself with various aspects of these problems and also how to estimate the thermodynamic properties of metastable solid solutions and compound phases, where similar problems arise when it is impossible to obtain data by experimental methods.

6.2. THERMOCHEMICAL ESTIMATIONS

6.2.1 General procedure for allotropic elements

Allotropic elements, which exhibit different crystal forms within accessible temperatures and pressures, allow the Gibbs energies to be measured for at least some of the possible alternative structures. Such data, although of necessity restricted to particular regimes of temperature and pressure, were therefore used as the platform for all the initial efforts in this field. The elements Mn (Weiss and Tauer 1958), Ti and Zr (Kaufman 1959a) and Fe (Kaufman *et al.* 1963) thus provided the basis for obtaining the first lattice stabilities for phase-diagram calculations. However, in order to make the CALPHAD approach universally applicable, a knowledge of the relative stability of the most frequently occurring crystal structures is also required for elements which do not manifest allotropy under normally accessible conditions of temperature and pressure. Kaufman's conviction that a reliable set of basic lattice stabilities could be assembled was undoubtedly the key concept for the whole CALPHAD approach.

As with all physical phenomena, an agreed reference state has to be established for each element or component. In order to have a firm foundation, this has generally been taken as the crystal structure in which that element exists at standard temperature and pressure.

This ensures that thermochemical (TC) lattice stabilities are firmly anchored to the available experimental evidence. Although the liquid phase might be considered a common denominator, this raises many problems because the structure of liquids is difficult to define; it is certainly not as constant as popularly imagined. It is therefore best to anchor the framework for lattice stabilities in the solid state.

Figure 6.1(a) shows the relative position of the experimentally determined Gibbs energy curves for Ti, which includes segments that refer to the stable low-temperature c.p.h. α-phase, the high temperature b.c.c. β-phase and the liquid phase. Figure 6.1(b) shows curves where properties have been extrapolated to cover regions where, although the phases are metastable, properties can be estimated by extrapolation from alloys which stabilise one or other of the phases in binary systems.

It is clearly desirable to see if the total curve can be de-convoluted into parts that can be identified with a specific physical property so that trends can be established for the many cases where data for metastable structures are not experimentally accessible. In principle, the TC lattice stability of an element in a specified crystal structure β relative to the standard state α can be comprehensively expressed as follows (Kaufman and Bernstein 1970):

$$\begin{aligned} G_{\text{Total}}^{\alpha\to\beta} = {} & H_{T=0}^{\alpha\to\beta} + G_{\text{Debye}}^{\alpha\to\beta} f(\theta_\alpha, \theta_\beta, T) + G_{\text{Elec}}^{\alpha\to\beta} f(\gamma_\alpha, \gamma_\beta, T) \\ & + G_{\text{Lambda}}^{\alpha\to\beta} f(\lambda_\alpha, \lambda_\beta, T) + G_{s1,s2}^{\alpha\to\beta} f(E_\alpha, E_\beta) + G_P^{\alpha\to\beta} f(V_\alpha, V_\beta, P). \quad (6.1) \end{aligned}$$

Here $H_{T=0}^{\alpha\to\beta}$ is the enthalpy of transformation at 0 K, $f(\theta, \gamma, \lambda, E)$ denote functions for the components associated with Debye temperature (θ), electronic specific heat (γ), lambda transitions such as magnetism (λ) and contributions (E) related to multiple electronic states, $s1$,$s2$..., while $f(P, V_\alpha, V_\beta)$ refers to the function which generates the pressure contribution.

It may of course be unnecessary to consider all these terms and the equation is much simplified in the absence of magnetism and multiple electronic states. In the case of Ti, it is possible to deduce values of the Debye temperature and the electronic specific heat for each structure; the pressure term is also available and lambda transitions do not seem to be present. Kaufman and Bernstein (1970) therefore used Eq. (6.2), which yields the results shown in Fig. 6.1(c).

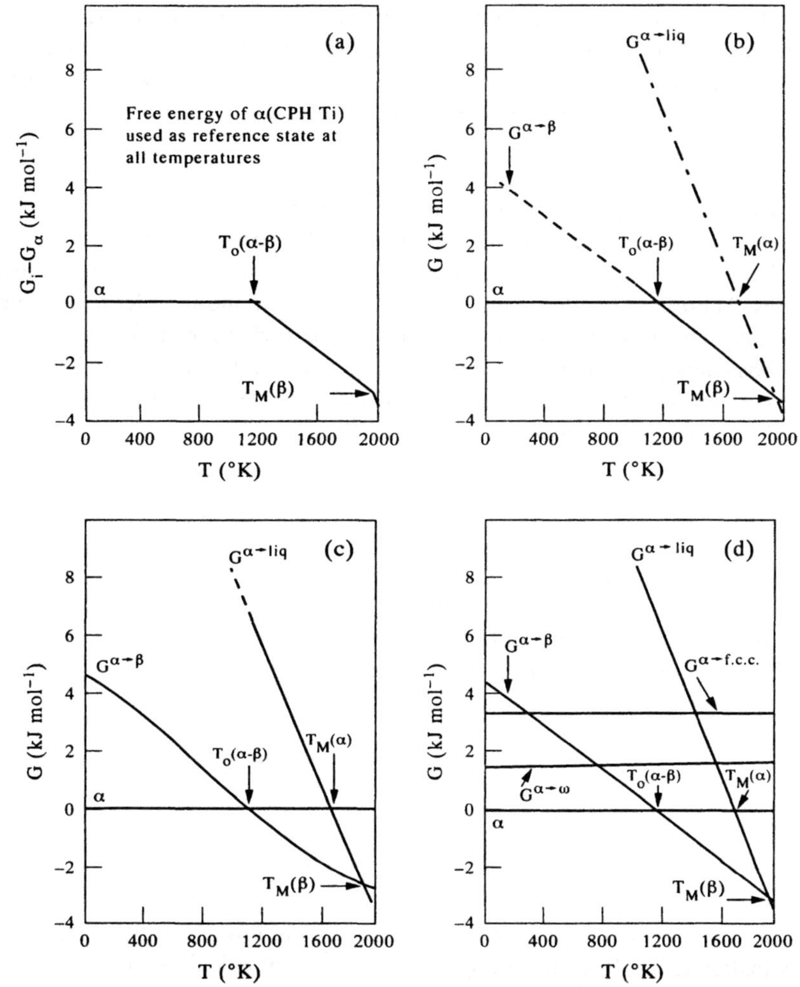

Figure 6.1. Gibbs free energy curves for Ti (a) α, β and liquid segments corresponding to the stable regions of each phase; (b) extrapolated extensions into metastable regions; (c) characterisation from Kaufman (1959a) using equation (6.2); and (d) addition of the G curves for ω and f.c.c. structures (from Miodownik 1985).

$$G_{\mathrm{Ti}}^{\alpha\to\beta} = H_{T=0}^{\alpha\to\beta} + \frac{9}{8}(\theta^\beta - \theta^\alpha) + \left[f\left(\frac{\theta^\beta}{T}\right) - f\left(\frac{\theta^\alpha}{T}\right)\right]$$
$$+ 10^{-4}\frac{R}{T}\left[\left(\frac{3T}{(2-\theta^\beta)}\right)^2 - \left(\frac{3T}{(2-\theta^\alpha)}\right)^2\right] - \frac{T^2}{2}(\gamma^\beta - \gamma^\alpha). \quad (6.2)$$

The right-hand side can be separated into five parts. The first part is the enthalpy at 0 K, the second represents the zero point energy, the third is the Debye energy term, the fourth is an approximation for the $C_p - C_v$ correction while the last part arises from the difference in electronic specific heats.

6.2.2 General procedure for non-allotropic elements

With the exception of a few allotropic elements, the necessary input parameters to Eqs (6.1) or (6.2) are not available to establish the lattice stabilities of metastable structures. Therefore an alternative solution has to be found in order to achieve the desired goal. This has evolved into a standard format where the reference or ground state Gibbs energy is expressed in the form of general polynomials which reproduce assessed experimental C_p data as closely as possible. An example of such a standard formula is given below (Dinsdale 1991):

$$G_m[T] - H_m^{\mathrm{SER}} = a + bT + cT\ln(T) + \sum_2^n d_n T^n. \tag{6.3}$$

The left-hand term is defined as the Gibbs energy relative to the standard element reference (SER) state where H_m^{SER} is the enthalpy of the element or substance in its defined reference state at 298.15 K, a, b, c and d_n are coefficients and n represents a set of integers, typically taking the values of 2, 3 and −1. From Eq. (6.3) further thermodynamic properties of interest can be obtained.

$$S = -b - c - c\ln(T) - \sum n d_n T^{n-1} \tag{6.3a}$$

$$H = a - cT - \sum (n-1)\, d_n T^n \tag{6.3b}$$

$$C_p = -c - \sum n(n-1)\, d_n T^{n-1}. \tag{6.3c}$$

Other phases are then characterised relative to this ground state, using the best approximation to Eq. (6.1) that is appropriate to the available data. For instance, if the electronic specific heats are reasonably similar, there are no lambda transitions and $T >> \theta_D$, then the entropy difference between two phases can be expressed just as a function of the difference in their Debye temperatures (Domb 1958):

$$S_{T>>\theta}^{\alpha\to\beta} \approx \log_e \frac{\theta_\beta}{\theta_\alpha}. \tag{6.4}$$

Combined with the 0 K enthalpy difference, the Gibbs energy of metastable phases can then be obtained by adding terms of the form (A–BT) to the reference state value or, more specifically:

$$G^{\alpha\to\beta} = H^{\alpha\to\beta} - TS^{\alpha\to\beta} \tag{6.5}$$

where $H^{\alpha\rightarrow\beta}$ and $S^{\alpha\rightarrow\beta}$ are taken as constant with respect to temperature. A linear model as shown in Fig. 6.1(b) is therefore a reasonable approximation at temperatures above θ_D. If the Debye temperatures are close enough, then the linear model will also give a reasonable description of the lattice stabilities below the Debye temperature (Miodownik 1986) and can be used to estimate a value of $H^{\alpha\rightarrow\beta}_{T=0}$, since this will be equal to the enthalpy of transformation measured at high temperature. Such linear expressions form the basis of many listed metastable lattice stability values but, in the longer term, it is desirable to return to a mode of representation for the unary elements which re-introduces as many physically definable parameters as possible. This is currently being pursued (Chase *et al.* 1995) by using either the Debye or Einstein equation where

$$C_{\text{Debye}} = 9R\left(\frac{T}{\theta_D}\right)^3 \int_0^{\theta_D} \frac{x^4\,\mathrm{e}^x}{(\mathrm{e}^x - 1)^2}\,\mathrm{d}x \tag{6.6}$$

and $x = h\nu/RT$, or

$$C_{\text{Einstein}} = 3R\left(\frac{\theta_E}{T}\right)^2 \frac{\mathrm{e}^{\theta_E/T}}{[\mathrm{e}^{\theta_E/T} - 1]^2}. \tag{6.7}$$

Empirical fitting coefficients can be added $(a_i(i = 1, 2 \ldots))$ so that the C_p is given by

$$C_{\text{fit}} = C_{\text{Debye, Einstein}} + \sum_i a_i T^i. \tag{6.8}$$

The excess term should allow the total Gibbs energy to be fitted to match that of Eq. (6.3) while at the same time incorporating a return to the inclusion of $f(\theta)$ and $f(\gamma)$ in the lattice stabilities. With the increased potential for calculating metastable Debye temperatures and electronic specific heats from first principles (Haglund *et al.* 1993), a further step forward would be to also replace Eq. (6.5) by some function of Eq. (6.8).

Expressions for the liquid phase are complicated by the presence of a glass transition temperature (T_g). A simple linear model for the liquid should not, therefore, be used close to and below T_g and the curve representing the properties of the liquid phase in Fig. 6.1(b) has subsequently not been extended to 0 K. Further algorithms can be added to the high temperature linear treatment to provide for the effect of the liquid → glass transition if required (Ågren *et al.* 1995). However, regardless of what happens close to and below T_g, an extrapolated curve based on the concept of a retained liquid phase at 0 K will certainly yield the correct Gibbs energy at high temperatures. A similar situation occurs with relation to a magnetic transition, where it is physically impossible to retain the high-temperature paramagnetic configuration below the corresponding critical Curie (or Neel) temperature. Nevertheless, the procedure that has been generally adopted is to define a Gibbs energy curve for the paramagnetic state which covers the whole range of temperatures, and then add a magnetic Gibbs energy to it.

In addition to the melting point of the β phase and the α/β allotropic transformation temperature in Fig. 6.1(b), there is a further intersection between the Gibbs energy of α and liquid phases. This corresponds to the metastable melting point of the α phase. A linear model will then dictate that the entropy of melting for α is defined by the entropies of melting and transformation at the two other critical points (Ardell 1963),

$$S^{\alpha \to L} = S^{\alpha \to \beta} + S^{\beta \to L}. \tag{6.9}$$

This is a reasonable assumption providing the metastable melting point is well above both the θ_D of the solid and the T_g of the liquid. Eq. (6.9) has the following corollary:

$$H^{\alpha \to \beta} = T_f^{\alpha}\, S^{\alpha \to L} - T_f^{\beta}\, S^{\beta \to L}. \tag{6.10}$$

If the procedure adopted to determine the melting characteristics of the metastable phase is reversed, this provides a possible route for establishing TC lattice stabilities in cases where the metastable allotrope is not experimentally accessible. Moreover this is so even where the Gibbs energy curve for the metastable crystal structure does not intersect the Gibbs energy curve of the ground state structure (Fig. 6.1(d)). At first sight this does not seem particularly useful, because there is still a requirement for two unknown parameters T_f^{Φ} and $\Delta S^{\Phi \to L}$, where Φ is the experimentally inaccessible allotrope. However, the next two sections show how meaningful estimates can be obtained for these two parameters.

6.2.2.1 The van Laar technique for estimating melting points. T_f^{Φ} can be obtained by extrapolation from the liquidus of alloy systems which contain a phase of structure Φ by a technique originally suggested by van Laar (1908) but not considered seriously until resuscitated by Hume-Rothery and Raynor (1940). It is described succinctly in the following extract from their paper:

> *"... the b.c.c. structure of the beta phase (in copper alloys) may be regarded as derived from a hypothetical b.c.c. form of copper, which in pure copper is less stable than the f.c.c. structure, but which becomes more stable when sufficient solute is present to raise the electron concentration. From this point of view, which was already discussed by Rosenhain in 1926, the actual phase diagram may be regarded as arising in the way shown in Fig. 6.2, where the broken lines represent the extrapolation of the beta phase solidus and liquidus into the region where the phase is no longer stable."*

It was therefore appropriate that the first attempt to produce lattice stabilities for non-allotropic elements dealt with Cu, Ag and Zn (Kaufman 1959b). It is also significant that, because of the unfamiliarity of the lattice stability concept, this paper did not appear as a mainstream publication although the work on Ti and Zr (Kaufman 1959a) was published virtually at the same time. It was also realised that the reliability of metastable melting points derived by extrapolation were best

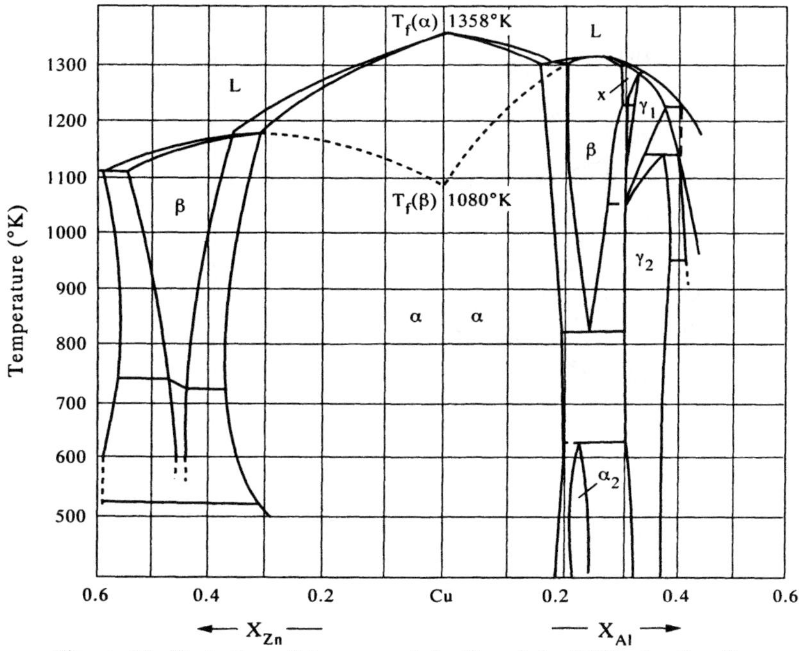

Figure 6.2. Illustration of the proposal by Rosenhain (1926) for Cu alloys containing α- and β-phases. The broken lines represent the extrapolation of the transus lines into the region where the β-phase is no longer stable, terminating at the metastable melting point of b.c.c. Cu (from Miodownik 1986).

obtained by combining the results from a number of different systems. Figure 3.6 from Kaufman and Bernstein (1970) shows this philosophy in action for the case of b.c.c. Re.

6.2.2.2 The estimation of metastable entropies of melting. Many of the lattice stabilities listed in the first embryonic database (Kaufman 1967) were obtained by combining the extrapolated values of melting points with an estimate for the entropy of fusion (S^f) based on Richard's rule. This dates to the last century and states that all metals should exhibit an entropy of fusion approximately equal to the gas constant R. However, with the accumulation of new experimental data for a wider range of elements, this was seen to be an over-simplification. Different crystal structures appear to have characteristic entropies of fusion with b.c.c. structures generally showed the lowest values of S^f, with higher values for f.c.c., c.p.h. and various structures of increasingly lower symmetry (Chalmers 1959, Gschneider 1961). An additional trend relates to the variation of S^f with melting point (Miodownik 1972a, Sawamura 1972). This was initially taken to be linear but, when more recent values for the higher melting-point elements were included, this

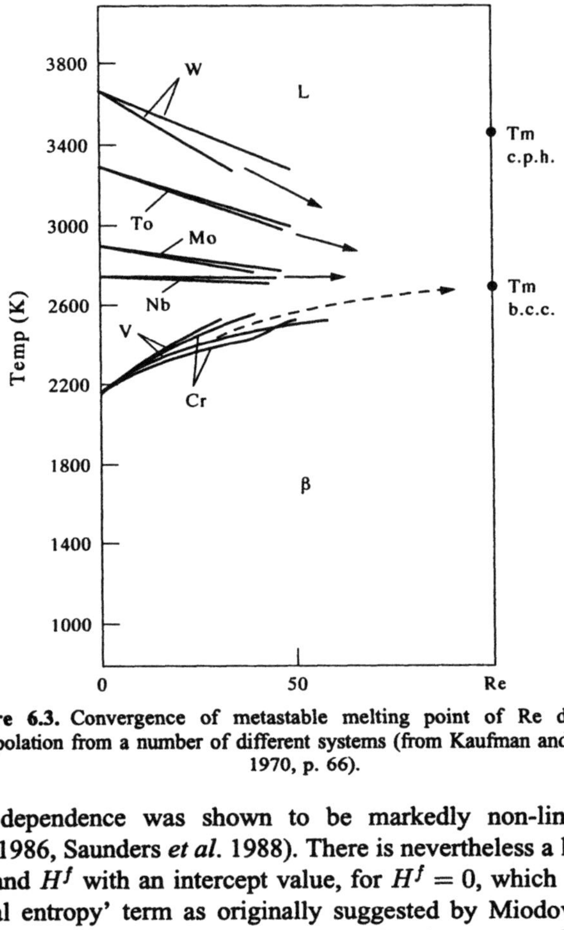

Figure 6.3. Convergence of metastable melting point of Re derived by extrapolation from a number of different systems (from Kaufman and Bernstein 1970, p. 66).

temperature dependence was shown to be markedly non-linear (Fig. 6.4(a)) (Miodownik 1986, Saunders *et al.* 1988). There is nevertheless a linear relationship between S^f and H^f with an intercept value, for $H^f = 0$, which is consistent with the 'structural entropy' term as originally suggested by Miodownik (1972a). As might be expected, there is a similar but smaller intercept value in the corresponding plot that relates S and H for allotropic transformations.

In order to use such empirical correlations as a general method for estimating lattice stabilities, it is necessary to establish whether properties of metastable phases will follow a similar trend to their stable counterparts. In the case of many transition metals (Miodownik 1972a) this certainly seems to be a useful working assumption. However, it is apparent that S^f for the high-temperature b.c.c. structures of the rare earths are anomalously low (Gschneider 1961, 1975) and some of the actinides show a similar effect; S^f for *β-Pu* is, for example, only 3.09 J mol^{-1} K^{-1} (Dinsdale 1991). It has been suggested that electrons in partially occupied *f*-levels can contribute an extra entropy component to the b.c.c. phase in rare earths and actinides (Kmetko and Hill 1976), while Tiwari (1978) suggested that the cumulative entropy of all prior transformations is a more constant quantity

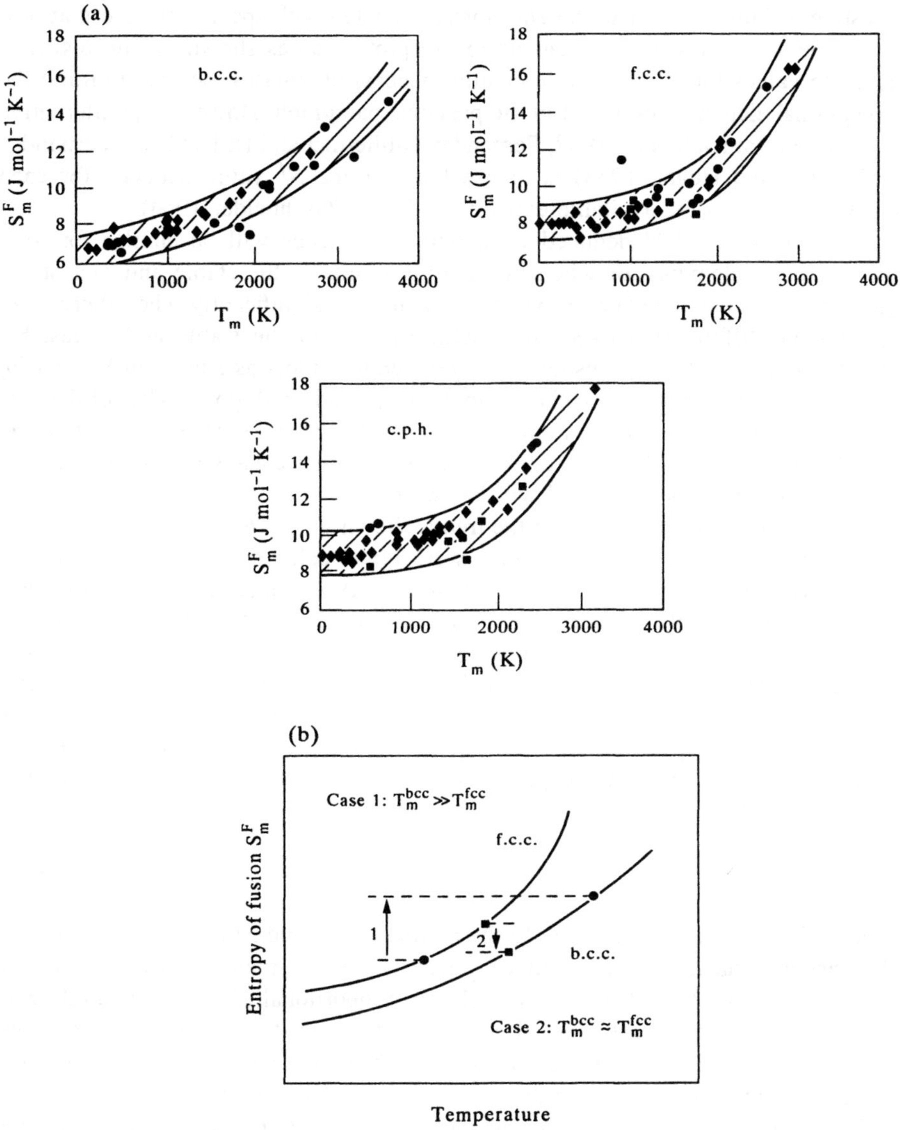

Figure 6.4. (a) Variation of the entropy of fusion with melting point for different crystal structures (from Saunders *et al.* 1988) and (b) schematic illustration of the possibility of a change in value and sign for the entropy of transformation if the metastable structure has a low melting point (from Miodownik 1992).

than the individual component entropies of transformation. But this begs the question of how to estimate the entropies associated with specific transformations.

On the whole it seems reasonable to use plots such as shown in Fig. 6.4(a) to estimate values for S^f of experimentally inaccessible phases, bearing in mind the exceptional elements discussed in the previous paragraph. However, an alternative procedure (Grimwall *et al.* 1987, Fernandez Guillermet and Hillert 1988, Fernandez Guillermet and Huang 1988) is to use the *average* S^f value deduced for each crystal class from the available experimental data (Gschneider 1964).

The values of $S^{\alpha\rightarrow\beta}$ deduced from both approaches will be similar for low-melting-point elements, or when the melting points of the stable and metastable elements are close together. However, they diverge significantly when there is an appreciable difference between the melting points of the stable and metastable structures. This is because using S^f vs temperature plots as shown in Fig. 6.4(b) predicts that there would be a change in the magnitude and, potentially, of the sign of $S^{\alpha\rightarrow\beta}$ when differences in melting temperatures become large (Saunders *et al.* 1988), while the use of an average S^f for the different classes of crystal structure automatically defines that $S^{\alpha\rightarrow\beta}$ will be constant.

It is worth considering whether the trends shown in Fig. 6.4 have any theoretical basis. Originally the intercept values were empirically rationalised in terms of numbers of nearest neighbours (Miodownik 1972a). An electron structure approach used by Friedel (1974) yields the right sign for $S^{\text{f.c.c.}\rightarrow\text{b.c.c.}}$, but the predicted magnitude is too high and applying this approach to the other crystal structures does not yield meaningful results. Relating S^f to differences in θ_D (see Eq. (6.4)) is a more fruitful approach (Grimwall and Ebbsjo 1975, Grimwall 1977). A relation between the intercept values of S^f in Fig. 6.4(a) and the Debye temperature can be invoked using the structural dependence of the constant (C_L) in the well-known Lindemann equation

$$\theta_D = C_L V_a^{-1/3} T_f^{1/2} M^{-1/2} \tag{6.11}$$

where V_a is the atomic volume, T_f the melting point and M the atomic weight. A dimensional analysis indicates that C_L^2 has the same dimensions as entropy, and Table 6.1 shows that S^f and C_L^2 are closely proportional (Achar and Miodownik 1974). This accounts for the empirical finding that a further set of constant values for C_L can be obtained if the allotropic transformation temperature is substituted for the melting point in the Lindemann equation (Cho and Puerta 1976).

However, most of the examples quoted in these earlier papers do not include the higher melting-point elements such as W, where a detailed treatment shows that the total entropy (at least of the solid phases) must include many other components such as the electronic specific heat, anharmonicity terms and the temperature dependence of θ_D (Grimwall *et al.* 1987, Moroni *et al.* 1996). An estimate for the Debye temperature of the high-temperature β phase was included in the seminal

References are listed on pp. 173–178.

Table 6.1. Correlation between the entropy of fusion S^f (extrapolated to $T_m = 0$) and the Lindemann constant (C_L) for various crystal structures (from Achar and Miodownik 1974)

Structure of solid phase	Lindemann constant (C_L)	Function of C_L ($10^{-4}\, C_L^2$)	Entropy of fusion (S^f)
b.c.c. (A2)	118	1.4	1.6
f.c.c. (A1)	138	1.9	1.9
c.p.h. (A3)	169	2.8	2.3
Bi (A7)	197	3.9	4.0
Diamond (A4)	215	4.6	4.6

paper on Ti and Zr (Kaufman 1959a), but the treatment did not take into account the inherent mechanical instability of the high temperature phase. More recent work has shown that, in the case of these elements, suitable anharmonic contributions can stabilise the b.c.c. phase in Ti and Zr and that the vibrational entropy contribution accounts for 70% of $S^{\alpha\to\beta}$ at the allotropic transformation temperature (Ho and Harman 1990, Petry *et al.* 1991). As, in this case, the mechanical instability is only marginal the treatment of Kaufman (1959a) may be considered a good approximation. However, when there is a much more marked instability, as in the case for tungsten (Fernandez Guillermet *et al.* 1995, Einarsdotter *et al.* 1997) any vibrational entropies calculated from Debye temperatures may be totally inappropriate.

6.2.2.3 Determination of transformation enthalpies in binary systems. Just as consistent values of T_m for elements can be obtained by back-extrapolation from binary systems, so it is possible to obtain values of $H^{\alpha\to\beta}$ by extrapolating the enthalpy of mixing vs composition in an alloy system where the phase has a reasonable range of existence. The archetypal use of this technique was the derivation of the lattice stability of f.c.c. Cr from the measured thermodynamic properties of the Ni-based f.c.c. solid solution (γ) in the Ni–Cr system (Kaufman 1972). If it is assumed that the f.c.c. phase is a regular solution, the following expression can be obtained:

$$RT \ln \frac{a_{\mathrm{Cr}}^{\gamma}}{x_{\mathrm{Cr}}} = G_{\mathrm{Cr}}^{\mathrm{b.c.c.}\to\mathrm{f.c.c.}} + \Omega\, x_{\mathrm{Ni}}^2 \qquad (6.12)$$

where the left-hand side refers to the activity coefficient of Cr in the Ni solid solution, γ, the right-hand side contains the 'lattice stability' value for f.c.c. Cr while Ω is the regular solution interaction parameter.

Plotting $a_{\mathrm{Cr}}^{\gamma}/x_{\mathrm{Cr}}$ vs x_{Ni}^2 then leads to a straight line with an intercept equal to the Gibbs energy difference between the f.c.c. and b.c.c. forms of Cr, at the temperature where measurements were made (Fig. 6.5), while the slope of the line yields the associated regular solution interaction parameter. The lattice stability and the interaction parameter are conjugate quantities and, therefore, if a different magnitude

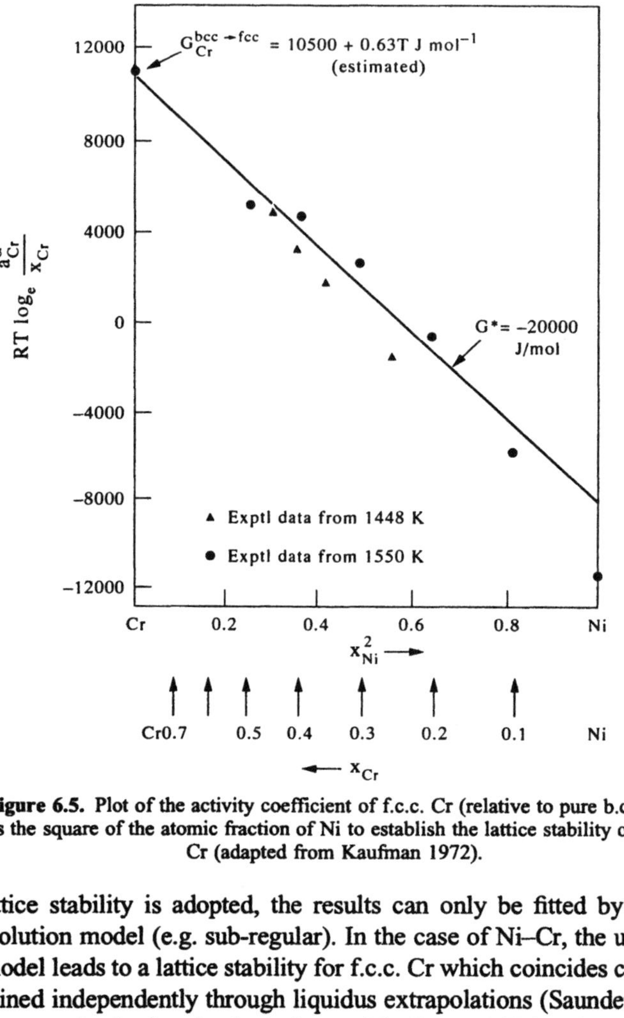

Figure 6.5. Plot of the activity coefficient of f.c.c. Cr (relative to pure b.c.c. Cr) vs the square of the atomic fraction of Ni to establish the lattice stability of f.c.c. Cr (adapted from Kaufman 1972).

for the lattice stability is adopted, the results can only be fitted by using a more complex solution model (e.g. sub-regular). In the case of Ni–Cr, the use of a regular solution model leads to a lattice stability for f.c.c. Cr which coincides closely with the value obtained independently through liquidus extrapolations (Saunders *et al.* 1988).

However, on the basis of calculations of lattice stabilities from spectroscopic data, Brewer (1967, 1979) has consistently maintained that interaction coefficients can change drastically with composition, and that extrapolated lattice stabilities obtained with simple models should therefore be considered as only 'effective' values. While this may indeed be true when mechanical instability occurs, many of the assumptions which underlie Brewer's methodology are questionable. A core principle of the spectroscopic approach is the derivation of 'promotion energies' which require the definition of both ground and excited levels. Assumptions concerning the relevant excited state have always been strongly coloured by adherence to the empirical views of Engel (1964) and Brewer (1967). By definition, the choice

References are listed on pp. 173–178.

of atomic ground states ignores all band-structure effects, and calculations by Moruzzi and Marcus (1988a) for the 3D transition elements highlight the omitted factors. It is interesting to note that there have been progressive changes in underlying assumptions over the years (Brewer 1975, Kouvetakis and Brewer 1993) and the lattice stabilities obtained by this route are now much closer to those currently produced by the more conventional TC method.

6.2.2.4 Utilisation of Stacking-Fault Energies. Experimental values of stacking-fault energies (SFE) offer a method of providing energy differences between stable and metastable close-packed structures. A rigorous relationship involves modelling the interface between the fault and the matrix, but a good working formula can be established by assuming that this interfacial energy term is constant for a given class of alloys (Miodownik 1978c).

$$\lambda = 2(G_s^{\text{f.c.c.}\rightarrow\text{c.p.h.}} + \sigma) \tag{6.13}$$

where

$$G_s^{\text{f.c.c.}\rightarrow\text{c.p.h.}} = \frac{10^7}{N^{1/3}}\left[\rho/M\right]^{2/3} G_m^{\text{f.c.c.}\rightarrow\text{c.p.h.}} \tag{6.14}$$

$G_s^{\text{f.c.c.}\rightarrow\text{c.p.h.}}$ is the Gibbs energy difference/unit area between the f.c.c. and c.p.h. phases in mJ m^{-2}, σ is the energy of the dislocation interface, ρ is the density in g cm^{-3}, M the molecular weight in grammes and $G_m^{\text{f.c.c.}\rightarrow\text{c.p.h.}}$ the Gibbs energy in J mol^{-1}. Eqs (6.13) and (6.14) were originally used to predict the SFE of a wide range of stainless steels, but they have also been used, in reverse, to estimate values of $G^{\text{f.c.c.}\rightarrow\text{c.p.h.}}$ of some f.c.c. elements (Saunders *et al.* 1988) where they provide values which are in excellent agreement (Miodownik 1992) with those obtained by FP methods (Crampin *et al.* 1990, Xu *et al.* 1991).

Although this method is essentially restricted to a particular sub-set of lattice stabilities, it nevertheless provides an additional experimental input, especially in cases where it is not possible to access the metastable phase by other methods. It is therefore disappointing that there are no *experimental* values of the SFE available for Ru or Os, which could provide confirmation of $G^{\text{c.p.h.}\rightarrow\text{f.c.c.}}$ obtained by other methods. High SFE values have, however, been both observed and predicted for Rh and Ir, which is indirect confirmation for a larger variation of $G_m^{\text{f.c.c.}\rightarrow\text{c.p.h.}}$ with *d*-shell filling than proposed by Kaufman and Bernstein (1970).

6.2.3 Summary of the current status of thermochemical estimates

While all extrapolative methods have some disadvantages, there is no doubt that remarkably consistent values have been obtained through combining extrapolations from various different TC methods. This has led to the consolidation of lattice stabilities of a large number of elements (Dinsdale 1991). The driving force for such a compilation is that, *if lattice stabilities are allowed to become floating parameters, characterisations of different sub-systems will become incompatible when joined in a multi-component database, even though the individual characterisations are internally self-consistent.* This leads to inconsistencies when such data bases are subsequently combined and invalidates subsequent calculations.

Individual lattice stabilities have nonetheless to be continuously scrutinised as there are many good scientific reasons to consider revisions at periodic intervals; some of these are discussed below:

(1) Although changes in stable melting temperature are now quite rare, and generally rather small, this is not the case for values of S^f, where relatively large changes have been reported in the last decade for a number of high-melting-point elements. This can cause large changes in values of lattice stabilities estimated from S^f in conjunction with extrapolated metastable melting points and using Eq. (6.10) (Miodownik 1986, Saunders *et al.* 1988).

(2) Estimations of metastable melting points are invariably problematical. They are inherently semi-quantitative in nature and depend on the extent of the required extrapolation of stable liquidus lines which can be far away from the element for which the extrapolation is being made.

(3) The use of linear models for many metastable allotropes, which incorporate temperature independent enthalpy and entropies of transformation, has proved adequate so far because calculations have tended to be made for elevated temperatures, where the effect of incorporating more complex models seems marginal. However, as requirements increase for handling non-equilibrium transformations at lower temperatures, it is clearly necessary to have more accurate information for extrapolated Gibbs energy curves. If this requires substantial revisions of lattice stabilities to be made, there is clearly a premium on adding any further effects (e.g., those arising from the liquid → glass transition) onto existing Gibbs energy expressions so as to avoid changes in well-substantiated high-temperature data (Agren *et al.* 1995). If high-temperature data were also to change this would inevitably require major re-calculation of many systems.

(4) The increasing availability of electron energy calculations for lattice stabilities has produced alternative values for enthalpy differences between allotropes at 0 K which do not rely on the various TC assumptions and extrapolations. Such calculations can also provide values for other properties such as the Debye temperature for metastable structures, and this in turn may allow the development of more physically appropriate non-linear models to describe low-temperature Gibbs energy curves.

6.3. *AB INITIO* ELECTRON ENERGY CALCULATIONS

Ideally, the 0 K values of the relative enthalpy for various crystal structures can be obtained by *ab initio* (*first-principles*) electron energy calculations, using merely atomic numbers and the desired atomic geometry as input. However, such methods did not have sufficient accuracy at the time lattice stabilities were first brought into use. An apocryphal analogy is that to obtain lattice stabilities by calculating the difference in total energy of two crystal structures, is like determining the weight of a ship's captain by first weighing the ship empty and then weighing the ship with the captain on the bridge!

Even as late as 1971 some methods were still unable to predict the correct ground state for elements such as Zn, while the scatter obtained from various calculation routes was far too high for *ab initio* phase stabilities to be introduced into a TC database (Kaufman 1972). However, owing to a combination of improved modelling and the availability of more powerful computers, results have become increasingly more consistent (Pettifor 1977, Skriver 1985, Watson *et al.* 1986, Paxton *et al.* 1990, Asada and Terakura 1993). In many cases, the values obtained from first principles (FP) have confirmed the values obtained by the thermochemical (TC) methods outlined in the previous section. The convergence of FP and TC values for the 0 K enthalpy is of great potential benefit to both the CALPHAD and physics communities for the following reasons.

1. Despite the success of phase-diagram calculations, there is still a considerable reluctance by sections of the scientific community to accept that TC lattice stabilities represent a real physical entity as distinct from an operational convenience. This inevitably creates doubts concerning the ultimate validity of the calculations. It is therefore important to verify that TC lattice stabilities, largely derived by extrapolation, can be verified by *ab initio* calculations and placed on a sound physical basis.
2. From the physicist's point of view, agreement with the CALPHAD figures for metastable allotropes represents one of the few ways of assessing the validity of their technique which is based on the principle that, *once a theoretical model is developed, there are no further adjustable input parameters other than the atomic number and a fixed geometry for relative atomic positions*. However, just as a number of different extrapolative techniques have been used in the TC approach, so a variety of assumptions have been used to solve the Schrödinger equation for a complex ensemble of atoms, combining different levels of accuracy with solutions that can be attained on a realistic time scale. A selection is given in Table 6.2. As many methods were developed in parallel, the order in which they are listed should not be considered important.

The most significant assumptions made in these various electron energy calculations are indicated in the various acronyms listed in Table 6.2. These can be permutated in many combinations and a proper comparison of these methods is beyond the scope of the present article. Excellent review articles (Pettifor 1977, Turchi and Sluiter 1993, de Fontaine 1996) are available if further detail is required. Other references of particular interest are those which compare the results

Table 6.2. Selection of various methods used to produce first-principles (FP) values for the relative stability at 0 K of different crystal structures

EPM	Empirical pseudo-potential method	Heine and Weaire (1971)
CPA	Coherent potential approximation	Ehrenreich and Schwartz (1976)
GPM	Generalised perturbation method	Ducastelle and Gautier (1976)
FRO	Frozen core approximation	Yin (1982)
GPT	Generalised pseudo-potential theory	Moriarty and McMahan (1982)
CFT	Concentration-functional theory	Gyorffy and Stocks (1983)
LMTO	Linear muffin tin orbital	Skriver (1983)
FLAPW	Full potential linearised augmented plane wave	Jansen and Freeman (1984)
ASA	Atomic spherical wave approximation	Skriver (1985)
LASTO	Linear augmented slater type orbital	Watson *et al.* (1986)
LSDA	Local spin density approximation	Moruzzi *et al.* (1986)
ECM	Embedded cluster method	Gonis *et al.* (1987)
GGA	Generalised gradient approach	Asada and Terakura (1993)
BOP	Bond order approximation	Pettifor *et al.* (1995)

of making different combinations for specific groups of elements, see for example Moriarty and McMahan (1982), Fernando *et al.* (1990) and Asada and Terakura (1993).

6.3.1 Comparison between FP and TC lattice stabilities

Despite the variety of assumptions that have been used, some general trends for the resultant lattice stabilities have been obtained for various crystal structures across the periodic table. The mean values of such (FP) lattice stabilities can therefore be compared with the equivalent values determined by thermochemical (TC) methods. Such a comparison shows the following important features (Miodownik 1986, Watson *et al.* 1986, Saunders *et al.* 1988, Miodownik 1992):

(1) In the main there is reasonable agreement for the sign and the magnitude of the lattice stabilities for elements whose bonding is dominated by sp electrons (Table 6.3) (see also Figs 6 and 7 of Saunders *et al.* 1988). In the case of elements such as Na, Ca and Sr, predictions have even included a reasonable estimate of their transition temperatures. Good agreement is also obtained for Group IVB elements such as Ge and Si (Table 6.4). The agreement for these elements is particularly striking because the values obtained by TC methods entailed large extrapolations, and for the case of Si(f.c.c.) a virtual, negative melting point is implied. At first sight such a prediction could be considered problematical. However, it can be interpreted in terms of the amorphous state being more stable at low temperatures than the competing f.c.c. crystalline configuration, and it does not contravene the third law as it is either metastable or unstable compared to the stable A4 structure. In the case of W and Mo there is now experimental evidence that supports this viewpoint (Chen and Liu 1997).

Table 6.3. Comparison of lattice stabilities (kJ mol^{-1}) obtained by TC and FP routes for elements whose bonding is dominated by sp electrons

Element	Method	b.c.c. → c.p.h.	f.c.c. → c.p.h.	Reference
Al	Pseudopotentials	−4.6	+5.5	Moriarty and McMahan (1982)
	TC	−4.6	+5.5	Kaufman and Bernstein (1970)
				Saunders *et al.* (1988)
Mg	Pseudopotentials	−2.7	−0.8	Moriarty and McMahan (1982)
	LMTO	−1.2	−0.9	Moriarty and McMahan (1982)
	TC	−3.1	−2.6	Saunders *et al.* (1988)
Zn	Pseudopotentials	−7.5	−1.7	Moriarty and McMahan (1982)
	LAPW	−9.2	−1.6	Singh and Papaconstantopoulos (1990)
	TC	−2.9	−3.0	Dinsdale (1991)
Be	Pseudopotentials	−8.1	−7.2	Lam *et al.* (1984)
	LMTO	−6.4	−5.9	Skriver (1982)
	TC	−6.8	−6.3	Saunders *et al.* (1988)

Table 6.4. Comparison of lattice stabilities for Si and Ge (kJ mol^{-1}) obtained by TC and FP routes for transformations from A4 to A1 (f.c.c.), A2 (b.c.c.), A3 (c.p.h.) and A5 (β-Sn)

		Structure				
Element	Method	A1	A2	A3	A5	Reference
Si	TC	51	47	49		Saunders *et al.* (1988)
	TC	50	42	51		Kaufman and Bernstein (1970)
	FP	55	51	53	26	Yin (1982)
	FP					Goodwin *et al.* (1989)
Ge	TC	36	34	35		Saunders *et al.* (1988)
	TC	36			28	Miodownik (1972b)
	FP	44	42	43	24	Yin (1982)
	FP					Goodwin *et al.* (1989)

(2) A sinusoidal variation of the 0 K energy difference between b.c.c. and close-packed structures is predicted across the transition metal series, in agreement with that obtained by TC methods (Saunders *et al.* 1988). For the most part magnitudes are in reasonable agreement, but for some elements FP lattice stabilities are as much as 3–10 times larger than those obtained by any TC methods (Fig. 6.6).
(3) FP methods inherently lead to a marked sinusoidal variation of $H^{\text{f.c.c.}\to\text{c.p.h.}}$ across the periodic table (Pettifor 1977) and for Group V and VI elements, electron energy calculations predict $H^{\text{f.c.c.}\to\text{c.p.h.}}$ of *opposite sign* to those obtained by TC methods. It is worth noting, however, that a sinusoidal variation is reproduced by one of the more recent TC estimates (Saunders *et al.* 1988) although displaced on the energy axis (Fig. 6.7).

Some reconciliation can be achieved by considering the effect of changing

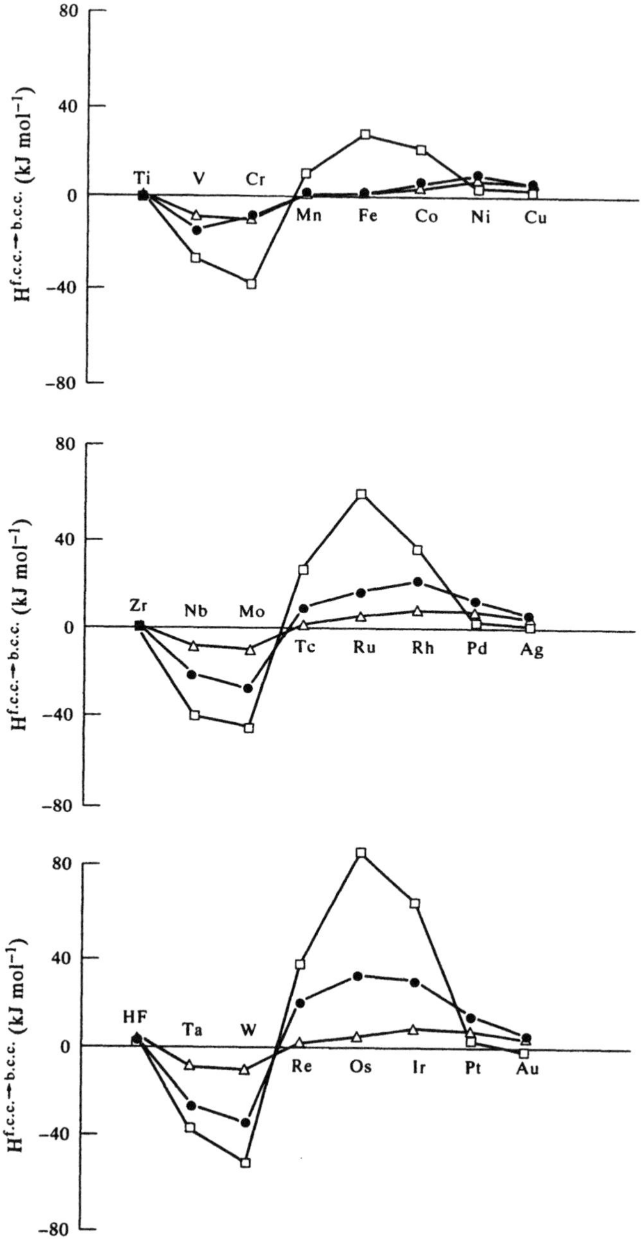

Figure 6.6. Variation of the enthalpy difference between f.c.c. and b.c.c. structures obtained by various TC and FP routes. • Saunders *et al.* (1988); Δ Kaufman and Bernstein (1970); □ Skriver (1985).

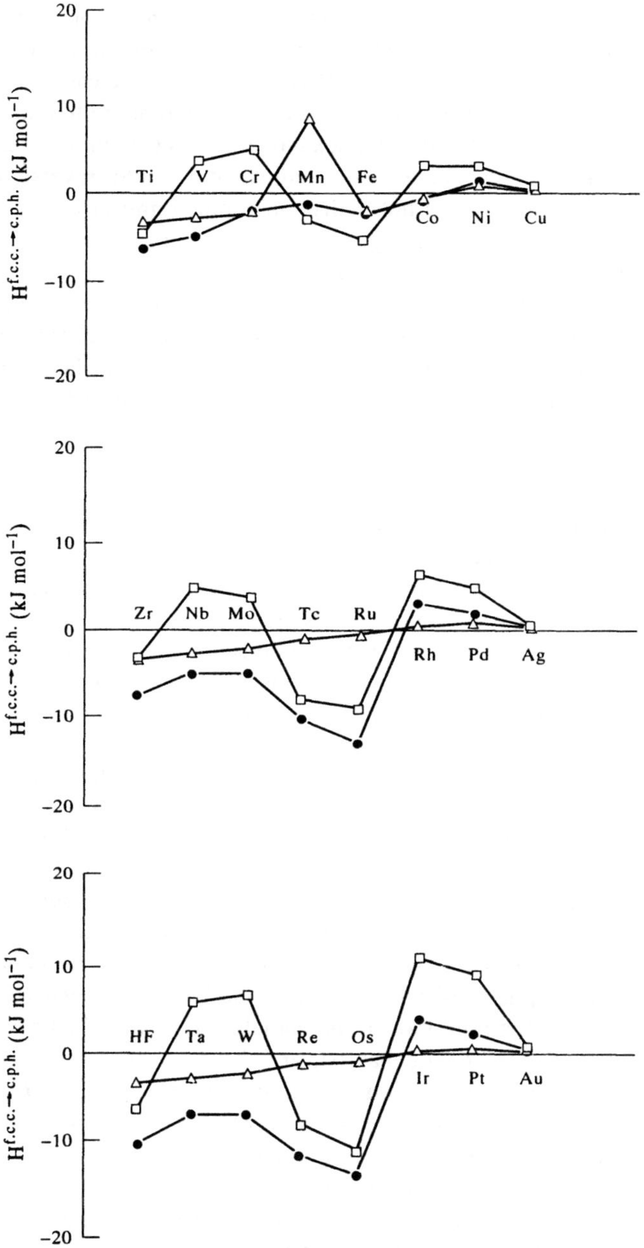

Figure 6.7. Variation of the enthalpy difference between f.c.c. and c.p.h. structures obtained by various TC and FP routes. • Saunders *et al.* (1988); Δ Kaufman and Bernstein (1970); □ Skriver (1985).

the *c/a* ratio (Fernando *et al.* 1990), as some of the hexagonal phases seem to be most stable at *c/a* ratios which depart substantially from the ideal close-packed value (~1.63). This factor is of significance in the case of elements like Cd and Zn where *c/a* ~ 1.8 and it is clearly not appropriate to equate the Gibbs energies of the ideal and non-ideal hexagonal forms (Singh and Papaconstantopoulus 1990). Making this kind of adjustment could reverse the sign of $H^{\text{f.c.c.}\rightarrow\text{c.p.h.}}$ for Nb but does not really make much impact on the situation for Mo, Ta and W. Likewise a change in volume within reasonable limits does not resolve the issue (Fernando *et al.* 1990).

(4) In the past, electron energy calculations have failed dramatically for magnetic elements since spin polarisation was not included. However, this can now be taken into account quite extensively (Moruzzi and Marcus 1988b, 1990, Asada and Terakura 1993) and calculations can reproduce the correct ground states for the magnetic elements.

6.3.2 Reconciliation of the difference between FP and TC lattice stabilities for some of the transition metals

Several approaches have been made in order to remove the outstanding conflicts between FP and TC lattice stabilities. Niessen *et al.* (1983) proposed a set of compromise values, which are also listed in de Boer *et al.* (1988), but these do not constitute a solution to this problem. These authors merely combined the two sets of values in relation to an arbitrary reference state in order to refine their predictions for heats of formation (see Section 6.6.1.1).

The accuracy with which a phase diagram can be fitted with competing values of phase stabilities appeared, at one time, to be a fairly obvious route to discover whether the TC or FP values were closer to reality. Tso *et al.* (1989) calculated the Ni–Cr phase diagram using a series of different values for the lattice stability of f.c.c. Cr. They were able to reasonably reproduce the phase diagram with energy differences close to those proposed by electron energy calculations. However, they could only do so by introducing a compensating change in the Gibbs energy of mixing for the f.c.c. phase, which had to become large and negative in sign. It is difficult to accept such a proposal for the enthalpy of mixing for f.c.c. alloys when the liquid exhibits almost ideal behaviour and the b.c.c. phase has mainly positive interactions. In addition, calculations of mixing energy by using *d*-band electron models (Colinet *et al.* 1985, Pasturel *et al.* 1985) supports values much closer to those already in use by CALPHAD practitioners.

A series of publications by various authors used a similar process to establish which lattice stability gave the best fit for particular phase diagrams (Anderson *et al.* 1987, Fernandez Guillermet and Hillert 1988, Fernandez Guillermet and Huang 1988, Kaufman 1993, Fernandez Guillermet and Gustafson 1985). Au–V formed a particularly useful test vehicle (Fernandez Guillermet and Huang 1988, Kaufman

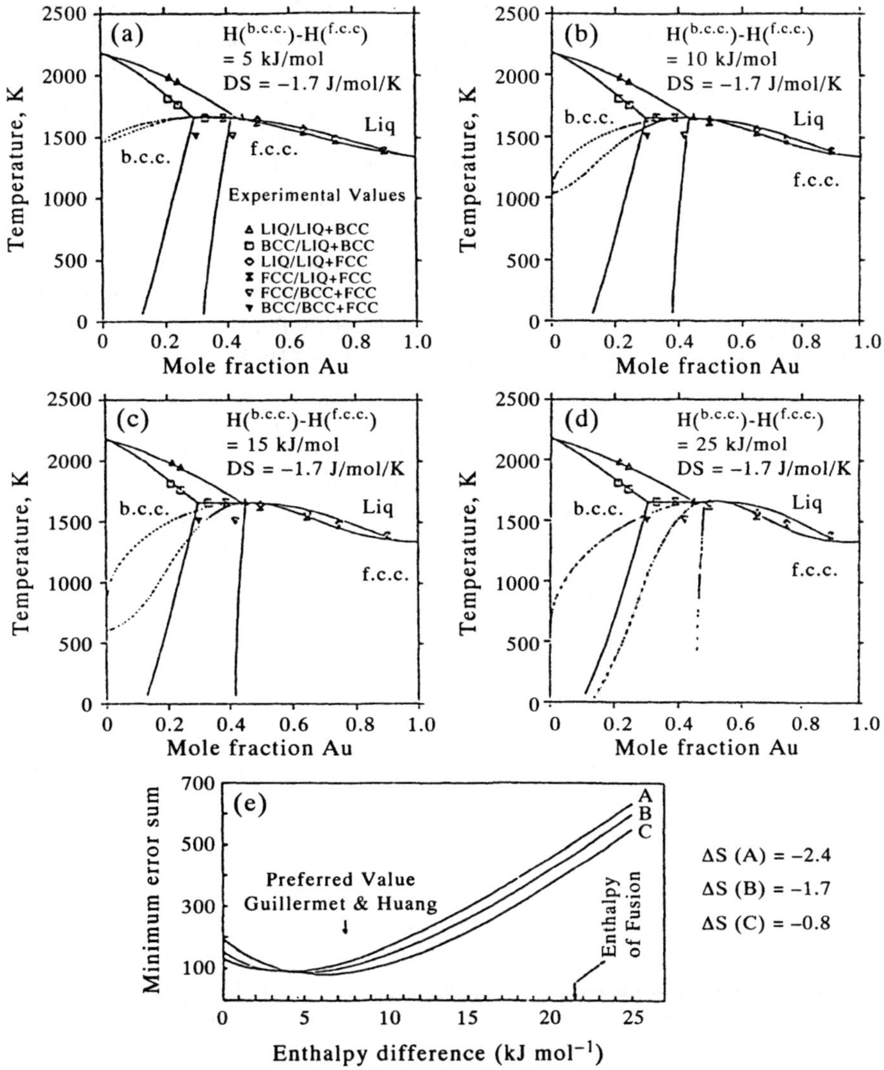

Figure 6.8. (a–d) The effect of varying the f.c.c.–b.c.c. V lattice stability on the matching of experimental and calculated phase boundaries in the Au–V system (after Fernandez Guillermet and Huang 1988) and (e) the effect of varying $\Delta S^{\text{b.c.c.}\rightarrow\text{f.c.c.}}$ on the minimum error sum, giving a preferred value of 7.5 kJ mol^{-1}.

1993) as V was one of the elements which showed a serious discrepancy (Fig. 6.8). Using a regular solution model and restricting the range of S^f, such an analysis (Fernandez Guillermet and Huang 1988) suggested that the optimum values of

$G^{\text{b.c.c.}\to\text{f.c.c.}}$ were markedly lower than those obtained by electron energy calculations (Fernandez Guillermet and Huang 1988) and also significantly lower than proposed by Saunders *et al.* (1988). The permitted range of lattice stability values can be increased by changing these constraints, in particular letting S^f vary as shown in Fig. 6.4(a), but any values still exclude FP predictions.

This impasse was eventually resolved by taking into consideration the calculated elastic constants of metastable structures in addition to their energy difference at 0 K. Craievich *et al.* (1994), Craievich and Sanchez (1995) and Guillermet *et al.* (1995), using independent calculations, have suggested that the difference between TC and *ab initio* predictions may be associated with mechanical instabilities in the metastable phase. This point had been raised earlier by Pettifor (1988) and has the following consequence as reported by Saunders *et al.* (1988):

> *"... the analogy with titanium and zirconium is useful. In these elements there is a reported lattice softening in the b.c.c. allotrope near the c.p.h. → b.c.c. transformation temperature. This instability is reflected by a substantial C_p difference between the c.p.h. and b.c.c. structures which effectively destabilises the b.c.c. phase with decreasing temperature. High-temperature extrapolations which ignore this, then overestimate the low temperature stability of the b.c.c. phase. As ab initio calculations give values for heats of transformation at 0 K, comparison with high-temperature phase diagram and thermodynamic extrapolations will produce different results."*

Furthermore, such instabilities will extend into the alloy system up to a critical composition and must therefore be taken into account by any effective solution-phase modelling. In the case of Ni–Cr , it is predicted that mechanical instability, as defined by a negative value of $c' = 1/2(c_{11} - c_{12})$, will occur between 60 and 70 at%Cr (Craievich and Sanchez 1995), so beyond this composition the f.c.c. phase cannot be considered as a competing phase.

While this concept of mechanical instability offers a potential explanation for the large discrepancies between FP and TC lattice stabilities for some elements, the calculations of Craeivich *et al.* (1994) showed that such instabilities also occur in many other transition elements where, in fact, FP and TC values show relatively little disagreement. *The key issue is therefore a need to distinguish between 'permissible' and 'non-permissible' mechanical instability.* Using the value of the elastic constant C' as a measure of mechanical instability, Craievich and Sanchez (1995) have found that the difference between the calculated elastic constant C' for f.c.c. and b.c.c. structures of the transition elements is directly proportional to the FP value of $H^{\text{b.c.c.}\to\text{f.c.c.}}$ (Fig. 6.9(a)).

The position of Ti and Zr is again important in this context. While the b.c.c. phase in these elements has long been known to indicate mechanical instability at 0 K, detailed calculations for Ti (Petry 1991) and Zr (Ho and Harmon 1990) show that it is stabilised at high temperatures by additional entropy contributions arising from low values of the elastic constants (soft modes) in specific crystal directions. This concept had already been raised in a qualitative way by Zener (1967), but the

References are listed on pp. 173–178.

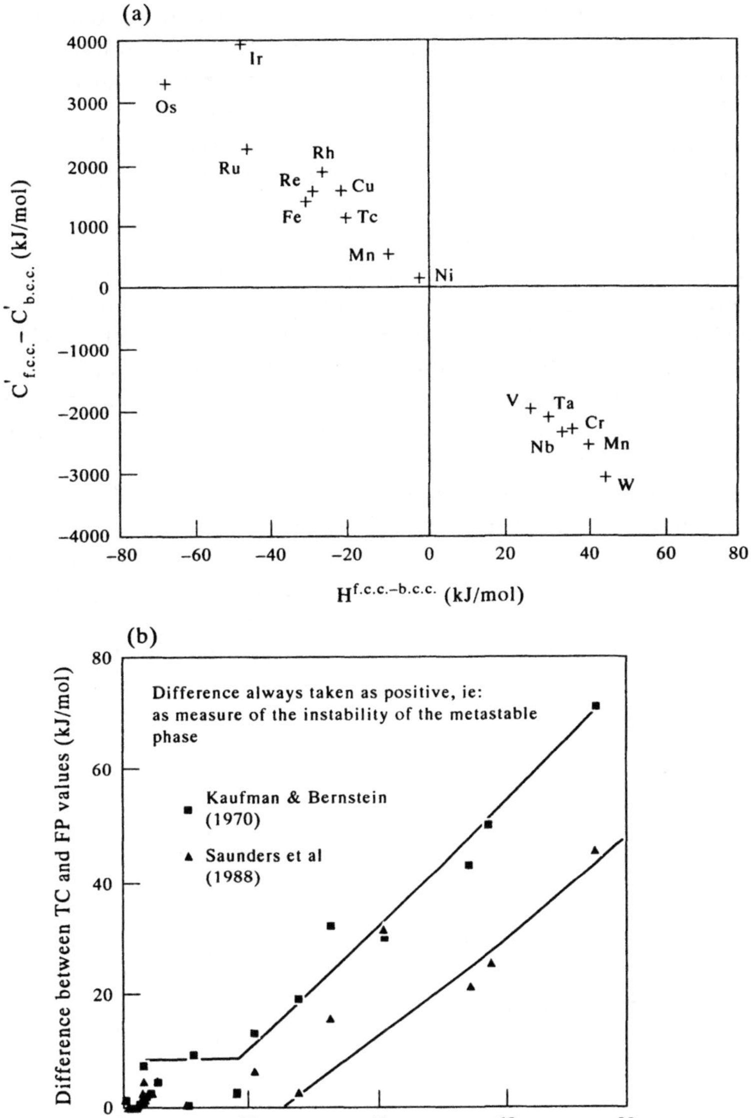

Figure 6.9. (a) Proportionality of calculated values of $\Delta H^{\text{f.c.c.-b.c.c.}}$ to the associated elastic instability as measured by the elastic constant C′ (Private Communication from Craievich and Sanchez 1995). (b) The difference between TC and FP lattice stabilities plotted vs the absolute value of the calculated FP lattice stability. When combined with Fig. 6.8 this suggests a critical value of the elastic instability that cannot be compensated for by thermal contributions.

key issue is to find out whether there is a maximum value to the additional entropy that is available to counterbalance specific values of mechanical instability at 0 K. This would be consistent with the observation that major discrepancies between FP and TC lattice stabilities only seem to arise when the mechanical instability rises beyond a critical value (Fig. 6.9(b)). This is also consistent with the findings of (Fernandez Guillermet *et al.* 1995), who could only reproduce the phase diagram for W–Pt by invoking an empirical and highly anomalous value of $S^{\mathrm{f.c.c.} \rightarrow \mathrm{b.c.c.}}$ which would also have to exhibit a strong temperature dependence to avoid the appearance of f.c.c. W as a stable phase below the melting point.

The effect of incorporating a more complex entropy function has also been examined by Chang *et al.* (1995) using the Ni–Cr system as a test vehicle. Here an anomalous entropy contribution was *simulated* by incorporating a dual Debye temperature function, linked to the critical composition at which the elastic constants had been calculated to change sign in this system. The resulting lattice stability for the f.c.c. phase nicely shows how this approach rationalises the extrapolated liquidus derived using the van Laar method and values that would be consistent with a FP approach. Figure 6.10 shows that the phase diagram in the stable region does not suffer inaccuracies by using high FP values as happened in the earlier attempts on Au–V (Fig. 6.8). It should, however, be emphasised that the equations used by Chang *et al.* (1995) to define the Gibbs energy of f.c.c. Cr are highly empirical and require a whole new set of adjustable parameters, such as a critical temperature which defines the onset of mechanical instability. This poses an

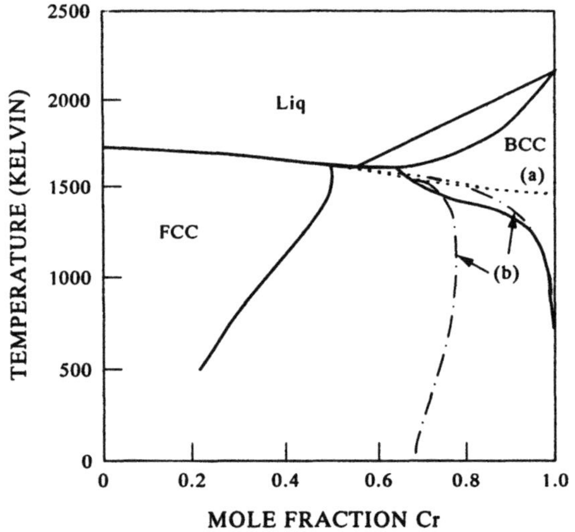

Figure 6.10. Comparison of the extrapolated liquidus/solidus lines relating to the f.c.c. phase in Ni–Cr alloys, derived by (a) using the van Laar method and (b) the trajectory obtained using the modified FP approach used by Chang *et al.* (1995).

exciting challenge to provide a sounder basis for a treatment that can encompass low-temperature mechanical instability. The present situation concerning FP and TC lattice stabilities can probably be described as follows.

1. Thermochemical methods generate lattice stabilities based on high-temperature equilibria that yield self-consistent multi-component phase-diagram calculations. However, as they are largely obtained by extrapolation, this means that in *some* cases they should only be treated as *effective lattice stabilities*. Particular difficulties may occur in relation to the liquid → glass transition and instances of mechanical instability.
2. By contrast, electron energy calculations have the inherent capability of yielding accurate values for many metastable structures at 0 K but have little or no capability of predicting the temperature dependence of the Gibbs energy, especially in cases where mechanical instabilities are involved.

Although the two methodologies are complementary, attempts at producing Gibbs energy curves which combine the two approaches are currently still empirical and would, in practice, be very difficult to incorporate in a general CALPHAD calculation. *A more fundamental treatment of various entropy contributions is required to achieve proper integration. Until this is realised in practice, the use of TC lattice stability values as well established operational parameters, valid at high temperatures and for most CALPHAD purposes, is likely to continue for the foreseeable future.*

6.4. THE BEHAVIOUR OF MAGNETIC ELEMENTS

The treatment outlined so far has not included the magnetic Gibbs energy contributions in Eq. (6.1) because the ground state of the majority of elements is paramagnetic. This, however, is certainly not the case for some key 3d transition elements, which exhibit various forms of magnetic behaviour not only in the ground state but also in one or more allotropes. Fe is a classic example and since ferrous metallurgy has been a major driving force in the development of phase diagram calculations, one of the first steps was to establish the magnitude of the magnetic component on the Gibbs energy. The basic factors that control this are detailed in Chapter 8, but it is worthwhile here to review some of the implications for specific elements, with particular emphasis on Fe.

6.4.1 Fe

Considerable information is available on the magnetic parameters associated with three different crystal structures of Fe which are b.c.c. and f.c.c. at ambient pressures and c.p.h. which is observed at high pressures. Table 6.5 gives the corresponding values of the maximum enthalpy and entropy contributions due to

Table 6.5. Maximum values of the magnetic enthalpy and entropy for various allotropes of Fe, Co and Ni based on data and methodology drawn from Miodownik (1977) and additional data from de Fontaine *et al.* (1995). Structures in brackets correspond to metastable forms that have not been observed in the *TP* diagram of the element

Element	Structure	H^{mag} (kJ mol^{-1})	S^{mag} (J mol^{-1} K^{-1})
Fe	b.c.c.	9.13	9.72
	f.c.c.	0.32	4.41
	c.p.h.	0	0
Co	(b.c.c.)	11.17	8.56
	f.c.c.	10.69	8.56
	c.p.h.	9.66	8.26
Ni	(b.c.c.)	1.91	3.37
	f.c.c.	2.27	4.01
	(c.p.h.)	0.94	3.48

magnetism, H^{mag}_{max} and S^{mag}_{max} respectively, and Fig. 6.11(a) shows how this affects the total Gibbs energy for the three structures in Fe. At high temperatures Fe behaves like many other allotropic elements, solidifying in a b.c.c. lattice, (δ-Fe). As the temperature is reduced, there is a transition to the f.c.c. γ-Fe phase which has both a lower energy and a lower entropy. However because of G^{mag} the b.c.c. phase reappears at lower temperature as α-Fe. While the Curie temperature of α-Fe, 770°C, is actually 140°C below the α/γ transformation temperature at 910°C, there is already sufficient short-range magnetic order to cause the transformation. A consequence of this unique behaviour is that the value of $G^{\text{b.c.c.}\rightarrow\text{f.c.c.}}$ remains exceedingly small in the region between the two transition points and never exceeds 50–60 J mol^{-1} (see Fig. 6.11(b)). This means that small changes in Gibbs energy, due to alloying, will substantially alter the topography of the γ-phase region in many Fe-base systems. If the alloying element also significantly affects the Curie temperature, the proximity of the latter to the α/γ transus will mean that this boundary is disproportionately altered in comparison to the γ/δ transus. This latter point can be shown to account for the otherwise puzzling asymmetric effects of alloying additions on the two transition points (Zener 1955, Miodownik 1977, Miodownik 1978b).

Because it is the basis of so many important commercial systems, the allotropy of Fe has been re-examined at frequent intervals. There is relatively more experimental information available for α-Fe, as this is the ground state, and relatively little controversy about characterising this phase. There is also little difference between the various proposals for $G^{\text{b.c.c.}\rightarrow\text{f.c.c.}}$ in the temperature range where γ-Fe is stable, because of the need to reproduce the two well-known transformation temperatures $T^{\alpha/\gamma}$ and $T^{\gamma/\delta}$ and their associated enthalpies of transformation (Fig. 6.11(b)). There are, however, different points of view regarding the extrapolation of the Gibbs energy of the γ-phase to low temperatures, which relate to different weightings given to various specific heat measurements, assumptions made

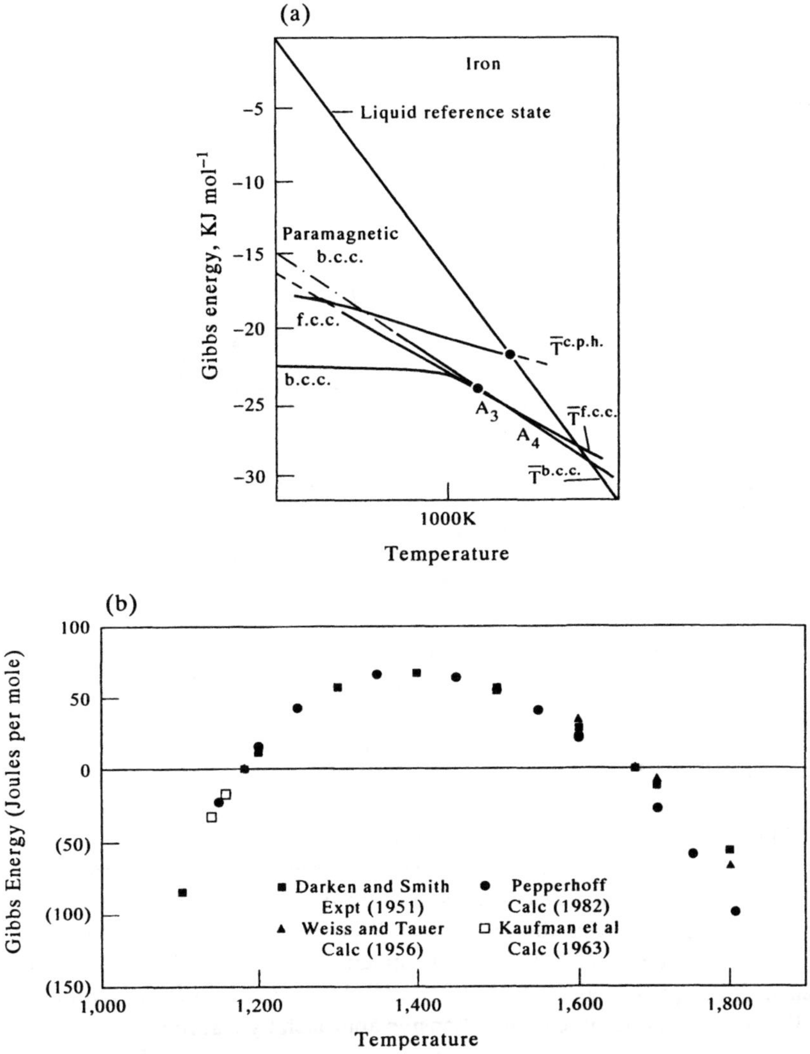

Figure 6.11. General overview of the relative stability of the f.c.c. and b.c.c. phases in pure Fe (a) over the whole temperature range (Miodownik 1978b) and (b) in the critical high-temperature region between the two critical points denoted A3 and A4, based on the work of Darken and Smith (1951), Weiss and Tauer (1956), Bendick and Pepperhoff (1982) and Kaufman *et al.* (1963).

regarding the magnetic description of γ-Fe, and the significance placed on data derived from low-temperature martensitic transformations (Fig. 6.12).

Johannson (1937) is generally credited with the earliest calculations of $G^{b.c.c. \rightarrow f.c.c.}$, which were based on the specific heat measurements of Austin (1932).

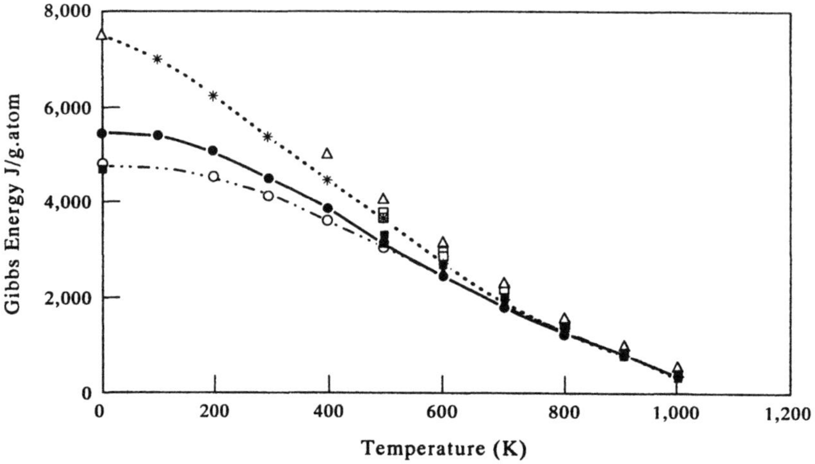

Figure 6.12. Effect of different models on the Gibbs energy difference between the f.c.c. and b.c.c. phases in pure iron. Data (□) of Darken and Smith (1951), (Δ) Weiss and Tauer (1956), (•) Kaufman *et al.* (1963), (⊡) Orr and Chipman (1967), (▲) Ågren (1979), (o) Bendick and Pepperhoff (1982), (∗) Fernandez Guillermet and Gustafson (1985), (■) Chang *et al.* (1985).

Relevant data extracted from activity measurements in the Fe–C phase were later included in the assessment by Zener (1946) and Darken and Smith (1951). All this work followed a traditional approach solely based on thermodynamic data derived under equilibrium conditions. By contrast, Kaufman and Cohen (1958) showed that the data of Johannson (1937) were more consistent with information derived from Fe-based martensite transformations than the interpretation used by Darken and Smith (1951). Together with the work of Svechnikov and Lesnik (1956) *this was a notable attempt to combine thermodynamic information derived from low-temperature metastable transformations with those from more traditional sources.*

Another such departure was the attempt by Weiss and Tauer (1956) to deconvolute the global value of $G^{\text{b.c.c.}\to\text{f.c.c.}}$ into magnetic, vibrational and electronic components. This represented the first attempt to obtain *a physical explanation* for the overall effect. An even more comprehensive approach by Kaufman *et al.* (1963) led to the further inclusion of two competing magnetic states for γ-Fe. Since the computer programmes available at that time could not handle such a sophisticated approach, the relevant values of $G^{\text{b.c.c.}\to\text{f.c.c.}}$ were converted into polynomial form and subsequently used for the calculations of key Fe-base diagrams such as Fe–Ni, Fe–Co and Fe–Cr (Kaufman and Nesor 1973).

Subsequent reassessments now took divergent routes. Ågren (1979) used thermodynamic data largely drawn from Orr and Chipman (1967) and re-characterised the magnetic component of the α-phase with the Hillert and Jarl model (see Chapter 8). The concept of two competing states in the γ-phase was abandoned as Orr and

Chipman (1967) favoured a high value for $H^{\text{b.c.c.}\rightarrow\text{f.c.c.}}$ and a suitable fit could be obtained without invoking the added complications of this model. The Ågren treatment has been further refined in the most recent assessment of the *T–P* properties of Fe made by Fernandez Guillermet and Gustafson (1985). By contrast, both Miodownik (1978b) and Bendick and Pepperhoff (1982) pursued further evidence for the existence of two gamma states from other physical properties and built alternative assessments for Fe around this concept.

Although attention has tended to concentrate on the equilibrium between α- and γ-Fe, the hexagonal variant, ε, has also to be considered. This can only be accessed experimentally in pure Fe when the b.c.c. phase is destabilised by pressure or alloying additions. The extrapolated properties of ε-Fe are consistent with it being paramagnetic, or very weakly anti-ferromagnetic, and so magnetism will provide a negligible contribution to its stability. Nevertheless, the extrapolation of SFE versus composition plots confirms that the ε-Fe becomes more stable than the γ-Fe at low temperatures, despite the magnetic contributions in the γ-phase (see Fig. 8.11 of Chapter 8). Hasegawa and Pettifor (1983) developed a model based on spin-fluctuation theory which accounts for the *T–P* diagram of Fe without invoking any differences in vibrational entropy or multiple magnetic states, but their theory predicts that both close-packed states should exhibit a temperature-induced local moment. The f.c.c. phase then behaves in a way which is rather similar to that predicted by the phenomenological two-γ-state model (Kaufman *et al.* 1963). The prediction of a high-temperature moment for the ε-phase is, however, at variance with the assumptions or predictions of most other workers who consider that the ε-phase has a zero or negligibly low moment. Various other models have been proposed which break down the total Gibbs energy in different ways, for instance by placing a different emphasis on the contribution of vibrational and electronic terms (Grimwall 1974). Both the Pettifor–Hasegewa and the Grimwall models can account for the some of the necessary qualitative features exhibited by the allotropy of Fe, but the intrinsic assumptions of these two approaches are incompatible with each other.

The various approaches were again reviewed by Kaufman (1991) who quoted further evidence for the two-gamma-state concept, but virtually all current databases incorporate the data generated by Fernandez Guillermet and Gustafson (1985) (Fig. 6.12). This has certainly been well validated in phase-diagram calculations, and it is hard to see how it could be improved at elevated temperatures. However, the differences that appear at low temperatures may be important in relation to metastable equilibria such as martensite transformations which, since the early work of Kaufman and Cohen (1956, 1958), have tended to be excluded from any optimisation procedure. Certainly there seems to be no doubt about the existence of many competing combinations of crystal structure and magnetic moments (Asada and Terakura 1993) (Fig. 6.13) which suggest that there must be room for an improved model.

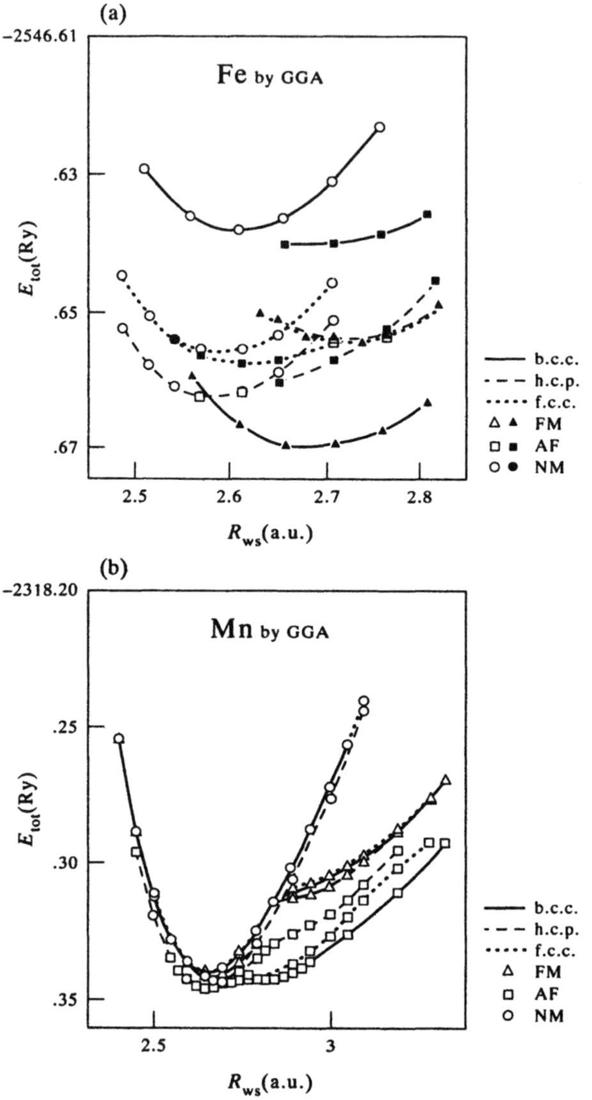

Figure 6.13. Enthalpy vs volume plots for f.c.c., b.c.c. and c.p.h. for (a) Fe and (b) Mn, incorporating the effect of assuming different magnetic interactions. Δ ferromagnetic, □ anti-ferromagnetic, ○ non magnetic (from Asada and Terakura 1993).

6.4.2 Co

Table 6.5 shows that $Co^{c.p.h.}$ has a high value of H^{mag}_{max}, so it is tempting to say that, as with Fe, the low-temperature allotrope forms are due to strong stabilisation by magnetic forces. However this is not the case, as $Co^{f.c.c.}$ is also ferromagnetic with very similar values of β and T_c. The magnetic contributions virtually cancel each other out in this case, although the f.c.c./c.p.h. transformation temperature will clearly be sensitive to small changes in magnetic parameters on alloying. The magnetic properties of metastable b.c.c. Co were derived indirectly by Inden (1975) as part of a calculation involving ordering in b.c.c. Fe–Co alloys. Interestingly, these values are substantially confirmed by more recent FP calculations. Table 8.2 and Fig. 8.11 of Chapter 8 can be consulted for further details.

6.4.3 Ni

Ni does not exhibit any allotropy with respect to either temperature or pressure, which implies that its f.c.c. structure must be significantly more stable than the nearest competitors. This is consistent with values for $G^{f.c.c. \rightarrow c.p.h.}$ derived from SFE measurements and changes in SFE with alloying also suggest that there is a magnetic component in $G^{f.c.c. \rightarrow c.p.h.}$. Since the magnetic enthalpy is much smaller than for Fe or Co, this in turn implies that that $Ni^{c.p.h.}$ is paramagnetic or has a low T_c (Miodownik 1978a). However, FP calculations imply a value of β not much lower than for $Ni^{f.c.c.}$, which would indicate that Ni behaves much like Co. Further experimental information on the magnetic properties of $Ni^{c.p.h.}$, possibly from thin films, would clearly be desirable.

6.4.4 Mn

Experimentally Mn solidifies as the b.c.c. δ-Mn phase, which then transforms to f.c.c. γ-Mn on cooling. This is entirely analogous to the behaviour of Fe. However, it then undergoes two further low-temperature transitions to β-Mn and α-Mn which are complex crystal structures with large unit cells. De-convoluting the experimental data into separate contributions is difficult. Three of the phases are anti-ferromagnetic with widely differing values of the saturation magnetisation at 0 K, and there are also significant differences in the Debye temperatures and electronic specific heat. It is therefore difficult to make a consistent characterisation which matches all these observations. By concentrating on matching phase transition temperatures and Debye temperature data, Fernandez Guillermet and Huang (1990) were forced to use *effective* magnetisation values which could be as low as ⅓ of those experimentally observed. This makes it impossible to estimate the real role of any magnetic factors in their treatment.

By contrast, in an earlier assessment, Weiss and Tauer (1958) decided to use the measured magnetisation values as a primary input over the whole temperature range, but then had to depart from the experimentally observed Debye temperatures. Interestingly, their treatment led to the conclusion that the G curves for β-Mn and γ-Mn would also intersect twice so that the behaviour of Fe could no longer be considered unique. However in this scenario it is γ-Mn that has the largest magnetic component, while the stability of β-Mn is dictated by a larger electronic specific heat. The finding by Gazzara *et al.* (1964) that the value of the saturation magnetisation for α-Mn (and other anti ferromagnetic materials) can be temperature dependent may have to be taken into account in order to finally reconcile what are otherwise different interpretations of the same data. The relative Gibbs energy curves for b.c.c., f.c.c. and c.p.h. Mn as calculated by Asada and Terakura (1993) are shown in Fig. 6.13(b).

6.5. THE EFFECT OF PRESSURE

6.5.1 Basic addition of a $P\Delta V$ term

It is commonplace to assume a form of the Gibbs energy function which excludes the pressure variable for solid-state phase transformations, as the magnitude of the $P\Delta V$ term is small at atmospheric pressures. This is of course not the case in geological systems, or if laboratory experiments are deliberately geared to high-pressure environments. Klement and Jayaraman (1966) provide a good review of the data available at the time when some of the earliest CALPHAD-type calculations were made (Kaufman and Bernstein 1970, Kaufman 1974). Much work was also carried out on specific alloy systems such as Fe–C (Hilliard 1963) and the Tl–In system (Meyerhoff and Smith 1963).

The extra pressure term to be added to the Gibbs energy can be expressed as

$$G_m^{\alpha\to\beta} = \int_0^P V_m^{\alpha\to\beta} \mathrm{d}P \tag{6.15}$$

where $V_m^{\alpha\to\beta}$ is the change in molar volume due to the transformation of α to β. Although this can be considered as constant to a first approximation, this will no longer be true at high pressures, and several empirical descriptions have been developed, depending on the pressure range in question. Bridgman (1931) used a simple second-power polynomial to define the effect of pressures up to 3 GPa:

$$V_m = V_0 + V_1 P + V_2 P^2 \tag{6.16}$$

where $V_{0,1,2}$ are temperature dependent and have to be determined by experiment. If the entropy, enthalpy and volume differences between these phases are assumed independent of temperature and pressure is not excessive, then Eq. (6.15) will obviously reduce to:

$$G_m^{\alpha\to\beta} = P V_m^{\alpha\to\beta} \tag{6.17}$$

where P and $V_m^{\alpha\to\beta}$ are given in GPa and m^3 respectively. This simple treatment also leads to the familiar Clapeyron equation where the slope of the temperature vs

Table 6.6. Linear Gibbs energy equations for some non-transition elements that include a pressure term taken (Kaufman and Bernstein 1970). Values are applicable only at temperatures >300 K

Element	Transformation	ΔH (J mol^{-1})	$-T\Delta S$ (S = J mol^{-1}; $T = K$)	$+P\Delta V$ ($P = GPa$; $V = m^3$)
Ca	f.c.c.→b.c.c.	243	−0.33T	+0.01P
Sr	f.c.c.→b.c.c.	837	−0.96T	−0.10P
Be	c.p.h.→b.c.c.	4602	−3.01T	−0.18P
Yb	f.c.c.→b.c.c.	3180	−2.97T	−0.80P
Ba	f.c.c.→c.p.h.	4351	+3.35T	−0.74P
Pb	f.c.c.→c.p.h.	2510	−0.42T	−0.18P
C	A4→Graphite	1255	−4.77T	+1.76P

Note: Although current values of ΔH and ΔS are different in later references this will not affect the pressure term.

pressure plot for the transformation $\alpha \rightarrow \beta$ is given by:

$$\frac{\mathrm{d}T}{\mathrm{d}P} = \frac{V^{\alpha\rightarrow\beta}}{S^{\alpha\rightarrow\beta}}. \tag{6.18}$$

Some examples of equations based on adding a simple $P\Delta V$ term to the definition of lattice stabilities are given in Table 6.6 (adapted from Kaufman and Bernstein 1970).

6.5.2 Making the volume a function of T *and* P

These earlier treatments have now been superseded by a more general approach, where the molar volume of each phase as a function of temperature and pressure is expressed as a function of the compressibility χ (Fernandez Guillermet *et al.* 1985).

$$V_m^{T,P} = V_0(1 + nP\chi)^{-1/n} \exp\left(\alpha_0 T + \tfrac{1}{2}\alpha_1 T^2\right) \tag{6.19}$$

where V_0 is an empirically fitted parameter with the dimensions of volume and α_0 and α_1 are parameters obtained from fitting the experimental lattice parameter and dilatometric data such that

$$\alpha_{P=0}^{T} = \alpha_0 + \alpha_1 T \tag{6.20}$$

where $\alpha_{P=0}^{T}$ is the thermal expansivity at zero pressure as a function of temperature. χ is also expressed as a function of temperature using the polynomial

$$\chi = \chi_0 + \chi_1 T + \chi_2 T^2. \tag{6.21}$$

At constant temperature, Eq. (6.19) reduces to the simpler expression suggested much earlier by Murnaghan (1944).

$$V_m^{T,P} = \frac{V_m^{T,P=0}}{(1 + nP\chi^{T,P=0})^{1/n}} \tag{6.21}$$

where $V_m^{T,P=0}$ and $\chi^{T,P=0}$ are, respectively, the molar volume and isothermal compressibility at zero pressure. G_m as a function of both temperature and pressure can then be obtained by adding the following expression to Eq. (6.3):

$$G_m^{T,P} - H_m^{\mathrm{SER}} = a + bT + cT\log_e(T) + \sum \mathrm{d}T^n + \int_0^P V_m^{T,P}\mathrm{d}P \tag{6.22}$$

where

$$\int_0^P V_m^{T,P}\mathrm{d}P = V_0\left\{(1 + nP\chi)^{(n-1)/n} - 1\right\}\frac{\exp\left(\alpha_0 T + \frac{1}{2}\alpha_1 T^2\right)}{\chi(n-1)}. \tag{6.23}$$

Available experimental data for various solid–liquid and solid–solid state transformations has been successfully fitted for C, Mo and W (Fernandez Guillermet *et al.* 1985) and Fe (Fernandez Guillermet and Gustafson 1985) using the above expression. It gives better results than Eq. (6.16) at high pressures and also has the advantage that Eq. (6.19) can be inverted to give an expression for P^{T,V_m}. This in turn allows an explicit function to be written for the Helmholtz energy. Another approach that has been considered recently is the Rose equation of state (Rosen and Grimwall 1983). This however requires the additional input the expansion coefficient and the pressure derivative of the bulk modulus. With the increasing availability of such quantities from first-principles calculations, this may well become the basis for future formulations (Burton *et al.* 1995).

Depending on the data available, Eqs (6.17)–(6.23) reproduce experimental pressure effects with considerable accuracy in many cases. In particular, Eq. (6.18) can be used to confirm entropy data derived using more conventional techniques and can also provide data for metastable allotropes. Ti again provides a leading example, as pressure experiments revealed that the ω-phase, previously only detected as a metastable product on quenching certain Ti alloys, could be stabilised under pressure (Fig. 6.14). Extrapolation of the β/ω transus line yields the metastable allotropic transformation temperature at which the β-phase would transform to ω in the absence of the α-phase, while the slope of the transus lines can be used to extract a value for the relevant entropy via Eq. (6.18).

6.5.3 Effect of competing states

All of the above approaches assume that missing pressure data can be estimated from consistent periodic trends and that individual phases do not exhibit anomalous behaviour. This is reasonable when the alternative crystal structures are each associated with specific energy minima, but there can be some important exceptions

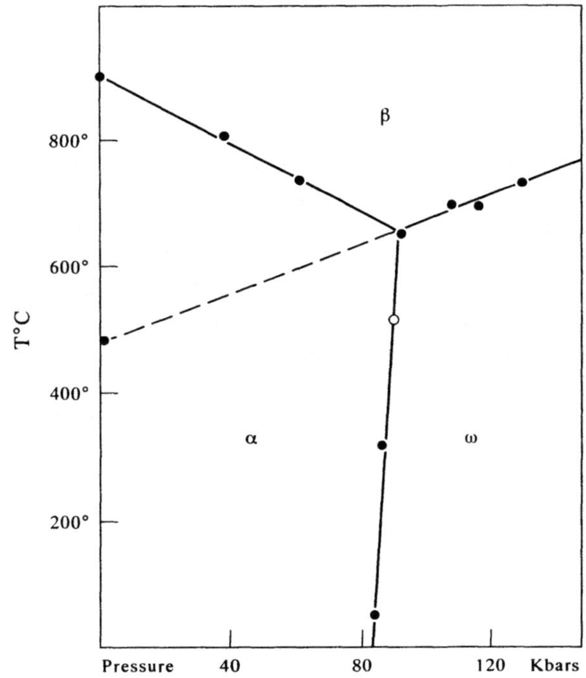

Figure 6.14. T–P diagram for Ti showing experimental data from Bundy (1965b) and Jayaraman (1966) and the extrapolation of the β–ω transus to yield the metastable $T^{\beta\to\omega}$ at atmospheric pressure (from Vanderpuye and Miodownik 1970).

to this general rule. For example, in magnetic systems it is possible to obtain markedly different combinations of exchange and correlation forces in the *same* crystal structure (Fig. 6.13) although the two structures can differ in energy by only a small amount (Asada and Terakura 1993). The possibility of Schottky excitations between alternative states can then produce an anomalous changes in volume with temperature and pressure. Although it is always possible to handle such a situation by a choice of suitable coefficients in Eq. (6.23), an alternative treatment is to explicitly determine the effect on the Gibbs energy of having a combination of states. This was attempted for γ-Fe using the following additional relationships (Kaufman *et al.* 1963, Blackburn *et al.* 1965):

$$V^{\gamma_1\to\gamma_2} = f^{\gamma_2}V_T^{\gamma_2} - f^{\gamma_1}V_T^{\gamma_1} \tag{6.24}$$

$$\frac{f_{T,P}^{\gamma_2}}{f_{T,P}^{\gamma_1}} = \alpha \exp^{-\Delta G/RT} \tag{6.25}$$

$$E_P^{\gamma_1\to\gamma_2} = E_{P=0}^{\gamma_1\to\gamma_2} + PV^{\gamma_1\to\gamma_2} \tag{6.26}$$

$$G_{T,P}^{\gamma_1\to\gamma_2} = 100PV_T^{\gamma_1} + RT\ln(1 - f^{\gamma_2}). \tag{6.27}$$

Here $V^{\gamma_1}, V^{\gamma_2}$ and $f^{\gamma_1}, f^{\gamma_2}$ refer to the molar volume and fractions of each state, $E^{\gamma_1\to\gamma_2}$ is the difference in energy between the two states and $G^{\gamma_1\to\gamma_2}$ is the extra increment of Gibbs energy resulting from the existence of the two states. Including this term, the γ/ε temperature–pressure transus for pure iron *was then quantitatively predicted before this was verified experimentally* by Bundy (1965a) (Fig. 6.15). The fit is clearly not as good as that which can be obtained by Fernandez Guillermet and Gustafson (1985) using Eqs (6.3) and (6.23), but on the other hand these authors *included the experimental pressure data in their optimisation* and did not make any specific provision for the possibility of multiple states. Which treatment is to be preferred depends on whether priority is to be given to the development of simple universally applicable algorithms or to more physically realistic models.

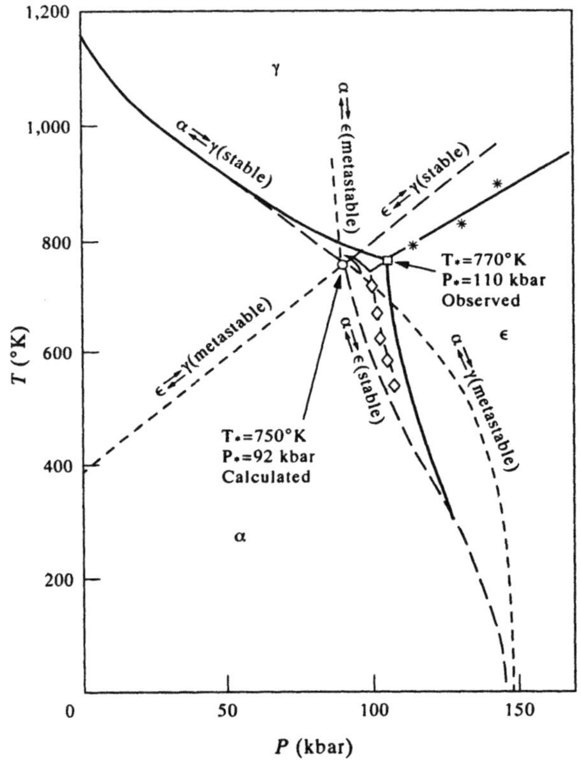

Figure 6.15. Comparison of experimental T–P diagram for Fe (Bundy 1965) * with calculated phase boundaries from Blackburn *et al.* (1965) and Fernandez Guillermet and Gustafson (1985) ◇.

Data derived from pressure experiments on semi-conducting elements by Klement and Jayamaran (1966) and Minomura (1974) have also been useful in obtaining confirmation that the entropies associated with transitions in Si, Ge and Sn form a consistent pattern, supporting the concept that each crystallographic transformation tends to have a characteristic associated entropy change (Miodownik 1972a, 1972b). Similarly, extrapolations from pressure data on alloys can be used to obtain estimates of lattice stabilities at $P = 0$, which can then be compared with estimates obtained by other routes, such as SFE measurements.

Calculation of the critical pressure required to cause a phase transformation at 0 K can also be obtained from first-principle calculations. Assuming the various phases exhibit normal elastic behaviour the tangency rule can be applied to energy vs volume plots to yield values for the critical pressure that would generate a phase transformation:

$$P_{\text{crit}} = \frac{\delta G}{\delta V}. \tag{6.28}$$

However, if there are large variations in the elastic constants with pressure, this can seriously affect P_{crit} by as much as an order of magnitude (Lam *et al.* 1984).

6.6. DETERMINATION OF INTERACTION COEFFICIENTS FOR ALLOYS AND STABILITY OF COUNTER-PHASES

The determination of individual binary equilibrium diagrams usually only involves the characterisation of a limited number of phases, and it is possible to obtain some experimental thermodynamic data on each of these phases. However, when handling multi-component systems or/and metastable conditions there is a need to characterise the Gibbs energy of many phases, some of which may be metastable over much of the composition space.

This requires a methodology for characterising a large range of metastable solutions and compounds which, by definition are difficult, if not impossible, to access experimentally. The available methods involve various levels of compromise between simplicity and accuracy and can be categorised by the choice of atomic properties used in the process.

At one end of the spectrum are first-principles methods where the only input requirements are the atomic numbers Z_A, Z_B, ... the relevant mole fractions and a specified crystal structure. This is a simple extension to the methods used to determine the lattice stability of the elements themselves. Having specified the atomic numbers, and some specific approximation for the interaction of the relevant wave functions, there is no need for any further specification of attractive and repulsive terms. Other properties, such as the equilibrium atomic volumes, elastic moduli and charge transfer, result automatically from the global minimisation of the energy for the total assembly of electrons and ions (Zunger 1980a, 1980b, Pettifor 1992).

At the other end of the spectrum are more empirical methods which utilise secondary properties which reflect more conventional terms, such as the atomic volume, electronegativity or some function of the enthalpy of evaporation (Darken and Gurry 1953, Hildebrand and Scott 1956, Mott 1968, Pauling 1960, Kaufman and Bernstein 1970). In many cases the desire for simplicity has led to allocation of a set of constant parameters to each atomic species, even though such properties are known to vary substantially with the geometric or chemical environment in which they are placed. This is probably adequate if such parameters are to be fed into a correspondingly simple regular solution model or line compound model, but will not be sufficient if there is a need for sub-regular solution parameters or if substantial directional bonding is involved. Many treatments have been modified by feedback to give better agreement with experimental trends. In some cases this has been achieved by introducing additional terms and in others by merely altering the numerical values that were originally tied rigidly to a measured property (Miedema 1976, Niessen *et al.* 1983).

6.6.1 The prediction of liquid and solid solution parameters

6.6.1.1 Empirical and semi-empirical approaches. The problem of making theoretical estimates for the interaction coefficients for the liquid phase has been tackled in different ways by various authors. Kaufman and Bernstein (1970) considered that the liquid state would exhibit the lowest repulsive forces of all the states of condensed matter and that a description of the interaction parameters for the liquid state would be the best basis for the prediction of interaction parameters for various solid phases.

Their expression for a liquid interaction parameter (L) follows Mott (1968) and starts with the simple sum of an attractive term, e_o, and a repulsive term, e_p,

$$L = e_o + e_p. \tag{6.29}$$

Kaufman and Bernstein (1970) derived e_o via the electronegativity values, φ_i and φ_2 (Pauling 1960)

$$e_o = -100\,(\varphi_1 - \varphi_2)^2. \tag{6.30}$$

Estimates for e_p were based on Hildebrand's solubility parameters, δ_{ij}, (Hildebrand and Scott 1950), which are related to the enthalpy of vaporisation ($H_{v(ij)}$) and the molar volume $V_{m(ij)}$ of elements i and j respectively so that

$$\delta_{(i,j)} = \left(-\frac{H_{v(i,j)}}{V_{m(i,j)}}\right)^{1/2}. \tag{6.31}$$

Making an adjustment for the effect of temperature finally leads to

$$e_p = 0.3\left(V_{m(i)} + V_{m(j)}\right)\left\{\left(\frac{-H_{v(i)}}{V_{m(i)}}\right)^{1/2} - \left(\frac{-H_{v(j)}}{V_{m(j)}}\right)^{1/2}\right\}. \tag{6.32}$$

Further parameters are then added to determine the interaction parameters for various crystal structures, e.g., the interaction parameter, *B*, for b.c.c. structure becomes

$$B = L + e_1 + e_2. \tag{6.33}$$

Following Kaufman and Bernstein (1970) the first of these is a strain-energy term depending on the volume difference between *i* and *j* which will always be a positive contribution and assumed to be proportional to $GV(\Delta V/V)^2$, where G is a modulus which, in the absence of data, can be approximated as $H_{v(ij)}/V$. The second term is based on the variation of $H^{L\rightarrow B}$ with group number across the periodic table reflecting the relation between e_o and the heat of vaporisation. Similarly, the interaction parameters E, for c.p.h. solutions, and A, for f.c.c. solutions, become

$$E = B + e_3 \tag{6.34}$$

and

$$A = E + e_4. \tag{6.35}$$

The strain energy term is thus assumed to be independent of crystal structure for the three major types of substitutional solid solutions.

By contrast, the 'macroscopic atom' model of Miedema (Miedema *et al.* 1975) starts with a descriptions of the solid state which is then modified to describe the liquid state (Boom *et al.* 1976a, 1976b). In their model the enthalpy of formation at 0.5 mole fraction, $H_{c=0.5}$ is given to a first approximation by:

$$H_{c=0.5} \approx \left[-P(\Delta\phi^*)^2 + Q(\Delta n_{ws}{}^{1/3})^2 - R\right]. \tag{6.36}$$

Here n_{ws} is an electron density based on the volume of Wigner–Seitz atomic cells, V_m, and it is assumed that differences in electron density Δn_{ws} between different species of atoms will always lead to local perturbations that give rise to a positive energy contribution. On the other hand, ϕ^* is a chemical potential based on the macroscopic work function ϕ, so that differences, $\Delta\phi^*$, at the cell surfaces between different species of atoms lead to an attractive term, analogous to the gain in energy from the formation of an electric dipole layer. P, Q and R are empirically derived proportionality constants. P and Q were initially taken to be adjustable parameters, but turned out to be almost universal constants for a wide variety of atomic combinations, especially amongst the transition metals. The parameter R was introduced as an arbitrary way of adjusting for the presence of atoms with p

electrons and takes on values that increase regularly with the valency of the B-group element. In contrast to the Kaufman model (Kaufman and Bernstein 1970), Eq. (6.36) contains no parameters which refer to crystal structure and therefore the heats of solution of f.c.c., b.c.c. and c.p.h. structures are predicted to be, a priori, identical. The same conclusion can also be expected for liquid solutions, and indeed the suggested parameters for liquid and solid solutions are very similar. The constants P and Q are practically the same, although it was necessary to introduce different values of R.

***6.6.1.2* Ab initio *electron energy calculations*.** While there are many methods for predicting the relative stability of ordered structures, electron energy calculations for predicting interaction parameters for the liquid is still a major problem. Calculations of the heat of solution for disordered solutions falls in an intermediate category. It is a surprising fact that it is at present still impossible to calculate the melting point, or the entropy and enthalpy of fusion of the elements to any reasonable degree of accuracy, which poses a major challenge to a full calculation of any phase diagram by FP methods. One method to overcome this problem for liquids is to assume that it will exhibit *sro* parameters based on the predominant solid state structure(s) in the phase diagram (Pasturel *et al.* 1992).

Pasturel *et al.* (1985) and Colinet *et al.* (1985) have used a tight-binding model which considers the moments of the density of states. The model provides an estimate for the *d* electron transfer between the elements in an alloy/compound by using the concept of partial density of states. The model is, however, simplified so that neither atomic position or crystal structure is calculated in detail. The results can therefore be considered to provide a good general prediction for disordered phases but is independent of crystal structure. The reason why this works reasonably well is because the magnitude of the fundamental electronic *d*-band effect for many such alloys is substantially larger than the effect arising from differences in crystal structure. More accurate results probably require the incorporation of directional bonding (Pettifor *et al.* 1995) which can lead to highly composition-dependent interaction parameters.

6.6.2 The prediction of thermodynamic properties for compounds

6.6.2.1 The concept of counter-phases. When a stable compound penetrates from a binary into a ternary system, it may extend right across the system or exhibit only limited solubility for the third element. In the latter case, any characterisation also requires thermodynamic parameters to be available for the equivalent metastable compound in one of the other binaries. These are known as *counter-phases*. Figure 6.16 shows an isothermal section across the Fe–Mo–B system (Pan 1992) which involves such extensions for the binary borides. In the absence of any other guide-

References are listed on pp. 173–178.

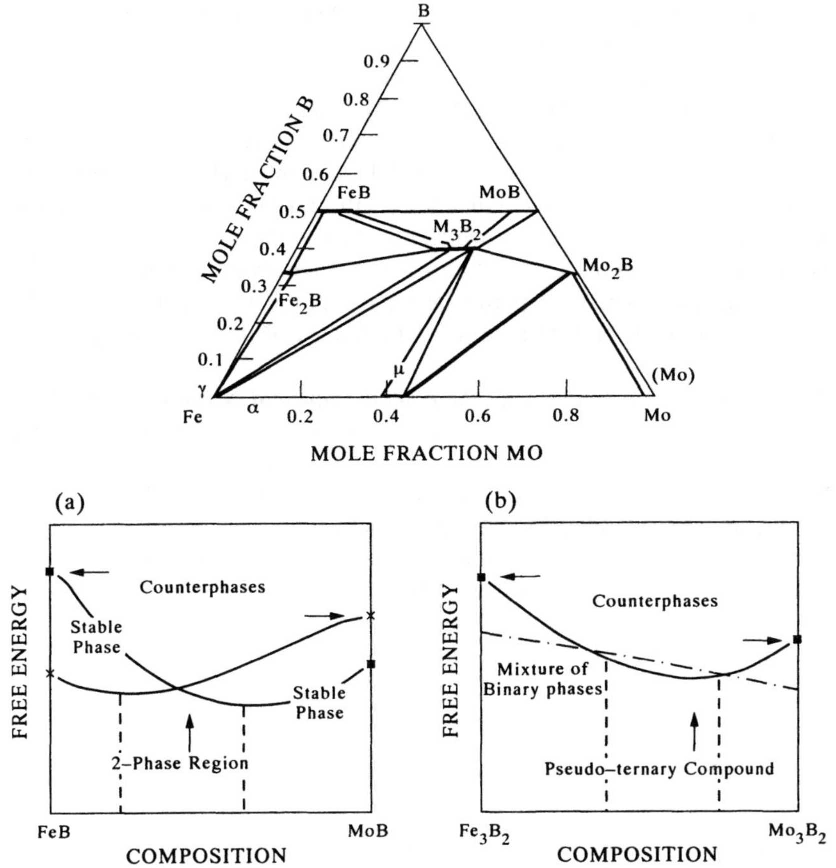

Figure 6.16. Counter phases in the Fe–Mo–B system: (a) the section FeB–MoB and (b) the section Fe_3B_2–Mo_3B_2 (from Pan 1992 and Miodownik 1994).

lines it is necessary to use empirical methods, which have been well documented in Kaufman and Bernstein (1970), and constitute a useful fitting procedure. Accurate results can be obtained if there are some prior phase-boundary data for the ternary system which can be used to validate the assumed thermodynamics of the counter-phase. However, problems arise when there is little or no experimental information and when the extension of the compound into the ternary is small. Further, when multiple sublattice modelling is used (see Chapter 5) it may be necessary to input thermodynamic properties for compounds which are compositionally far away from the area of interest in the CALPHAD calculation. It is therefore desirable to have a method of *predicting* the thermodynamic properties of as many counter-phases as possible.

6.6.2.2 Structure maps. Attempts have been made to chart the occurrence of particular structures in relation to familiar parameters such as size, electronegativity and electron concentration (Darken and Gurry 1953, Villars *et al.* 1989). Some of these schemes have only limited applicability to particular areas of the periodic table, but more recently there has been greater success through structure maps which are based on the concept of a Mendeleev number (Pettifor 1985a). These have been extended to ternary compounds (Pettifor 1985b) and give considerable insight into the choice of structures which may be significant competitors in a given multicomponent situation. Calculations which give more detailed energies of various competing structures in particular regions of the map, such as for the Laves phases (Ohta and Pettifor 1989) are now emerging with increased frequency.

6.6.2.3 The Miedema model and other semi-empirical methods. The Miedema model was originally devised as a tool for merely predicting the *sign* of the heat of solution at the equiatomic composition. Therefore Eq. (6.36) does not contain any concentration-dependent terms (Miedema 1973, Miedema *et al.* 1973). However, the treatment was extended in subsequent publications, and modifications were made to include ordering contributions and asymmetric effects (Miedema *et al.* 1975, de Boer *et al.* 1988). Additional functions $f(c_A^s, c_B^s)$ and $g(c_A, c_B, V_{mA}, V_{mB})$ were kept very simple and did not include additional parameters other than (V_m), which had already been used to determine (n_{ws}).

$$\Delta H(c_A, c_B) = f\left(c_A^s, c_B^s\right) g\left(c_A, c_B, V_{mA}, V_{mB}\right) \left[-P(\Delta\phi^*)^2 + Q\left(\Delta n_{ws}{}^{1/3}\right)^2 - R\right] \tag{6.37}$$

where $f(c_A^s, c_B^s) = c_A^s c_B^s(1 + 8(c_A^s c_B^s)^2)$ and c_A^s, etc., are surface concentrations. These arise from the concept that in their 'macroscopic atom' model the enthalpy effects are generated at the common interface of dissimilar atomic cells. Such an expression contains only a limited degree of asymmetry and essentially use a very primitive ordering model that does not even consider nearest-neighbour interactions but assumes that $H_{\mathrm{ord}}/H_{\mathrm{disord}}$ is virtually a constant. Since there is no provision for any crystallographic parameters it is impossible to make any predictions about the relative stability compounds with different crystal structures at the same stoichiometry. A quantum-mechanical rationale of the Miedema approach has been published by Pettifor (1979a), outlining its strengths and weaknesses.

Machlin (1974, 1977) developed a semi-empirical treatment which used a constant set of nearest-neighbour interactions and was one of the earliest semi-empirical attempts to obtain the relative enthalpies of formation between different crystal structures. This successfully predicted the correct ground states in a substantial number of cases, but the treatment was generally restricted to transition metal combinations and a limited number of crystal structures.

References are listed on pp. 173–178.

6.6.2.4 **Ab initio** ***electron energy calculations.*** Some of the earliest electron energy calculations were made for transition metals where it was noted that trends in cohesive energy could be approximated to a very reasonable degree by considering only the *d*-band electrons (Friedel 1969, Ducastelle and Cyrot-Lackman 1970). Various means of representing the density of states were then applied in an attempt to quantify this. Van der Rest *et al.* (1975) utilised the coherent potential approximation (CPA) method with off-diagonal disorder to compute the density of states curve. This method was employed by Gautier *et al.* (1975) and Ehrenreich and Schwartz (1976) for extensive calculations of transition metal alloys. Later work by Pettifor (1978, 1979b), Varma (1979) and Watson and Bennett (1981) extended the *d*-band concept further and comparison with experimental enthalpies of formation for equi-atomic compounds were quite reasonable, although there is a clear tendency for the predictions to be too exothermic. They are also limited because they are insensitive to crystal structure and, subsequently, give no predictions for metastable compounds. Pseudo-potential calculations are also a method by which heats of formation can be predicted (Hafner 1989, 1992). These work better for sp-bonded compounds such as the alkali metal Laves phases (Hafner 1987) where nearly free electron theory is well suited, but they are not so applicable to *d*-band calculations which are necessary for transition metal alloys.

Electron energy calculations utilise methods which solve the Schrödinger equation and start with the specification of the lattice space group taking into account a wide variety of interatomic forces. It is therefore possible to make calculations for a large group of ordered structures and calculate the difference in enthalpy that arise from any specified change in crystallography at a given composition for any combination of elements (Pettifor 1985a, Finnis *et al.* 1988). Such calculations incorporate the effects of changes in band width, the centre of gravity of the bands, various forms of hybridisation and can also include directional bonding (Pettifor 1989, Phillips and Carlsson 1990, Pettifor *et al.* 1995).

Although there are inherent approximations in any electron energy calculation, it is significant that this route now has a high success rate in predicting the correct ground state amongst a large range of competing structures (Gautier 1989, Sluiter and Turchi 1989, Nguyen-Manh *et al.* 1995). In order to shorten the calculation time these are often limited to those structures thought most likely to occur. This can potentially lead to the omission of other contenders but algorithms have recently been developed which can markedly extend the number of structures sampled and remove such arbitrary limitations (Lu *et al.* 1991).

Electron energy calculations now offer a coherent explanation of trends observed both across and down the periodic table and the grouping and overlaps observed in structure maps. Of particular importance are the marked changes that occur on moving to elements of higher atomic number, which means that some of the earlier assumptions concerning similarities of behaviour for compounds of the 3*d*, 4*d*, and 5*d* elements (Kaufman and Bernstein 1970) have had to be revised. Quantum mechanical effects also lead to significantly greater sinusoidal variations in stability with atomic number and hence also to markedly greater asymmetry in the variation of enthalpy of formation of similar structures in a given binary system.

6.7. SUMMARY

Despite a sometimes turbulent history a series of standard and effective lattice stability values are now readily available for the majority of elements. Most of the remaining areas of controversy concern structures which seem to be elastically unstable or where the magnetic structure is uncertain. The relative stability of different compounds seems to be taking a similar route, but clearly involves a much wider range of structures. The availability of many different calculation routes has led to numerous comparisons being made between the various calculated results and whatever experimental information is available for the stable phases (Kaufman 1986, Birnie *et al.* 1988, Colinet *et al.* 1988, Watson *et al.* 1988, de Boer *et al.* 1988, Aldinger *et al.* 1995). Figure 6.17 shows that there is often good general agreement for a given class of structures so that it is now possible to make a better estimate for the enthalpies of unknown metastable compounds.

The success of earlier empirical schemes in predicting the sign and magnitude of the enthalpy of formation can in part be attributed to making predictions for *AB* compounds. This can hide asymmetrical effects which are more apparent when other stoichiometries are considered. The empirical and semi-empirical schemes do

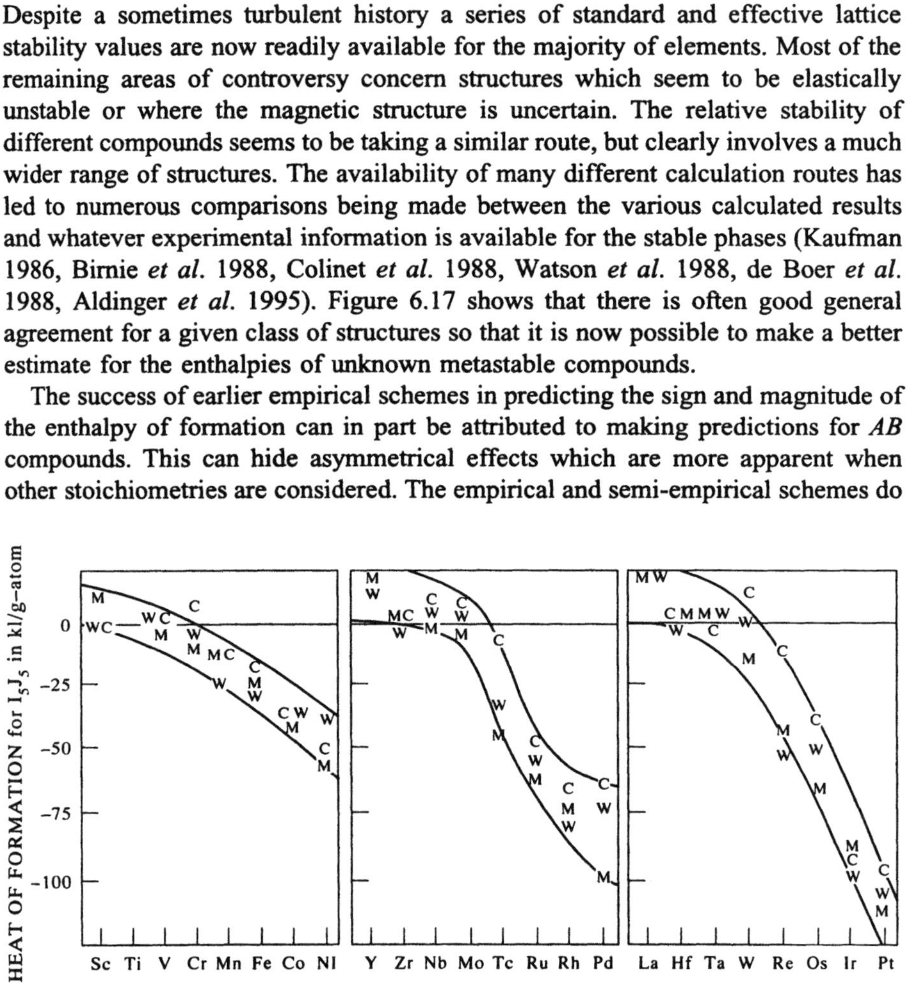

Figure 6.17. Comparison of enthalpies of formation for AB Titanium Alloys M=predictions of Miedema (de Boer *et al.* 1988), W=predictions of Watson and Bennett (1984) and C=predictions of Colinet *et al.* (1985). Figure from Aldinger *et al.* (1995).

References are listed on pp. 173–178.

not have the capacity to handle a wide range of bonding characteristics without the introduction of many additional parameters with arbitrary numerical values. This much diminishes their predictive value as far as defining the relative stability of counter-phases is concerned. The more sophisticated electron energy calculations are now undoubtedly a better guide for the relative enthalpy of a much wider range of structures and should be increasingly used as input data in any CALPHAD optimisation. However, such calculations cannot predict the relative entropies of the competing structures and this is still an area where an empirical approach is required for the present.

REFERENCES

Achar, B. S. and Miodownik, A. P. (1974) Report on Phase Equilibria in Alloys, Report to Science Research Council (UK) under Contract B/RG/19175.

Ågren, J. (1979) *Met. Trans.*, **10A,** 1847.

Ågren, J., Cheynet, B. Clavaguera-Mora, M. T., Hack, K., Hertz, J., Sommer, F. and Kattner, U. (1995) *CALPHAD,* **19,** 449.

Aldinger, F., Fernandez Guillermet, A., Iorich, V. S., Kaufman, L., Oates, W. A., Ohtani, H., Rand, M. and Schalin, M. (1955) *CALPHAD,* **19,** 555.

Anderson, J. O., Fernandez Guillermet, A. and Gustafson, P. (1987) *CALPHAD,* **11,** 365.

Ardell, A. J. (1963) *Acta. Met.*, **11,** 591.

Asada, T. and Terakura, K. (1993) in *Computer Aided Innovation of New Materials, Vol. 1,* eds Doyama, M. *et al.* (Elsevier Science, Amsterdam), p. 169.

Austin, J. B. (1932) *Ind. Chem. Engineering,* **24,** 1225.

Bendick, W. and Pepperhoff, W. (1982) *Acta. Met.*, **30,** 679.

Birnie, D., Machlin, E. S., Kaufman, L. and Taylor, K. (1982) *CALPHAD,* **6,** 93.

Blackburn, L. D., Kaufman, L. and Cohen, M. (1965) *Acta. Met.*, **13,** 533.

Boom, R., de Boer, F. R. and Miedema, A. R. (1976a) *J. Less Common Metals,* **45,** 237.

Boom, R., de Boer, F. R. and Miedema, A. R. (1976b) *J. Less Common Metals,* **45,** 271.

Brewer, L. (1967) in *Phase Stability in Metals and Alloys,* eds Rudman, P. S. *et al.* (McGraw-Hill, New York), p. 39.

Brewer, L. (1975) LBL-3720 Reprint on Contract W-7405-ENG-48.

Brewer, L. (1979) in *Calculation of Phase Diagrams and Thermochemistry of alloy phases* eds Chang, Y. A. and Smith, J. F. (AIME, Warrendale, PA), p. 197.

Bridgman, P. W. (1931) in *The Physics of High Pressure* (Bell, London).

Bundy, F. P. (1965a) *J. Appl. Physics,* **36,** 616.

Bundy, F. P. (1965b) ASTM Special Technical Publication Number 374.

Burton, B., Chart, T. G., Lukas, H. L., Pelton, A. D., Siefert, H. and Spencer, P. (1995) *CALPHAD,* **19,** 537.

Chalmers, B. (1959) in *Physical Metallurgy* (John Wiley), p. 85.

Chang, Y. A., Colinet, C., Hillert, M., Moser, Z., Sanchez, J. M., Saunders, N., Watson, R. E. and Kussmaul, A. (1995) *CALPHAD,* **19,** 481.

Chase, M. W., Ansara, I., Dinsdale, A. T., Eriksson, G., Grimvall, G., Höglund, L. and Yokokawa, H. (1995) *CALPHAD,* **19,** 437.

Chen, Y. G. and Liu, B. X. (1997) *J. Alloys & Compounds*, **261,** 217.
Cho, S. A. and Puerta, M. (1976) *J. Solid State Chem.*, **16,** 355.
Chuang, Y. Y., Schmid-Fester, R. and Chang, Y. A. (1985) *Met. Trans.*, **16A,** 153.
Colinet, C., Bessound, A. and Pasturel, A. (1988) *J. Phys. F*, **18,** 903.
Colinet, C., Pasturel, A. and Hicter, P. (1985) *CALPHAD*, **9,** 71.
Craievich, P. J., Weinert, M., Sanchez, J. and Watson, R. E. (1994) *Phys. Rev. Lett.*, **72,** 3076.
Craievich, P. J. and Sanchez, J. (1995) Unpublished research reported in Chang *et al.* (1995).
Crampin, S., Hampel, K., Vvedensky, D. D. and McLaren, J. M. (1990) *J. Mater. Res.*, **5,** 2107.
Darken, L. S. and Gurry, R. W. (1953) *The Physical Chemistry of Metals* (McGraw-Hill, New York).
Darken, L. S. and Smith, R. P. (1951) *Ind. Eng. Chem.*, **43,** 1815.
de Boer, F. R., Boom, R., Mattens, W. C. M., Miedema, A. R., and Niessen, A. K. (1988) in *Cohesion in metals: Cohesion and Structure, Vol. 1* (Elsevier Science, Amsterdam), p. 644.
de Fontaine, D. (1996) *MRS Bulletin*, **21,**(8), 16.
Dinsdale, A. T. (1991) *CALPHAD*, **15,** 317.
Domb, C. (1958) *Nuevo Cimento*, **9**(S1), 9.
Ducastelle, F. and Cyrot-Lackman, F. (1970) *J. Phys. Chem. Solids*, **31,** 1295.
Ducastelle, F. and Gautier, F. (1976) *J. Phys. F*, **6,** 2039.
Ehrenreich, H. and Schwartz, L. (1976) in *Solid State Physics*, eds Ehrenreich, H. *et al.* (Academic Press, New York), p. 149.
Einarsdotter, K., Sadigh, B., Grimwall, G. and Ozolins, V. (1997) Submitted to *Phys. Rev. Lett.*
Engel, N. (1964) *ASM Trans. Quarterly*, **57,** 619.
Fernandez Guillermet, A. and Gustafson, P. (1985) *High Temp.–High Press.*, **16,** 591.
Fernandez Guillermet, A. and Hillert, M. (1988) *CALPHAD*, **12,** 337.
Fernandez Guillermet, A. and Huang, W. (1988) *Z. Metallkde.*, **79,** 88.
Fernandez Guillermet, A. and Huang, W. (1990) *Int. J. Thermophys.*, **11,** 949.
Fernandez Guillermet, A., Gustafson, P. and Hillert, M. (1985) *J. Phys. Chem. Solids*, **46,** 1427.
Fernandez Guillermet, A., Ozolins, V., Grimwall, G. and Korling, M. (1995) *Phys. Rev. B*, **51,** 10,364.
Fernando, G. W., Watson, R. E., Weinert, M., Wang, Y. J. and Davenport, J. W. (1990) *Phys. Rev. B*, **41,** 11,813.
Finnis, M. W., Paxton, A. T., Pettifor, D. G., Sutton, A. P. and Ohta, Y. (1988) *Phil. Mag. A*, **58,** 143.
Friedel J. (1969) in *Physics of Metals*, ed. Ziman, J. (Cambridge Univ. Press), p. 340.
Friedel, J. (1974) *J. de Physique*, **75,** 59.
Gautier, F., Van der Rest, J. and Brouers, F. (1975) *J. Phys. F*, **5,** 1184.
Gautier, F. (1989) in *Alloy Phase Stability*, eds Stocks, G. M. and Gonis, A. (NATO ASI series E *Appl. Sci.*, **163,** Kluwer, Dordrecht), p. 321.
Gazzara, C. P., Middleton, R. M. and Weiss, R. J. (1964) *Phys. Lett.*, **10,** 257.
Gonis, A., Zhang, X. G, Freeman, A. J., Turchi, P. E. A., Stocks, G. M. and Nicholson (1987) *Phys. Rev. B*, **36,** 4630.

Goodwin, L., Skinner, A. J. and Pettifor, D. G. (1989) *Europhys. Lett.*, **9**, 701.
Grimwall, G. (1974) *Sol. State Comm.*, **14**, 551.
Grimwall, G. (1977) *Inst. Phys. Conf. Ser.* No 30.
Grimwall, G. (1997) Ringberg III to be published.
Grimwall, G. and Ebbsjo, I. (1975) *Physica Scripta*, **12**, 168.
Grimwall, G., Thiessen, M. and Fernandez Guillermet, A. (1987) *Phys. Rev. B*, **36**, 7816.
Gschneider, K. A. (1961) *Rare-Earth Alloys* (Van Nostrand, Amsterdam).
Gschneider, K. A. (1964) *Solid State Physics*, **16**, 275.
Gschneider, K. A. (1975) *J. Less Common Metals*, **43**, 179.
Gyorffy, B. L. and Stocks, G. M. (1983) *Phys. Rev. Lett.*, **50**, 374.
Hafner, J. (1987) *From Hamiltonians to Phase Diagrams* (Springer, Berlin).
Hafner, J. (1989) *Cohesion and Structure: Vol. 2*, eds Pettifor, D. G. and de Boer, F. R. (North Holland, Amsterdam).
Hafner, J. (1992) in *Electron Theory in Alloy Design*, eds Pettifor, D. G. and Cotttrell, A. H. (Inst. Materials, London), p. 44.
Haglund, J., Grimwall, G. and Jarlborg, T. (1993) *Phys. Rev. B*, **47**, 9279.
Hasegawa, H. and Pettifor, D. G. (1983) *Phys. Rev. Lett.*, **50**, 130.
Heine, V. and Weaire, D. (1971) *Solid State Physics*, **24**, 247.
Hildebrand, J. H. and Scott, R. L. (1956) *The Solubility of Non-Electrolyes* (Reinhold, New York).
Hilliard, J. (1963) *Trans. A.I.M.E.*, **223**, 429.
Ho, K. M. and Harmon, B. N. (1990) *Mat. Sci. Eng. A*, **127**, 155.
Hume-Rothery, W., Reynolds, P. W. and Raynor, G. V. (1940) *J. Inst. Metals*, **66**, 191.
Inden, G. (1975) *Z. Metallkunde*, **66**, 577.
Jansen, J. F. and Freeman, A. J. (1984) *Phys. Rev.*, **B30**, 561.
Johannson, C. H. (1937) *Arch. Eisenhuttenw.*, **11**, 241.
Kaufman, L. (1959a) *Acta. Met.*, **7**, 575.
Kaufman, L. (1959b) *Bull. Amer. Phys. Soc.*, **4**, 181.
Kaufman, L. (1967) in *Phase Stability of Metals and Alloys*, eds Rudman, P. S. *et al.* (McGraw-Hill, New York), p. 125.
Kaufman, L. (1972) in *Metallurgical Thermochemistry* (NPL-HMSO, London), p. 373.
Kaufman, L. (1974) in *Materials under Pressure*, Honda Memorial Series on Materials Science No. 2 (Maruzen, Tokyo), p. 66.
Kaufman, L. (1986) in *Computer Modelling of Phase Diagrams*, ed. Bennett, L. H. (Met. Soc. AIME, Warrendale, PA), p. 237.
Kaufman, L. (1991) *Scand. J. Met.*, **20**, 32.
Kaufman, L. (1993) *CALPHAD*, **17**, 354.
Kaufman, L. and Bernstein, H. (1970) *Computer Calculations of Phase Diagrams* (Academic Press, New York).
Kaufman, L. and Cohen, M. (1956) *Trans. AIME*, **206**, 1393.
Kaufman, L. and Cohen, M. (1958) *Progr. Metal Physics*, **7**(3), 165.
Kaufman, L. and Nesor, H. (1973) *Z. Metallkde.*, **64**, 249.
Kaufman, L., Clougherty, E. V. and Weiss, R. J. (1963) *Acta. Met.*, **11**, 323.
Klement, W. and Jayaraman, A. (1966) *Progress in Solid State Chemistry*, **3**, 289.
Kmetko, E. A. and Hill, H. H. (1976) *J. Phys. F*, **6**, 1025.
Kouvetakis, J. and Brewer, L. (1993) *J. Phase Equilibria*, **14**, 563.

Lam, P. K., Chou, M. Y. and Cohen, M. L. (1984) *J. Physics C*, **17**, 2065.
Lu, Z. W., Wei, S. H., Junger, Frota-Pessao, A. S. and Ferreira, L. G. (1991) *Phys. Rev. B*, **44**, 512.
Machlin, E. S. (1974) *Acta. Met.*, **22**, 95.
Machlin, E. S. (1977) *CALPHAD*, **1**, 361.
Meyerhoff, R. W. and Smith, J. F. (1963) *Acta. Met.*, **11**, 529.
Miedema, A. R. (1973) *J. Less Common Metals*, **32**, 117.
Miedema, A. R. (1976) *J. Less Common Metals*, **46**, 67.
Miedema, A. R., Boom, R. and de Boer, F. R. (1975) *J. Less Common Metals*, **41**, 283.
Miedema, A. R., de Boer, F. R. and de Chatel, P. F. (1973) *J. Phys. F*, **3**, 1558.
Minomura, S. (1974) in Honda Memorial Series on Materials Science, Monograph No. (2), p. 104.
Miodownik, A. P. (1972a) in *Metallurgical Thermochemistry* (NPL-HMSO, London), p. 233.
Miodownik, A. P. (1972b) in *Metallurgical Thermochemistry* (NPL-HMSO, London), p. 352.
Miodownik, A. P. (1977) *CALPHAD*, **1**, 133.
Miodownik, A. P. (1978a) in Honda Memorial Volume on Metal Science and Metallurgy, in *Physics and Application of Invar Alloys*, ed. Saito (Maruzen, Tokyo) **3**(18), p. 429.
Miodownik, A. P. (1978b) ibid. **3**(12), p. 288.
Miodownik, A. P. (1978c) *CALPHAD*, **2**, 207.
Miodownik, A. P. (1986) in *Computer Modelling of Phase Diagrams* ed. Bennett, L. H. (Met. Soc. AIME, Warrendale, PA), p. 253.
Miodownik, A. P. (1994) in *Statics and Dynamics of Alloy Phase Transformations*, ed. Turchi, P. E. A. and Gonis, A. (NATO ASI Series 8 *Physics* **319**, 45).
Miodownik, A. P. (1992) in *Structural and Phase Stability of Alloys*, eds Moran-Lopez, J. L. *et al.* (Plenum Press, New York), p. 65.
Moriarty, J. A. and McMahan, A. K. (1982) *Phys. Rev. Lett.*, **48**, 809.
Moroni, E. G., Grimwall, G. and Jarlborg, T. (1996) *Phys. Rev. Lett.*, **76**, 2758.
Moruzzi, V. L. and Marcus, P. M. (1988a) *Phys. Rev. B*, **38**, 1613.
Moruzzi, V. L. and Marcus, P. M. (1988b) *J. Appl. Phys.*, **64**, 5598.
Moruzzi, V. L. and Marcus, P. M. (1990) *Phys. Rev. B*, **42**, 8361.
Moruzzi, V. L., Marcus, P. M., Schwartz, K. and Mohn, P. (1986) *Phys. Rev. B*, **34**, 1784.
Mott, B. W. (1968) *J. Mat. Sci.*, **3**, 424.
Murnaghan, F. D. (1944) *Proc. Natl. Acad. Sci. (USA)*, **30**, 244.
Nguyen-Manh, D., Bratkovsky, A. M. and Pettifor, D. G. (1995) *Phil. Trans. R. Soc. A*, **351**, 529.
Niessen, A. K., de Boer, F. R., Boom, R., de Chatel, P. F., Mattens, W. C. M. and Midema, A. R. (1983) *CALPHAD*, **7**, 51.
Ohta, Y. and Pettifor, D. G. (1989) *J. Phys.: Condens. Matter*, **2**, 8189.
Orr, R. L. and Chipman, J. (1967) *Trans. AIME*, **239**, 630.
Pan, L.-M. (1992) Ph.D. Thesis, University of Surrey, Guildford, U.K.
Pasturel, A., Colinet, C. and Hicter, P. (1985) *CALPHAD*, **9**, 349.
Pasturel, A., Colinet, C., Paxton, A. T. and van Schilfgaarde, M. (1992) *J. Phys. Cond. Matter.*, **4**, 945.
Pauling, L. (1960) *The Nature of the Chemical Bond*, (Cornell Univ. Press, Ithaca).
Paxton, A. T., Methfessel, M. and Polatoglon, H. M. (1990) *Phys. Rev. B*, **40**, 425.

Petry, W. *et al.* (1991) *Phys. Rev. B*, **43,** 10,933.
Pettifor, D. G. (1977) *CALPHAD*, **1,** 305.
Pettifor, D. G. (1978) *Solid State Comm.*, **31,** 621.
Pettifor, D. G. (1979a) *Phys. Rev. Letters*, **42,** 846.
Pettifor, D. G. (1979b) *Phys. Rev. Letters*, **43,** 1130.
Pettifor, D. G. (1985a) *J. Less Common Metals*, **114,** 7.
Pettifor, D. G. (1985b) *Mat. Sci. Tech.*, **4,** 2480.
Pettifor, D. G. (1988) Private communication to Saunders *et al.* (1988).
Pettifor, D. G. (1989) *Phys. Rev. Lett.*, **63,** 2480.
Pettifor, D. G., Aoki, M., Gumbsch, P., Horsfield, A. P., Nguyen-Manh, D. and Vitek, V. (1995) *Mat. Sci. Eng.*, **A192/193,** 24.
Phillips, R. and Carlsson, A. E. (1990) *Phys. Rev. B*, **42,** 3345.
Rosen, J. and Grimwall, G. (1983) *Phys. Rev. B*, **27,** 7199.
Rosenhain, W. (1926) *J. Inst. Metals*, **35,** 353.
Saunders, N., Miodownik, A. P. and Dinsdale, A. T. (1988) *CALPHAD*, **12,** 351.
Sawamura, H. (1972) *Nippon Kinzoku Kakkai Laiho*, **11**(6), 413.
Shao, G., Miodownik, A. P., Nguyen-Manh, D. and Pettifor, D. (1996) to be published, abstract in summary of proceedings for Calphad XXV (A. T. Dinsdale, *CALPHAD*, **21**(1), 115).
Singh, D. and Papaconstantopoulos, D. A. (1990) *Phys. Rev. B*, **37,** 8885.
Skriver, H. L. (1982) *Phys. Rev. Lett.*, **49,** 1768.
Skriver, H. L. (1983) *The LMTO Method* (Springer, Heidelberg).
Skriver, H. L. (1985) *Phys. Rev. B*, **31,** 1909.
Sluiter, M. and Turchi, P. (1989) in *Alloy Phase Stability*, eds Stocks, G. M. and Gonis, A. (NATO ASI series E *Appl. Sci.*, **163,** Kluwer, Dordrecht), p. 521.
Svechnikov, V. N. and Lesnik, A. G. (1956) *Fisika Met. Metalloved.*, **3**(1), 87.
Tiwari, G. P. (1978) *Metal Sci.*, **12,** 317.
Tso, N. C., Kosugi, M. and Sanchez, J. (1987) *Acta Met.*, **37,** 121.
Turchi, P. and Sluiter, M. (1993) in *Computer Aided Innovation of New Materials, Vol. 1*, eds Doyama, M. *et al.* (Elsevier Science, Amsterdam), p. 1449.
Vanderpuye, N. A. and Miodownik, A. P. (1970) in *The Science, Technology and Applications of Titanium*, eds Jaffee, R. *et al.* (Pergamon Press, Oxford), p. 719.
Van der Rest, J., Gautier, F. and Brouers, F. (1975) *J. Phys. F*, **5,** 2283.
van Laar, J. J. (1908) *Z. Physik. Chem.*, **63,** 216; **64,** 257.
Varma, C. M. (1979) *Solid State Comm.*, **31,** 295.
Villars, P., Mathis, K. and Hulliger, F. (1989) in *Cohesion and Structure, Vol. 2*, eds de Boer, F. R. and Pettifor, D. G. (Elsevier Science, Amsterdam), p. 1.
Watson, R. E. and Bennett, L. E. (1981) *CALPHAD*, **5,** 25.
Watson, R. E. *et al.* (1988) *Scripta Met.*, **22,** 1285.
Watson, R. E., Bennett, L. H., Davenport, J. W. and Weinert, M. (1986) in *Computer Modelling of Phase Diagrams*, ed. Bennett, L. H. (Met. Soc. AIME, Warrendale), p. 273.
Weiss, R. J. and Tauer, K. J. (1956) in *Theory of Alloy Phases* (ASM, Metals Park, OH), p. 290.
Weiss, R. J. and Tauer, K. J. (1958) *J. Phys. Chem. Solids*, **4,** 135.
Xu, J. H., Lin, W. and Freeman, J. (1991) *Phys. Rev.*, **B43,** 603.
Yin, M. T. (1982) Ph.D. Thesis, Univ. of California, Berkeley.

Zener, C. (1946) *Trans. AIME*, **167,** 513.
Zener, C. (1955) *Trans. AIME*, **203,** 619.
Zener, C. (1967) in *Phase Stability in Metals and Alloys*, eds Rudman, P. S. *et al.* (McGraw-Hill, New York), p. 31.
Zunger, A. (1980a) *Phys. Rev. Lett.*, **44,** 582.
Zunger, A. (1980b) *Phys. Rev. B*, **22,** 5839.

Chapter 7

Ordering Models

Chapter 7

Ordering Models

7.1. INTRODUCTION

In most phase-diagram calculations there is a tendency for different phases to be characterised thermodynamically using separate models, irrespective of their crystal structure. However, in systems where the structures of several phases have common symmetry elements and similar bonding, it is possible to describe the system using an ordering model which integrates the descriptions of the various phases into a single modelling entity. By doing this it is also possible to both reduce the number the parameters used to characterise the system and give these terms a sounder physical meaning. Although the mathematical procedures are inherently simple, the number of computational steps for any rigorous treatment tends to rise geometrically with the number of possible ordered configurations. This imposes a high penalty when such ordering calculations are incorporated into phase-diagram calculations. In addition, the coupling of parameters in various phases can reduce the degree of freedom with which particular phase boundaries can be described.

A number of different approaches have been adopted for the description of ordered systems. One route is to start with a fundamental description of bonding and crystallography, establish the ordering energy of various phases at 0 K and then add suitable expressions for the structural entropy. The second route is to start with a description of the high-temperature disordered phase and derive expressions for the ordering energy, which is then added to derive the properties of the ordered phase(s). Each route can then be subdivided according to the level of approximation used and the desired restraints on the computing time involved. Before beginning a detailed examination of ordering treatments it is useful to make a number of critical definitions.

7.1.1 Definition of long-range order

A structure can be defined as possessing long-range order if at least two sets of positions can be distinguished by a different average occupation. These classes are usually called *sub-lattices*. The simplest example of an order/disorder transformation occurring in a b.c.c. lattice may be described in terms of two interpenetrating simple cubic arrays. If the occupation probability of each species is the same on both sub-lattices, then this is equivalent to a fully disordered b.c.c. structure, A2 (Fig. 7.1).

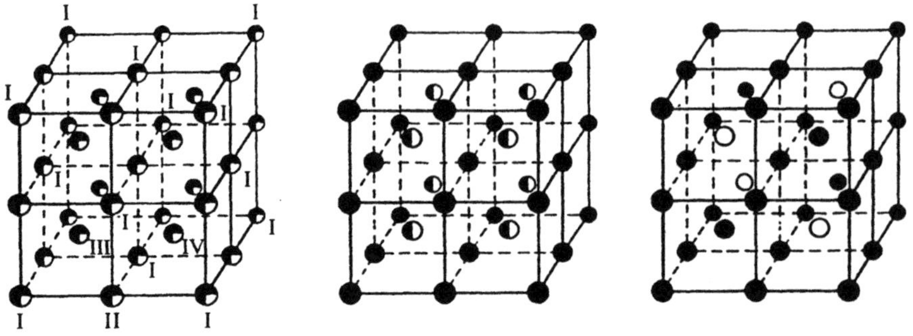

Figure 7.1. Relative occupation of sub-lattices for the A2, B2 and DO3 variants of a basic body-centred cubic arrangement from Inden (1982).

By contrast, if the two sub-lattices are totally occupied by different species the structure exhibits full long-range order, (*lro*). In the binary case, and for the equiatomic composition, this is then represented by a separate symbol, B2 (Fig. 7.1). Between 0 K and the ordering temperature (T^{ord}) there is a finite degree of order expressed in terms of the occupation probabilities (η) of the two sub-lattices α and β

$$\eta_i = x_i^{\beta} - x_i^{\alpha}. \tag{7.1}$$

In the case of a binary alloy containing only A and B atoms there is no need to use a subscript notation; x_i becomes the number of A atoms and $\eta_i = \eta_A = \eta$. Then for a b.c.c. lattice containing N atoms, the degree of *lro* can then be defined by letting the number of A atoms on α sites be $(1+\eta)N/4$. All possible states of *lro* are then represented as η ranges from 0 to ± 1. With $\eta = 0$ the lattice is disordered, while for a degree of order η, the concentrations on the two sub-lattices are $c = (1+\eta)/2$ and $(1-c) = (1-\eta)/2$ respectively. The entropy can then be derived as

$$S = -Nk_B\left[\left(\frac{1+\eta}{2}\right)\log_e\left(\frac{1+\eta}{2}\right) + \left(\frac{1-\eta}{2}\right)\log_e\left(\frac{1-\eta}{2}\right)\right] \tag{7.2}$$

and minimising the Gibbs energy with respect to (η) then provides the following relationship at high temperatures:

$$\eta = \frac{2k_BT}{z(V_{AA} + V_{BB} - 2V_{AB})}\log_e\left(\frac{1+\eta}{1-\eta}\right) \tag{7.3}$$

where z is the nearest-neighbour co-ordination number.

7.1.2 Definition of short-range order

By definition, *lro* falls to zero at T^{ord}, but it is still possible for there to be local short-range-ordered (*sro*) configurations above T^{ord}. This can be expressed mathematically in terms of the two-point correlation function of adjacent atoms, or in terms of the pair probability (ζ^*) of having an A atom at site m relative to a B atom at site n. It is common practice to use a short-range-order parameter (α) which is defined in such a way that it can be directly correlated with experimental X-ray information such as the Warren–Cowley coefficients (Ducastelle 1991)

$$\alpha = 1 - \left[\frac{\zeta^*}{x_A x_B}\right] \tag{7.4}$$

Ordering of unlike atoms corresponds to negative values of α where $\zeta^* > x_A x_B$, while $\zeta^* < x_A x_B$ and positive values of α correspond to the clustering of like atoms. When $\zeta^* = x_A x_B$, α is zero and the solution is perfectly random.

7.1.3 Magnetic ordering vs structural ordering

If magnetic behaviour is treated entirely in terms of localised spins, it is possible to use the same formalism as for configurational ordering and also treat mixtures of the two types of ordering. In the Ising model (Ducastelle 1991) a magnetic moment is assumed to have only two orientations in a given direction at each site and a simple nearest-neighbour coupling is then introduced so that parallel pairs have an energy –J, while anti-parallel pairs have an energy +J.

Even by using such simple assumptions, it is only possible to obtain exact solutions for a limited number of cases. In fact such assumptions are not necessarily well suited for real magnetic systems, because moments are vectors and not two-valued quantities. In many metallic alloys the magnetic moments exhibit non-integral values and are also extremely sensitive to atomic volume. In general it is therefore difficult to produce a rigorous treatment of magnetic order and extend it into a multi-component system. The two major approaches that are used in practice either assume a simplified spin state for each magnetic species, or alternatively describe magnetic order by an empirical treatment of the associated changes in specific heat. The magnitude of the Gibbs energy change associated with magnetic ordering can be surprisingly large and incorporation of the magnetic Gibbs energy is therefore very important, especially in ferrous alloys (see Chapter 8). This means that residual short-range magnetic ordering above the critical temperature can also provide a significant Gibbs energy contribution. The interplay of chemical and magnetic ordering is left to a later section.

7.1.4 Continuous vs discontinuous ordering

In alloys, structural ordering transformations can proceed with either a continuous or discontinuous change in entropy at the transformation temperature, while in unary systems, magnetic ordering is always associated with a continuous, second-order transformation. However, a rigid separation into first- and second-order transformations is not always practical, as the two types of transformation can merge in binary and higher-order systems, depending on the precise variation of the Gibbs energy curves with temperature and composition (Inden 1981a).

7.2. GENERAL PRINCIPLES OF ORDERING MODELS

7.2.1 Interaction parameters

All ordering models, whether simple or complex, require the input of suitable interaction parameters for each set of nearest neighbours, which create a hierarchy of terms for each crystal structure. This is frequently truncated to first- and second-nearest neighbours, but an increasing number of nearest neighbours need to be used as the symmetry of a given structure decreases. A more important issue is whether central or point interactions, no matter how far they extend from the point of reference, need to be expanded to include interactions between atoms in a group or cluster. This marks the essential difference between the treatment by Bragg and Williams (1934, 1935a, 1935b), also derived by Gorsky (1928) and generally known as the BWG treatment, and the more sophisticated *cluster variation method* (CVM) proposed by Kikuchi (1951).

Although the formal treatment of CVM includes pairs and triplets, the most popular cluster is the tetrahedron (Fig. 7.2a). It is possible to consider larger clusters of atoms (Fig. 7.2b) as well as taking into account as many nearest-neighbour interactions as desired. In the case of the b.c.c. lattice, the basic tetrahedral cluster conveniently takes into account second-nearest neighbours. The utilisation of more complex clusters is only curtailed by the inevitable increase in computing time. *Effective cluster interaction* (ECI) parameters can be derived using methods developed by Connolly and Williams (1983). This leads to a set of *composition-independent* parameters from which the energy of a set of related structures based on some basic symmetry such as the f.c.c. or b.c.c. lattice can be reproduced. More recently this has been extended by Lu *et al.* (1991) to make the ECIs more independent of the initial choice of basic symmetries. Most models yield the same enthalpy at 0 K and the same entropy at $T = \infty$, so the crucial point of interest is the change in properties such as G or C_p in the vicinity of T^{ord}. Differences between the models then arise from the permitted degrees of freedom when moving from a condition of full order to one of complete randomness (Fig. 7.3a).

7.2.2 Hierarchy of ordering models

Given a finite time for a given calculation, it is unrealistic to take into account all possible atomic configurations. It is, however, possible to consider the smaller

References are listed on pp. 222–226.

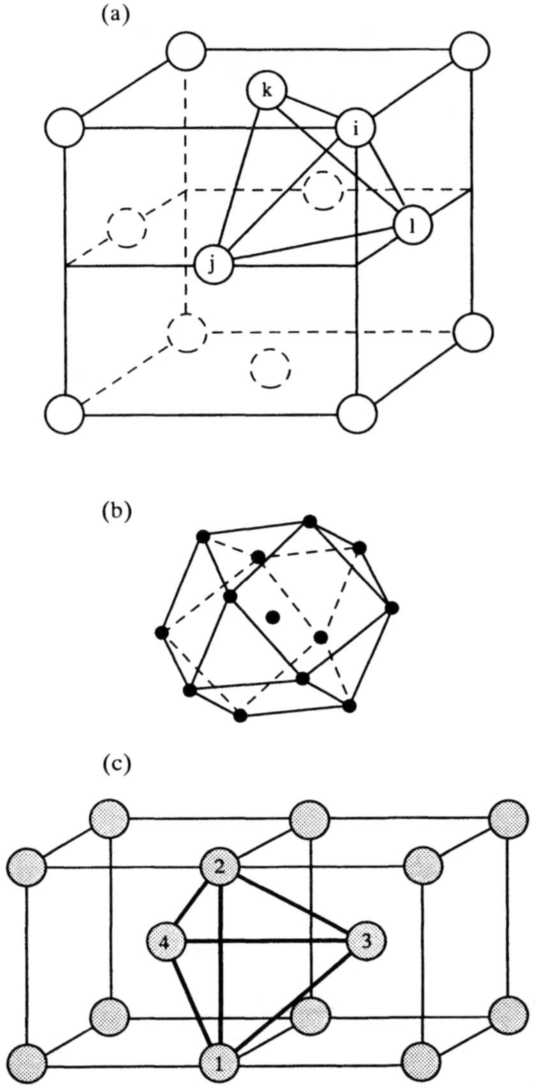

Figure 7.2. Typical clusters used in CVM calculations. (a) Tetrahedral (4-point) and (b) cube-octahedral (13-point) clusters for an f.c.c. lattice, (c) the irregular tetrahedral cluster used in b.c.c. lattices (adapted from Finel 1994 and Inden and Pitsch 1991).

number of degrees of freedom associated with a more limited region of space (e.g., a suitable cluster of atoms) and then use a mean field approximation for any interaction with other (similar) regions. The Monte Carlo method (MC) is assumed to

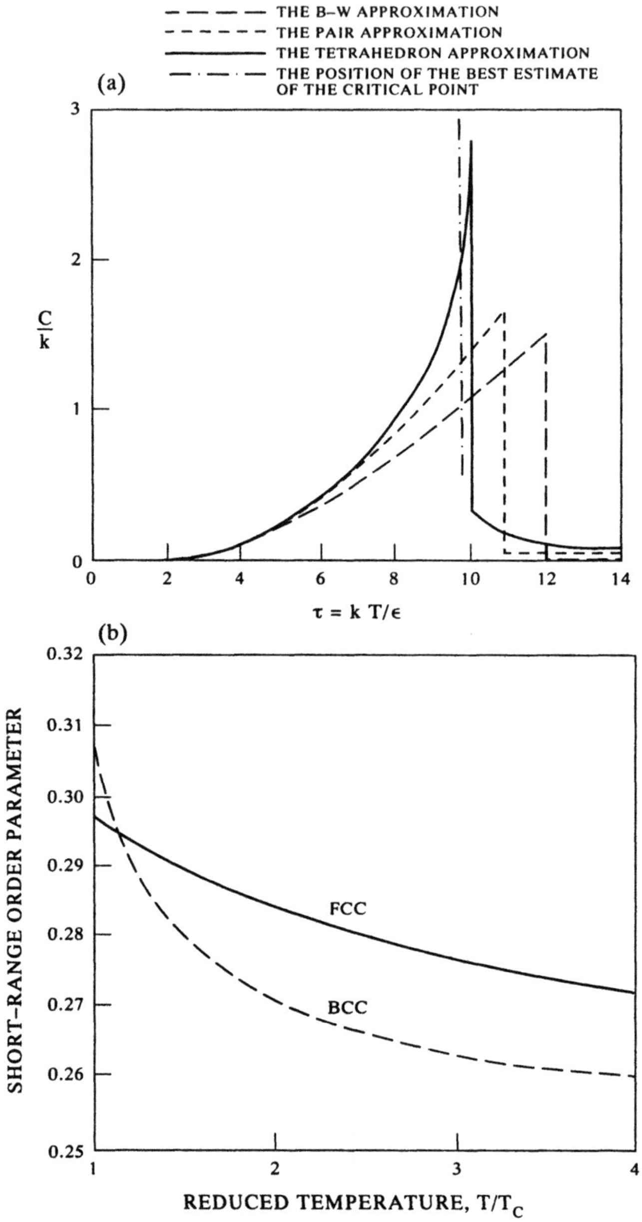

Figure 7.3. Variation of short-range order (*sro*) (a) in an f.c.c. lattice above the critical temperature, as evidenced by the specific heat predicted by various ordering models (Kikuchi 1977) and (b) *sro* characteristics for b.c.c. and f.c.c. structures at the equiatomic composition (Kikuchi and Saito 1974).

yield the closest approximation to reality, as it effectively takes into account the largest number of neighbours (Binder 1986). It is therefore considered the standard against which the results of other more approximate treatments can be measured. By contrast, the BWG treatment only deals with the statistically averaged occupation probabilities on different sub-lattices. As this specifically excludes any correlation between atoms in local clusters, this makes the BWG treatment incapable of handling any form of *sro*, which is particularly important above T^{ord}.

In between these extremes lie a large number of CVM treatments which use combinations of different cluster sizes. The early treatment of Bethe (1935) used a pair approximation (i.e., a two-atom cluster), but this cluster size is insufficient to deal with frustration effects or when next-nearest neighbours play a significant role (Inden and Pitsch 1991). A four-atom (tetrahedral) cluster is theoretically the minimum requirement for an f.c.c. lattice, but clusters of 13–14 atoms have been used by de Fontaine (1979, 1994) (Fig. 7.2b). However, since a comprehensive treatment for an $[n]$-member cluster should include the effect of all the component smaller $(n-1,\ n-2\ldots)$ units, there is a marked increase in computing time with cluster size. Several approximations have been made in order to circumvent this problem.

Oates and Wenzl (1996) have proposed a *cluster-site approximation* (CSA) that gives a two-term entropy by restricting the way tetrahedra are allowed to mix and neglecting the effects of the subsidiary pair interactions. Their approach is mathematically related to an earlier quasi-chemical model used by Guggenheim (1935). Various interpretations exist of both the quasi-chemical and CSA approaches, but they can all be considered part of the spectrum of approximate treatments that come under the umbrella of a total CVM hierarchy (Kikuchi and Saito 1974). The accuracy of the various approaches can vary depending on the level of approximation, and it is interesting to compare predictions for the critical temperature, as defined by the maximum rate of change of ordering with temperature, using the BWG and various CVM models. This is shown in Table 7.1 where the critical temperatures were calculated using a single set of model-independent interaction energies and a simple spin-up/spin-down f.c.c. Ising model.

All the models converge at very low and very high temperatures but, because they take into account varying degrees of short-range order, they give a different

Table 7.1. Ratio of the critical temperature calculated for an f.c.c. Ising lattice by different methods, assuming only constant nearest-neighbour interactions and using MC as the standard (after Kikuchi 1977)

BWG	1.23
Pair	1.13
CSA	1.10
Tetrahedron	1.024
Octahedron+tetrahedron	1.022
Oct + double tetrahedron	1.017
Monte Carlo ('Exact')	1.000

value of critical temperature. The exact variation in thermodynamic properties as a function of temperature will depend on the specific crystal structure, but the general effect is well illustrated in Fig. 7.3(a) and Table 7.1.

This makes it clear why the tetrahedron approximation is so popular, but it should be noted that the ratios in Table 7.1 were obtained by considering only nearest-neighbour interactions. Because a higher accuracy is usually associated with longer computational times, it is important to establish the degree of approximation that can be safely used in particular circumstances. This requires considerable judgement, as it will also depend on the bonding features of the particular alloy system and whether longer range forces are important. The main ordering models will now be considered in greater detail.

7.3. FEATURES OF VARIOUS ORDERING MODELS

7.3.1 The Monte Carlo method

The Monte Carlo (MC) method has been well documented by Binder (1986) and Inden and Pitsch (1991). Average thermodynamic properties are calculated for a series of finite three-dimensional arrays of increasing size and the results extrapolated to an infinitely large array. The effect of changing the occupancy of a particular site, on the change in energy of the whole system, is tested systematically and the occupancy changed or retained according to an algorithm which represents the effect of temperature.

However, it should be noted that this method is *as dependent on the availability of suitable interaction parameters as any of the other more approximate methods.* Moreover, even relatively small arrays require large amounts of computing time. For instance, it is possible to handle arrays of 10^4 to 10^5 sites but the typical numbers of MC steps required to reach equilibrium can be as high as 10^3 to 10^4 per site, even for the case of a binary system. This number can be reduced by restricting calculations to only those configurations with a relatively high probability of occurrence, but this essentially assumes prior knowledge of the end-point configuration. A faster approximation to the MC method has been used in Pd–V which samples only configurations on small clusters around each lattice point (Ceder *et al.* 1994).

Other problems associated with the MC method have been raised by Ducastelle (1991), who has pointed out that boundary conditions need to be specified correctly in order to properly reflect the crystallography of the system and that it is difficult to know whether the system has been caught in a metastable configuration. The MC method also tends to round off critical singularities and is, therefore, not ideal for second-order transformations. Despite all these factors, it is nonetheless accepted as the standard against which most other methods can be judged. The lengthy computing time has mitigated against its use as a sub-routine in more extensive phase-diagram calculations. An alternative *entropy Monte Carlo* functional (EMC) has recently been proposed, which may achieve an accuracy normally associated with MC but with the speed of CVM (Ferreira *et al.* 1997).

7.3.2 The BWG approximation

The advantage of the BWG approximation is that it leads to simple associated formulae for the heat of solution and other related properties. It can also lead to reasonable values for the ordering energy at 0 K and its inability to deal with *sro* at high temperatures is not important in that case.

7.3.2.1 BWG enthalpies. The enthalpy of formation of a random b.c.c._A2 solid solution, ΔH^{A2}, is given by Inden (1991) as

$$\Delta H^{\mathrm{A2}} = \frac{-N}{2} x_A x_B \left[z^{(1)} W^{(1)} + z^{(2)} W^{(2)} + \cdots \right] \tag{7.5}$$

where $W^{(k)}$ are interaction parameters associated with, 1st, 2nd ... k-th nearest neighbours and defined as $W^{(k)} = (V_{AA} + V_{BB} - 2V_{AB}^{(k)})$ and $z^{(k)}$ is the co-ordination number of the k-th shell. For the ordered B2 structure at 0 K, and when $x_B \leq 0.5$, the corresponding expression is

$$\Delta H_{T=0}^{B2} = \frac{-N}{2} x_B \Big[(z^{(1)} W^{(1)} + z^{(4)} W^{(4)} + \cdots) + (x_A - x_B)(z^{(2)} W^{(2)} + z^{(3)} W^{(3)} + z^{(5)} W^{(5)} + \cdots) \Big]. \tag{7.6}$$

General equations, including magnetic terms for various b.c.c. based structures have been tabulated (Inden 1975a); a typical application to a non-magnetic system is found in Inden (1975b). Analogous equations for some f.c.c. structures are given by Inden (1977b) and their implications elaborated further by Buth and Inden (1982). A selection is given below:

$$0 \leq x_B \leq 1 \quad \Delta H^{\mathrm{A2}} = -N[x_B(1 - x_B)\,(4W^{(1)} + 3W^{(2)})] \tag{7.6a}$$

$$0 \leq x_B \leq 0.5 \quad \Delta H_{T=0}^{\mathrm{B2}} = -N[x_B(4W^{(1)} + 3W^{(2)}) - 6x_B^2 W^{(2)}] \tag{7.6b}$$

$$0 \leq x_B \leq 0.25 \quad \Delta H_{T=0}^{\mathrm{DO_3}} = -N x_B(4W^{(1)} + 3W^{(2)}) \tag{7.6c}$$

$$0 \leq x_B \leq 0.5 \quad \Delta H_{T=0}^{\mathrm{B32}} = -N[x_B(4W^{(1)} + 3W^{(2)}) - 4x_B^2 W^{(1)}] \tag{7.6d}$$

$$0 \leq x_B \leq 1 \quad \Delta H^{\mathrm{A1}} = -N[x_B(1 - x_B)\,(6W^{(1)} + 3W^{(2)})] \tag{7.6e}$$

$$0 \leq x_B \leq 0.5 \quad \Delta H_{T=0}^{\mathrm{L1_0}} = -N[x_B(6W^{(1)} + 3W^{(2)}) - x_B^2(4W^{(1)} + 6W^{(2)})] \tag{7.6f}$$

$$0 \leq x_B \leq 0.25 \quad \Delta H_{T=0}^{\mathrm{L1_2}} = -N[x_B(6W^{(1)} + 3W^{(2)}) - 12x_B^2 W^{(2)}]. \tag{7.6g}$$

While the previously mentioned B2 structure, as shown in Fig. 7.1(b), can be interpreted as two interpenetrating simple cubic sub-lattices (L = I, II), it is necessary to consider four sublattices (L = I, II, III and IV) in order to handle variants of b.c.c. ordering such as DO_3 (Fig. 7.1(c)) and B32. These can be distinguished through different combinations of the occupation probabilities (x, y, z) as defined by Inden (1974):

$$x = \left(p_A^{\mathrm{I}} + p_A^{\mathrm{II}} - p_A^{\mathrm{III}} - p_A^{\mathrm{IV}}\right)/4 \tag{7.7}$$

$$y = \left(p_A^{\mathrm{III}} - p_A^{\mathrm{IV}}\right)/2 \tag{7.8}$$

$$z = \left(p_A^{\mathrm{I}} - p_A^{\mathrm{II}}\right)/2. \tag{7.9}$$

A combination of $x = 1$ with $y = z = 0$ corresponds to full B2 ordering, while DO_3 ordering requires that y or $z > 0$ in addition to $x > 0$. B2 ordering is replaced by B32 when $W^{(1)} < 1.5W^{(2)}$ (Inden 1974) and other lattices such as $F\bar{4}3m$ have their own unique combination of site occupation probabilities (Richards and Cahn 1971, Allen and Cahn 1972, Inden and Pitsch 1991). With the assumption of *composition-independent* values of W, *symmetrical* ground-state diagrams are formed which indicate the relative stability of different structures as a function of the ratio $W^{(1)}/W^{(2)}$ (Figs 7.4(a) and (b)). However, when the interaction energies are determined via first principles calculations (Zunger 1993), they are usually dependent on composition and the resultant ground-state diagrams are markedly asymmetric (Fig. 7.4(c)). This figure also demonstrates the effect of superimposing the ground states of different parent structures as distinct from just looking at the derivatives of one parent structure.

7.3.2.2 Approximate derivation of $T^{ord}_{(BWG)}$. Given values of $W^{(1)}$ and $W^{(2)}$, the basic BWG treatment also leads to explicit equations for the ordering temperature, T^{ord}, but the omission of *sro* inevitably leads to calculated values that are appreciably higher than shown by experiment. If the simplicity of the BWG method is to be retained, an empirical correction factor (χ) has to be included, where $\chi = T^{\mathrm{ord}}/T^{\mathrm{ord}}_{\mathrm{BWG}}$. Typical equations for various structural transitions are given below:

$$kT^{\mathrm{ord}}_{\mathrm{B2 \to A2}} = \chi x_B(1 - x_B)\left(8W^{(1)} - 6W^{(2)}\right) \tag{7.10}$$

$$kT^{\mathrm{ord}}_{\mathrm{DO3 \to A2}} = \chi x_B(1 - x_B)\, 6W^{(2)} \tag{7.11}$$

$$kT^{\mathrm{ord}}_{\mathrm{L12 \to A1}} = \chi x_B(1 - x_B)\left(4W^{(1)} - 6W^{(2)}\right). \tag{7.12}$$

Working values of χ can be obtained by plotting $T^{\mathrm{ord}}_{\mathrm{BWG}}$ against values of T^{ord} obtained by more accurate approximations, e.g., CVM and MC. For b.c.c. alloys such values have been plotted against $W^{(1)}/W^{(2)}$ so that an appropriate choice of (χ) can be made (Inden 1975a) (Fig. 7.5). Although inferior to more sophisticated

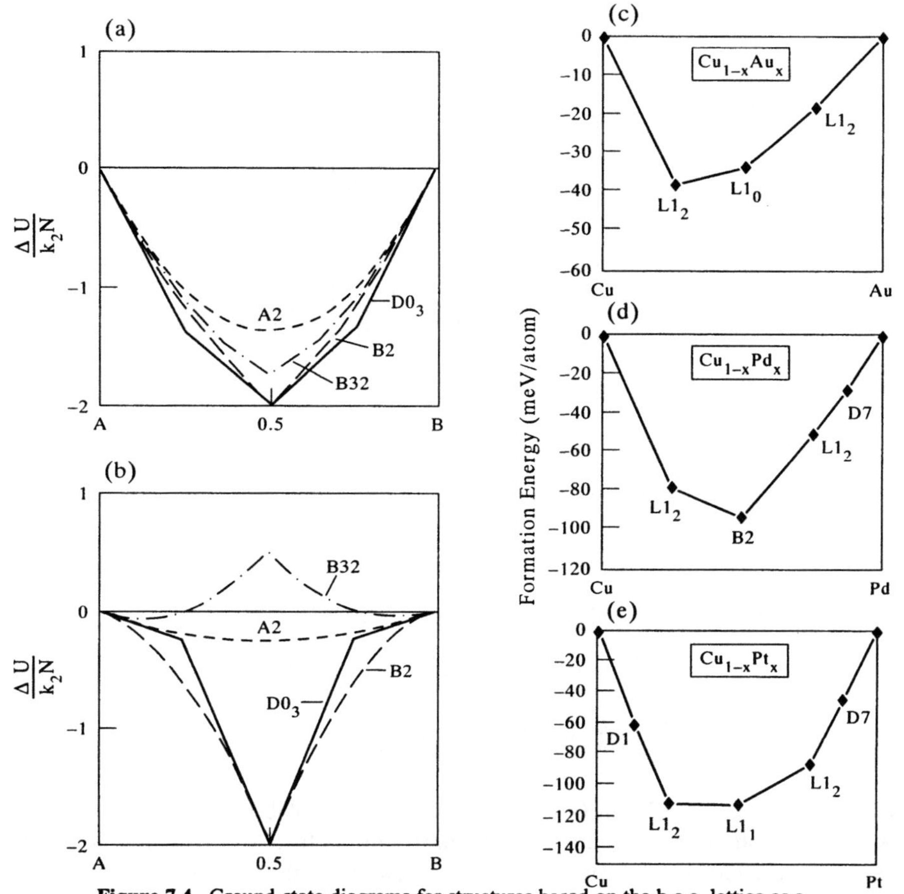

Figure 7.4. Ground-state diagrams for structures based on the b.c.c. lattice as a function of composition corresponding to (a) $W^{(1)} = 2W^{(2)} > 0$, (b) $W^{(1)} = -W^{(2)} \geq 0$ (Inden and Pitsch 1991) and (c)–(e) examples from first-principles calculations for selected systems (Zunger 1993).

treatments, this modified BWG treatment gives surprisingly good answers for many b.c.c.-ordering systems (see later sections) but can be misleading for close-packed systems. There is also an inherent inconsistency, in that experimental heats of solution for the disordered phase actually include a finite degree of *sro* but it has to be treated as being completely random.

7.3.2.3 Magnetic interactions in the BWG treatment. The BWG treatment has been applied extensively to b.c.c. Fe-base alloys, but since these are magnetic it is necessary to include some extra terms. The configurational Gibbs energy per mole for a b.c.c. solid solution is then given by the following expression:

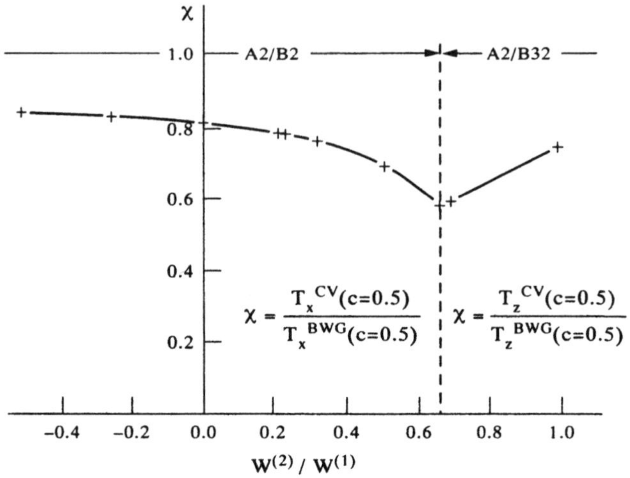

Figure 7.5. Variation of the (χ) factor with the $W^{(1)}/W^{(2)}$ ratio for b.c.c. alloys (Inden 1975a).

$$\begin{aligned} G^{\text{b.c.c.}} &= G^{\text{o}} - Nx_A x_B(4W^{(1)} + 4\sigma^2 M^{(1)} + 3W^{(2)}) \\ &\quad - \frac{N}{2}\left[(8W^{(1)} + 8\sigma^2 M^{(1)} - 6W^{(2)})\,x^2 + 3W^{(2)}(y^2 + z^2)\right] \\ &\quad + \frac{NkT}{4}\sum_L (p_A^L \log_e p_A^L + p_B^L \log_e p_B^L) - T[\alpha x_A + \beta x_B]\, S_{\text{mag}}. \end{aligned} \tag{7.13}$$

The first term in Eq. (7.13), G^{o}, allows for any phase stability correction if one or both of the components do not occur naturally in the b.c.c. form. The second term represents the enthalpy of mixing, the third term gives the energy contribution due to ordering, the fourth term gives the atomic configuration entropy, while the last term is the contribution arising from the magnetic entropy and assumes a linear variation of the Curie temperature, T_c, with composition. Setting the coefficients α and β to either 1 or 0 then allows for various combinations of magnetic and non-magnetic elements to be treated; Inden (1975a) should be consulted for a full derivation. Obviously the equation is much simplified in the case of non-magnetic alloys when $M^{(1)}$ and S_{mag} are equal to zero. $M^{(1)}$ is the magnetic interaction energy between nearest neighbours analogous to $W^{(1)}$ and σ is the magnetic ordering parameter. Applications have been made to both binary (Inden 1977a) and ternary systems (Inden 1979).

7.3.2.4 BWG and anti-phase boundary energies. The bond-breaking method first described by Flinn (1960) allows an estimate to be made for the structural component of *anti-phase boundary* (APB) energies. As good agreement can be obtained between calculated and experimental APB energies in the cases of both the non-magnetic Cu–Zn, B2 phase (Inden 1975b) and magnetic Fe–Co, B2 alloys (Inden 1977a), it would appear that the total fault energy in these cases is dominated by the structural (chemical plus magnetic) component. For the A2/B2 case, the APB energy is given by:

$$\gamma_1 = \gamma_{a/4<111>\{110\}}$$
$$= \frac{16}{a^2}\sqrt{h^2+k^2+l^2}\left[x^2(W^{(1)}+\sigma^2 M^{(1)}) - x^2 W^{(2)} + \tfrac{1}{4}(y^2+z^2)W^{(2)}\right] \quad (7.14)$$

where a = the lattice spacing of the sublattice and h, k, l are the Miller indices. For b.c.c.{110} planes the (h, k, l) function reduces to $\sqrt{2}$. The equivalent formula for f.c.c., $L1_2$ structures is given by

$$\text{APB}[111](L1_2) = a^{-2}(1/3)^{-1/2}[W^{(1)} - 3W^{(2)} + 4W^{(3)} - \cdots] \quad (7.15)$$

$$\text{APB}[100](L1_2) = a^{-2}(1)\,[-W^{(2)} + 4W^{(3)} - \cdots]. \quad (7.16)$$

Inden *et al.* (1986) and Khachaturyan and Morris (1987) can be consulted for equations that include longer-range terms. At least third-nearest-neighbour interaction energies need to be taken into account for $L1_2$ structures and, in general, there is insufficient experimental data to derive all the necessary parameters. In order to be able to use a BWG formalism, Miodownik and Saunders (1995) therefore assumed a constant $W^{(3)}/W^{(2)}$ ratio, based on FP calculations by Sluiter *et al.* (1992), and were able to derive APB energies for multicomponent nickel-base $L1_2$ (γ') phases that matched experimental values. Reasonable estimates for metastable ordering temperatures were also obtained by assuming a specific ratio for *sro* and *lro* at T^{ord} despite the known problems of applying the BWG treatment to close-packed phases. APB energies have also been calculated using CVM with a combination of the tetrahedron approximation and nearest-neighbour interaction energies for B2 (Beauchamp *et al.* 1992) and $L1_2$ ordered compounds (Inden *et al.* 1986).

7.3.3 The cluster variation method (CVM)

CVM is a more powerful formalism which can include the *mutual interaction of all the atoms in sets of clusters*, so as to properly reflect a greater variety of atomic interactions (de Fontaine 1979, 1994). The smallest cluster that should be used to describe a three-dimensional lattice is the tetrahedron (Kikuchi 1951), but the complete calculation of entropy using CVM requires a consideration of all subsidiary clusters (van Baal 1973). This means pairs and planar triangles have to be included for the tetrahedron approximation, and all these in turn have to be included when larger clusters such as octahedra are considered. Various mathematical techniques are available to maintain self-consistency between contiguous clusters and to avoid double counting (Inden and Pitsch 1991, de Fontaine 1994).

CVM gained widespread acceptance when it was shown that the tetrahedron approximation could reproduce the observed f.c.c. Cu–Au phase boundaries, especially if asymmetric four-body interactions are included (Figs 7.6(a–b)). This may be contrasted with the results previously obtained by the BWG method which failed to produce even a qualitative resemblance to the observed phase diagram (Fig. 7.7). The effect of various approximations including the Monte Carlo method is shown in Fig. 7.8(a). As might be expected, the CSA approximation produces an intermediate result (Fig. 7.8(b)). The CVM approach also clearly gives much more information in cases where *sro* is important. For instance, Kikuchi and Saito (1974) showed that the decay of *sro* with temperature was significantly different in f.c.c. and b.c.c. structures (Fig. 7.3(b)). Nonetheless, residual differences in results obtained by the CVM and MC technique led to the CVM procedure being extended to ever larger clusters. This forced the development of alternative cluster expansion routes; but it is still an open question whether faster convergence can offset the large increase in the number of equations which arise as the size and complexity of clusters is increased. However, it is a bonus that the cluster expansion route can also be used for any other property that is a function of configurational variables (Sanchez 1992).

Although the parameters that enter into the CVM procedure follow some simple mathematical principles, the procedures are undoubtedly much more complex than in the BWG approximation. Further complications arise when additional variables such as the volume change as a result of atomic interchange are considered. Global volume relaxation can be incorporated by minimising G with respect to changes in inter-atomic distance (Mohri *et al.* 1989, 1992), but this assumes that local volumes are independent of configuration, which is physically implausible when the constituent atoms have large size differences. However, taking local relaxation into account leads to even more complex algorithms (Lu *et al.* 1991, Laks *et al.* 1992, Zunger 1994). The Ni–Au system has recently been re-examined as a good testing ground for various ordering theories (Wolverton and Zunger 1997) as high-temperature measurements indicate the presence of *sro*, but the system exhibits phase separation at low temperatures.

Developments in CVM have been reviewed at regular intervals (Kikuchi 1977, Khachaturyan 1978, de Fontaine and Kikuchi 1978, de Fontaine 1979, Inden and Pitsch 1991, de Fontaine 1994, Zunger 1994). The succeeding sections should be taken as merely defining some of the relevant key concepts.

7.3.3.1 Site-occupation parameters. A complete description of a system requires detailed knowledge about the occupancy of all available sites by all the participating species. However, with K constituents and N sites, this leads to K^N different configurations, which is computationally unrealistic. A useful approximation is to use *average site-occupation parameters*, ξ, for equivalent crystallographic sites.

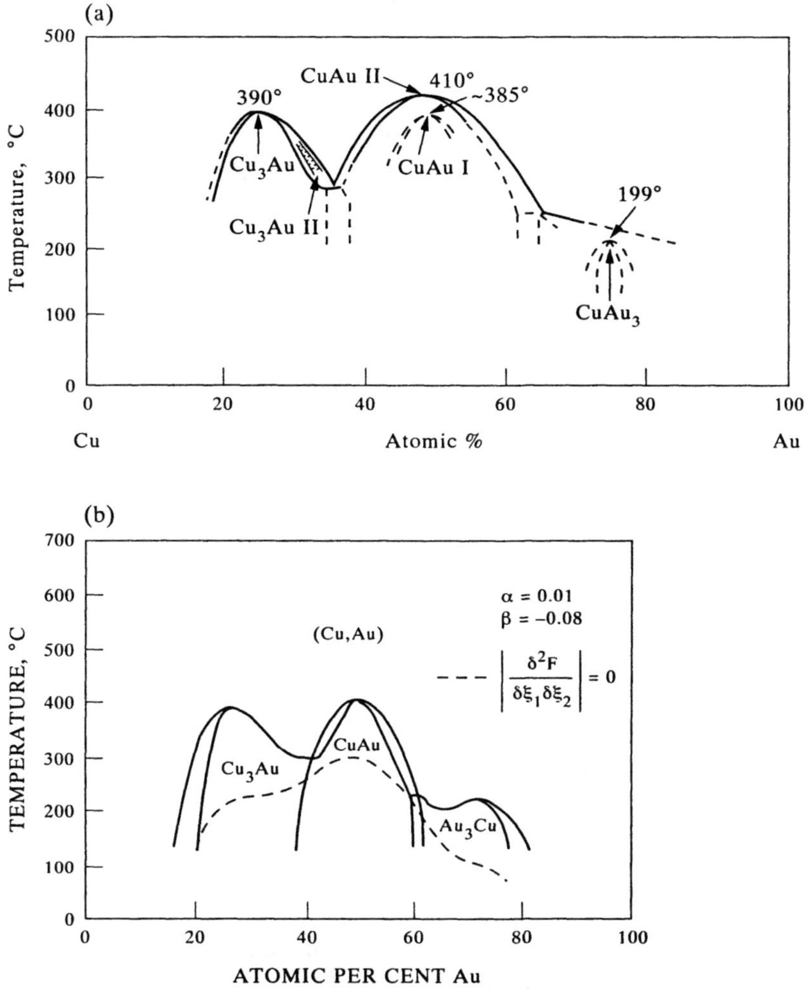

Figure 7.6. Comparison of (a) experimental phase diagram for the Cu–Au system (Hansen 1958) with (b) predictions for the Cu–Au system calculated using the tetrahedron approximation but including asymmetric four-body interactions (de Fontaine and Kikuchi 1978).

Working with clusters defined by r lattice points reduces the number of configurations by a further factor of r. *A multi-site correlation parameter* (ξ^*) can then be constructed from the sum of all such values and used to define the overall configuration.

The required thermodynamic functions then depend on the sum of all the cluster probabilities. This is simplified by expressing the probability of each higher-order

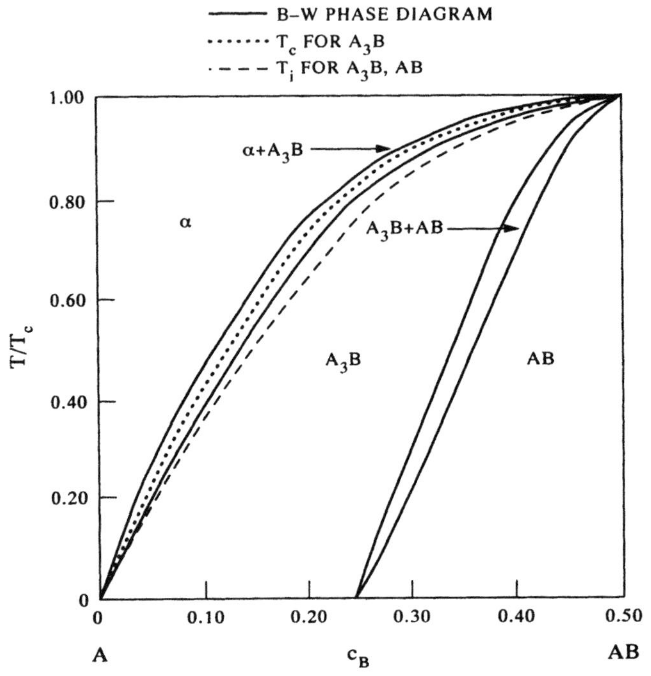

Figure 7.7. Earlier prediction for Cu–Au using the BWG formalism (Shockley 1938).

cluster in terms of its lower components, e.g., pair probabilities (y_{ij}) as products of point variables. This approach can be extended to include higher-order systems, for example in ternary systems one then has combinations of y_{11}, y_{21} and y_{31}. It is important to note that it does not matter whether the added species are vacancies, chemically different atoms, or atoms of the same species with different spin configurations, but all the pair variables need to be normalised to unity at some stage.

To complete the picture it is useful to define distribution variables, x_i, which are atomic fractions normalised to a mole of atoms as related to site-occupation parameters. The x and y variables are related geometrically through the equation:

$$x_i = \sum_j y_{ij}. \tag{7.17}$$

The formalism can then be extended to as many co-ordination shells (n-th nearest neighbours) as required.

7.3.3.2 Effective cluster interactions. The second important parameter in all cluster models is a series of interaction coefficients that define the bond strengths. An

References are listed on pp. 222–226.

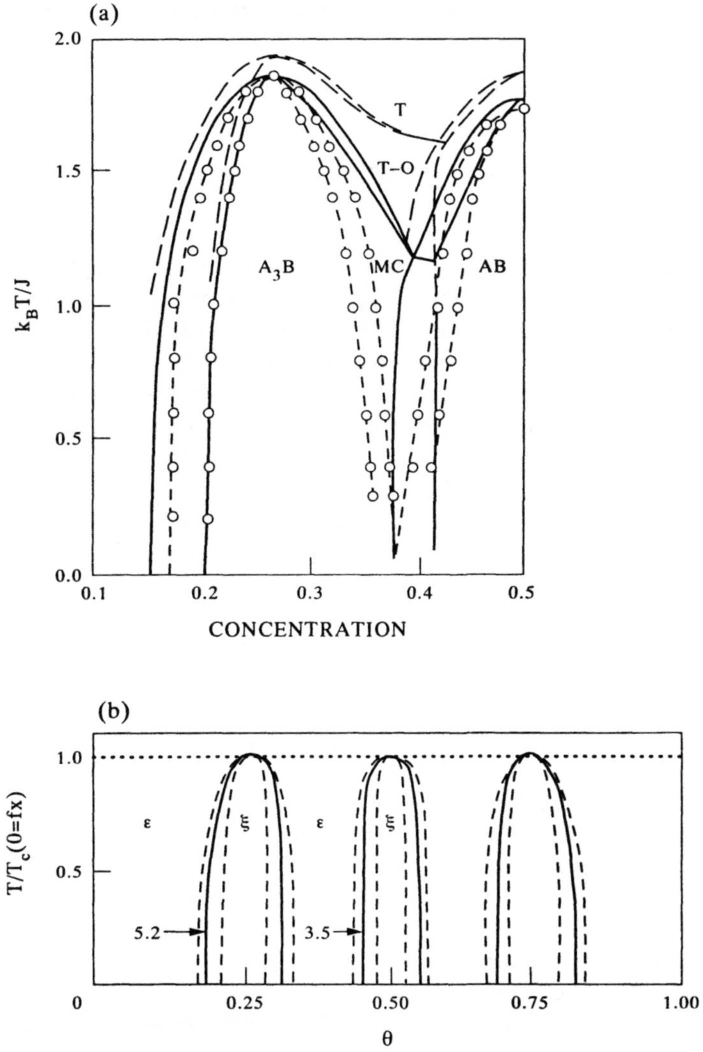

Figure 7.8. (a) enlarged region for $0 \leq x \leq 0.5$ calculated according to CVM in the tetrahedron approximation (*T*), the tetrahedron–octahedron approximation (T–O) and using a Monte Carlo method (MC) (Sanchez *et al.* 1982). (b) results obtained from using the quasi-chemical or CSA approximation (Kikuchi 1977).

important assumption is that *all* the members of a particular 'family' of related ordered structures can be described by combining *a single set of effective cluster interaction* (ECI) *energies* $E_\gamma(r)$. The ECI values relate in turn to the total energy, $E^\alpha_{\text{tot}}(r)$, of a structure, α, through an appropriate set of site correlation functions, ξ^α_γ.

Equations (7.18) and (7.19) give the relationships that hold for a binary system:

$$E^{\alpha}_{\text{tot}}(r) = \sum_{\gamma}^{\gamma_{\max}} N_{\gamma}\, E_{\gamma}(r)\, \zeta^{\alpha}_{\gamma} \tag{7.18}$$

$$\zeta^{\alpha}_{\gamma} = \frac{1}{N_{\gamma}} \sum_{|n_j|} \sigma_{n_1}\, \sigma_{n_2} \ldots \sigma_{n_{\gamma}} \tag{7.19}$$

$\sigma_n = (1 - 2p_n)$ takes the values +1 or −1 depending on the occupancy of the site, n, N_{γ} is the total number of γ-type clusters and the sum runs over all such clusters that can be formed by combining sites for the entire crystal.

Providing the bonding does not change dramatically across the system, the derived ECIs should be *composition independent.* Connolly and Williams (1983) were interested in calculating the ordering energies of a variety of structures from a set of constant ECI values, but more importantly their method can be reversed (Carlsson 1989) to extract ECI values from ordering energies that have been calculated by electron energy calculations (Pasturel *et al.* 1992). The number of structures for which the energy needs to be calculated has, clearly, to be at least equal to the number of unknown values of $E_{\gamma}(r)$. This is a relatively small number if only first- and second-neighbour interactions need to be considered, but can rise substantially with the inclusion of longer-range forces.

The relevant number of interaction coefficients can sometimes be inferred directly from the electron energy calculations. For example, if structures such as DO_{22} are found to be more stable than $L1_2$, reference to ground-state diagrams shows that at least second-nearest-neighbours have to be taken into account. Likewise if the ECIs cannot be extracted without showing a strong composition dependence, this suggests that longer-range forces need to be taken into account. In the case of Pd–V it was found that interactions up to the 4th-nearest neighbours had to be included to obtain good convergence (Ceder *et al.* 1994) and structures such as L60 and B11 in Nb–Al indicate that it might be necessary to include the 5th-nearest neighbours in this system (Colinet *et al.* 1997).

There are, therefore, some reservations about the use of the inverse Connolly method if ECIs are determined from an insufficient number of base structures (Ducastelle 1991). On the other hand, arbitrarily increasing the number of structures to be calculated is an open-ended recipe for unrealistic increases in computing time. This problem was solved for a series of semi-conductor compounds by using lattice theory to map the energies of 10–20 selected A_mB_n compounds (Wei *et al.* 1990), and then extended to AB intermetallic compounds by Lu *et al.* (1991). The procedure has been reviewed by Zunger (1994). It efficiently screens a large number of symmetries (~65,000 configurations) with a reasonable number of calculations and then produces a set of cluster interactions that minimises the predictive error in the energy of all the structures used in the analysis.

References are listed on pp. 222–226.

7.3.3.3 Effective pair interaction parameters. In the CVM formalism, effective pair interaction coefficients, defined by the symbol $V_{\rm pair}^{(n)\rm eff}$, are given in Eq. (7.20) which clearly resembles the equivalent BWG point interaction coefficients, W, of Eq. (7.21)

$$V_{\rm pair}^{(n)\rm eff} = \frac{1}{4}\left(V_{AA}^{(n)} + V_{BB}^{(n)} - 2V_{AB}^{(n)}\right) \quad (7.20)$$

$$W_{\rm point}^{(n)} = \left(W_{AA}^{(n)} + W_{BB}^{(n)} - 2W_{AB}^{(n)}\right) \quad (7.21)$$

where $V_{IJ}^{(n)}$ are the pair potentials for atoms I and J at the n-th nearest-neighbour distance which can be related to the previously derived parameters through a re-normalisation technique developed by Carlsson (1987). Examples from Pasturel (1992) are given below:

$$V_2^{(1)\rm eff}(\text{f.c.c.}) = E_2^1 + 4E_3^1\zeta_1 + 2E_4^1(\zeta_1)^2 + \cdots \quad (7.22)$$

$$V_2^{(1)\rm eff}(\text{b.c.c.}) = E_2^1 + 6E_3^1\zeta_1 + 6E_4^1(\zeta_1)^2 + \cdots \quad (7.23)$$

$$V_2^{(2)\rm eff}(\text{b.c.c.}) = E_2^2 + 4E_3^1\zeta_1 + 4E_4^1(\zeta_1)^2 + \cdots \quad (7.24)$$

Here E_2^1 represents the first-neighbour pair energy and E_2^2 the second-neighbour pair energy. E_3^1 represents the energy in a first-neighbour triangle, which for an f.c.c. lattice contains only first-neighbour pairs, but both first- and second-neighbour pairs in the case of a b.c.c. lattice.

7.3.3.4 Use of the general perturbation method. All of the above treatments have essentially started with the assumption that one should first calculate the energy of an ordered phase and then describe the effect of temperature on the disordering process. It is, however, equally valid to use the high-temperature random state as a reference point and extract the ordering energy by difference. This can be implemented through the *general perturbation method* (GPM), first suggested by Ducastelle and Gautier (1976). In this case the ordering energy is written as an expansion of concentration-dependent n-th order effective clusters (Gonis *et al.* 1987, Turchi *et al.* 1988, Ducastelle 1989). Unfortunately, a large number of band-structure calculations have to be made in order to establish the heat of formation of the disordered solution over the whole composition range. The calculations also have to be of high accuracy, so truncating the series expansion is not an option. An extensive computing effort is, therefore, required even before interaction coefficients are extracted from the ordering enthalpy and introduced into a CVM procedure.

The interaction parameters derived in this way *are usually concentration-dependent*, making for lengthy computations. It is therefore unlikely that this route will ever be adopted for multi-component calculations unless more efficient algorithms can be developed. Apart from this aspect, the method is attractive because many other associated properties such as APB energies and the variation of surface energy with orientation can also be derived (Turchi and Sluiter 1992, 1993) and the method clearly predicts important trends in the properties of both ordered and disordered transition metal compounds (Sluiter *et al.* 1987, Turchi *et al.* 1994). Some applications of the method to the calculation of binary phase diagrams are discussed in a later section.

7.3.3.5 General form of the CVM enthalpy. The enthalpy associated with clusters is given by the sum of expressions that cover all the possible combinations of points, pairs, triplets, tetrahedra and whatever other clusters are being considered:

$$\Delta E = -\left[\sum_{n,m..}\sum_{i,j} p_i^n\, p_j^m\, V_{i,j}^{n,m} + \sum_{n,m,q.}\sum_{i,j,k} p_i^n\, p_j^m\, p_k^q\, V_{i,j,k}^{n,m,q} + \cdots\right] \tag{7.25}$$

where p_i^n represents the site (n) occupancy and equals 1, site (n) is occupied by (i), otherwise it equals zero. For a given crystal structure there are constraints on the occupation probabilities of various types of clusters, e.g., tetrahedra, triangles, pairs and points. Reference should be made to de Fontaine (1994) and Cacciamani *et al.* (1997) for explicit values of the enthalpy derived with various different assumptions.

7.3.3.6 Relation of BWG, pair and CVM enthalpies. If the more general expression of Eq. (7.25) is restricted to pair interactions, the result is:

$$E(y_{ij}) = \frac{z}{2} N \sum_{i=1}^{2}\sum_{j=1}^{2} \varepsilon_{ij}\, y_{ij} \tag{7.26}$$

which is equivalent to the approximation first suggested by Bethe (1935). As pointed out by Kikuchi (1977), the BWG treatment involves a further reduction of the four (2^2) variables y_{11}, y_{12}, y_{21} and y_{22} which are implicit in Eq. (7.26), to just the two variables x_1 and x_2 and then

$$E(x_1, x_2) = \frac{z}{2} N \sum_{i=1}^{2}\sum_{j=1}^{2} \varepsilon_{ij}\, x_i\, x_j. \tag{7.27}$$

This equation effectively imposes the condition that $p_{ij} = p_{ji} = x_A x_B$, *with the important result that sro* is totally excluded, because, when the pair probability ξ is equal to $x_A x_B$, the short-range-order parameter α becomes zero (see Section 7.1.2).

7.3.4 CVM entropy

It must be clear that although the CVM approach is much more accurate than the BWG treatment, the corresponding entropy expressions are now complicated, and this escalates markedly as more distant neighbours are also included. In this section, examples are restricted to the pair and tetrahedron approximations to give a feeling for the basic features and allow a comparison with the cluster site approximation discussed in Section 7.3.5.

However, in all cases care has to be taken to avoid double-counting various permutations. A good correction (Kikuchi 1977) is to divide by a factor which numerically describes the nearest-neighbour (n) connectivity (c) of the lattice, which for pairs leads to

$$\rho_2 = \frac{\prod\limits_{<ij>} \rho_{ij}}{\prod\limits_{<i>} \rho_i^{c-1}}. \tag{7.28}$$

If the site correlations are extended to triangular clusters, then

$$\rho_3 = \frac{\prod\limits_{<ijk>} \rho_{ijk} \prod\limits_{<i>} \rho_i}{\prod\limits_{<ij>} \rho_{ij}}. \tag{7.29}$$

The rationale for this is that each first-neighbour pair has been counted twice because it belongs to two triangles. Each point belongs to six triangles and six pairs, so the correction factors here cancel out and a further term representing the point probabilities has to be reintroduced (Finel 1994).

7.3.4.1 Criteria for judging CVM approximations. Finel (1994) has addressed several important issues such as the *criteria that should be used to judge the quality of a CVM approximation* and what sequence of clusters lead to a rapid convergence to an exact solution. This includes the interesting point that use of a given cluster size yields less precise answers for the disordered phase in comparison to its ordered counterpart. Fortunately, the higher symmetry of a disordered lattice also reduces the number of variational parameters for a given cluster size. It is therefore sensible to use a pair approximation for $T < T^{\mathrm{ord}}$ combined with the tetrahedron approximation for $T > T^{\mathrm{ord}}$, or the tetrahedron approximation for the ordered phase and the tetrahedron/octahedron approximation for the disordered phase. Other combinations have also been used, such as the MC method to obtain enthalpies and cluster probabilities and then using CVM in the tetrahedron approximation to calculate the entropy (Bichara and Inden 1991). Likewise a mixture of MC plus a 13-point cluster has also been used (Ceder *et al.* 1994). A comparison between BWG, CVM and MC in the more complex case of the A15 structure can be found in Turchi and Finel (1992) (Fig. 7.9).

7.3.4.2 Entropy on the pair interaction model. The entropy of the system in a pair

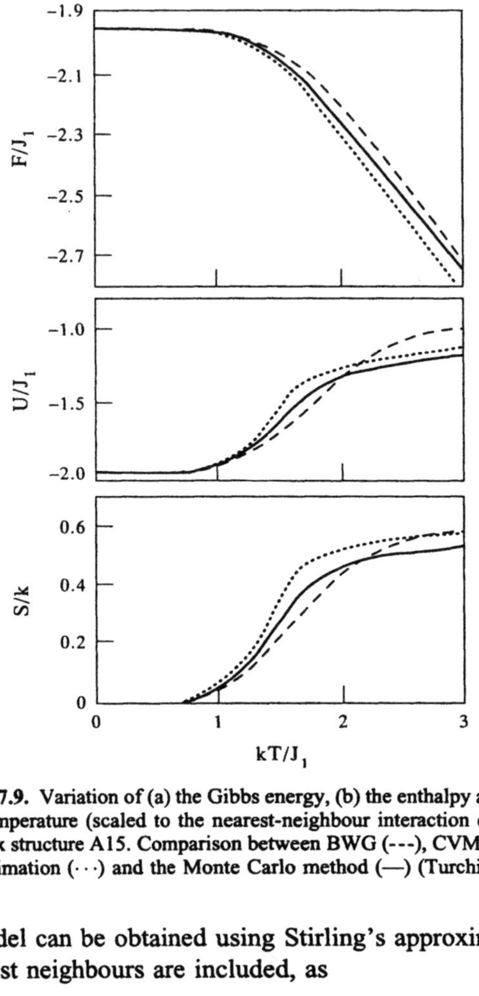

Figure 7.9. Variation of (a) the Gibbs energy, (b) the enthalpy and (c) the entropy with temperature (scaled to the nearest-neighbour interaction energy J_1) for the complex structure A15. Comparison between BWG (---), CVM in the tetrahedron approximation (···) and the Monte Carlo method (—) (Turchi and Finel 1992).

interaction model can be obtained using Stirling's approximation, if it is assumed that only nearest neighbours are included, as

$$S_N = kN\left[(z-1)\sum_i x_i \ln x_i - \frac{z}{2}\sum_i\sum_j y_{ij} \ln y_{ij}\right]. \qquad (7.30)$$

7.3.4.3 Entropy on the tetrahedron approximation. If the treatment is extended to the case of a tetrahedral cluster, this immediately raises the number of required equations to (2^4) variables. As already indicated, the pair approach is not sufficient to adequately describe the f.c.c. lattice, where an additional variable z_{ijkl} is needed where i, j, k, l can take values 1, 2, 3... corresponding to various atoms or vacancies. The energy expression corresponding to the tetrahedra is then given by

References are listed on pp. 222–226.

$$E_N = 2N \sum_{i,j,k,l} \varepsilon_{ijkl}\, z_{ijkl} \tag{7.31}$$

where $2N$ is the total number of tetrahedra formed from N lattice points and ε_{ijkl} contains the four-body interactions. Even this may not produce a fully accurate answer, since an analysis of several f.c.c. systems shows that their behaviour necessitates taking into account next-nearest neighbours (Inden 1977a). In that case, an octahedral plus tetrahedral cluster is really required (de Fontaine and Kikuchi 1978). Fortunately the CVM treatment of the b.c.c. structure is much simpler than for f.c.c. as it can be handled with an irregular tetrahedron that already includes first and second neighbours (van Baal 1973). The many CVM reviews already cited should be consulted for the entropy expressions relating to more elaborate clusters.

7.3.4.4 Implementation of CVM. The implementation of any improved ordering algorithms must take into account both a minimisation of computing time and reliable convergence to the equilibrium result. The Newton–Raphson method of steepest descent is not convenient when there is a large number of variables, as it can oscillate and produce negative values for probability variables. By contrast the natural iteration method (Kikuchi 1974, de Fontaine and Kikuchi 1978) was specifically designed to handle the complexities of CVM, and gives only positive probability variables which converge smoothly to the state of lowest Gibbs energy. Although 'safer', the natural iteration method is, unfortunately, much slower than the method of steepest descent.

7.3.5 The cluster site approximation (CSA)

The emphasis in the previous sections has been on the *accuracy* with which the Gibbs energy, particularly the entropy component above T^{ord}, can be calculated. However, as the number of components, C, and the number of atoms in the chosen cluster, n, increases, the number of simultaneous equations that have to be solved is of the order of C^n. This number is not materially reduced by redefining the equations in terms of multi-site correlation functions (Kikuchi and Sato 1974). The position may be eased as extra computing power becomes available, but a choice will inevitably have to be made between supporting a more complex model or extending a simpler model to a greater number of components.

The cluster site approximation, CSA, has the great advantage that the required number of equations to handle the necessary variables is given by the product Cn (Oates and Wentl 1996) as compared to the exponential value C^n aassociated with a full CVM treatment (Inden and Pitsch 1991). The CSA is a variant of the quasi-chemical model proposed by Li as early as 1949, where the number of clusters that are considered to contribute to the entropy are reduced by excluding all clusters that share edges or bonds. Kikuchi (1977) has deduced the consequential changes in entropy for an f.c.c. structure (Eqs 7.32 and 7.33), which places CSA intermediate between BWG and CVM in the overall hierarchy of CVM techniques (see Table 7.1, p. 187).

$$\text{(CVM)}\qquad S\{w_{i,j,k,l}\} = R\ln\left(\frac{\left[\prod_{i,j}(y_{ij}N)!\right]^6 N!}{\left[\prod_{i,j,k,l}(w_{ijkl}N)!\right]^2\left[\prod_{i}(x_iN)!\right]^5}\right) \tag{7.32}$$

$$\text{(CSA)}\qquad S\{w_{i,j,k,l}\} = R\ln\left(\frac{\left[\prod_{i,j}(x_iN)!\right]^3}{\left[\prod_{i,j,k,l}(w_{ijkl}N)!\right] N!^2}\right). \tag{7.33}$$

An alternative way of expressing these differences is:

$$\begin{aligned} S^{\text{f.c.c.}}[\text{CVM}] &= 2S_t - 6S_{p1} + 5S_s \\ S^{\text{b.c.c.}}[\text{CVM}] &= 6S_t - 12S_{tr} + 3S_{p2} + 4S_{p1} - S_s \\ S^{\text{f.c.c.}}[\text{CSA}] &= S^{\text{b.c.c.}}[\text{CSA}] = S_t - 3S_s \end{aligned} \tag{7.34}$$

where the subscripts t, tr, p and s refer to tetrahedral, triangular, pair and single-site contributions respectively. As an example, the expanded form of the entropy contribution for S_t is given by:

$$S_t = -Nk\sum_{ijkl} w_{ijkl}\ln w_{ijkl}. \tag{7.35}$$

The single-site entropy term in the CSA version arises from the normalisation of the partition function (Oates and Wentl 1996), while the relative complexity of the b.c.c., CVM case occurs because the tetrahedron is asymmetric in this case. There is then a need to take into account both nearest- and next-nearest-neighbour interactions (Inden and Pitsch 1991).

One result of the simplification inherent in the CSA treatment is that the same expression is obtained for the entropy of both f.c.c. and b.c.c. lattices which clearly distinguishes it from the differences noted in Fig. 7.3(b). However, Fig. 7.10 shows that the overall variation of Gibbs energy derived from the CSA method agrees well with CVM, falling between the pair approximation, which overestimates the number of AB bonds, and the point approximation, where these are underestimated. As might be expected, if larger clusters are admitted to the CSA approximation the results become closer to the CVM result. However, this is counterproductive if the object is to increase the speed of calculation for multi-component systems.

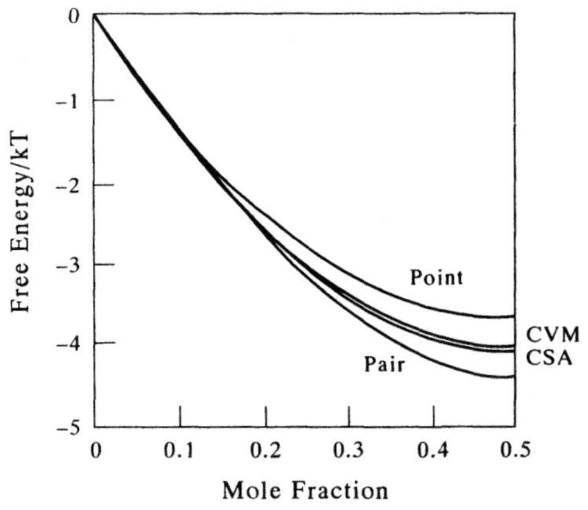

Figure 7.10. Variation of free energy with composition for an f.c.c. lattice as calculated by the point, pair and tetrahedron approximations for both CSA and CVM just above the critical temperature (Oates and Wenzl 1996).

It remains to be seen whether the CSA approximation will prove viable when applied to multi-component systems, but it does appear to be a useful compromise. So far the method has also been restricted to only first- and second-neighbour interactions; clearly any extension to longer-range forces will start to offset its current advantages in terms of computing time.

7.3.6 Simulation of CVM in the framework of a sub-lattice model

Sundman and Mohri (1990) have made a number of simplifying assumptions in order to implement the cluster variation method within the framework of a sub-lattice model and produced a method that can handle varying degrees of short-range order within a random solid solution. Using the specific example of Cu–Au, clusters of specific compositions (e.g., Cu, $Cu_{.75}Au_{.25}$, $Cu_{.5}Au_{.5}$, $Cu_{.25}Au_{.75}$ and Au) are treated as separate composition variables. Although this has superficial analogies with the treatment of molecular species in a gas phase, there are constraints on the positioning of such clusters in the solid-state analogue (as indeed there are in the full CVM treatment). Suitable correction terms have therefore to be introduced with the further constraint that the final expression reduces to the normal regular solution model when the *sro* becomes zero.

This treatment was used to determine the high-temperature equilibrium in the system and combined with a standard four sub-lattice treatment of the ordered phase, equivalent to the BWG treatment for a binary f.c.c. lattice. The Cu–Au diagram generated by this means (Fig. 7.11) is quite close to that obtained by an early MC calculation (Binder 1980), but the latter result is not now universally accepted (Ducastelle 1991). Sundman and Mohri (1990) suggested that their hybrid

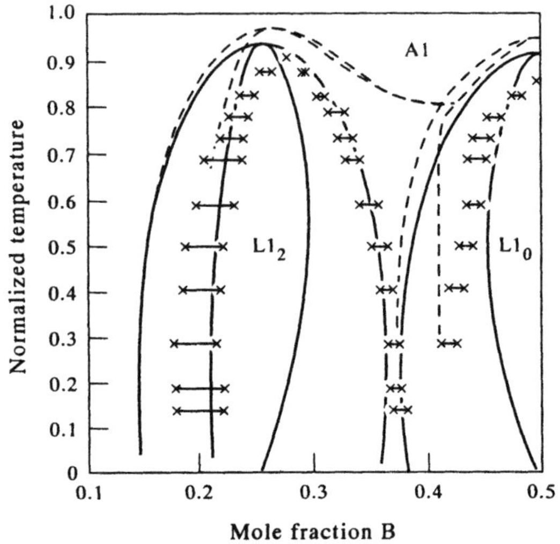

Figure 7.11. Comparison of an ordering treatment in an f.c.c. lattice using a sub-lattice approximation for tetrahedral clusters for disordered alloys, combined with a BWG treatment for the ordered phases (—) with a conventional tetrahedron-CVM calculation (---) (Sundman and Mohri 1990). The tie lines are from a Monte Carlo calculation that has since been superseded.

approach would produce some self-cancellation of the errors inherent in both treatments, but the accuracy of this approach remains an open question, particularly as it has only been tested in the context of nearest-neighbour interactions.

It will be interesting to see how such treatments based on a sub-lattice model can be made more general. So far it has only been used for quite simple systems with ordered structures such as $L1_2$ and $L1_0$. It may be necessary to include more than four sub-lattices for complex ordered phases which are superstructures of these types, and to consider more than 1st or 2nd neighbour energies. Furthermore, the choice of clusters for the *sro* part must relate back to the sub-lattice model itself and it is difficult to see how the more complex clusters routinely handled by CVM models can be reconciled with the sub-lattice models used so far.

7.4. EMPIRICAL ROUTES

7.4.1 Specific heat (C_p) *approximation*

Integration of experimental C_p data with temperature can be used to bypass the plethora of models described in the previous sections, as it provides both the critical ordering temperature and the ratio of *sro* to *lro*. This route has been extensively used to describe magnetic transformations (Inden 1977a, Hillert and Jarl 1978, Nishizawa *et al.* 1979) and generalised through the development of a series of approximate treatments. In the magnetic case these all require a priori knowledge of the critical ordering temperature and the number of Bohr magnetons which are uniquely related to the magnetic ordering energy and entropy of the fully (magnetically) disordered state (Chapter 8). Analogous functions can be constructed for the equivalent chemical or structural ordering transformation and have been used to describe the configurational ordering of the β-phase in the Cu–Zn system (Inden 1981), in Fe–Al (Inden 1983) and for both configurational and magnetic ordering in Fe–Si (Lee *et al.* 1987).

The relative reluctance to apply this method more extensively to structural ordering can be ascribed to several factors. Firstly, the main objective in the magnetic case has been to calculate a ΔG_{mag} contribution *from* the critical ordering temperature; the latter is therefore an *input* and not an output from the calculations. By contrast, the *prediction* of T^{ord} is a major objective in the case of structural ordering. A second factor is that there is a general lack of experimental C_p data for configurational ordering transformations, which has made it difficult to establish general trends for the degree of *sro* associated with different crystal structures. This is an important input parameter and has led to the use of values that have only been really validated for the magnetic case. Thirdly, Inden (1991) has shown that unacceptable singularities can arise in the Gibbs energy functions if input data are derived from the terminal entropy of the disordered state. However, the same article shows that satisfactory results can be obtained with negligible loss of accuracy by alternative routes (Fig. 7.12). This is a good example of the fact that small changes in the way parameters are derived can make significant differences to the final phase diagram.

The main advantage of the C_p route is that the Gibbs energy contribution of a structural ordering transformation can be incorporated using algorithms that are already available in a number of software packages to describe the magnetic Gibbs energy. The ordering Gibbs energy is automatically partitioned into long-range and

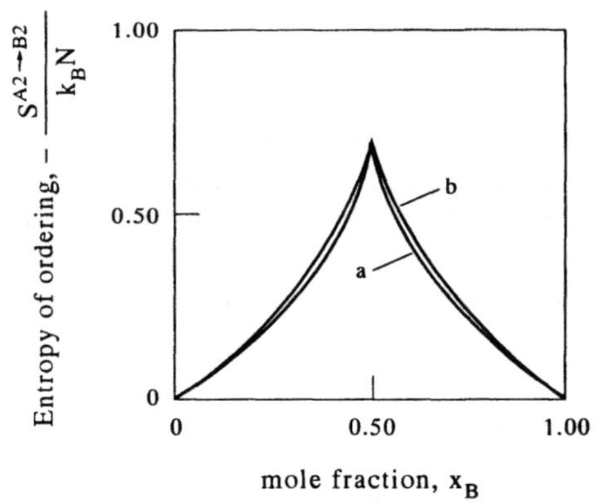

Figure 7.12. Ordering entropy as a function of composition: (a) exact values derived from the terminal entropy for a completely random solution and (b) using an empirical route via the enthalpy The difference between these two curves is sufficient to generate spurious miscibility gaps (Inden 1991).

short-range components and asymptotes smoothly to the function representing the disordered state. An associated, and not insignificant advantage, is that these algorithms are designed to handle asymmetrical values of the ordering temperature. The main disadvantage is clearly the need to have a value of T^{ord} as an input. One possibility is to derive T^{ord} from a simple BWG approximation together with an appropriate χ factor (see Section 7.3.2.2), but this only a reasonable option in the case of b.c.c. systems. Nevertheless, there are systems, such as Nb–Al, where electron energy calculations have shown a rapid variation of ordering energy with composition (Colinet *et al.* 1997). The C_p method may therefore find a niche for handling ordering in certain multi-component b.c.c. alloys.

7.4.2 General polynomial approximation

The possibility of simulating the actual BWG ordering energy, rather than C_p, using a polynomial approximation was also examined by Inden (1976) using the disordered solid solution as a reference state. The following expression was suggested for a continuous second-order transformation such as A2/B2:

$$G_{x,T}^{\text{dis}\rightarrow\text{ord}} = \varepsilon + (\theta T_0)\tau - (3\varepsilon + 2\theta T_0)\tau^2 + (2\varepsilon + \theta T_0)\tau^3 \qquad (7.36)$$

where

$$\tau = T/T^{\text{ord}}, \varepsilon = H_{x,0\text{K}}^{\text{dis}\rightarrow\text{ord}} \quad \text{and} \quad \theta = S_{(x)}^{\text{dis}} - S_{x,0\text{K}}^{\text{ord}}. \qquad (7.37)$$

Assuming these three parameters are available, it is then possible to obtain explicit values for

$$S_{x,T}^{\text{dis}\rightarrow\text{ord}} = -\theta + (6\varepsilon/T_0 + 4\theta)\tau - (6\varepsilon/T_0 + 3\theta)\tau^2 \qquad (7.38)$$

and

$$H_{x,T}^{\text{dis}\rightarrow\text{ord}} = G_{x,T}^{\text{dis}\rightarrow\text{ord}} + TS_{x,T}^{\text{dis}\rightarrow\text{ord}}. \qquad (7.39)$$

A good match was obtained with the results obtained from the standard BWG treatment, both for the A2/B2 and the B2/DO3 transformation, and the expression was extended for use in the Cu–Zn–Mn system (Chandrasekaran 1980). However, as these equations are based on the BWG approximation, which does not take into account *sro*, this approach was abandoned when the CVM formalism became more established.

7.5. ROLE OF LATTICE VIBRATIONS

7.5.1 Interaction of ordering and vibrational entropy

All the approaches described so far concern themselves only with the calculation of Gibbs Energy changes that are caused by changing atomic configurations *per se* and do not necessarily take into account effects associated with the elastic modulus, vibrational entropy or atomic volume. *Changes of these properties with composition will also affect the Gibbs energy, and therefore, the matching of experimental and calculated phase diagrams.* The proper inclusion of vibrational with electronic and magnetic excitations may indeed reduce the number and magnitude of arbitrary parameters introduced into various ordering models (Cacciamani *et al.* 1997). This is a general problem which affects all manner of phase transformations, not just ordering, and is the reason why there is now increasing emphasis on entering specific functions for parameters such as Debye temperatures into the next generation of Gibbs energy algorithms (Sundman and Aldinger 1995).

However, it is also possible for *changes in the vibrational spectrum to occur with changes in degree of order itself,* producing a further change in the entropy. Experiments indicate that, in the case of Ni_3Al, the vibrational component, 0.2–0.3 $R\,mol^{-1}$, can be of comparable magnitude to the configurational entropy, 0.56 $R\,mol^{-1}$ (Anthony *et al.* 1993, Fultz *et al.* 1995). Other examples are a change of 0.10 $R\,mol^{-1}$ for Fe_3Al and 0.12–0.23 $R\,mol^{-1}$ for Cu_3Au (Anthony *et al.* 1994). To appreciate what effect this may have on the ordering temperature the mean free approximation can be extended (Anthony *et al.* 1993) to include different frequency factors, ω, for the various bonds:

$$\eta = \frac{2k_BT}{z(V_{AA} + V_{BB} - 2V_{AB})} \ln\left[\left(\frac{1+\eta}{1-\eta}\right)\left(\omega_{AB}^2/\omega_{AA}\,\omega_{BB}\right)^{3/2}\right]. \qquad (7.40)$$

This is essentially the same expression already given for the BWG approximation (Eq. (7.3)) with an additional function that combines the various vibrational frequencies of different bonds. In an ordered alloy the A–B bonds are expected to be stiffer than those of the A–A and A–B bonds, so the vibrational entropy of the ordered state will be lower than that of the disordered state, thus *lowering the critical ordering temperature.*

In the absence of experimental information on vibrational frequencies, a first estimate can be made via Debye temperatures derived from First Principle calculations. However, the Debye model does not necessarily give an accurate enough answer because it does not include the high-frequency part of the vibrational spectrum. Garbulsky and Ceder (1996) have provided an estimate of the possible fractional change in critical ordering temperature that could be caused by changes in lattice vibrations (Fig. 7.13). This can clearly be quite significant, and the change can be of either sign, depending on the combination of size-mismatch and chemical affinity. A redistribution of masses *per se* as a result of ordering does not seem to produce any major effect (Grimwall 1996).

7.5.2 Kinetic development of ordered states

The structures that are actually observed in practice depend on the kinetics of

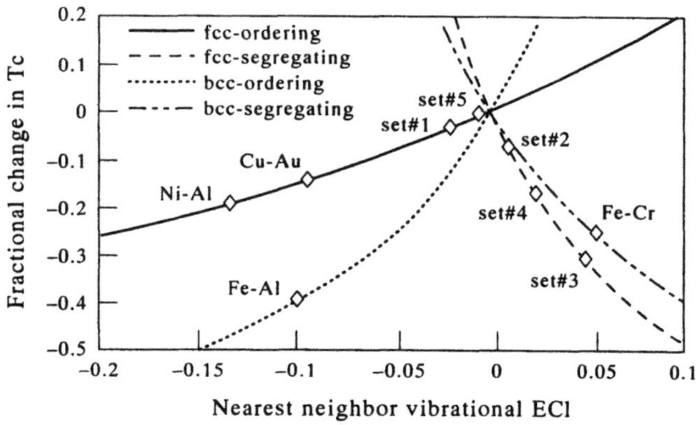

Figure 7.13. Estimated fractional change in the ordering temperature resulting from incorporating the effect of ordering on the vibrational spectrum by an additional effective cluster interaction (ECI_{vib}) (Garbulsky and Ceder 1996).

transformation, and a time-dependent version of CVM called the *path probability method* was already proposed in the early stages of CVM (Kikuchi 1966, 1977). More recently, Mohri *et al.* (1992) have outlined some of the implications in relation to an f.c.c. lattice including a switch from nucleation and growth to spinodal decomposition, a concept first proposed by de Fontaine (1975). Implementation of MC versions of such techniques have shown that there can be transient metastable ordered states before the equilibrium ground state is finally achieved (Fultz 1992, 1994). Thus Hong *et al.* (1995) showed that small regions having the B32 structure can form on the way to attaining the stable B2 structure, even though the selected ratio of first- and second-neighbour bond strengths make B32 technically unstable. Calculations by Turchi *et al.* (1994a) for Fe–Cr alloys indicate that, if the sigma phase was excluded, a B2 phase would have a Gibbs energy intermediate between that of the random A2 high-temperature solution and separation into two Fe- and Cr-rich solutions at low temperatures. The conclusion that a virtual ordered phase can exist even when the heat of mixing is positive has considerable consequences for the prediction of metastable states during rapid cooling.

7.6. INTEGRATION OF ORDERING INTO PHASE DIAGRAM CALCULATIONS

The previous sections have largely concentrated on the relative accuracy with which a given order–disorder reaction can be described and the associated computing time. However, one of the ultimate tests is to see whether these procedures can be integrated into calculations for real, rather than model, systems.

7.6.1 Predictions restricted to phases of related symmetry

Table 7.2 gives selected examples of applications of various models to various systems, covering the range through model calculations in binary systems to ternary applications. Bearing in mind the scale of the subject, the list cannot be considered exhaustive and other reviews should also be consulted (Ducastelle 1991, Inden 1991, de Fontaine 1994, Turchi 1994). The earliest applications either concentrated on simple systems such as the f.c.c.-based transformations in Cu–Au, or only tried to reproduce a restricted region of a real alloy system; for example the b.c.c. family of structures in the Fe-rich region of the Fe–Si system.

Ordering in the b.c.c. phase of Fe–Co alloys (Figs 7.14(a) and (b)) is a good example of the relatively small difference between using the BWG and CVM methods in the case of b.c.c. systems, even where magnetic effects have to be included. Similar comments apply to the ternary Fe–Co–Al system (Figs 7.15(a)–(c)). This may be contrasted with the large variations obtained in close-packed systems, as already evidenced by the various phase diagrams produced for Cu–Au (Figs 7.6–7.8). It is therefore not surprising that many applications to b.c.c. ternary systems tend to concentrate on a description of a limited region in phase space where a particular family of ordering reactions based on b.c.c. derivatives are known to take place, and have not tried to integrate these into a characterisation of the whole diagram.

7.6.2 Predictions using only first-principles plus CVM

It is possible that a *complete solid-state phase diagram* can be predicted from a combination of first-principles (FP) and CVM without the introduction of *any* arbitrary adjustable parameters. However, the present state of the art cannot, as yet, fully deal with the liquid phase. Examples of this kind include Fe–Cr (Turchi *et al.* 1994a), Ti–Rh (Sluiter *et al.* 1988), Ru–Nb (Becker *et al.* (1991) and Al–Li (Sigli and Sanchez, 1985). For a useful review see Turchi (1994). Given the severe restrictions placed on this route, the reproduction of many qualitative features in complex diagrams must be considered a *tour de force*. However, the resulting phase boundaries have not been generally accurate enough to allow their incorporation into any subsequent calculation for industrial alloys.

7.6.3 Methods which maximise the FP input

To be useful from a CALPHAD point of view, it is mandatory to have a characterisation of the liquid phase, and the insistence on working totally from FP has to be relaxed. The simplest solution is to add data for the melting points and enthalpies of fusion for the elements from conventional CALPHAD sources. As for heat of mixing for the liquid phase, one approximation is to assume it will follow a suitably weighted behaviour of the f.c.c. and b.c.c. phases. This kind of treatment

Table 7.2. List of Methods applied to Selected Ordering Systems

System	Lattice	Model	Nearest neighbours	Sub-lattice/ clusters	References
A–B	f.c.c.	BWG	$W^{(1)}$ and $W^{(2)}$	4 sublattices	Inden (1977b)
A–B	f.c.c.	BWG	$W^{(1)}$ and $W^{(2)}$	8 sublattices	Buth and Inden (1982)
A–B	f.c.c. ($L1_0/L1_2$)	CVM	variable $W^{(1)}/W^{(2)}$	T, T/O and MC	Mohri *et al.* (1985)
A–B	(A2/B2/B32)	CVM	variable $W^{(1)}/W^{(2)}$	T and MC	Ackermann *et al.* (1989)
A–B	f.c.c.	CSA	$W^{(1)}$ only	Pt, Tr, T	Oates and Wentl (1996)
A–B	f.c.c./b.c.c.	CVM	$W^{(1)}$ only	Pt, Tr, O, T	Kikuchi and Sato (1974)
A–B	f.c.c.	CVM	$W^{(1)}$ only	Pt, Tr, O, T	Kikuchi (1977)
A–B	($L1_0/L_{12}$)	MC	$W^{(1)} - W^{(2)} > 0$	T/O and MC	Inden and Pitsch (1991)
A–B	c.p.h.	MC	anistr. $W^{(11)}/W^{(12)}$		Crusius and Inden (1988)
A–B	c.p.h./DO_{19}	MC	function $W^{(12)}$		Kikuchi and Cahn (1987)
A–B	f.c.c	MC		mixed" CVM	Finel (1994)
Au–Cu	f.c.c.	CVM		T	Kikuchi (1974)
Au–Cu	f.c.c.	CVM		assym. T	Van Baal (1973)
Ni–Ga	B2 + vacancies				Neumann and Chang (1979)
Co–Ga	B2 + vacancies				Neumann and Chang (1979)
A–B	A15 alloys	CVM	$W^{(1)} - W^{(7)}$	+BWG +MC	Turchi and Finel (1992)
Fe–Co	b.c.c.	CVM	$W^{(1)}$ and $W^{(2)}$	T	Inden and Pitsch (1991)
Fe–Co	b.c.c.	BWG	$W^{(1)}$ and $W^{(2)}$		Inden (1977a)
Fe–Co	b.c.c.	CVM	$W^{(1)}(W^{(2)} = 0)$	T	Colinet *et al.* (1993)
Al–Co	b.c.c.	MC			Ackermann (1988)
Cu–Zn	b.c.c.	GPM			Turchi *et al.* (1991)
Cu–Zn	b.c.c.	BWG	$W^{(1)}$ and $W^{(2)}$	4 sublattices	Inden (1975b)
Fe–Al	A2/B2	MC	up to $W^{(5)}$		Bichara and Inden (1991)
Fe–Al–Co	B2	BWG	Fe, Co, mag spin 1		Kozakai and Miyazaki (1994 a,b)
Fe–Al–Co	B2	BWG	Fe, Co, mag spin 1		Miyazaki *et al.* (1987)
Fe–Al–Co	b.c.c. variants	CVM		T	Colinet *et al.* (1993)
Fe–Al–Co	b.c.c. variants	MC			Ackerman (1988)
Ni–Al–Ti	f.c.c./b.c.c.	CVM		T+T/O	Sanchez (1992)
Fe–Si–Co	B2/DO_3/$L2_1$	BWG	Fe, Co, mag spin 1		Inden (1979)
Fe–Si–Co	A2/B2/DO_3	BWG	Fe, Co, mag spin 1		Fukaya *et al.* (1991)
Fe–Si–Al	A2/B2/DO_3	BWG	Fe, Co, mag spin 1		Fukaya *et al.* (1991)
Fe–Al–Ge	A2/B2/DO_3	BWG			Kozakai and Miyasaki (1993)
Ti–Al–Mo		CVM		T	Rubin and Finel (1993)
Ti–Al–Nb		CVM		T	Rubin and Finel (1993)
Ti–Al–W		CVM		T	Rubin and Finel (1993)
Ga–In–P	III–V	CVM			Silverman *et al.* (1994)
Ga–As–Sb	Zn–blende	CVM	up to $W^{(4)}$		Zunger (1994)
Cu–In–Se	Chalcopyrite	CVM			Osorio *et al.* (1993)
In–P–Sb	Zn–Blende	CVM		T+T/O	Nakamura *et al.* (1991)

Abbreviations: BWG = Bragg–Williams–Gorsky, CVM = Cluster Variation method, T = Tetrahedron approximation, T/O = Tetrahedron/Octahedron approximation, MC = Monte Carlo, SC = Simple cubic approximation, SP = simple prism approximation, Pt = point approximation, Tr = triangle approximation

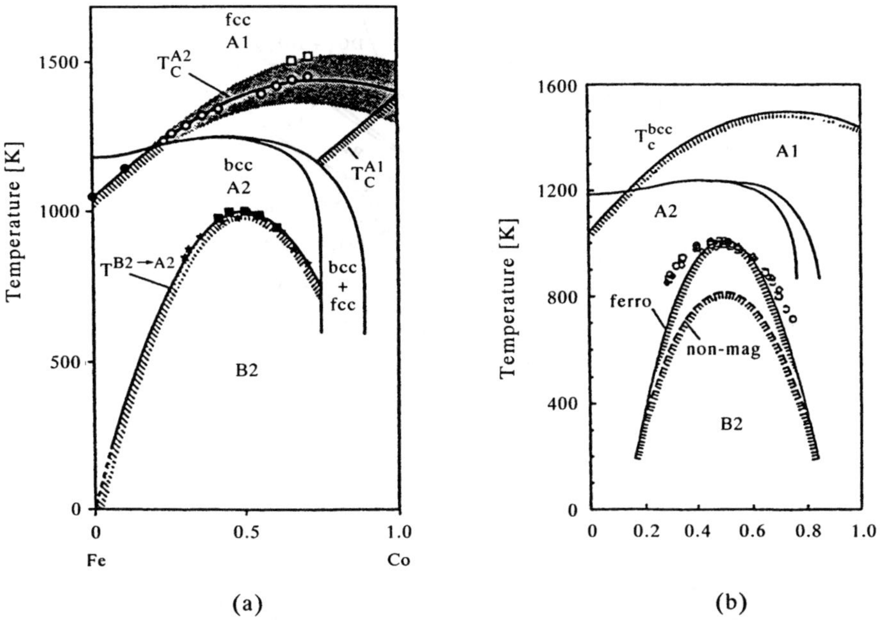

Figure 7.14. Comparison of ordering in binary b.c.c. Fe–Co alloys (a) according to the BWG approximation (Inden 1977) and (b) using CVM (Colinet *et al.* 1993).

has been applied to Ni–Ti (Pasturel *et al.* 1992) and Nb–Al (Colinet *et al.* 1997), but it is of course possible to also import such data from CALPHAD assessments if these are available.

While this can produce diagrams that are closer to reality, certain critical features can still be in error by a considerable margin (several hundred degrees in some instances). One way of improving the agreement is to use a simple scaling factor for the initial set of calculated interaction coefficients so as to obtain a better fit with observed critical temperatures. Such discrepancies might of course simply reflect the fact that vibrational terms were often ignored. It is possible to partially include the effect of the phonon spectrum, without adding further empirical parameters, by deriving vibrational frequencies from FP calculations of the bulk modulus (Sanchez 1992).

It is difficult to assess the relative merits of the different approaches as they have not necessarily been applied to the same systems. They are further characterised by other differences such as the CVM cluster size, different characteristics of the liquid phase and, notably, by the adoption of different values for the lattice stabilities. The Ni–Al system is useful in this respect as it contains both ordered b.c.c. and ordered f.c.c. phases as well as having been tackled by many

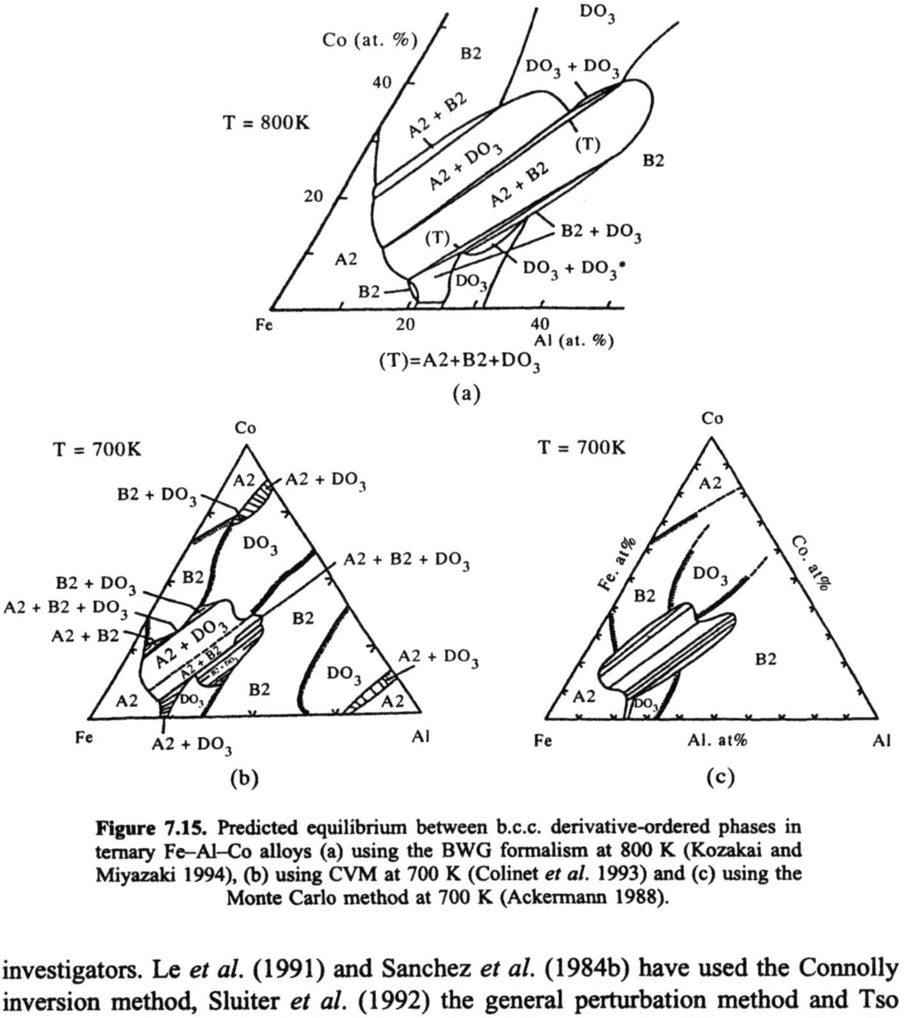

Figure 7.15. Predicted equilibrium between b.c.c. derivative-ordered phases in ternary Fe–Al–Co alloys (a) using the BWG formalism at 800 K (Kozakai and Miyazaki 1994), (b) using CVM at 700 K (Colinet *et al.* 1993) and (c) using the Monte Carlo method at 700 K (Ackermann 1988).

investigators. Le *et al.* (1991) and Sanchez *et al.* (1984b) have used the Connolly inversion method, Sluiter *et al.* (1992) the general perturbation method and Tso (1992) the mixed CVM–CALPHAD method. More recently, this system has also been used as a test vehicle (Cacciamani *et al.* 1997) to compare other combinations of CVM and CALPHAD methods (see next section).

7.6.4 The mixed CVM-CALPHAD approach

Even with the incorporation of CALPHAD data for the liquid phase, the insistence on using the electron energy lattice stabilities for the end-members can make it virtually impossible to integrate any resulting FP phase diagram with other systems based on standard CALPHAD assessments. A totally different approach is necessary

References are listed on pp. 222–226.

to achieve such a possibility. A mixed CVM–CALPHAD technique (Sigli 1986, Tso 1992) essentially accepts the CALPHAD view of placing a premium on accurately fitting the experimental phase boundaries rather than on fixing all the parameters from FP. Once the connection with FP calculations becomes less rigid, it also becomes possible to use standard thermochemical lattice stabilities, thus allowing a more seamless integration with other CALPHAD assessments. However, the crucial element of this approach is the incorporation of a suitable excess term to remove any discrepancies between a CVM calculation of the ordering component and the observed phase equilibria.

$$G^{\phi} = \sum_{i=1...n} x_i\, G_i^{\phi} + G^{\phi}_{\text{CVM}} + G^{\phi}_{\text{excess}}. \tag{7.41}$$

Here the first term represents the lattice stability components of the phase ϕ, the second term the Gibbs energy contribution arising from cluster calculations and the third term is the excess Gibbs energy expressed in the form of a standard Redlich–Kister polynomial (see Chapter 5).

Once the aim has been shifted away from maximising the FP input, it seems sensible to extend the range of input parameters that are used to define the cluster interactions and include all the available experimental data such as enthalpies of formation. It is then possible to derive ECI parameters entirely from experimental data (Oates *et al.* 1996), although in that case they may well turn out to be concentration- and temperature-dependent. The Ni–Al phase diagram resulting from the CVM–CALPHAD treatment of Tso (1997) is shown in Fig. 7.16(a). The results show that the CVM–CALPHAD route leads to phase boundaries of reasonable accuracy but with potentially fewer parameters than a CALPHAD approach, such as the specific form of sub-lattice model used by (Ansara *et al.* 1995) Fig. 7.16(b) (see also Chapter 5). The work reported by Cacciamani *et al.* (1997) should be consulted for the effect of making other assumptions, and work involving a combination of FP, cluster expansion and vibrational energies for this system is in progress.

7.6.5 Applications of FP–CVM calculations to higher-order metallic alloys

There have been relatively few attempts to determine ordering in a ternary system solely using a combination of FP calculations and CVM. Computing time obviously rises substantially with the number of components, so some simplifications may have to be made. One suggested technique (Sanchez 1992) is to perform the main calculation by using the tetrahedron approach but use re-normalised parameters which reproduce a T^{ord} previously deduced from the more accurate tetrahedron–octahedron approximation. This allows both first- and second-neighbour interactions in the binary systems to be taken into account, but how

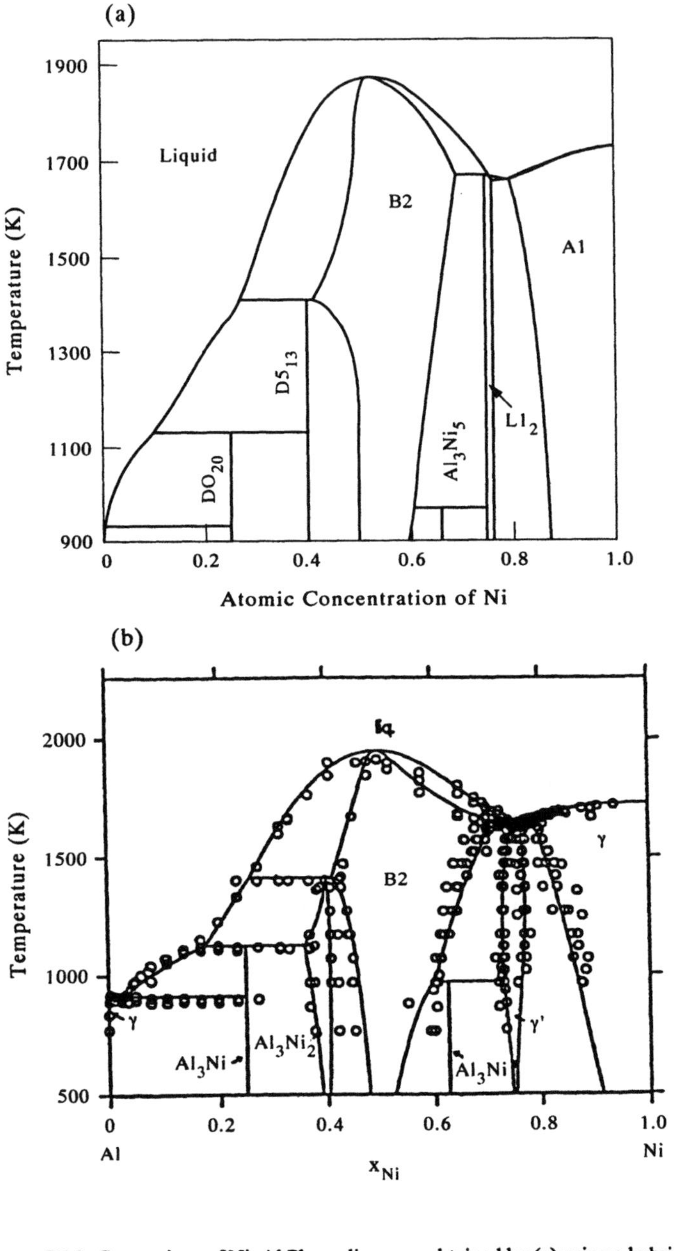

Figure 7.16. Comparison of Ni–Al Phase diagrams obtained by (a) using a hybrid CVM–CALPHAD approach (Tso 1992, Cacciamani 1997) and (b) CALPHAD approach incorporating a sub-lattice ordering model (Ansara *et al.* 1995).

far it will be possible to extend this to a multi-component situation is still an open question. Some results for restricted regions of the Ni–Al–Ti are given in Sanchez (1992), while Asta *et al.* (1995) have performed an FP–CVM study of intermetallic phase stability in Ti–Al–Nb alloys. Because of interest in exotic intermetallic compounds, Becker and Sanchez (1993) have also explored the Ru–Nb–Zr system. Some further general aspects of modelling can be found in the work of Tso *et al.* (1989) and Tso (1992).

Earlier work on systems such as Ni–Al–Cr reported in Sanchez *et al.* (1984b) used FP methods to obtain information on phases for which there was no experimental information. In the case of Ni-base alloys, the results correctly reproduced the main qualitative features of the $\gamma - \gamma'$ equilibrium but cannot be considered accurate enough to be used for quantitative alloy development. A closely related example is the work of (Enomoto and Harada 1991) who made CVM predictions for order/disorder $(\gamma - \gamma')$ transformation in Ni-based superalloys utilising Lennard–Jones pair potentials.

7.6.6 Applications to more complex structures

Many semiconductors and ceramics contain phases with more complex variations based on larger unit cells and with marked deviations from cubic symmetry. This involves an enormous increase in the number of potential structures that have to be screened in order to find the ground-state configuration. In addition, a higher accuracy is required because the relevant Gibbs energy differences become smaller, which makes the BWG approximation inappropriate. Applying the CVM technique is also difficult because conventional cluster expansions only converge quickly in the absence of appreciable atomic relaxation and the computational effort also becomes much higher as the number of atoms in the unit cell increases. These problems have been overcome through the methods developed by Zunger and his co-workers and can be applied not only to conventional order–disorder transformations but also to APB boundaries and interfacial energies. Systems that have been studied include the quasi-binary system AlAs/GaAs (Magri *et al.* 1991), many other III–V quasi-binary systems (Wei *et al.* 1990), pseudo-ternary phase diagrams of chalcopyrite–zinc blende alloys (Osorio *et al.* 1993) (Fig. 7.18) and surface-stabilised ternary semi-conductors (Zunger 1993). The GaSb–InSb and GaP–InP systems have been studied by Mohri *et al.* (1989) and InP–InSb by Nakamura *et al.* (1991). The importance of lattice relaxation has been emphasised by Zunger (1994). Technically all these examples are restricted to solid-state applications since, as has already been mentioned, FP calculations cannot as yet handle the liquid phase. There is therefore significant scope for combining these treatments with a standard CALPHAD treatment of the liquidus and solidus (Ishida *et al.* 1988, Ohtani *et al.* 1992).

In the field of ceramics the CVM technique has been applied to the system

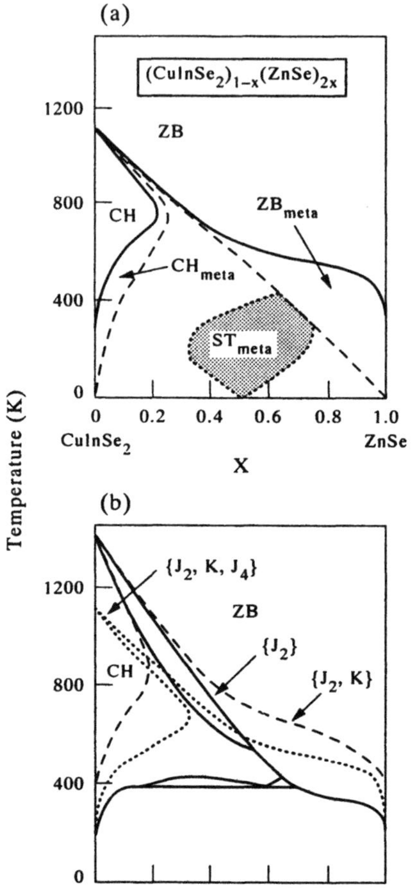

Figure 7.17. Predicted phase diagram for the system $(CuInSe_2)_{1-x}(ZnSe)_{2x}$ (a) stable and metastable equilibrium between chalcopyrite (CH) and zinc-blende (ZB) phases using volume-dependent interaction parameters; (b) using a hierarchy of less accurate approximations (Osorio *et al.* 1993).

$CaCO_3$–$MgCO_3$, using the tetrahedron approximation, and compared with the result by using BWG (Burton and Kikuchi 1984a). As might be expected, CVM gives better results than BWG because of the inclusion of *sro* (Fig. 7.18). Good agreement was obtained for the sub-system $CaCO_3$–$[(CaCO_3)_{0.5}$–$(MgCO_3)_{0.5}]$ (calcite–dolomite), but interestingly the features of the dolomite–magnesite ($[(CaCO_3)_{0.5}$–$(MgCO_3)_{0.5}]$–$MgCO_3$) segment do not match even qualitatively. The authors therefore suggested that an additional transition based on the rotation of CO_3 groups might have to be included to give an extra degree of freedom. In the case of hematite Fe_2O_3, the corundum structure necessitates the use of a prismatic

References are listed on pp. 222–226.

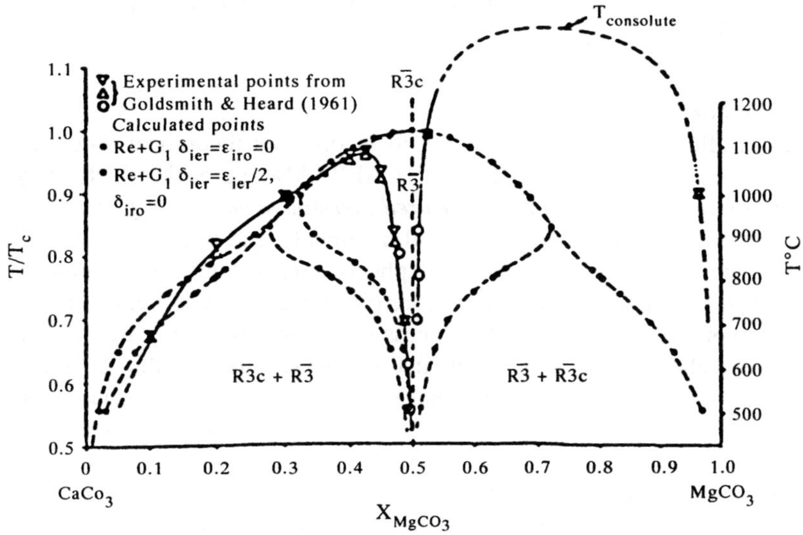

Figure 7.18. Comparison of experimental and predicted phase equilibria in the system $CaCO_3$–$MgCO_3$ using CVM in the tetrahedron approximation for a trigonally distorted f.c.c. Ising lattice. Semi-quantitative agreement is achieved for the calcite–dolomite segment but the Mg-rich side of the diagram indicates the need to include a more complex model (Burton and Kikuchi 1984b).

cluster (Burton and Kikuchi 1984b). Better agreement for the calculated Neel temperature was obtained than with simpler pair-models, but the calculation was still considerably in excess of experimental results. The order–disorder transition in $FeTiO_3$ shows that structural information is important in systems which have anisotropic coupling (Burton 1985a). This becomes even more necessary in carbonates and silicates where intra-layer interactions favour phase separation, while inter-layer interactions favour ordering (Burton 1985b). Only for highly ionic materials, such as CaO–MgO, does the charge state remain invariant on mixing; but even in this case discrepancies may arise due to the neglect of vibrational entropy (Tepesch *et al.* 1996).

A distinctive feature of many oxides is the presence of lattice defects, requiring the evaluation of defect interaction energies (Tetot *et al.* 1994). These can be of vital significance in relation to the superconductivity of materials such as $YBa_2Cu_3O_x$ whose basic ordering behaviour has been studied by Kikuchi and Choi (1989) and de Fontaine *et al.* (1987). Vacant sites can also have a significant effect in intermetallic compounds, effectively adding an extra component, since the sites have to be treated as a separate species and given suitable interaction energies. An early example is the paper by Neumann and Chang (1979) on defect structures in B2 transition metal intermetallic compounds.

7.7. COMMENTS ON THE USE OF ORDERING TREATMENTS IN CALPHAD CALCULATIONS

7.7.1 General comments

Difficulties often arise when attempting to incorporate theoretical ordering treatments within the so-called 'traditional' CALPHAD approach which obeys the paradigm of *accurate coupling of measured thermodynamics and phase diagrams*. This is because the majority of theoretical treatments tend to be associated with an exclusive sub-set of particular physical phenomena and may not include other, equally important effects within their theoretical basis. For example, the BWG treatment does not include *sro* while the excess Gibbs energy arising from vibrational entropy is rarely considered in standard CVM procedures. Within a 'traditional' CALPHAD optimisation, such additional effects are empirically included through the various excess terms. Added to this, the greater number of parameters which can be optimised means that accurate calculations for binary and ternary phase diagrams are possible. Without this required degree of accuracy it is impossible to make accurate predictions for phase equilibria in multi-component alloys.

This accounts for the moves to combine a theoretical CVM treatment with a CALPHAD optimisation technique which include an excess term (see Section 7.6.4). However, CVM then becomes just another optimisable model and it is not yet clear whether there is any gain in predictability in the higher-order systems when adopting this procedure for multi-component alloys, even if vibrational entropy can be properly included.

At the present time the strength of the more theoretically based ordering models is that they offer the potential for *predicting ordering diagrams of many different types*, which is not possible within the more empirical CALPHAD framework and, very importantly, they provide the possibility of *predicting material properties which are more physically based*. If these approaches are adapted too much merely to achieve high levels of accuracy in the coupling of phase diagrams and thermodynamic properties, the essence of predictability becomes lost and it also may be that their theoretical basis is submerged. For example, recent attempts with sub-lattice models for ordering (Ansara *et al.* 1988, 1995), while being faster than either CVM or MC, are limited to basic ordering diagrams. They are also empirically based in that parameters are *fitted* as part of an optimisation process and, therefore, any predictive capability is compromised.

There are other factors to be considered. In a number of systems ternary phases exist whose stability indicates that there are additional, and as yet unknown, factors at work. For example, in the Ni–Al–Ta system η, which has the Ni_3Ti structure, *only exists as a stable phase in ternary alloys, although it can only be fully ordered in the binary system where it is metastable*. However, the phase competes successfully with *equilibrium* binary phases that have a substantial extension into the ternary system. The existence of the η-phase, therefore, is almost certainly due to ternary interactions within the ordered phase itself which would require knowledge of the interaction energies for higher-order clusters containing all three elements. The chances of providing a CVM optimisation which is more accurate than achieved through more empirical approaches may then be small.

The increased computing power necessary to include just binary CVM models (without even considering any extension to multi-component alloys) then raises serious question-marks concerning the philosophy of combining CVM–CALPHAD as a general tool for predicting equilibria in higher-order systems. There may be a real danger that attempts to combine the *current*, physically based theoretical models with more empirically based CALPHAD calculations *will not allow the full strengths of either approach to be properly realised*. This prognosis is, of course, tempered by the fact that there will, inevitably, be advances in theoretical models, and it may be that in future years CALPHAD will be predominantly based on models whose basis lies in electronic energy calculations tied to very exact and fundamental ordering models. However, such a pleasant prospect will probably only come to fruition in the next decades rather than the next few years.

In the meantime, an excellent example of how ordering models can be used in practice is given by Enomoto *et al.* (1991) (Section 7.6.5). These authors showed that it was possible, using the tetrahedron approximation within CVM, to make predictions for ordering in multi-component, Ni-based superalloys which could subsequently be used in the design of γ/γ'-based single-crystal alloys. Although within this framework it is not possible to predict more general phase equilibria, one can envisage a fully integrated FP–CVM approach which would further allow the possibility of predicting APB and stacking fault energies in the γ'-phase and therefore provide vital input parameters for predicting yield strength or creep resistance in such alloys. Further to this, it may also be possible to predict the formation of other, important ordered phases found in Ni-based superalloys such as γ''.

7.7.2 The prediction of ordering temperatures

As a final comment it is clearly necessary to make a distinction between progress in evaluating ordering energies and progress in predicting ordering temperatures. Despite the fact that ordering energies can now, for the most part, be evaluated accurately from FP, the *accurate prediction of ordering temperatures* is still proving elusive in many cases. Although the problem is not as difficult as predicting melting points, there are still a large number of contributing factors that have been found to enter into the evaluation of the entropy of the relevant competing structures. Just as the BWG method has been largely superseded by CVM and MC methods because BWG *per se* cannot handle *sro*, so it now appears that a merely configurational MC approach, which does not take into account changes in the vibrational spectrum *due to* ordering, may also prove insufficient.

This means that the predictability of ordering temperatures is going to vary from system to system and the accuracy will depend markedly on the relevant structure and bonding characteristics. If the structure is b.c.c., and bonding forces are symmetrical across the system, it may well be that ordering temperatures can be predicted quite accurately by using the BWG method with a scaling factor. For f.c.c. systems of reasonable symmetry an equivalent accuracy can only be attained by using a suitable cluster size within a CVM treatment. However, there will be both b.c.c. and f.c.c. systems which show asymmetrical bonding behaviour with respect to composition and/or a temperature dependence of the interaction parameters. In these cases going to larger cluster sizes, or using the MC method with assumed constant interaction parameters, will lead to a potential error in predicted ordering temperatures comparable to the effect of omitting *sro* in the BWG approximation.

When a suitable algorithm can be found that takes all the relevant factors into consideration, it will obviously become possible to make more accurate predictions of the ordering temperature for unknown systems. However, this is likely to be at the expense of even more computing time than the conventional MC method and therefore may not lend itself for incorporation into a CALPHAD methodology for handling multi-component alloys. The possibility of extending T^{ord} values from binary systems into a ternary or higher-order systems will therefore depend on how far it is possible to make meaningful calculations without knowing the exact site occupancies for each component, and this is an area which needs further examination and validation.

REFERENCES

Ackermann, H. (1988) Ph.D. Thesis, RWTH Aachen.
Ackermann, H., Inden, G. and Kikuchi, R. (1989) *Acta Met.*, **37**, 1.
Allen, S. M. and Cahn, J. W. (1972) *Acta Met.*, **20**, 423.
Ansara, I., Sundman, B. and Willemin, P. (1988) *Acta Met.*, **36**, 977.
Ansara, I., Dupin, N., Lukas, H. L. and Sundman, B. (1995) in *Applications of Thermodynamics in the Synthesis and Processing of Materials*, eds Nash, P. and Sundman, B. (TMS, Warrendale), p. 273.
Anthony, L., Okamoto, J. K. and Fultz, B. (1993) *Phys. Rev. Lett.*, **70**, 1128.
Anthony, L., Nagel, L. J., Okamoto, J. K. and Fultz, B. (1994) *Phys. Rev. Lett.*, **73**, 3034.
Asta, M., Ormecci, A., Wills, J. M. and Albers, R. C. (1995) *Mat. Res. Soc. Symp. Proc.*, **364**, 157.
Beauchamp, P., Dirras, G. and Veyssiere, P. (1992) *Phil. Mag. A.*, **65**, 477.
Becker, J. D, Sanchez, J. M. and Tien, J. K. in *High-Temperature Ordered Intermetallic Alloys IV*, eds Johnson, L. A. *et al.* (1991) *Mat. Res. Soc. Symp. Proc.*, **213**, 113.
Becker, J. D and Sanchez, J. M. (1993) *Mat. Sci. Eng.*, **A170**, 161.
Bethe, H. A. (1935) *Proc. Roy. Soc.*, **150A**, 552.
Bichara, C. and Inden, G. (1991) *Scripta Met.*, **25**, 2607.
Binder, A. R. (1986) *Monte-Carlo Methods in Statistical Physics*, ed. Binder, A. R. (Springer, Berlin).

Bragg, W. L. and Williams, E. J. (1934) *Proc. Roy. Soc.*, **145A,** 699.

Bragg, W. L. and Williams, E. J. (1935a) *Proc. Roy. Soc.*, **151A,** 540.

Bragg, W. L. and Williams, E. J. (1935b) *Proc. Roy. Soc.*, **152A,** 231.

Burton, B. and Kikuchi, R. (1984a) *Phy. Chem. Minerals*, **11,** 125.

Burton, B. and Kikuchi, R. (1984b) *Amer. Mineralogist*, **69,** 165.

Burton, B. (1985a) in *Computer Modelling of Phase Diagrams*, ed. Bennett, L. H. (AIME, Warrendale, PA), p. 81.

Burton, B. (1985b) in *Computer Modelling of Phase Diagrams*, ed. Bennett, L. H. (AIME, Warrendale, PA), p. 129.

Buth, J. and Inden, G. (1982) *Acta Met.*, **30,** 213.

Cacciamani, G., Chang, Y. A., Franks, P., Grimwall, G., Kaufman, L., Miodownik, A. P., Sanchez, J. M., Schalin, M. and Sigli, C. (1997) *CALPHAD*, **21,** 219.

Carlsson, A. E. (1987) *Phys. Rev. B*, **35,** 4858.

Carlsson, A. E. (1989) *Phys. Rev. B*, **40,** 912.

Ceder, G., Tepesch, P. D., Wolverton, C. and de. Fontaine, D. (1994) in *Statics and Dynamics of Alloy Phase Transformations*, eds Turchi, P. E. A. and Gonis, A. (Plenum Press, New York), p. 571.

Chandrasekaran, L. (1980) ''Ordering and martensitic transformations in Cu–Zn–Mn shape memory alloys'', Ph.D. Thesis, University of Surrey, Guildford, UK.

Colinet, C., Pasturel, A., Nguyen Manh, D, Pettifor, D. G. and Miodownik, A. P. (1997) *Phys. Rev. B*, **56,**(2).

Colinet, C., Inden G. and Kikuchi, R. (1993) *Acta Met.*, **41,** 1109.

Connolly, J. W. and Williams, A. R. (1983) *Phys. Rev. B.*, **27,** 5169.

Crusius, S. and Inden, G. (1988) in *Proc. Int. Symp. Dynamics of Ordering Processes in Condensed Matter, Kyoto 1987*, eds Komura, S. and Furukawa, H. (Plenum Press, New York), p. 139.

de Fontaine, D. (1975) *Acta. Met.*, **23,** 553.

de Fontaine, D. (1979) *Solid State Physics*, **34,** 73.

de Fontaine, D. (1994) *Solid State Physics*, **47,** 33.

de Fontaine, D. and Kikuchi, R. (1978) in *Applications of Phase Diagram in Metallurgy and Ceramics*, ed. Carter, G. (NBS Special Publication SP 496, Gaithesburg), p. 999.

de Fontaine, D., Willie, L. T. and Moss, S. C. (1987) *Phys. Rev. B*, **36,** 5709.

Ducastelle, F. (1989) in *Alloy Phase Satability*, eds Stocks, G. M. and Gonis, A. (NATO ASI series E *Appl. Sci.* **163,** Kluwer Acad. Publ.), p. 2329.

Ducastelle, F. (1991) *Order and Phase Stability in Alloys, Cohesion and Structure, Vol. 3*, eds de Boer, F. R. and Pettifor, D. G. (Elsevier, Amsterdam).

Ducastelle, F. and Gautier, F. (1976) *J. Phys. F*, **6,** 2039.

Dunweg, B. and Binder, K. (1987) *Phys. Rev.*, **B36,** 6935.

Enomoto, M. and Harada, H. (1991) *Met. Trans. A*, **20A,** 649.

Enomoto, M., Harada, H. and Yamazaki, M. (1991) *CALPHAD*, **15,** 143.

Ferreira, L. G., Wolverton, C. and Zunger, A. (1997) submitted to *J. Chem. Phys.*.

Finel, A. (1994) in *Statics and Dynamics of Alloy Phase Transformations*, ed. Turchi, P. E. A. and Gonis, A. (Plenum Press, New York), p. 495.

Flinn, P. A. (1960) *Trans. AIME*, **218,** 145.

Fowler, R. H. and Guggenheim, E. A. (1939) in *Statistical Mechanics* (Cambridge University Press, Cambridge), p. 162.

Fukaya, M., Miyazaki, T. and Kozakai, T. (1991) *J. Mat. Sci.*, **26,** 5420.
Fultz, B. (1992) *Phil. Mag. B*, **193,** 253.
Fultz, B. (1994) in *Statics and Dynamics of Alloy Phase Transformations*, eds Turchi, P. E. A. and Gonis, A. (Plenum Press, New York), p. 669.
Fultz, B., Anthony, L., Nagel, L. J., Nicklow, R. M. and Spooner, S. (1995) *Phys. Rev. B*, **52,** 3315.
Garbulsky, G. D. and Ceder, G. (1996) *Phys. Rev. B*, **53,** 8993.
Gonis, A., Zhang, X. G., Freeman, A. J., Turchi, P., Stocks, G. M. and Nicholson, D. M. (1987) *Phys. Rev. B*, **36,** 4630.
Golosov, N. S., Pudan, L. Ya., Golosova, G. S. and Popov, L. E. (1972) *Sov. Phys. Sol. State*, **14,** 1280.
Gorsky, V. S. (1928) *Z. Phys.*, **50,** 64.
Grimwall, G. (1996) *J. Alloys and Compounds*, **233,** 183.
Guggenheim, E. A. (1935) *Proc. Roy. Soc.*, **A148,** 304.
Hansen, M. and Anderko, K. (1958) Constitution of Binary Alloys: 2nd ed. (McGraw-Hill, New York).
Hillert, M. and Jarl, M. (1978) *CALPHAD*, **2,** 227.
Hong, L. B., Anthony, L. and Fultz, B. (1995) *J. Mat. Res.*, **10,** 126.
Inden, G. (1974) *Z. Metallkde.*, **65,** 94.
Inden, G. (1975a) *Z. Metallkde.*, **66,** 577.
Inden, G. (1975b) *Z. Metallkde.*, **66,** 648.
Inden, G. (1976) *Proc. CALPHAD V*, paper III.1.
Inden, G. (1977a) *Z. Metallkde.*, **68,** 529.
Inden, G. (1977b) *J. de Physique*, Coll. C.7, Suppl., **38,** 373.
Inden, G. (1979) *Phys. Stat. Sol.*, **56,** 177.
Inden, G. (1981) *Physica*, **103B,** 82.
Inden, G. (1982) *Bulletin of Alloy Phase Diagrams*, **2,** 406.
Inden, G. (1983) *Mat. Res. Soc. Symp.*, **19,** (Elsevier, New York), p. 175
Inden, G. (1991) *Scand. J. Metallurgy*, **20,** 112.
Inden, G. and Pitsch, W. (1991) in *Materials Science and Technology, Vol. 5*, ed. Haasen, P. (VCH Verlagsgesellschaft, Weinheim), p. 497.
Inden, G., Bruns, S. and Ackermann, H. (1986) *Phil. Mag. A*, **53,** 87.
Ishida, K., Shumiya, T., Nomura, T., Ohtani, H. and Nishizawa, T. (1988) *J. Less. Comm. Metals*, **142,** 135.
Khachaturyan, A. G. (1978) *Prog. Mater. Sci.*, **22,** 1.
Khachaturyan, A. G. and Morris, J. W. (1987) *Phil. Mag. A.*, **56,** 517.
Kikuchi, R. (1951) *Phys. Rev.*, **81,** 988.
Kikuchi, R. (1966) *Suppl. Prog. Theor. Physics*, **35,** 1.
Kikuchi, R. (1974) *J. Chem. Phys.*, **60,** 1071.
Kikuchi, R. (1977) *J. de Physique*, Coll. C7, Suppl., **38,** 307.
Kikuchi, R. and Sato, H. (1974) *Acta. Met.*, **22,** 1099.
Kikuchi, R. and von Baal, C. M. (1974) *Scripta Met.*, **8,** 425.
Kikuchi, R. and Cahn, J. W. (1987) in *User Applications of Phase Diagrams*, ed. Kaufman, L. (ASM, Metals Park, OH), p. 19.
Kikuchi, R, and Choi, J. S. (1989) *Physica C*, **160,** 347.
Kozakai, T. and Miyasaki, T. (1993) in *Computer Aided Innovation of New Materials II*, eds Doyama, M. *et al.* (Elsevier, Amsterdam), p. 767.

Kozakai, T. and Miyazaki, T. (1994a) *ISIJ International*, **34,** 373.
Kozakai, T. and Miyazaki, T. (1994b) *J. Mat. Sci.*, **29,** 652.
Laks, D. B. *et al.* (1992) *Phys. Rev. B*, **46,** 12,587.
Le, D. H., Colinet, C., Hicter, P. and Pasturel, A. (1991) *J. Phys. Cond. Matter*, **3,** 7895; *ibid* 9965.
Lee, B.-J., Lee, S. K. and Lee, D. N. (1987) *CALPHAD*, **11,** 253.
Li, Y. (1949) *Phys. Rev.*, **76,** 972.
Lu, Z. W., Wei, S. H., Zunger, A., Frota-Pessao, S. and Ferreira, L. G. (1991) *Phys. Rev. B*, **44,** 512.
Magri, R., Bernard, J. E. and Zunger, A. (1991) *Phys. Rev. B.*, **43,** 1593.
Miyazaki, T., Isobe, K., Kosakai, T. and Doi, M. (1987) *Acta Met.*, **35,** 317.
Miodownik, A. P. and Saunders, N. J. (1995) in *Application of Thermodynamics in the Synthesis and Processing of Materials*, eds Nash, P. and Sundman, B. (TMS, Warrendale, PA), p. 91.
Mohri. T, Sanchez, J. M., and de Fontaine, D. (1985) *Acta Met.*, **33,** 1171.
Mohri, T. Kobayashi, C. and Watanabe, K. (1988) *Mem. Fac. Eng. Hokkaido Univ.*, **XVII**(3), 76.
Mohri, T., Koyanagi, K., Ito, T. and Watanabe, K. (1989) *Jap. J. Appl. Phys.*, **28,** 1312.
Mohri, T., Sugawara, Y., Watanabe, K. and Sanchez, J. M. (1992) *Mat. Trans. Jap. Inst. Met.*, **33,** 558.
Nakamura, K., Mohri, T. and Ito, T (1991) *Bull. Fac. Eng. Hokkaido Univ.*, **155,** 19.
Neumann, J. P. and Chang, Y. A. (1979) *Z. Metallkde.*, **70,** 118.
Nishizawa, T., Hasebe, M. and Ko, M. (1979) *Acta Met.*, **27,** 817.
Oates, W. A., and Wentl, H. (1996) *Scripta Mater.*, **35,** 623.
Oates, W. A., Spencer, P. J. and Fries, S. G. (1996) *CALPHAD*, **20,** 481.
Ohtani, H., Kojima, K., Ishida, K. and Nishizawa, T. (1992) *J. Alloys and Compounds*, **182,** 103.
Osorio, R., Lu, Z. W., Wei, S.-H., and Zunger, A. (1993) *Phys. Rev. B*, **47,** 9985.
Pasturel, A., Colinet, C., Paxton, A. T. and van Schilfgaarde, M. (1992) *J. Phys. Cond. Matter*, **4,** 945.
Richards, M. J. and Cahn, J. W. (1971) *Acta Met.*, **19,** 1263.
Rubin, G, and Finel, A. (1993) *J. Phys. Condensed Matter*, **5,** 9105.
Sanchez, J. M., Ducastelle, F. and Gratias, D. (1984a) *Physica*, **128A,** 334.
Sanchez, J. M., Jarrett, R. N., Sigli, C. and Tien, J. K. (1984b) in *HighTemperature Alloys: Theory and Design*, ed. Stiegler, O. R. (TMS, Warrendale, PA), p. 83.
Sanchez, J. M. (1992) in *Structure and Phase Stability of Alloys*, eds Moran-Lopez *et al.* (Plenum Press, New York), p. 151.
Sanchez, J. M. and de Fontaine, D. 1980 *Phys. Rev.*, **B21,** 216.
Sanchez, J. M., de Fontaine, D. and Teitler, W. (1982) *Phys. Rev. B.* **26,** 1465.
Sigli, C. (1986) "On the electronic structure and thermodynamics of alloys", Ph.D. Thesis, Columbia University.
Sigli, C. and Sanchez, J. M. (1985) *Acta Met.*, **34,** 1021.
Silverman, A., Zunger, A., Kalish, R. and Adler, J. (1995) *J. Phys.* **7**(6), 1167.
Sluiter, M. and Turchi, P. E. A. (1989) *Phys. Rev. B*, **40,** 11,215.
Sluiter, M., Turchi, P. E. A. and de Fontaine, D. (1987) *J. Phys. F*, **17,** 2163.
Sluiter, M., Turchi, P. E. A., Pinski, F. J. and Stocks, G. M. (1992) *J. Phase Equilibria*, **13,** 605.

Sluiter, M., Turchi, P. E. A., Zezhong, F. and de Fontaine, D. (1988) *Phys. Rev. Lett.*, **60**, 716.
Sundman, B. and Mohri, T. (1990) *Z. Metallkde.*, **81**, 251.
Sundman, B. and Aldinger, F. (1995) Ringberg (I) Workshop, *CALPHAD*, **19**, 433.
Tepesch, P. D., Kohan, A. F., Garbulsky, G. D., Ceder, G., Coley, C., Stokes, H. T., Boyer, L. L., Mehl, M. J., Burton, B. P., Cho, K. and Joannopoulos, J. (1996) *J. Am. Cer. Soc.*, **79**, 2033–2040.
Tetot, R., Giaconia, C., Nacer, B. and Boureau. G (1994) in *Statics and Dynamics of Phase Transformations*, eds Turchi, P. E. A. and Gonis, A. (Plenum Press, New York), p. 577.
Tso, N. C, Sanchez, J. M. and Tien, J. K (1989) in *Superalloys, Supercomposites and Superceramics*, eds Tien, J. K. and Caulfield, T. (Academic Press, New York). p. 525.
Tso, N. C. (1992) "Thermodynamic modelling of ternary alloy phase solubility", Ph.D. Thesis, Columbia University.
Turchi, P. E. A. (1994) in *Intermetallic Compounds, Vol 1*, eds Westbrook, J. H. and Fleischer, R. L. (John Wiley, New York), p. 21.
Turchi, P. E. A. and Finel, A. (1992) *Phys. Rev. B*, **46**, 702.
Turchi, P. E. A. and Sluiter, M. (1992) *Mat. Res. Soc. Symp.*, **253**, 227.
Turchi, P. E. A. and Sluiter, M. (1993) in *Computer Aided Innovation of New Materials II*, eds Doyama, M. *et al.* (Elsevier, Amsterdam), p. 1449.
Turchi, P. E. A, Reinhard, L. and Stocks, G. M. (1994a) *Phys. Rev. B.*, **50**, 15,542.
Turchi, P. E. A., Singh, P. P., Sluiter, M. and Stocks, G. M. (1994b) in *Metallic Alloys: Experimental and Theoretical perspectives*, eds Faulkner, J. S. and Jordan, R. G. (Kluwer Acad.), p. 177.
Turchi, P. E. A., Sluiter, M., Pinski, F. J., Johnson, D., Nicholson, D. M., Stocks, G. M. and Staunton, J. B. (1991) *Phys. Rev. Lett.*, **67**, 1779.
Turchi, P. E. A., Stocks, G. M., Butler, W. H., Nicholson, D. M. and Gonis, A. (1988) *Phys. Rev. B*, **37**, 5982.
van Baal, C. M. (1973) *Physica*, **64**, 571.
Wei, S.-H., Ferreira, L. G. and Zunger, A. (1990) *Phys. Rev. B*, **41**,(12), 8240.
Wolverton, C. and Zunger, A. (1997) *J. Comp. Mat. Sci* **8** (1–2), 107.
Yang, C. N. and Li, Y. (1947) *Chinese J. Phys.*, **7**, 59.
Zunger, A. (1993) *Japanese J. Appl. Phys.*, **32**(Supplement 3), 14.
Zunger, A. (1994) in *Statics and Dynamics of Phase Transformations*, eds Turchi, P. E. A. and Gonis, A. (Plenum Press, New York), p. 361.

Chapter 8

The Role of Magnetic Gibbs Energy

Chapter 8
The Role of Magnetic Gibbs Energy

8.1. INTRODUCTION

In paramagnetic materials there is no polarisation of electron spins and therefore it is unnecessary to consider a magnetic contribution to the Gibbs energy if this condition is taken to be the standard state. However in ferromagnetic, anti-ferromagnetic and ferri-magnetic materials, there is also competition between different spin arrangements and the enthalpy of certain transition metals, rare earths and their associated alloys and compounds are lowered by specific forms of spin polarisation. This includes the technologically important elements Fe, Ni and Co and their alloys and the effect can be of sufficient magnitude to have a major effect on phase transformation.

There are many different forms of spin polarisation. The *ferromagnetism* exhibited by b.c.c. Fe is probably the best-known variant. Here the coupling between atoms favours parallel spin configurations between nearest neighbours and the critical ordering temperature is known as the Curie temperature (T_c). In other materials the exchange forces favour anti-parallel spins between nearest neighbours. If there is only one species of atoms this is known as *anti-ferromagnetism* and the critical ordering temperature is known as the Neel temperature (T_n). If the anti-parallel coupling involves more than one species of atoms with different values of spin, there will be a net spin in one particular direction. Such a material, e.g., Fe_3O_4, is considered to be *ferri-magnetic* and again behaves like a ferromagnetic with a Curie temperature.

Many other subtle differences can occur, such as periodic changes in the crystallographic direction in which the spins are orientated, but these details are rarely considered in thermodynamic terms. In the treatments which follow, it is sufficient to note that the quantum number $s = +\frac{1}{2}$ or $-\frac{1}{2}$ defines the spin-up or spin-down direction for individual electrons, while the magnitude of any magnetic effects depends on the number of electrons that are being polarised. The unit magnetic moment associated with spin $\frac{1}{2}$ is equal to a Bohr magneton (μ_B) and the number of magnetic electrons/atom is defined by the symbol β, so that $\beta = 2s$ (Bozorth 1956).

Unlike chemical ordering, which can only occur in alloys, magnetic ordering can occur in unary systems and the magnetic Gibbs energy turns out to be sufficiently large to cause fundamental changes in structure. For example, the high-temperature form of solid Fe is δ which has a b.c.c. structure and transforms to γ-Fe with the more close-packed f.c.c. structure at 1394°C. Both of these phases are paramagnetic at these elevated temperatures, but the onset of ferromagnetism in the b.c.c. phase causes this phase to reappear at 912°C as α-Fe. This makes Fe a unique element in that the high-temperature allotrope reappears at low temperatures, and it should be noted that the thermodynamic effect due to short-range magnetic order is already sufficient to cause the re-appearance of the b.c.c. form; α-Fe only becomes ferromagnetic below its Curie temperature at 770°C. At 0 K the magnetic enthalpy ($\approx$9000 J mol^{-1}) is an order of magnitude larger than was involved in the $\delta-\gamma$ high temperature transformation ($\approx$900 J mol^{-1}).

The change in Gibbs energy with temperature of Fe, with respect to the paramagnetic form δ-Fe, is shown in Fig. 8.1(a). This can be correlated directly with the value of the magnetic specific heat (Fig. 8.1(b)) (Nishizawa *et al.*. 1979), which in turn is related to a corresponding change in the saturation magnetisation, β, (Fig. 8.1(c)) (Nishizawa 1978). The critical magnetic ordering temperature, the Curie temperature (T_c), is defined either by the peak in C_p^{mag} or by the maximum rate of change in the variation of β with T. The situation is made more complex on alloying as both T_c and β vary with composition.

It has already been mentioned in the previous chapter that the Ising model, which underlies the formalism used for most chemical ordering treatments, was originally used to describe magnetic transitions. It is, however, technically only valid if the various magnetic states correspond to multiples of the unit magnetic spin, s. Early attempts to describe the associated thermodynamic functions within formal ordering theory were by Saito *et al.* (1959), Arita (1978) and Moran-Lopez and Falicov (1979). More recent examples of the combination of chemical and magnetic ordering, as applied to b.c.c. Fe-base alloys, have been given by Inden and Pitsch (1991) and Kosakai and Miyazaki (1994). The BWG treatment appears to be a reasonable assumption in such systems because the saturation magnetisation of Fe (2.2 μ_B) is close to the integral value $\beta = 2$ which is equivalent to $s = 1$. However, the concentration dependence of the saturation magnetisation, and the important role played by magnetic short-range order, makes it difficult to use such a treatment in the general case and it has been necessary to devise alternative methodologies to describe the magnetic Gibbs energy of alloy systems.

8.1.1 Polynomial representation of magnetic Gibbs energy

While acknowledging that the 'Gibbs energy anomaly' was associated with magnetism, early attempts to characterise the behaviour of Fe, did not derive G^{mag} explicitly from ferromagnetic parameters such as β. Instead, G^{mag} was derived either by the direct graphical integration of C_p^{mag} or simply incorporated into the overall Gibbs energy difference between the γ- and α-phases in Fe by means of a global polynomial expression such as used by Kaufman and Nesor (1973)

References are listed on pp. 256–258.

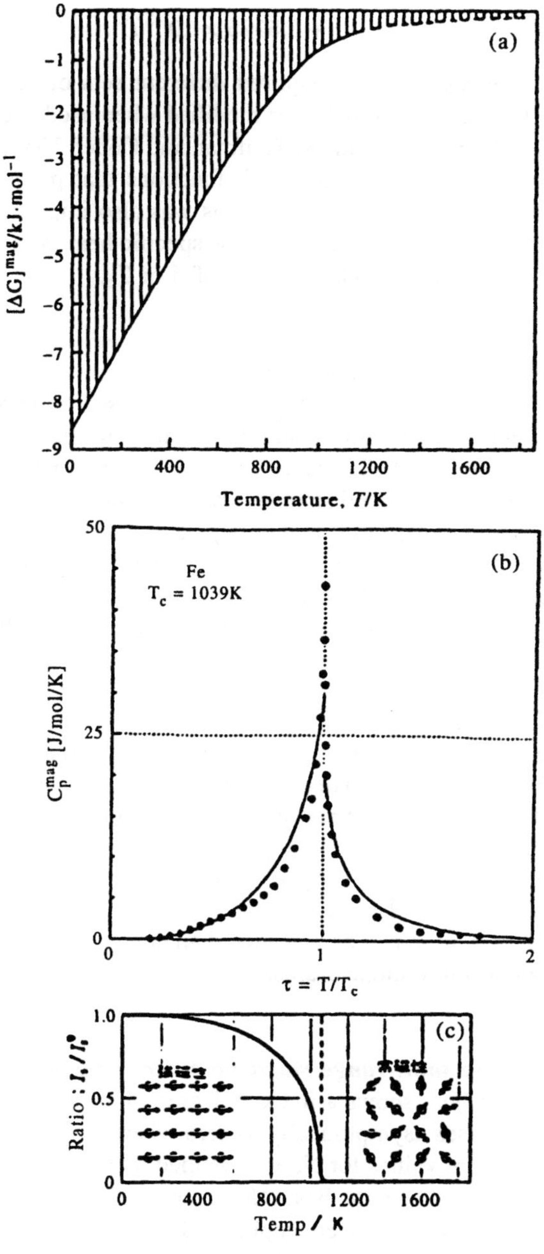

Figure 8.1. Correlation between (a) magnetic Gibbs energy, (b) magnetic specific heat (from Nishizawa 1992) and (c) change of saturation magnetisation for pure Fe (from Nishizawa 1978).

$$300 < T < 1100K \qquad G^{\gamma \rightarrow \alpha} = 6109 - 3.462T - 0.747217 \times 10^{-2}T^2 + 0.5125 \times 10^{-5}T^3. \tag{8.1}$$

Different temperature regions then required different sets of coefficients and, in specific systems, the coefficients had to be modified when used to describe the low-temperature martensitic transformation (Kaufman and Cohen 1956). It soon became evident that it was preferable to work with functions that properly reflected the physical origin of the extra Gibbs energy. This was also important when moving from pure Fe to increasingly complex steels, as specific heat measurements are not often as generally available as measurements of β or T_c.

8.1.2 Consideration of the best reference state

The first step is to consider the methods which can be used to represent the change of magnetisation with temperature, and how this can be translated into the corresponding change of G^{mag}. By analogy with configurational ordering, it would be reasonable to choose the ground state at 0 K as the fundamental reference point. This view was taken by Zener (1955) who may be considered the first author to try and obtain an explicit description of G^{mag}. However, many important phase changes are associated with the high-temperature behaviour, especially the region where short-range magnetic order is present. Starting at 0 K therefore presents certain problems. Firstly, an accurate representation of the temperature dependence of the magnetic parameters is required, and this can be highly non-linear. Secondly, anomalies can arise if there is a rapid variation of the magnetic ordering temperature with composition (Inden 1991). This is particularly true if T_c reaches zero in the middle of a system. Finally, it is necessary to devise a system which can take care of mixtures of magnetic and non-magnetic components.

In contrast to most treatments of configurational ordering, the high-temperature paramagnetic state has therefore been adopted as the best reference state. Objections to using this state on the grounds that it cannot be retained below T_c are unwarranted, as the situation can be considered analogous to using the liquid phase as a reference state at high temperatures.

8.1.3 Magnitude of the short-range magnetic order component

The magnetic specific heat shows a marked degree of short-range magnetic order above T_c which decreases asymptotically to zero when $T >> T_c$. Figure 8.2 illustrates how the equivalent values for Ni are obtained by careful subtraction of other components from the total specific heat (Hofmann *et al.* 1956). The importance of standardising this procedure has been emphasised by de Fontaine *et al.* (1995). The fraction (ϕ) of the total magnetic enthalpy retained above T_c is clearly always an important quantity and, by analysing experimental results, Inden (1976) obtained a

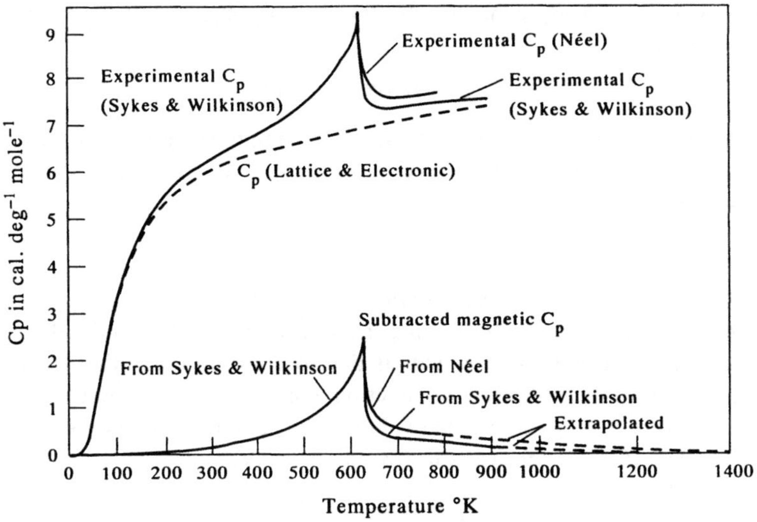

Figure 8.2. Extraction of magnetic specific heat from the total specific heat for pure Ni (from Hoffman *et al.* 1956).

value of $\phi = 0.27$ for both f.c.c. Ni and Co while the larger value of $\phi = 0.43$ was found for b.c.c. Fe. These values were subsequently assumed equally valid for all other f.c.c. and b.c.c. phases (Hillert and Jarl 1978, Inden 1981a). While this is a reasonable assumption for steels, it may not hold for atoms with larger magnetic moments (de Fontaine *et al.* 1995). This has been confirmed by a CVM treatment applied to Gd (Schon and Inden 1996). In order to incorporate this result into a semi-empirical treatment suitable for multi-component systems it may be worthwhile to revert to an earlier suggestion (Paskin 1957) that ϕ is a function of s and the nearest-neighbour co-ordination number, (z),

$$\phi = \frac{s+1}{s(z-1)}. \tag{8.2}$$

8.2. DERIVATION OF THE MAGNETIC ENTROPY

While it is important to partition the long-range and short-range magnetic components correctly, the maximum entropy contribution due to magnetism, $S_{\max}^{\text{mag}}$, is an equally crucial factor.

8.2.1 Theoretical value for the maximum magnetic entropy

Despite the intrinsic complexity of magnetic phenomena, there is general agreement that the maximum entropy generated by the de-coupling of spins between atoms can be expressed by:

$$S^{\rm mag}_{\rm max} = R\log_{\rm e}(2s+1). \tag{8.3}$$

Technically, the values of s are theoretically restricted to integral multiples of s and one should consider the sum of each contributing species, but these values are not always available.

8.2.2 Empirical value for the maximum magnetic entropy

The saturation magnetisation β, in μ_B per atom, of many materials often corresponds to non-integral values of s. This is due to contributions other than s being involved, for example polarised conduction electrons. It is, therefore, general practice to substitute the experimental value of the saturation magnetisation at 0 K, β_0, for $2s$ in Eq. (8.3) which leads to (Miodownik 1977)

$$S^{\rm mag}_{\rm max} = R\log_{\rm e}(\beta_0+1). \tag{8.4}$$

8.2.3 Explicit variation in entropy with magnetic spin number and temperature

The magnetic entropy may be rigorously specified if the BWG mean-field approximation is combined with specific values of the magnetic spin. The expression for the magnetic entropy corresponding to $s = ½$, as a function of the degree of order η, is identical to that already given in Eq. (7.2) of the previous chapter on configurational ordering

$$S^{\rm mag}_{\rm BWG} = -Nk_B\left[\left(\frac{1+\eta}{2}\right)\log_{\rm e}\left(\frac{1+\eta}{2}\right)+\left(\frac{1-\eta}{2}\right)\log_{\rm e}\left(\frac{1-\eta}{2}\right)\right]. \tag{8.5}$$

However, the permutation of permissible spins rapidly becomes more complicated with higher values of s. The equivalent expression for $s = 1$, as derived by Semenovskaya (1974) and used by Inden (1975, 1981), is given by

$$\begin{aligned} S^{{\rm mag},s=1}_{\rm BWG} = &- Nk_B\Big[\log_{\rm e}(8-6\eta+2\sqrt{4-3\eta^2}) - (1-\eta)\log_{\rm e}(2(1-\eta)) \\ &- (1+\eta)\log_{\rm e}(\eta+\sqrt{4-3\eta^2})\Big]. \end{aligned} \tag{8.6}$$

Expressions corresponding to $s = 3.2$ and $s = 2$ are listed in Inden (1981).

8.3. DERIVATION OF MAGNETIC ENTHALPY, $H^{\rm MAG}$

8.3.1 Classical derivation

The simplest case is to start with a situation where all the atoms have the same spin.

The general expression for the BWG (mean field) approximation then gives

$$H^{\text{mag}} = \frac{N}{2}\sum_k z^{(k)} J^{(k)} \eta^2 s^2. \tag{8.7}$$

This allows for a summation for magnetic interactions in successive (k-th) neighbouring shells but, unless J values can be derived from first-principles calculations, there is generally insufficient experimental data to allow anything other than a single nearest-neighbour interaction parameter to be used. This, then, must be taken to incorporate any other longer-range effects, and application to real alloy systems is also technically restricted by having to use multiples of $s = 1/2$. Much therefore depends on whether this realistically simulates experimental values of β, which fortunately seem to be the case in Fe-rich, Co-rich and Ni-rich alloys. However, this is less secure at higher solute concentrations, especially if there are transitions between different forms of magnetism in the system (see Fig. 8.4).

An equation which includes the effect of adding a magnetic term to the enthalpy of an ordered system has already been given in Section 7.3.2.3 of Chapter 7, but no details were given of how to determine the value of the magnetic interaction energy $M^{(1)}_{A,B}$. For a binary alloy $A_{1-x}B_x$ this involves the introduction of three magnetic interaction parameters $J^{(1)}_{AA}$, $J^{(1)}_{BB}$ and $J^{(1)}_{AB}$ which describe the strength of the coupling between nearest-neighbour atoms whose (magnetic) electrons have parallel spins as compared to anti-parallel spins. The magnetic interchange energy $M^{(1)}_{AB}$ is then obtained by combining these parameters to yield

$$M^{(1)}_{AB} = J^{(1)}_{AA} + J^{(1)}_{BB} - 2J^{(1)}_{AB}. \tag{8.8}$$

This is entirely analogous to the treatment of the chemical interaction energies V_{ij} treated in Chapter 7. Negative values of $J_{i,j}$ correspond to ferromagnetism while positive values correspond to anti-ferromagnetism. Positive values of $M_{i,j}$ imply that the strength of the ferro-magnetic coupling between unlike atoms is stronger than between like atoms. Most treatments do not consider magnetic interactions beyond the nearest-neighbour coordination shell.

The most convenient way of determining the interchange energies is from experimental values of the critical ordering temperature, which is designated the Curie temperature for ferromagnetic ordering The Curie temperature for alloys with two magnetic components having the same spin (1/2 or 1) can be written as follows:

$$T_c = \mu T_c^{\text{BWG}} =$$
$$\left\{\frac{-8K_s}{\alpha(1-x)+\beta x}\left[(1-x)\frac{J^{(1)}_{AA}}{k} + x\frac{J^{(1)}_{BB}}{k} - x(1-x)\frac{M^{(1)}_{A,B}}{k} - \eta^2\frac{M^{(1)}_{A,B}}{k}\right]\right\}. \tag{8.9}$$

In this equation, μ is the magnetic analogue to the χ factor used in the chemical ordering case to compensate for the higher critical temperatures always generated by the BWG formalism (see Chapter 7). Comparing the results from the BWG treatment with those of other more sophisticated treatments leads to a value of $\mu = 0.8$ (Inden 1975) and the long-range ordering parameter η takes values between 0 and 1 as in the case of chemical ordering. In the magnetic case η equals the ratio of the mean spin value per atom at temperature T to the maximum value of s (at 0 K). K_s is a parameter which varies with the magnitude of the magnetic spin s such that $K_s = 1$ for $s = 1/2$ and $K_s = 2/3$ for $s = 1$. The coefficients α and β can be set either to 0 or 1 to cover various combinations of magnetic and non-magnetic components.

8.3.2 Empirical derivation

Given that S^{mag}_{max} can be described using Eq. (8.4), it may be asked whether a corresponding value for H^{mag}_{max} can be obtained by combining S^{mag}_{max} with T_c. If magnetic ordering was a first-order transformation with a critical temperature, T^*, it would then follow that (Fig. 8.3).

$$H^{mag}_{max} = T^* S^{mag}_{max}. \tag{8.10}$$

However, since magnetic ordering is a second-order transformation, this has to be

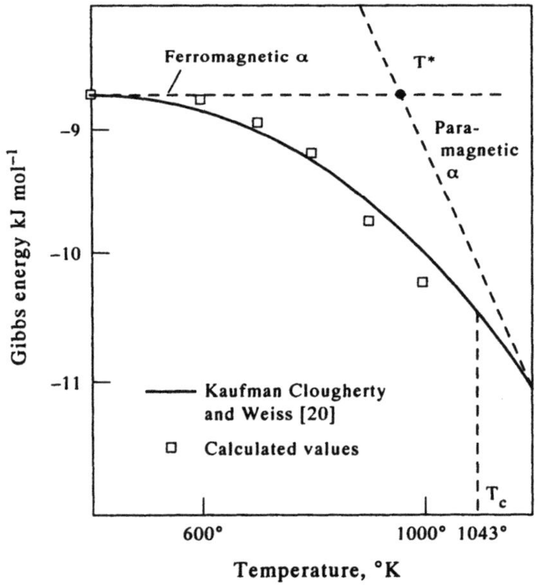

Figure 8.3. Relation of an effective first-order transformation temperature (T^*) to the real second-order Curie temperature (T_c) (from Miodownik 1977).

modified to take into account the proportion of short- and long-range order on either side of T_c (see Section 8.1.3). Miodownik (1977) assumed that $T^* \approx 0.9T_c$ for Fe, Co, Ni and their alloys. It then follows that

$$H_{\max}^{\text{mag}} = 0.9T_c \log_e(\beta_0 + 1). \tag{8.11}$$

Weiss and Tauer (1958) made the alternative assumption that

$$H_{\max}^{\text{mag}} \approx RT_c s^{1/3}. \tag{8.12}$$

Such equations have the advantage of simplicity but comparison with Eq. (8.2) implies that ϕ is totally independent of s and z. Whichever expression is used, the next step is to incorporate the effect of temperature on the alignment of the magnetic spins. In view of the other simplifications which have already been adopted, this can be achieved (Miodownik 1977, 1978a) by approximating the Brillouin–Langevin formalism for the number of aligned spins (β^T) remaining at $T < 0.9T_c$ through the following expression:

$$\beta^T = \beta_0\left[1 - \tau^6\right] \tag{8.13}$$

where τ defines the ratio (T/T_c). Assuming that the magnetic enthalpy scales as (β^T/β_0) this can then be combined with an arbitrary power function to account for short-range magnetic order for $T > 0.9T_c$)

$$\beta^T = \beta_0(1/2)^{[2+10(\tau-1)]}. \tag{8.14}$$

Equation (8.14) leads to a value for $\phi = 0.25$ at T_c, which is of comparable magnitude to the values assumed by Inden (1981a), although it clearly does not include any dependence on the co-ordination number z.

8.4. DERIVATION OF MAGNETIC GIBBS ENERGY

Even if magnetic interaction energies are available to define a magnetic enthalpy, the development of a viable Gibbs energy expression is difficult. As already noted in the preceding section, the expressions for the magnetic entropy in a BWG treatment become increasingly complicated when $s > 1/2$ and, indeed, may become inexact if the distribution of individual spins is not known. In practice their application also requires a series of nested Gibbs energy minimisation calculations which become costly in computing time. Also, of course, the BWG approximation does not take into account any short-range order except through empirical corrections.

Despite these drawbacks remarkably good results have been obtained in the equiatomic region of the Fe-Co system (Inden 1977) which are virtually indistinguishable from those obtained by CVM calculations (Colinet 1993). However, this region deals only with A2/B2 equilibria where the BWG method is known to work well. It is otherwise inadequate to deal with f.c.c.–b.c.c. equilibria, notably in pure Fe or Fe-rich alloys, because of the very small differences in the free energies of the f.c.c. and b.c.c. allotropes. In such cases it is vitally important to include short-range magnetic order and find alternative means of defining the magnetic Gibbs energy.

8.4.1 General algorithms for the magnetic Gibbs energy

With expressions for the magnetic enthalpy and entropy it is now possible to build up an algorithm for G^{mag} at any temperature. The maximum magnetic entropy is taken as

$$S_{\mathrm{max}}^{\mathrm{mag}} = S_{T=\infty}^{\mathrm{para}} - S_{T=0}^{\mathrm{ferro}}. \tag{8.15}$$

In an analogous way the maximum magnetic enthalpy is

$$H_{\mathrm{max}}^{\mathrm{mag}} = H_{T=\infty}^{\mathrm{para}} - H_{T=0}^{\mathrm{ferro}}. \tag{8.16}$$

The magnetic entropy and enthalpy at a given temperature, S_T^{mag} and H_T^{mag}, respectively, are

$$S_T^{\mathrm{mag}}, H_T^{\mathrm{mag}} = f(T, T_c, \beta) \tag{8.17}$$

while the magnetic Gibbs energy G_T^{mag} at a given temperature is given by

$$G_T^{\mathrm{mag}} = G_T^{\mathrm{ferro}} - G_T^{\mathrm{para}}. \tag{8.18}$$

8.4.2 Magnetic Gibbs energy as a direct function of β and T_c

Combining the expressions for magnetic entropy and enthalpy and assuming that the magnetic enthalpy scales as β^T/β_0

$$G_T^{\mathrm{mag}} = -0.9RT_c \log_e(\beta_0 + 1)(\beta^T/\beta_0) - RT \log_e(\beta_0 - \beta^T + 1). \tag{8.19}$$

Here the first term represents H_T^{mag} and the second term the effect of S_T^{mag}. Eq. (8.19) represents one of the many empirical ways in which Eq. (8.18) can be made to work in practice (Miodownik 1977). However, this expression does not lend itself easily to a derivation of the associated value of C_p nor to an explicit formulation of $\delta\Delta G^{\mathrm{mag}}/\delta T$. Therefore, most alternative expressions have been based on approximate analytical expressions for the magnetic specific heat. However, all such expressions also incorporate the functions $\log_e(\beta + 1)$ and τ and are a far cry from the original methods of graphical integration.

8.4.3 Magnetic Gibbs energy as a function of C_p^{mag} *for ferromagnetic systems*

8.4.3.1 The model of Inden. This approach was pioneered by Inden (1976) who developed the following empirical equations:

for $\tau < 1$ $$C_p^{\mathrm{mag}} = K^{lro} R \log_e(1+\tau^3)/\log_e(1-\tau^3) \tag{8.20a}$$

for $\tau > 1$ $$C_p^{\mathrm{mag}} = K^{sro} R \log_e(1+\tau^5)/\log_e(1-\tau^5) \tag{8.20b}$$

where K^{lro} and K^{sro} are empirically derived coefficients. At first sight these equations appear to differ substantially from Eq. (8.19), but the values of K^{lro} and K^{sro} are constrained by the need to correctly reproduce both the total entropy and take into account experimental values of ϕ. Since the total entropy is an explicit function of β, the net result is that this treatment also has a mixed dependence on the two key magnetic parameters β and T_c. Further, a relationship between K^{sro} and K^{lro} can be obtained by considering H_{mag}:

$$K^{sro} = 518/675(K^{lro} + 0.6K^{sro}) = \log_e(\beta+1) \tag{8.21}$$

$$K^{lro} = 474/497[(1-\phi)/\phi]K^{sro}. \tag{8.22}$$

A dependence on crystallographic parameters is introduced through making ϕ a function of z, as already indicated in Section 8.1.3. Introducing fixed values of ϕ appropriate to each crystal structure is a useful simplification as it is then possible to uniquely define the values of K^{lro} and K^{sro} (de Fontaine *et al.* 1995).

8.4.3.2 Model of Hillert and Jarl. In his original treatment, Inden (1976) used a complicated but closed expression for the enthalpy, but had to use a series expansion in order to calculate the entropy. Hillert and Jarl (1978) therefore decided to convert the C_p expression directly through a series expansion which substantially simplifies the overall calculation and leads to a maximum error of only 1–2 J/mol at the Curie temperature of Fe. The equivalent equations to those used by Inden (1976) are given by

for $\tau < 1$ $$C_p^{\mathrm{mag}} = 2K^{lro} R(\tau^m + \tfrac{1}{3}\tau^{3m} + \tfrac{1}{5}\tau^{5m}) \tag{8.23a}$$

for $\tau > 1$ $$C_p^{\mathrm{mag}} = 2K^{sro} R(\tau^{-n} + \tfrac{1}{3}\tau^{-3n} + \tfrac{1}{5}\tau^{-5n}). \tag{8.23b}$$

8.4.3.3 Alternative* C_p *models. Chuang *et al.* (1985) have developed alternative exponential functions to describe the C_p curves and applied this to various Fe-base alloys (Chuang *et al.* 1986)

for $\tau < 1$ $$C_p = K' R\tau \exp(-p(1-\tau)) \tag{8.24a}$$

for $\tau > 1$ $$C_p = K'' R\tau \exp(-q(1-\tau)). \tag{8.24b}$$

These equations have the advantage of greater mathematical simplicity, but it is still necessary to evaluate the constants K', K'', p and q.

8.4.3.4 Comparison of models for the ferromagnetic Gibbs energy. It is difficult to compare the various treatments described in Sections 8.4.3.1 to 8.4.3.3, because simultaneous changes were made in both the models and input parameters during their formulation. This is a perennial problem if different weightings are attached to magnetic and thermodynamic data (de Fontaine *et al.* 1995). The suggestion by Inden (1976) that values for m and n in Eqs (18.18a) and (18.18b) should be taken as 3 and 5 respectively has been preserved by Hillert and Jarl (1978). This is still the core assumption as far as $3d$ elements are concerned in current models. However, it may be necessary to relax this assumption in systems with larger magnetic spin numbers (de Fontaine *et al.* 1995)

8.4.4 Anti-ferromagnetic and ferri-magnetic systems

Providing there is no change in value of β_0 with temperature, Eq. (8.4) can also be used to determine the maximum magnetic entropy in anti-ferromagnetic and ferri-magnetic materials (Miodownik 1978a, Smith 1967, Hofmann *et al.* 1956). A comparison of predicted and experimental magnetic entropies and energies (Table 8.1) indicates that this is a reasonable assumption in most cases. If serious discrepancies remain between theoretical expectations and experimental results, it may be necessary to consider the existence of multiple magnetic states (see Section 8.7).

8.5. THE EFFECT OF ALLOYING ELEMENTS

The treatment of compounds and other end-members of fixed composition does not differ from that of the elements, but the next vital step is to consider the representation of these parameters for solid solutions.

Table 8.1. A comparison of predicted and experimental magnetic entropies and enthalpies.

Element	Exp. S^{mag}_{max} (J/mol/K)	Calc. (Eq. (8.4)) (J/mol/K)	Exp. H^{mag}_{max} (J/mol)	Calc. (Eq. (8.11)) (J/mol)
Ni	3.39	3.47	1757	1987
Fe	8.99	9.20	8075	8556
Gd	17.74	17.41	3372	4602
Dy	23.09	23.05	—	—
Ho	23.56	23.56	—	—
MnO_2	10.87	11.51	816	900
NiF_2	9.46	9.12	590	657
Cr_2O_3	22.72	23.22	5314	6255

References are listed on pp. 256–258.

8.5.1 Effect of changes in T_c *and* β *with composition*

The earliest derivation of G_T^{mag} was made by Zener (1955) who postulated that the effect of an alloying element was proportional to its effect on T_c. With the assumption that $\delta Tc/\delta x$ is a constant, it is then possible to write:

$$G_{\rm alloy}^{\rm mag}(T) = G_{\rm Fe}^{\rm mag}(T) - x\left(\frac{\delta T}{\delta x}\right)S_{\rm Fe}^{\rm mag} \qquad (8.25)$$

where $T = T_c - x(\delta T/\delta x)$. The Zener method therefore depends totally on the assumption of a linear variation of $\delta T_c/\delta x$ and any associated changes in β are reflected in the change of T_c. It can, nevertheless, give reasonable results for dilute solutions and was extensively used by Hillert *et al.* (1967) in the period before the Inden formulation was more generally adopted in Europe. It, however, remained popular in Japan for the characterisation of Fe alloys, which was its original field of application (Nishizawa *et al.* 1979, Hasebe *et al.* 1985).

Zener's method does not explicitly take into account the effect of changes in β with composition but implicitly assumes that any changes in β are reflected in the changes in T_c. Clougherty and Kaufman (1963) considered how changes in β with composition might affect the terminal magnetic entropy of f.c.c. Ni–Zn alloys, but this did not lead to a generalised approach which was applicable across the whole system. Miodownik (1977) showed that values of T_c and β for alloys could be introduced into the same algorithm developed for the elements, but this was superseded by the advent of the Hillert–Jarl formalism. With the need to handle multi-component calculations it was then a natural step to develop (separate) Redlich–Kister polynomials for the variation of β and T_c with composition (Chin *et al.* 1987).

8.5.2 Systems whose end-members exhibit different forms of magnetism

8.5.2.1 Ferromagnetic to anti-ferromagnetic transition. When end-members exhibit different forms of magnetism there are often significant deviations from linearity in both T_c and β and these parameters may become zero in the middle of the system. One example is associated with a change from ferromagnetic to anti-ferromagnetic behaviour which occurs in a number of industrially important systems, such as Fe–Ni (Fig. 8.4(a)) and the related ternary system Fe–Ni–Mn (Fig. 8.4(b)). b.c.c. Fe–Cr, rare-earth and oxide systems can also exhibit similar complexity. In such situations special measures need to be taken to allow a Redlich–Kister-type polynomial to be used to describe T_c and β. Since ferromagnetism and anti-ferromagnetism are associated with values for $J_{i,j}$ of opposite sign, Weiss and Tauer (1956) suggested that a continuous change between these two forms of magnetic ordering could best be described by attaching a negative sign to Neel temperatures. These authors also suggested a semi-empirical

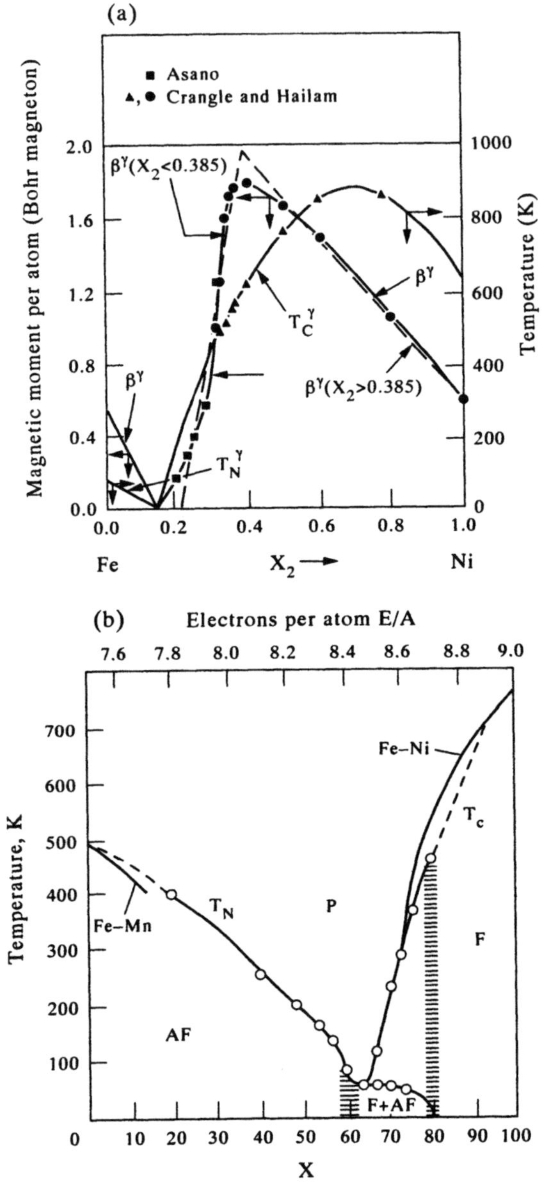

Figure 8.4. Variation of magnetic parameters T and β with composition in (a) Fe–Ni (from Chuang *et al.* 1986), (b) Fe–Ni–Mn alloys (from Ettwig and Pepperhoff 1974).

relation between T_{crit} and β which is useful if one or other of these parameters is not experimentally available.

$$T_{\text{crit}} = 113.5(Z\uparrow - Z\downarrow)\ln(\beta + 1). \tag{8.26}$$

Here the two parameters, $Z\uparrow$ and $Z\downarrow$, refer to the relative number of spins of opposite sign, which allows this equation to be used for both ferromagnetic and anti-ferromagnetic materials. One implication of Eq. (8.26) is that, for a given value of β and type of magnetism, the critical temperature is expected to be higher for f.c.c. materials than for b.c.c. materials, which accounts for the observation that $T_c^{\text{Co}} > T_c^{\text{Fe}}$ although $\beta^{\text{Co}} < \beta^{\text{Fe}}$. There are, however, problems when applying this simple empirical equation to anti-ferromagnetic BCC materials such as Cr and to structures with lower symmetry.

A more important conclusion of Eq. (8.26) is that, for a given value of β in an f.c.c. lattice, the value of T_n derived for an anti-ferromagnetic alloy will be 1/3 of the equivalent value of T_c in the ferromagnetic case. The utility of this approach has been verified by Chin *et al.* (1987) who, in order to optimise parameters for their Redlich–Kister polynomials for the critical temperatures of various binary alloy systems, converted positive values of T_n to negative values of T_c and were able to reproduce the critical temperature variation arising from the ferromagnetic/anti-ferromagnetic transition. These authors also used the same convention to differentiate ferromagnetic and anti-ferromagnetic values of β when adapting the Hillert–Jarl formalism into Thermo-Calc, although this must be considered purely as a mathematical convenience and has no theoretical justification.

Another simplification that has been made with respect to the implementation of magnetic algorithms into software packages, such as Thermo-Calc, relates to the composition at which T_c and β reach zero. One might expect that this would happen at one particular composition but Figs. 8.4(a) and (b) show that the situation may be much more complicated. In order to be consistent with a simplified magnetic model Chin *et al.* (1987) chose a set of parameters which effectively smoothed out the differences in the two critical compositions for binary alloys. This however precludes a proper prediction of regions of mixed magnetism, which is important in Invar alloys (Miodownik 1978b) as well as making it impossible to reproduce effects such as shown in Figs. 8.4(a) and (b).

8.5.2.2 Ferromagnetic–paramagnetic transition. In other cases, a zero T_c may be associated with the transition between ferromagnetism and paramagnetism. Even the use of higher-order Redlich–Kister polynomials cannot properly reproduce such a sudden change in slope and some smoothing of the measured variation of T_c and β has to be accepted in the vicinity of 0 K. As both enthalpy and entropy contributions scale approximately with T_c, at low temperatures, this will only lead to small errors on the overall Gibbs energy at such compositions. However, under certain circumstances spurious miscibility gaps have been predicted at high temperatures in systems that exhibit abrupt changes in T_c with composition (Inden 1981). Care has, therefore, to be taken in deriving the input parameters to the Hillert–Jarl formalism in such cases. The fact remains that such miscibility gaps have never been observed in practice and it may, therefore, be that as T_c approaches zero some other effects, such as itinerant ferromagnetism (Wohlfarth 1974), may occur which essentially smooth out the variation of magnetic entropy. In this instance, the Redlich–Kister formalism may actually be a better approximation to reality than would be indicated by trying to reproduce a T_c or T_n obtained by linear extrapolation of experimental results to 0 K.

8.6. THE ESTIMATION OF MAGNETIC PARAMETERS

8.6.1 Magnetic versus thermochemical approaches to evaluating the magnetic Gibbs energy

In principle, the value of G_T^{mag} should be the same whether it is derived from knowledge of β or from the magnetic specific heat. Since the latter is derived by subtraction of other major terms from the total specific heat, such an agreement is a useful confirmation that the deconvolution process has been properly conducted. However, this is not always achieved. In some studies, e.g., for Cr (Andersson 1985), Co (Fernandez-Guillermet 1987) and Ni (Dinsdale 1991), the listed β values are 'effective thermochemical moments' which have been derived on the basis of a global thermochemical assessment rather than from magnetic measurements. Small differences will always arise when considering ways of averaging data and truncating series, but one should be cautious of a tendency for β to become an adjustable parameter. Such discrepancies will hopefully be reduced in the future by the incorporation of better magnetic models (de Fontaine *et al.* 1995).

8.6.2 Values of the saturation magnetisation, β

It should be emphasised that it is the rule rather than the exception for β to change markedly with crystal structure (Table 8.2). It is therefore unwise to assume that various metastable allotropes can be given the same value of β as for the stable structure. In some cases values of β can be extrapolated from stable or metastable alloys with the requisite crystal structure, but in others this is not possible. A significant development is that it is now possible to include spin polarisation in electron energy calculations (Moruzzi and Marcus 1988, 1990a, b, Asada and Terakura 1995). This allows a calculation of the equilibrium value of β_0 to be made in any desired crystal structure. More importantly, such values are in good accord with known values for equilibrium phases (Table 8.2). It has also been shown that magnetic orbital contributions play a relatively minor role (Eriksson *et al.* 1990), so calculated values of β for metastable phases should be reasonably reliable.

References are listed on pp. 256–258.

Table 8.2. Comparison of SGTE (*) first-principle and experimental magnetic moments for selected elements (adapted from de Fontaine *et al.* 1995)

		Ferromagnetic moment (μB)			Anti-ferromagnetic moment (μB)		
Element	Source	b.c.c.	f.c.c.	c.p.h.	b.c.c.	f.c.c.	c.p.h.
Cr	Chin *et al.* 1987*			0	0.4	0.82	0
Cr	Dinsdale 1991*			0	0.008	0.82	0
Cr	Asada 1993 (FP)	0	0	0	0/0.5	0/3.0	0
Cr	Moruzzi 1990 (FP)	0			0/0.6		
Cr	Weiss 1979				0/0.7		
Cr	Exp (β)				0.4 $f[T]$		
Cr	Exp (C_p)				0.0+		
Mn	Chin *et al.* 1987*						
Mn	Dinsdale 1991*				0.09	0.62	0
Mn	Asada 1993 (FP)	1.0	0/4	2.0	2.76	2.25	0.52
Mn	Moruzzi 1990 (FP)	0.9					
Mn	Weiss 1979		4.5			2.4/0	
Mn	Exp (β)	(1)			(1)	2.3	
Fe	Chin *et al.* 1987*	2.22		0		0.70	0
Fe	Dinsdale 1991*	2.22		0		0.70	0
Fe	Asada 1993 (FP)	2.32	2.56	2.56	1.55	1.21	0
Fe	Moruzzi 1990 (FP)	2.2	2.7		1.75	0.5/0	
Fe	Weiss 1979		2.6			0.40	
Fe	Exp (β)	2.22				0.70	
Fe	Chuang 1985 (C_p)	2.05				0.57	
Fe	Lytton 1964 (C_p)	1.03					
Co	Chin *et al.* 1987*	1.80	1.70	1.70			
Co	Dinsdale 1991*	1.35	1.35	1.35			
Co	Asada 1993 (FP)	1.80	1.70	1.61	1.0	0	1.2
Co	Moruzzi 1990 (FP)	1.70	1.80				
Co	Miodownik 1978	1.70	1.80			0	
Co	Exp (β)	2.00	1.80	1.70			
Co	Chuang 1985 (C_p)			0.89			
Co	Lytton 1964 (C_p)			1.21			
Ni	Chin *et al.* 1987*	0.85	0.62				
Ni	Dinsdale 1991*	0.85	0.52	0.52			
Ni	Asada 1993 (FP)	0.50	0.66	0.60	0	0.1	0.25
Ni	Moruzzi 1990 (FP)	0.30					
Ni	Weiss 1956		0.85				
Ni	Exp (β)		0.62				
Ni	Chuang 1985 (C_p)		0.25				
Ni	Lytton 1964 (C_p)		0.58				
Ni	Meschter 1981 (C_p)	0.53					

In addition to offering a comparison between theoretical and experimental values, Table 8.2 also indicates that there may be a significant difference between the β values that have been independently obtained by the deconvolution of C_p measurements and from magnetic measurements. There is also a significant spread between the β values obtained by using different methods to extract the magnetic specific heat (Lytton 1964, Meschter *et al.* 1981, Chuang *et al.* 1985). Some of these differences can be rationalised by the occurrence of mixed magnetic states resulting from increasing temperatures, as most of the data in Table 8.2 essentially refers to 0 K. This is almost certainly the case for Cr and Mn (Weiss 1972, 1979, Moruzzi and Marcus 1990a, b). In the case of Co, the c.p.h. form, ε-Co, is a close competitor to the f.c.c. form, α-Co, near the Curie temperature. High stacking-fault densities may therefore also lead to unusual effects (Miodownik 1977). In the case of α-Mn, there is documented evidence for a considerable decrease in β on passing through T_n which could be associated with a transition from an anti-ferromagnetic to a 'non-magnetic state' (Gazzara *et al.* 1964, Weiss 1979). Nevertheless, a single temperature-independent value of β seems a useful starting point in the majority of cases, especially when no other data is available to predict values for metastable elements and solutions. It is noted that β is very sensitive to atomic volume (Moruzzi and Marcus 1990a, b) which is certainly one of the reasons for the non-linearity observed for the variation of β even in simple solid solutions.

8.7. MULTIPLE MAGNETIC STATES

It is not generally appreciated that there are many competing forms of magnetic ordering. When one particular magnetic ground state is substantially more stable than other alternatives, the conventional disordering of magnetic spins, as described in the previous sections, is the only scenario which needs to be considered. However, additional excitations at high temperatures can arise if another magnetic configuration with a comparable ground-state energy exists. Such a hypothesis was used to explain a number of anomalous experimental results in the case of metastable phases in various Fe, Ni and Co alloys and particularly in the case of Invar alloys (Weiss 1963, Chikazumi and Matsui 1978). Together with the switch from ferromagnetic to anti-ferromagnetic behaviour (Fig. 8.4) this makes compelling experimental evidence for the co-existence of two magnetic states in Fe–Ni and Fe–Co alloys (Miodownik 1978b). Weiss (1979) later extended this concept to other 3*d* elements.

8.7.1 Treatments of multiple states

The simplest way to describe the equilibrium between various competing states is to use a Shottky model (Weiss 1963) where

$$\alpha = \frac{f_{\gamma_1}}{f_{\gamma_2}} = \frac{g_2}{g_1} \exp\left(-\Delta E/RT\right). \tag{8.27}$$

Here α is the ratio of the fraction of atoms, f, corresponding to the two magnetic states, γ_1 and γ_2, and g_1 and g_2 are the corresponding degeneracies of the two states. Clearly if ΔE (the difference in energy between the two states) is large and/or the temperature is low, there is effectively only one state and one magnetic moment. However, as ΔE becomes smaller there can be changes in the effective magnetic moment, especially in the case of γ-Fe. Here the two states correspond to a ferromagnetic moment of $\approx 2.8\,\mu B$ and an anti-ferromagnetic moment of $\approx 0.5\,\mu B$, which leads to the *effective* moment being given by

$$\beta_T^{\max} = [2.8/(1+\alpha) + 0.5\alpha/(1+\alpha)]. \tag{8.28}$$

This contrasts with the assumption made in virtually all other magnetic models that the value of β is independent of temperature. Several variants of the Shottky model have been developed by Miodownik (1977, 1978a) to take into account the situation where one of the two states subsequently undergoes magnetic ordering (Fig. 8.5). In such cases it may also be necessary to consider a temperature-dependent ΔE (Miodownik and Hillert 1980).

8.7.2 Thermodynamic consequences of multiple states

One of the consequences of accepting the presence of multiple magnetic states is an additional contribution to the entropy and, therefore, several authors have considered the inclusion of multiple states in their description of low-temperature phase transformations in Fe and its alloys (Kaufman *et al.* 1963, Miodownik 1970, Bendick and Pepperhoff 1978). However, most authors have, in the end, preferred to describe the magnetic effects in Fe using more conventional temperature-independent values for the magnetic moments of the relevant phases. This is partly linked to the absence of any provision for the necessary formalism in current

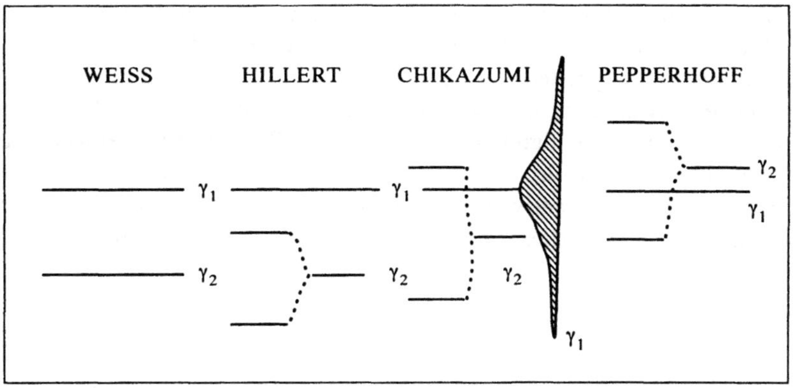

Figure 8.5. Schematic comparison of various two-state models (from Miodownik 1979).

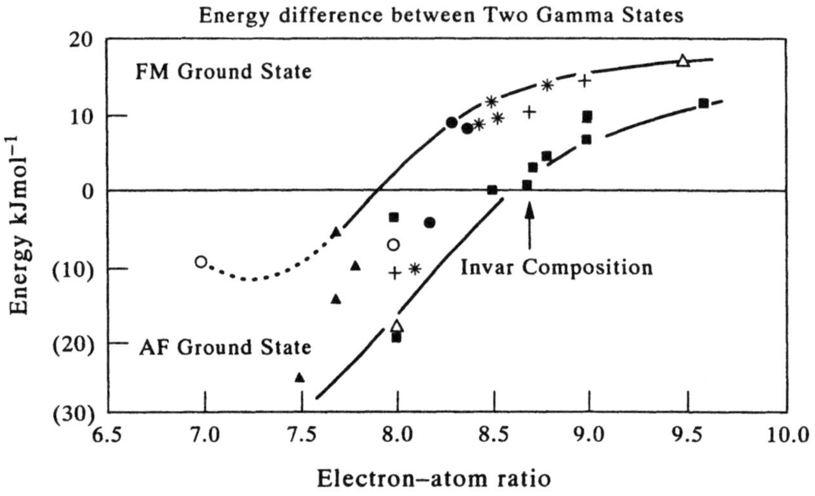

Figure 8.6. Variation of δE between two competing magnetic states with electron/atom ratio for some 3d elements (from de Fontaine *et al.* 1995). ■ Weiss (1963), ▲ Miodownik (1978b), • Bendick *et al.* (1977), * Bendick and Pepperhoff (1978), + Bendick *et al.* (1978), □ Moruzzi and Marcus (1990a), △ Moroni and Jarlsberg (1990), ○ Asada and Terakura (1995).

software packages for phase-diagram calculations, and also to the fact that, with the exception of Roy and Pettifor (1977), there was little theoretical backing for the concept of multiple states in its early stages of development.

It is interesting to note that recent electron energy calculations (Moroni and Jarlsberg 1990, Moruzzi and Marcus 1990a, b, Asada and Terakura 1995) have not only confirmed the necessary energetics for the existence of multiple states, but have also confirmed both the values of the moments (Table 8.2) and the energy gaps which had been previously inferred from experimental evidence (Fig. 8.6). It has therefore been suggested (de Fontaine *et al.* 1995) that future developments should make provision for the inclusion of multiple states. While having only a marginal effect on ΔG at high temperatures (Fig. 8.7a) such a model would lead to significant changes in the driving force for low-temperature (martensitic) transformations and, more importantly, should lead to better modelling of many associated changes in physical properties (Bendick and Pepperhoff 1978, Bendick *et al.* 1977).

8.8. CHANGES IN PHASE EQUILIBRIA DIRECTLY ATTRIBUTABLE TO G^{MAG}

The energy of magnetic transformations can be deceptively large, often exceeding that released by ordinary phase transformations. The following effects have been

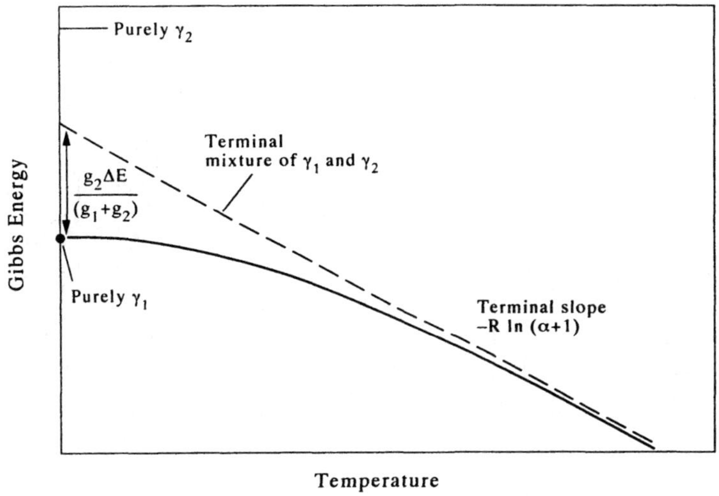

Figure 8.7. Variation of ΔG with temperature for a two-state model (from Miodownik 1977).

observed in various systems and confirmed by Gibbs energy calculations (Miodownik 1982).

A marked change in solid solubility. This occurs at the point where the solubility limit is intersected by the locus of magnetic transformation temperatures (Fig. 8.8). The magnitude of such effects obviously scales with the value of $G^{\mathrm{mag}}/\mathrm{d}T$ and are most marked in Fe- and Co-rich alloys (Nishizawa *et al.* 1979, Hasebe *et al.* 1985). An interesting recent example is given by the Fe–Co–Zn system as the latter two elements have opposite effects on the Curie temperature of Fe (Takayama *et al.* 1995). In some early diagrams, such abrupt changes were inexplicable and deemed to be due to 'experimental error'.

Distortion of miscibility gaps. In the case of some miscibility gaps such intersections are accompanied by an even more marked distortion (Fig. 8.9). This is now often called the Nishizawa Horn, due to the extensive work of Nishizawa and co-workers (1979, 1992) on this effect, but it is interesting to note that the effect had previously been noted by Meijering (1963). Here, too, the apparent presence of more than one maximum in a miscibility gap was believed to represent experimental error before it was shown to have a sound theoretical foundation.

Continuous transition between first- and second-order transformations. When examined more closely, the Nishizawa Horn represents a situation where there is a continuous transition between a first- and second-order transformation. This remarkable situation is not restricted to systems which exhibit a miscibility gap (Inden 1981a) (Fig. 8.10), and it therefore remains to be seen whether it is possible to maintain a hard and fast distinction between these two types of transformation

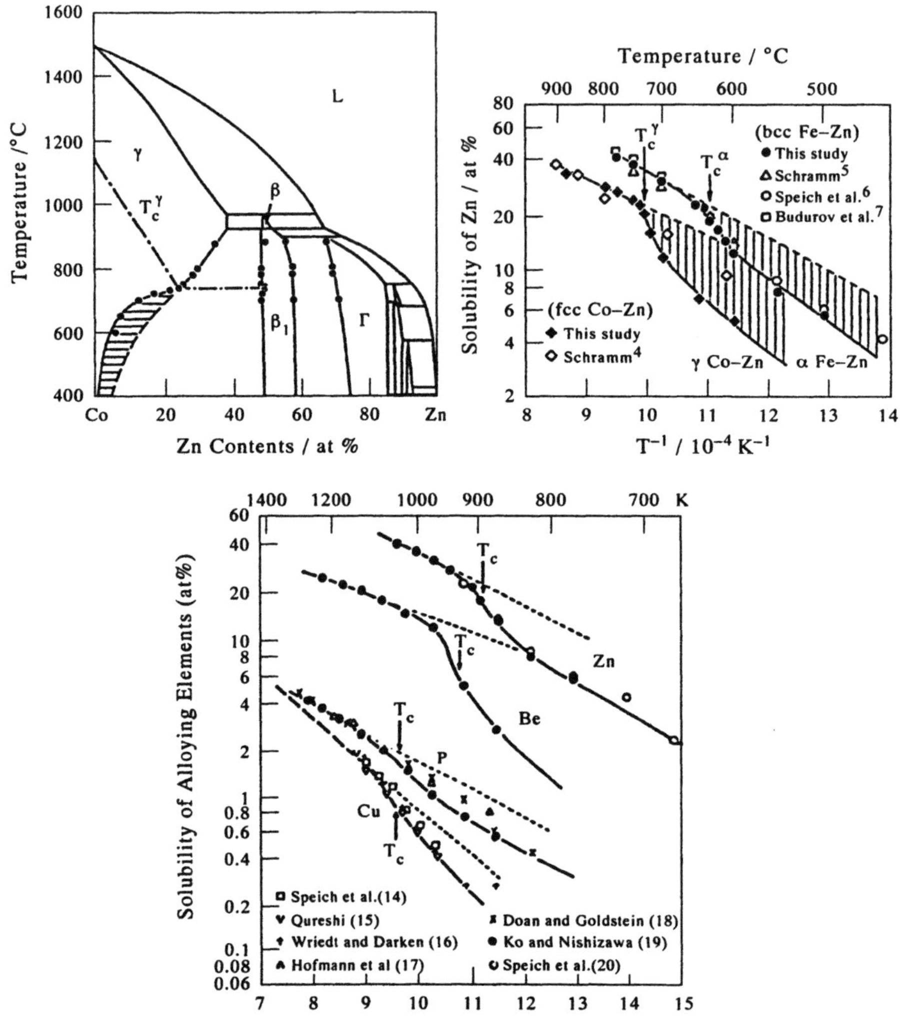

Figure 8.8. Change in solubility where a solubility transus is intersected by T_c (from Takeyama *et al.* 1995).

when faced with such well-documented effects (Hillert 1996).

Stabilisation of metastable phases by the magnetic Gibbs energy contribution. The allotropy of Fe represents an important example where, in the γ region of pure Fe, the Gibbs energy difference between b.c.c. Fe and f.c.c. Fe is small (50–60 J/mol). As a corollary, small changes in G_{mag} have an apparently disproportionate

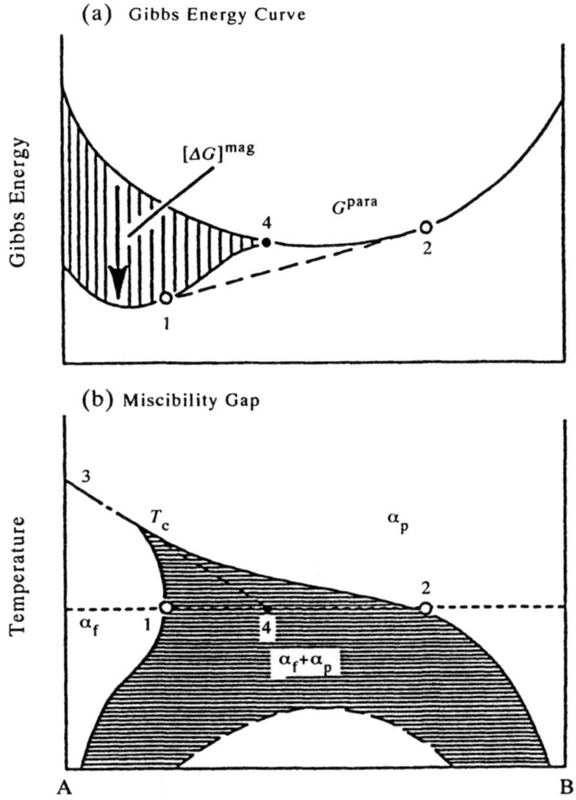

Figure 8.9. The formation of a Nishizawa Horn from the intersection of a magnetic transition and a miscibility gap (from Nishizawa 1992).

effect on the topography of the α/γ region in Fe-base alloys. Because of the location of the Curie temperature in pure Fe, the effect of magnetic forces on the A3 is much more pronounced than on the A4 and alloying additions can produce asymmetric effects on the two transitions (Zener 1955, Miodownik 1977, 1978a, b). Interestingly, although the Curie temperature of Co is much higher, the value of $G_{mag}^{f.c.c.-c.p.h.}$ is not as affected because the magnetic properties of both phases are similar (Miodownik 1977). However, because $S^{f.c.c.-c.p.h.}$ is also small, the allotropic transition temperature can be substantially reduced (Fig. 8.11).

Magnetic effects on metastable transformations. The underlying factor in all the above effects is the magnitude of the ratio (G^{mag}/G^{total}) and especially its variation with temperature. It follows that there can also be a substantial effect on the driving force for phase transformations, including shear transformations. Thus the martensite start temperature, $M_s^{\gamma\rightarrow\alpha}$, in most Fe alloys is dominated by the

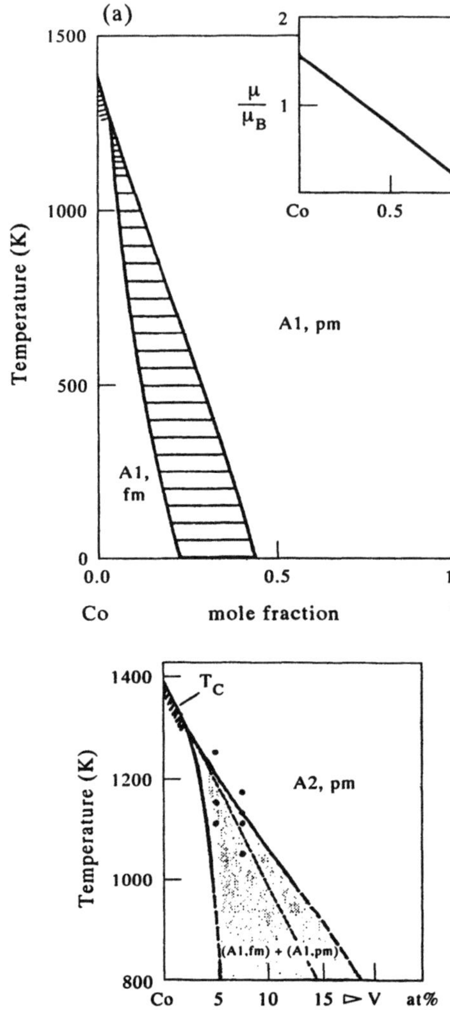

Figure 8.10. Continuous transition between first- and second-order transformations in Co–V alloys: (a) prediction and (b) experiment (from Inden 1985 and 1982).

ferromagnetism of the α phase while the $M_s^{\gamma\to\epsilon}$ in Fe–Mn alloys is controlled by the anti-ferromagnetism of the γ phase (Miodownik 1982). Complex interactions can be expected in systems where the critical ordering temperatures for both chemical and magnetic ordering intersect (Inden 1982, Inden 1991, Skinner and Miodownik 1979), Fe–Si being a good example (Fig. 8.12).

Magnetic effects on stacking fault energy. As the width of a stacking fault in

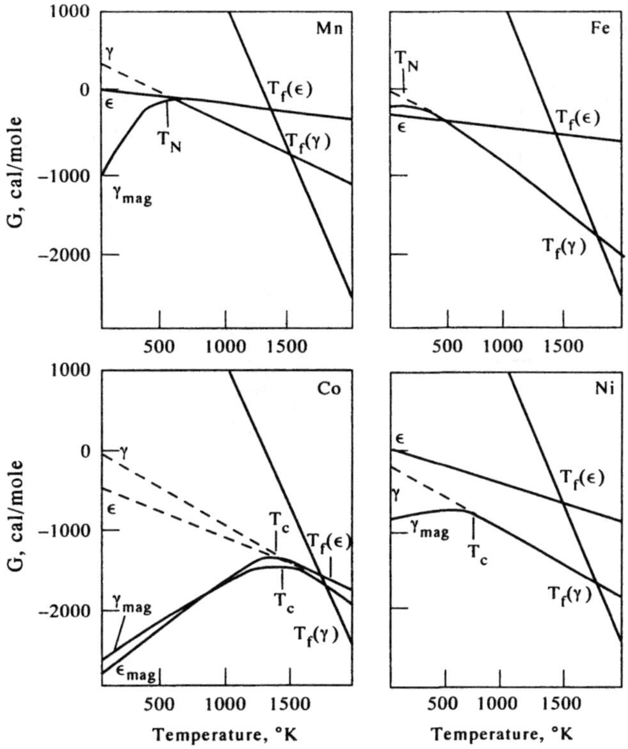

Figure 8.11. Effect of magnetic *G* on f.c.c.–c.p.h.-phase transitions in certain 3*d* elements (from Miodownik 1977).

f.c.c. lattices is a function of the Gibbs energy difference between f.c.c. and c.p.h. structures, this will also be affected by any magnetic component in either or both these structures (Fig. 8.11). This has been analysed by both Ishida (1975) and Miodownik (1978c). Although some of the parameters in both papers need to be re-examined in the light of more recent data, it is difficult to account for the observed trends without taking the magnetic Gibbs energy into consideration.

8.9. INTERACTION WITH EXTERNAL MAGNETIC FIELDS

All the effects noted above can be considered to arise from an *internal* magnetic field. But a Gibbs energy contribution, G_H^{mag}, is also generated if two phases differing in saturation magnetisation by an amount, ΔJ, are placed in an *external* magnetic field, **H**, and is given by

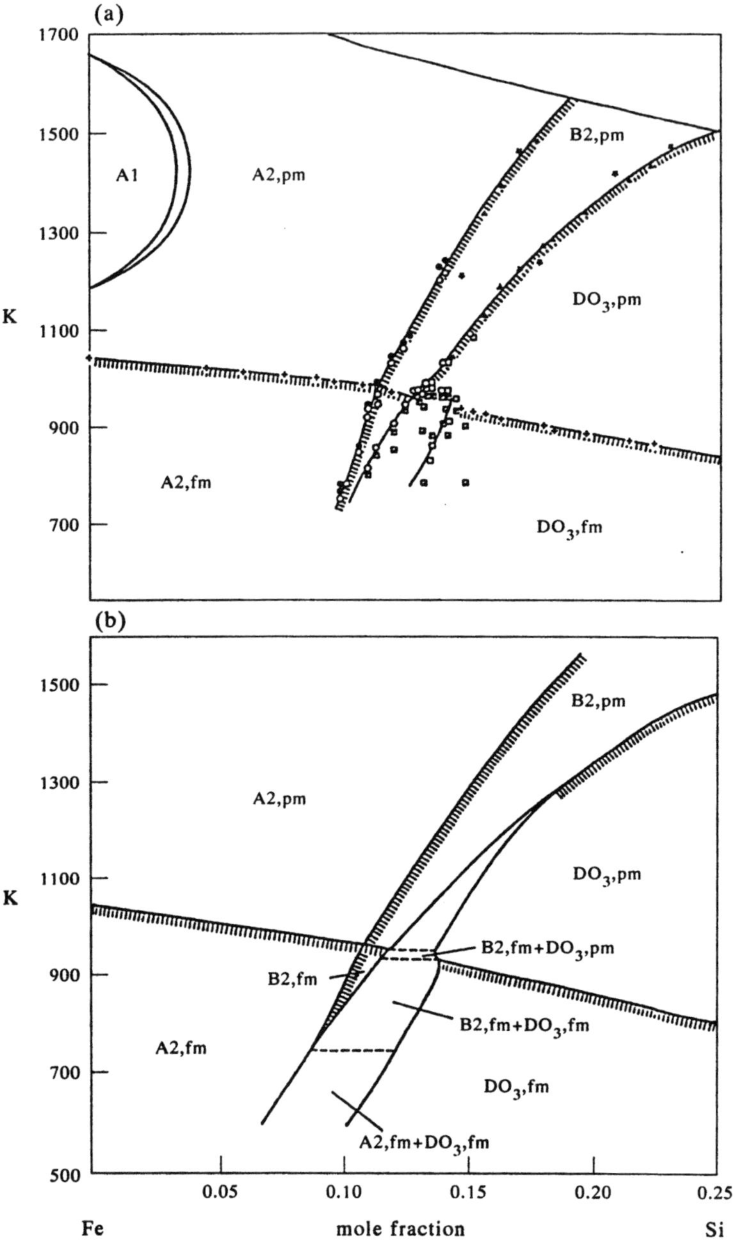

Figure 8.12. Equilibrium between magnetic and non-magnetic-ordered derivatives of b.c.c. Fe–Si alloys: (a) experimental and (b) calculated values (from Inden 1982).

$$G_{\mathbf{H}}^{\mathrm{mag}} = \mathbf{H}\Delta J. \tag{8.24}$$

This term is entirely analogous to the more commonly recognised product $P\Delta V$ and is yet another constituent of the expression for the total Gibbs energy. The effect has been confirmed by many experiments and an external magnetic field can destabilise austenite during martensitic transformations, causing a rise in M_S temperature of approximately 3°K Tesla^{-1} (Sadovskiy *et al.* 1961, Fields and Graham 1976). This effect turns out to be directly proportional to the difference in entropy between the two phases (Satyanarayan *et al.* 1972). A magnetic field can also markedly alter the rate of transformation in isothermal martensite reactions (Peters *et al.* 1972, Kakeshita *et al.* 1993). While changes in M_S due to deformation are well known, the presence of high magnetic fields generated by superconductors at cryogenic temperatures can produce similar effects. Diffusion-controlled transformations are also prone to magnetic-field effects as the driving force for transformation is affected. Thus Fe–Co alloys can be converted from 100% austenite to 100% ferrite at elevated temperatures (Fig. 8.13) under the influence of a field generated by a conventional bench magnet (Peters and Miodownik 1973). The expansion of a miscibility gap and the precipitation of the σ phase in Fe–Cr alloys (Chen *et al.* 1983) are another example of effects that can be induced by an external field.

Although a magnetic field is not commonly applied, it has certain advantages in allowing a precise increment of Gibbs energy to be added instantaneously to a

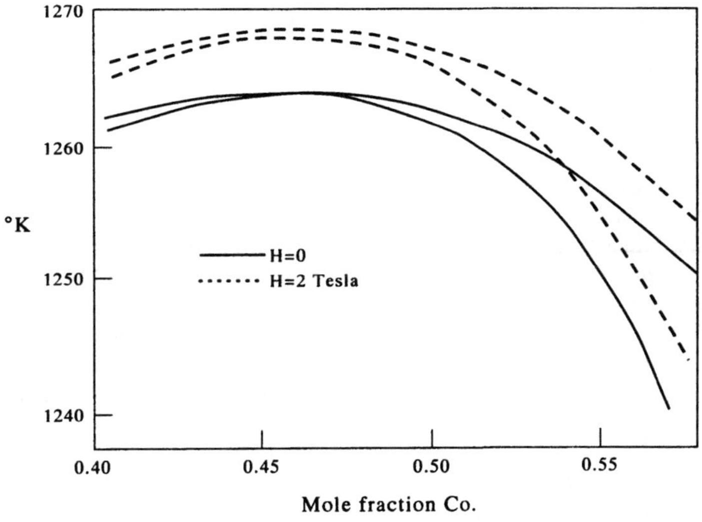

Figure 8.13. Effect of an external magnetic field on phase transformations in Fe–Co (from Peters and Miodownik 1973).

system simply by means of a field and without any requirement for direct contact with the specimen. It is also a salutary reminder that the Gibbs energy expressions in common use do not contain the complete range of terms implicit in Gibbs original formulation.

REFERENCES

Andersson, J.-O. (1985) *Int. J. Thermophysics*, **6**, 411.

Arita, M. (1978) *Acta Metall.*, **26**, 259.

Asada, T. and Terakura, K. (1995) in *Computer Aided Innovation of New Materials, Vol. 1*, eds Doyama, M. *et al.* (Elsevier Science, Amsterdam), p. 169.

Bendick, W. and Pepperhoff, W. (1978) *J. Phys. F*, **8**, 2535.

Bendick, W., Ettwig, H. H. and Pepperhoff, W. (1978) *J. Phys. F*, **8**, 2525.

Bendick, W., Ettwig, H. H., Richter, F. and Pepperhoff, W. (1977) *Z. Metallkde.*, **68**, 103.

Bozorth, R. M. (1956) in *Ferromagnetism* (Van Nostrand).

Chikazumi, S. and Matsui, M. (1978) *J. Phys. Soc. Japan*, **45**, 458.

Chin, C-P., Hertzman, S. and Sundman, B. (1987) TRITA-MAC 0203, Dept. Mat. Sci. Eng, Royal Institute of Technology, Sweden.

Chen, I-Wei, Faillace, E. and Miodownik, A. P. (1983) in *Ferritic Alloys for use in the Nuclear Energy Technologies*, eds Davis, J. W. and Michel, D. J. (AIME, Warrendale), p. 517.

Chuang, Y. Y., Schmid-Fester, R. and Chang, Y. A. (1985) *Met. Trans.*, **16A**, 153.

Chuang, Y. Y., Lin, J-C., Schmid-Fester, R. and Chang, Y. A. (1986) *Met. Trans.*, **17A**, 1361.

Clougherty, E. V. and Kaufman, L. (1963) *Acta. Met.*, **11**, 1043.

Colinet, C., Inden, G. and Kikuchi, R. (1993) *Acta. Metall.*, **41**, 1109.

de Fontaine, D., Fries, S. G., Inden, G., Miodownik, A. P., Schmid-Fetzer, R. and Chen, S. L. (1995) *CALPHAD*, **19**, 499.

Dinsdale, A. T. (1991) *CALPHAD*, **15**, 317.

Eriksson, O., Johansson, B., Albers, R. C. and Boring, A. M. (1990) *Phys. Rev. B*, **42**, 2707.

Ettwig, H. H. and Pepperhoff, W. (1974) *Phys. Sat. Solidi*, **23**, 105.

Fernandez Guillermet, A. (1987) *Int. J. Thermophysics*, **8**, 481.

Fields, R. and Graham, C. D. (1976) *Metall. Trans. A*, **7A**, 719.

Gazzara, C. P., Middleton, R. M. and Weiss, R. J. (1964) *Physics Letters*, **10**, 257.

Hasebe, M., Ohtani, H. and Nishizawa, T. (1985) *Metall. Trans.*, **16A**, 913.

Hillert, M. and Jarl, M. (1978) *CALPHAD*, **2**, 227.

Hillert, M., Wada, T. and Wada, H. (1967) *J. Iron and Steel. Inst.*, **205**, 539.

Hillert, M. (1996) Verbal communication at CALPHAD XXV, Erice, Sicily.

Hofmann, J. A., Paskin, A., Tauer, K. J. and Weiss, R. J. (1956) *J. Phys. Chem. Solids*, **1**, 45.

Inden, G. (1975) *Z. Metallkde.*, **66**, 577.

Inden. G. (1976) Proc. Calphad V (Dusseldorf) III-(4)-1.

Inden, G. (1977) *Z. Metallkde.*, **68**, 529.

Inden, G. (1981a) *Physica. B*, **103**, 82.

Inden, G. (1981b) *Scripta Metall.*, **15**, 669.

Inden, G. (1982) *Bull. Alloy. Phase. Diag.*, **2**, 412.

Inden, G. (1991) *Scand. J. Met.*, **20,** 112.

Inden, G. and Pitsch, W. (1991) in *Materials Science and Technology, Vol. 5*, ed. Haasen, P. (VCH Verlagsgesellschaft, Weinheim), p. 497.

Ishida, K. (1975) *Phil. Mag.*, **32,** 663.

Kakeshita, T., Kuroiwa, K., Shimizu, K., Ikeda, T., Yamagishi, A. and Date, M. (1993) *Mat. Trans. Jap. Inst. Met.*, **34** (5), 415.

Kaufman, L. and Cohen, M. (1956) *Trans. AIME*, **206,** 1393.

Kaufman, L. and Cohen, M. (1958) *Progress in Metal Physics*, **7** (3), 165.

Kaufman, L. and Nesor, H. (1973) *Z. Metallkde.*, **64,** 249.

Kaufman, L., Clougherty, E. V. and Weiss, R. J. (1963) **11,** 323.

Ko, M. and Nishizawa, T. (1979) *J. Japan. Inst. Metals*, **43,** 126.

Kosakai, T. and Miyazaki, T. (1994) *Iron and Steel Inst. Japan*, **34,** 373.

Lytton, J. L. (1964) *J. Appl. Phys.*, **35,** 2397.

Meschter, P. J., Wright, J. W. and Brooks, C. R (1981) *J. Phys. Chem. Sol.*, **42,** 861.

Meijering, J. L. (1963) *Philips Res. Report*, **18,** 318.

Miodownik, A. P. (1970) *Acta Met.*, **18,** 541.

Miodownik, A. P. (1977) *CALPHAD*, **1,** 133.

Miodownik, A. P. (1978a) in *Honda Memorial Volume on Metal Science and Metallurgy*, in *Physics and Application of Invar Alloys*, ed. Saito, H. (Maruzen) **3** (18), p. 429.

Miodownik, A. P. (1978b) in *Honda Memorial Volume on Metal Science and Metallurgy*, in *Physics and Application of Invar Alloys*, ed. Saito, H. (Maruzen) **3** (12), p. 288.

Miodownik, A. P. (1978c) *CALPHAD*, **2** (3), 207.

Miodownik, A. P. and Hillert, M. (1980) *CALPHAD*, **4,** 143.

Miodownik, A. P. (1982) *Bull. Alloy. Phase. Diag.*, **2,** 406.

Moran-Lopez, J. L. and Falicov, L. M. (1979) *Sol. Stat. Comm.*, **31,** 325.

Moroni, E. G. and Jarlsberg, T. (1990) *Phys. Rev. B*, **41,** 9600.

Moruzzi, V. L. and Marcus, P. L. (1988) *Phys. Rev. B*, **38,** 1613.

Moruzzi, V. L. and Marcus, P. L. (1990a) *Phys. Rev. B*, **42,** 8361.

Moruzzi, V. L. and Marcus, P. L. (1990b) *Phys. Rev. B*, **42,** 10,322.

Nishizawa, T. (1978) *Bull. Japanese Inst. Metals*, **17,** 780.

Nishizawa, T. (1992) *Mat. Trans. Japanese Inst. Metals*, **33,** 713.

Nishizawa, T., Hasebe, M., and Ko, M. (1979) *Acta. Met.*, **27,** 817.

Paskin, A. (1957) *J. Phys. Chem. Solids*, **2,** 232.

Peters, C. T. and Miodownik, A. P. (1973) *Scripta Met.*, **7,** 955.

Peters, C. T., Bolton, P. and Miodownik, A. P. (1972) *Acta Met.*, **20,** 881.

Roy, D. M. and Pettifor, D. G. (1977) *J. Phys. F*, **7,** 183.

Sadovskiy, V. D., Rodigin, L., Smirnov, L. V., Filonchik, G. M. and Fakidov, I. G. (1961) *Phys. Met. Metallography*, **12** (2), 131.

Saito, H., Arrott, A. and Kikuchi, R. (1959) **10,** 19.

Satyanarayan, K. R., Eliasz, W. and Miodownik, A. P. (1972) *Acta. Met.*, **16,** 877.

Schon, C. G. and Inden, G. (1996) *Scripta Mater. Met.*, **30,** 000.

Semenovskoya, S. V. (1974), *Phys. Stat. Sol. B*, **64,** 291.

Skinner, D. and Miodownik, A. P. (1979) in *Proc. Int. Symp. on Martensitic Transformations: ICOMAT II*, eds Cohen, M. and Owen, W. (MIT, Boston), p. 259.

Smith, D. (1967) *Scripta Met.*, **1,** 157.

Takayama, T., Shinohara, S., Ishida, K. and Nishizawa, T. (1995) *J. Phase Equilibria*, **16,** 390.

Weiss, R. J. (1963) *Proc. Phys. Soc.*, **82**, 281.
Weiss, R. J. (1972) *Phil. Mag.*, **26**, 261.
Weiss, R. J. (1979) *Phil. Mag. B*, **40**, 425.
Weiss, R. J. and Tauer, K. J. (1956) in *Theory of Alloy Phases* (ASM, Metals Park, OH), p. 290.
Weiss, R. J. and Tauer, K. J. (1958) *J. Phys. Chem. Solids*, **4**, 135.
Wohlfarth, E. P. (1974) *Phys. Stat. Sol. (a)*, **25**, 285.
Zener, C. (1955) *Trans. AIME*, **203**, 619.

Chapter 9
Computational Methods

Chapter 9
Computational Methods

9.1. INTRODUCTION

Most of this book concerns the development and application of theoretical thermodynamic models, as these are the basis of the CALPHAD method. However, none of this would be possible without the existence of the computational methods and software which allow these models to be applied in practice. In essence, the issues involved in computational methods are less diverse and mainly revolve around Gibbs energy minimisation. In addition, there are 'optimiser' codes which are used for the thermodynamic assessment of phase equilibria. The essential aim of these codes is to reduce the statistical error between calculated phase equilibria, thermodynamic properties and the equivalent experimentally measured quantities.

It is also worthwhile to make some distinctions between methods of calculating phase equilibrium. For many years, equilibrium constants have been used to express the abundance of certain species in terms of the amounts of other arbitrarily chosen species, see for example Brinkley (1946, 1947), Kandliner and Brinkley (1950) and Krieger and White (1948). Such calculations suffer significant disadvantages in that some prior knowledge of potential reactions is often necessary, and it is difficult to analyse the effect of complex reactions involving many species on a particular equilibrium reaction. Furthermore, unless equilibrium constants are defined for all possible chemical reactions, a true equilibrium calculation cannot be made and, in the case of a reaction with 50 or 60 substances present, the number of possible reactions is massive.

CALPHAD methods attempt to provide a true equilibrium calculation by considering the Gibbs energy of all phases and minimising the total Gibbs energy of the system (G). In this circumstance G can be calculated either from knowledge of the chemical potential ($\bar{G}_i$) of component i, by

$$G = \sum_i n_i \bar{G}_i \tag{9.1}$$

where n_i is amount of component i, or alternatively by

$$G = \sum_\phi N^\phi G_m^\phi \tag{9.2}$$

where N^ϕ is the amount of phase ϕ and G^ϕ_m its Gibbs energy. The number of unknowns is now considerably reduced in comparison to an equilibrium constant approach.

This chapter will describe aspects of the computation of phase equilibria by starting with the simple case of two-phase equilibria in binary and ternary systems. This provides a good example by which the process of Gibbs energy minimisation can be conceptualised and the principles are equally applicable to multi-component systems. The extension to more general cases will then be discussed. One of the most significant advantages of the CALPHAD methodology is that, because the total Gibbs energy is calculated, it is possible to derive all of the associated functions and characteristics of phase equilibria, e.g., phase diagrams, chemical potential diagrams, etc. Examples of some graphical forms of representing such functions will also be given. Finally, 'optimiser' programmes will be discussed with particular emphasis on the two major codes currently used world-wide, the Lukas programme (Lukas *et al.* 1977) and PARROT (Jansson 1984b).

9.2. CALCULATION OF PHASE EQUILIBRIA

9.2.1 Introduction

The actual calculation of phase equilibria in a multi-component, multi-phase system is a complex process involving a high level of computer programming. Details of programming aspects are too lengthy to go into this chapter, but most of the principles by which Gibbs energy minimisation is achieved are conceptually quite simple. This section will therefore concentrate on the general principles rather than go into detail concerning the currently available software programmes which, in any case, often contain proprietary code.

9.2.2 Binary and ternary phase equilibria

9.2.2.1 Analytical solutions. It is possible to calculate certain two-phase equilibria using analytical equations. The simplest such case is a eutectic system where there is negligible solubility in the solid state of the terminal phases and there is ideal mixing in the liquid (Fig. 9.1). The composition of the liquidus of α at any given temperature can then be directly calculated by finding the composition where the partial Gibbs energy of A in the liquid, $\bar{G}^{\rm liq}_A$ is equal to the Gibbs energy of the pure solid component A (Fig. 9.2). Thus

$$\bar{G}^{\rm liq}_A = G^{\rm liq\to\alpha}_A \tag{9.3}$$

or

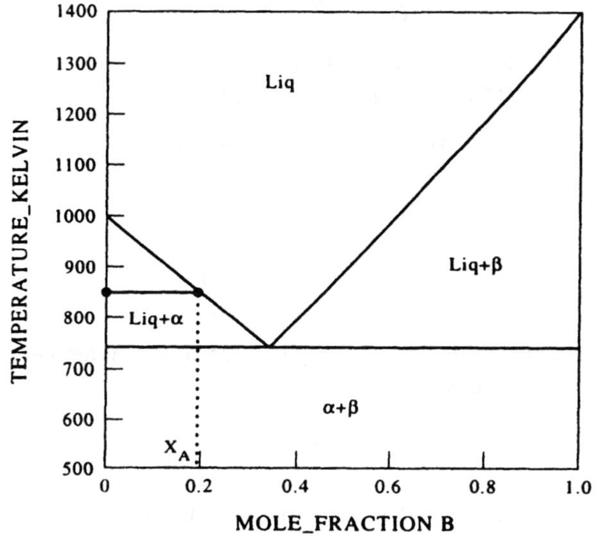

Figure 9.1. Simple eutectic system with ideal mixing in the liquid and negligible solid solubility in the terminal solid phases, α and β.

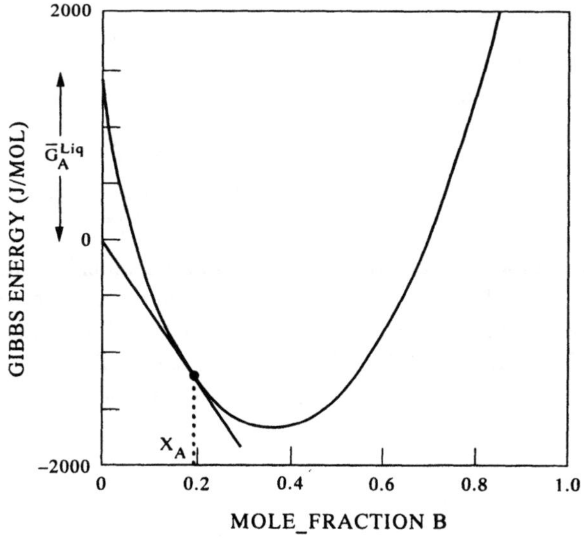

Figure 9.2. G/x diagram at 850 K for eutectic system in Fig. 9.1.

$$\bar{G}_A^{\mathrm{liq}} - G_A^{\mathrm{liq}\to\alpha} = 0 \tag{9.4}$$

where $G_A^{\mathrm{liq}\to\alpha}$ is the Gibbs energy of fusion of solid A with respect to α. In an ideal solution $\bar{G}_A^{\mathrm{liq}} = RT \log_e x_A$ and Eq. (9.4) becomes

$$RT \log_e x_A - G_A^{\mathrm{liq}\to\alpha} = 0 \tag{9.5}$$

which can be rewritten as

$$x_A = \exp\left(\frac{G_A^{\mathrm{liq}\to\alpha}}{RT}\right) \tag{9.6}$$

where x_A is the composition of the liquidus at a given temperature T. More generally Eq. (9.4) can be written as

$$\left(\bar{H}_A^{\mathrm{liq}} - T\bar{S}_A^{\mathrm{liq}}\right) - \left(\bar{H}_A^{\mathrm{liq}\to\alpha} - T\bar{S}_A^{\mathrm{liq}\to\alpha}\right) = 0 \tag{9.7}$$

where $\bar{H}_A^{\mathrm{liq}}$ and $\bar{S}_A^{\mathrm{liq}}$ are the partial enthalpy and entropy of solution of A in the liquid at x_A and $H_A^{\mathrm{liq}\to\alpha}$ and $S_A^{\mathrm{liq}\to\alpha}$ are the enthalpy and entropy of fusion of α in pure A. This can be rearranged to give an expression for the liquidus temperature T^{liq}, as

$$T^{\mathrm{liq}} = \frac{\bar{H}_A^{\mathrm{liq}} - \bar{H}_A^{\mathrm{liq}\to\alpha}}{\bar{S}_A^{\mathrm{liq}} - \bar{S}_A^{\mathrm{liq}\to\alpha}}. \tag{9.8}$$

If $\bar{H}_A^{\mathrm{liq}}$, $\bar{S}_A^{\mathrm{liq}}$, $\bar{H}_A^{\mathrm{liq}\to\alpha}$ and $\bar{S}_A^{\mathrm{liq}\to\alpha}$ are temperature-independent, the liquidus temperature can also be defined exactly. This can be extended for the case of an intermetallic compound of fixed stoichiometry (Kaufman 1970, Ansara 1979) so that

$$T^{\mathrm{liq}} = \frac{x_A^\phi \bar{H}_A^{\mathrm{liq}} + x_B^\phi \bar{H}_B^{\mathrm{liq}} + x_A^\phi H_A^{\mathrm{fus}} + x_B^\phi H_B^{\mathrm{fus}} - H^\phi}{x_A^\phi \bar{S}_A^{\mathrm{liq}} + x_B^\phi \bar{S}_B^{\mathrm{liq}} + x_A^\phi \dfrac{H_A^{\mathrm{fus}}}{T_A^{\mathrm{fus}}} + x_B^\phi \dfrac{H_B^{\mathrm{fus}}}{T_B^{\mathrm{fus}}} - S^\phi} \tag{9.9}$$

where x_A^ϕ and x_B^ϕ are the compositions of A and B in the stoichiometric compound ϕ, H^ϕ and S^ϕ are the enthalpy and entropy of formation of ϕ, given with respect to the A and B in their standard states, and H_A^{fus}, S_A^{fus}, H_B^{fus} and S_B^{fus} are the enthalpies and entropies of fusion of A and B in their standard states. This expression can be further extended for any number of components providing the compound, ϕ, is stoichiometric, and becomes

$$T^{\mathrm{liq}} = \frac{\sum_i x_i^\phi \bar{H}_i^{\mathrm{liq}} + \sum_i x_i^\phi H_i^{\mathrm{fus}} - H^\phi}{\sum_i x_i^\phi \bar{S}_i^{\mathrm{liq}} + \sum_i x_i^\phi \dfrac{H_i^{\mathrm{fus}}}{T_i^{\mathrm{fus}}} - S^\phi}. \tag{9.10}$$

Again, providing the various H and S terms are temperature-independent, the solution remains exact and provides a rapid method of calculating the liquidus temperature. Equation (9.10) is generally applicable to any phase boundary between a solution phase and stoichiometric compound, so could equally well be used for solid-state solvus lines.

9.2.2.2 General solutions. In the case of a solution phase and stoichiometric compound which have temperature-dependent enthalpies and entropies, an iterative process by which T is varied can be utilised to find the liquidus by using the following equation

$$\Delta G^{1\phi} = \sum_i x_i^\phi \bar{G}_i^{\rm liq} - G^\phi = 0 \tag{9.11}$$

where x_i^ϕ is the mole fraction of component i in the compound ϕ and $\bar{G}_i^{\rm liq}$ is the partial Gibbs energy of i in the liquid. Figure (9.3) shows a schematic diagram for how $\Delta G^{1\phi}$ varies with temperature and becomes zero at the liquidus. This temperature can be found by first choosing an arbitrary temperature and then iteratively stepping the temperature until $\Delta G^{1\phi}$ has a value which is below a small, defined, accuracy limit. Rapid convergence can be achieved by using a Newton–Raphson method. An initial arbitrary temperature (T_1) is chosen and both $\Delta G^{1\phi}$ and $\Delta G^{1\phi}/\mathrm{d}T$ are calculated and then used to estimate a new temperature (T_2) where

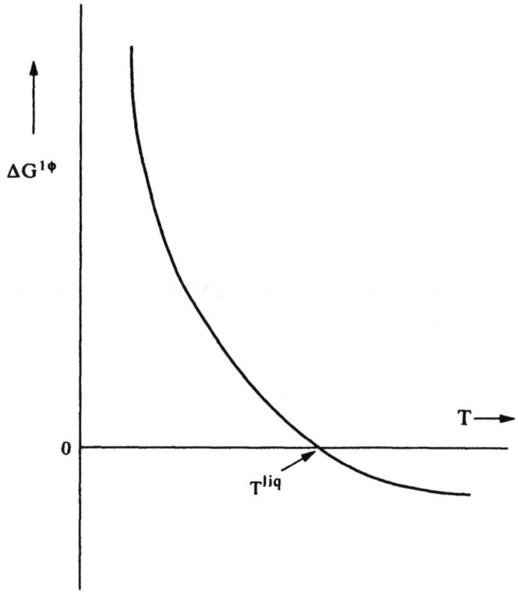

Figure 9.3. Schematic diagram showing variation in $\Delta G^{1\phi}$ as a function of temperature.

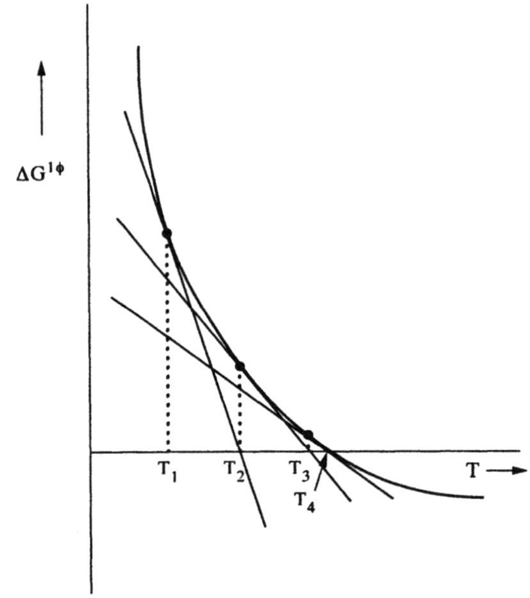

Figure 9.4. Schematic diagram showing method used to calculate liquidus temperature for compound ϕ from curve $\Delta G^{1\phi}/T$.

$\Delta G^{1\phi}$ is expected to be zero. If the new value of $\Delta G^{1\phi}$ does not satisfy the convergence criterion, a new temperature (T_3) is estimated. The process is repeated if necessary, until the convergence criterion is satisfied. Figure (9.4) shows a schematic diagram representing this process.

In the more general case both phases could exhibit significant stoichiometric deviation, and so a further degree of freedom is added. A method for dealing with this in the binary system Ni–Cu was suggested in Chapter 3.7 (Fig. 9.5). In this case the temperature is set as a constant and the Gibbs energy where liquid and f.c.c. phases are equal is taken as the start point of the minimisation process (Figs. 9.6(a,b)). A solution must be found such that the Gibbs energy of a mixture of the f.c.c. and liquid phases gives the lowest Gibbs energy and the compositions of the f.c.c. and liquid phases are $x^e_{\text{f.c.c.}}$ and x^e_{liq} respectively. This can be achieved by changing the compositions of the liquid and f.c.c. phases away from the start point and calculating a new Gibbs energy based on the constraints of Eq. (9.2) which is re-stated below:

$$G = \sum_{\phi} N^{\phi} G^{\phi}_m \tag{9.2}$$

with general mass balance equations where

References are listed on pp. 294–296.

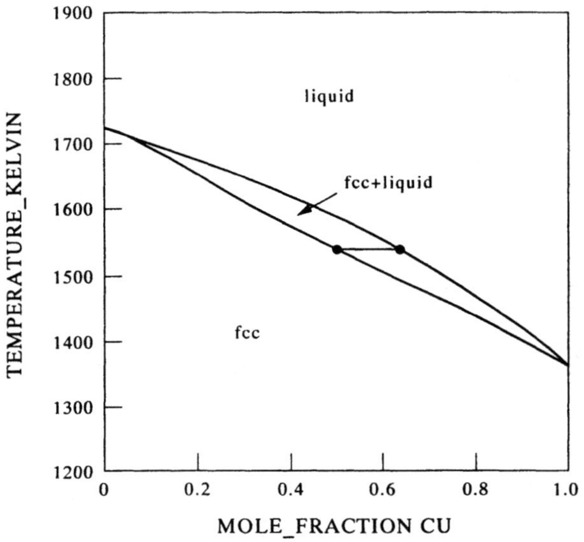

Figure 9.5. Ni–Cu phase diagram.

$$N_i = \sum_{\phi} N_i^{\phi} \tag{9.12}$$

and

$$\sum_{\phi} N^{\phi} = M \tag{9.13}$$

where N_i is the total number of moles of component i in the system and N_i^{ϕ} is the number of moles of component i in the phase ϕ and M is the total number of moles in the system.

It is not necessary to use x_0 as a starting point and any composition x_i can be used which satisfies the lever rule, where

$$N^{\phi_1} = \frac{|(x_i - x_i^{\phi_2}|}{|(x_i^{\phi_1} - x_i^{\phi_2})|} \tag{9.14}$$

N^{ϕ_1} is the amount of phase ϕ_1 in moles and $x_i^{\phi_1}$ and $x_i^{\phi_2}$ are the compositions of phases ϕ_1 and ϕ_2 respectively and x_i is the composition of the alloy. A general procedure for minimisation could then follow the process described below.

The alloy is first assumed to be either single-phase liquid or f.c.c., (in this case f.c.c.), and an arbitrary amount of liquid introduced. To retain mass balance, a corresponding change in composition of the f.c.c. phase is made and G is calculated.

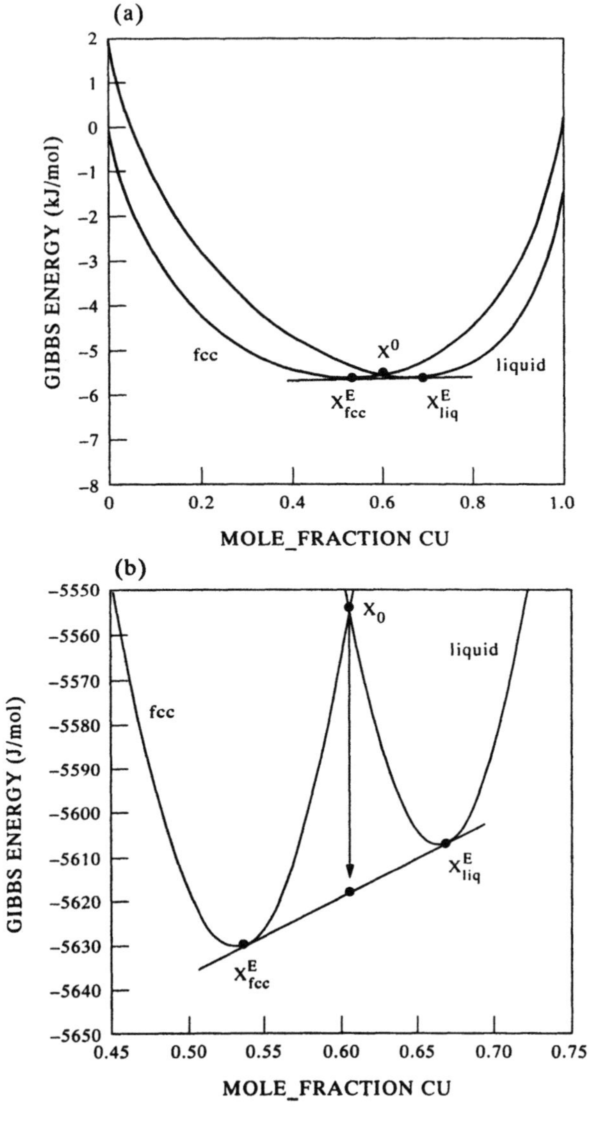

Figure 9.6. (a) G/x curves for Ni–Cu at 1523 K and (b) expanded to show a two-phase region.

The composition of the f.c.c. phase is then kept constant and the amount of liquid changed, with a corresponding change in its composition to maintain mass balance, so that the Gibbs energy is reduced. The process is iterated (Fig. 9.7) until G reaches a minimum value. Figure 9.8 shows a graph of G vs N^{liq} and the minimum

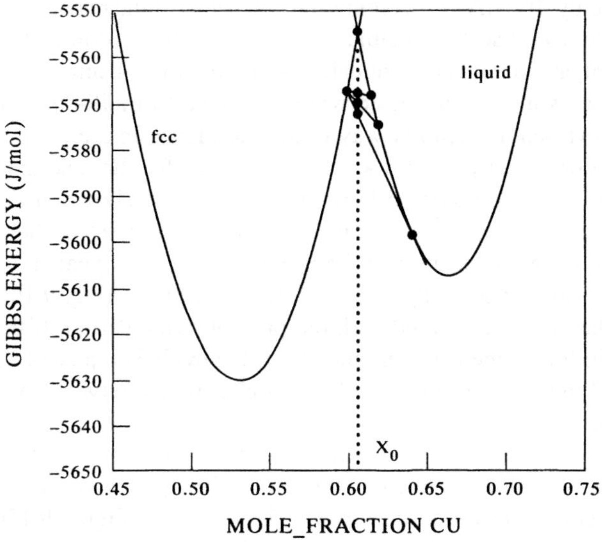

Figure 9.7. Schematic diagram showing first iteration stage in the Gibbs energy minimisation process of an alloy with composition x_0 in Ni–Cu at 1523 K.

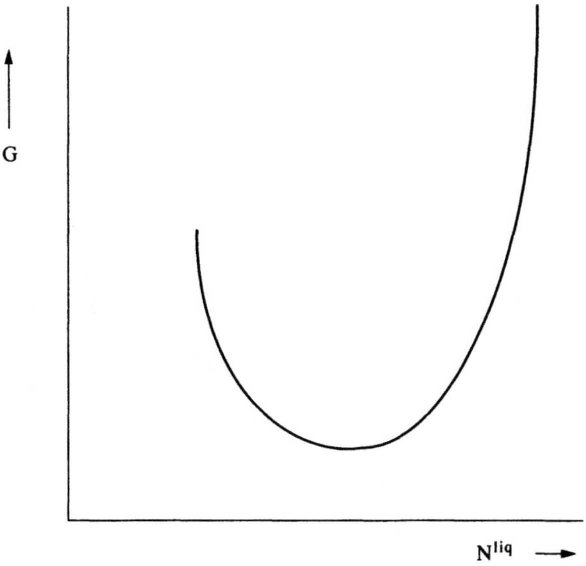

Figure 9.8. Schematic diagram showing the Gibbs energy as N^{liq} is varied in a two-phase mixture as shown in Fig. 9.7.

is characterised by dG/dN^{liq} becoming zero. Because of the mass balance relationship it would, of course, also be possible to vary the composition of the liquid, rather than its amount, and obtain an identical answer. However, using the amount of phase makes the process more generalised when extended to multi-component systems.

N^{liq} can be varied in a similar way to that described in the previous section but, as the lowest Gibbs energy is not known, the guess for the next amount of N^{liq} is made by calculating the second differential to the curve shown in Fig. 9.8 so that a composition where $dG/dN^{liq} = 0$ can be estimated (Fig. 9.9). The composition of the liquid is now held constant and the previous process repeated, but this time by changing the amount and composition of the f.c.c. phase (Fig. 9.10). A new value of G is calculated and compared with the previous calculation. If the difference in G is not within the defined convergence limit the cycle is repeated again (Figs 9.11 and 9.12) until the difference in calculated Gibbs energy between cycles falls below the defined convergence limit.

The method described above is quite general. If an alloy is chosen outside the equilibrium two-phase field, this will be automatically recognised and convergence to the correct single phase equilibrium will be achieved. Figure 9.13(a) shows such a case, again for Ni–Cu, where the composition x_i is in the single-phase f.c.c. region. An amount of liquid is introduced and G is calculated. The amount of liquid is then changed until a minimum value of G is obtained. The process is repeated but with the amount of the f.c.c. phase being changed (Fig. 9.13(b)). In this case G is

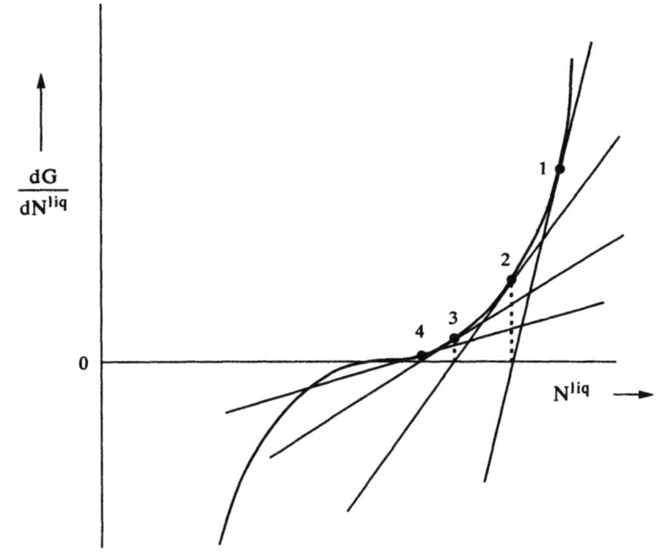

Figure 9.9. Schematic diagram of the first derivative of the G/N^{liq} curve in Fig. 9.8 showing the calculation of the minimum in the Gibbs energy as a function of N^{liq}.

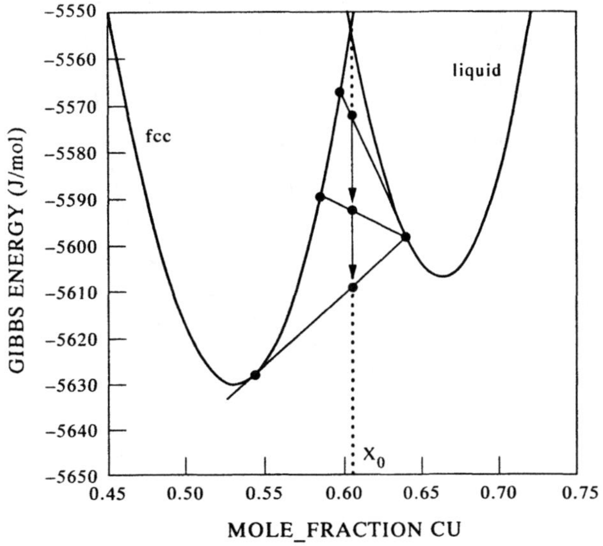

Figure 9.10. Schematic diagram showing seconnd iteration stage in the Gibbs energy minimisation process of an alloy with composition x_0 in Ni–Cu at 1523 K.

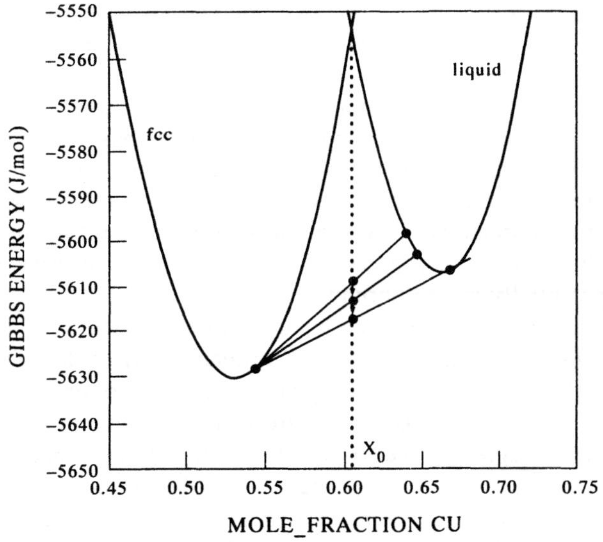

Figure 9.11. Schematic diagram showing third iteration stage in the Gibbs energy minimisation process of an alloy with composition x_0 in Ni–Cu at 1523 K.

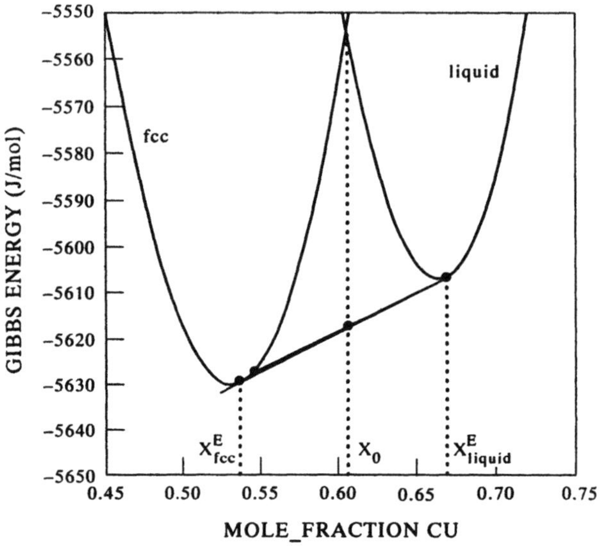

Figure 9.12. Schematic diagram showing fourth iteration stage in the Gibbs energy minimisation process of an alloy with composition x_0 in Ni–Cu at 1523 K.

minimised by the composition of the f.c.c. phase reaching x_1 and the amount of liquid is therefore reduced to zero to maintain mass balance. Furthermore it is possible to check the stability of *three* phases to find which one, or which combination, is the equilibrium state. In this case an estimate for compositions of the three phases is made which satisfies the mass balance and a Gibbs energy calculated, giving point A in Fig. 9.14. With the constraint of the other two phases being considered in the total Gibbs energy calculation, the Gibbs energy of the system decreases as the amount of the unstable third phase is decreased until a zero amount is present, at point B. Finally the normal two-phase minimisation provides the final equilibrium solution at C (Fig. 9.14). This process will also work with numerous phases being considered as potentially in equilibrium.

The process has been described for a binary alloy but, essentially, this method of minimisation holds for a ternary system (Kaufman 1970). In the binary case the amount of ϕ_1 or ϕ_2 is altered by changing the composition of one of the respective components, but in a ternary system the effect of the additional element must be included as a further degree of freedom is introduced. Although in a ternary system it is possible to have more than two phases in equilibrium, defining the two-phase fields can still be used to delineate a three-phase triangle because the sides of the three-phase field are defined by the tie-lines of the respective adjoining two-phase fields (Fig. 9.15). Therefore, to plot an isothermal section it is only necessary to calculate the possible two-phase equilibria and the three-phase fields are then easily

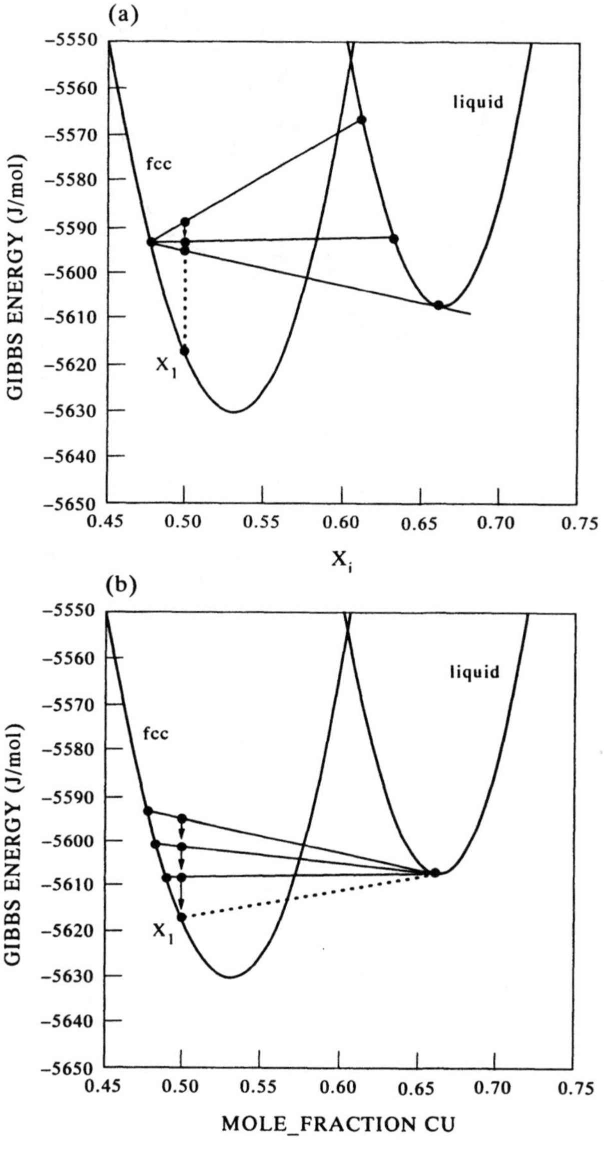

Figure 9.13. Schematic diagrams showing (a) first iteration stage and (b) second iteration stage in the Gibbs energy minimisation process of an alloy with composition x_1 in Ni–Cu at 1523 K.

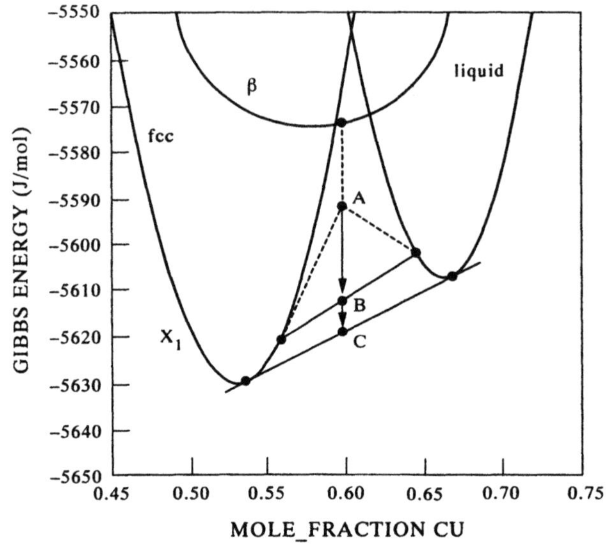

Figure 9.14. Schematic diagram showing Gibbs energy minimisation process for an alloy of composition x_0 in Ni–Cu at 1523 K including a metastable phase β.

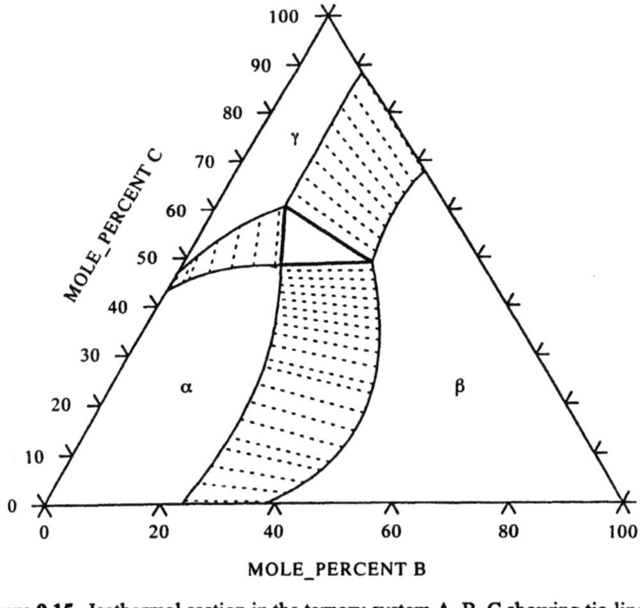

Figure 9.15. Isothermal section in the ternary system A–B–C showing tie-lines in the various two-phase fields and their intersection to form a three-phase field, $\alpha+\beta+\gamma$.

found where the three two-phase fields converge (Kaufman 1970). The calculation method can be further extrapolated to a multi-component system but, unless the problem specifically relates to two phases, cf. austenite/ferrite phase equilibria, it is not useful as a general tool.

9.2.3 Calculation methods for multi-component systems

Essentially, the calculation must be defined so that (1) the number of degrees of freedom is reduced, (2) the Gibbs energy of the system can be calculated, and (3) some iterative technique can be used to minimise the Gibbs energy. The number of degrees of freedom is reduced by defining a series of constraints, such as the mass balance, electroneutrality in ionic systems, or the composition range in which each phase exists. Preliminary estimates for equilibrium values must be given so that the process can begin and subsequently proceed smoothly. These values are called the start points and are critical in the minimisation process of finding *true* equilibrium (see Section 9.2.5).

One of the earliest examples of Gibbs energy minimisation applied to a multi-component system was by White *et al.* (1958) who considered the chemical equilibrium in an ideal gas mixture of O, H and N with the species H, H_2, H_2O, N, N_2, NH, NO, O, O_2 and OH being present. The problem here is to find the most stable mixture of species. The Gibbs energy of the mixture was defined using Eq. (9.1) and defining the chemical potential of species i as

$$\bar{G}_i = \bar{G}_i^{\mathrm{o}} + RT \log_{\mathrm{e}} a_i \qquad (9.15)$$

where $\bar{G}_i^{\mathrm{o}}$ is the standard chemical potential of species i. Standard mass balance equations were then introduced where

$$\sum_i a_{ij}\, x_i = n_j \quad (j = 1, 2 \ldots m) \qquad (9.16)$$

where a_{ij} represents the number of atoms of element j in the species i, x_i the number of moles of species i and n_j the total number of moles of j in the system. These authors presented two methods of Gibbs energy minimisation, one which utilised a linear programming method developed by Dantzig *et al.* (1957) and one based on the method of steepest descent using Lagrange's method of undetermined multipliers.

Their linear programming method used the following equation:

$$\frac{G}{RT} = \sum_i x_i \left[\left(\frac{\bar{G}_i^{\mathrm{o}}}{RT} \right) + \log_{\mathrm{e}} \frac{x_i}{x_{\mathrm{total}}} \right] \qquad (9.17)$$

where $x_{\mathrm{total}} = \sum_i x_i$. This can be rewritten as

$$\frac{G}{RT} = \sum_i x_i \left(\frac{\bar{G}_i^{\mathrm{o}}}{RT}\right) + x_{\mathrm{total}} \sum \left(\frac{x_i}{x_{\mathrm{total}}}\right) \log_{\mathrm{e}} \left(\frac{x_i}{x_{\mathrm{total}}}\right) \tag{9.18}$$

G is then calculated as a linear function of $(x_i/x_{\mathrm{total}}) \log_{\mathrm{e}}(x_i/x_{\mathrm{total}})$ and standard Simplex code was used to find the Gibbs energy minimum.

The method of steepest descent provides a rapid solution to the minimisation problem and was later used by Eriksson (1971, 1975) and Eriksson and Rosen (1973) in the software codes SOLGAS and SOLGASMIX. SOLGAS, the earlier code, treated a mixture of stoichiometric condensed substances and an ideal gas mixture, while SOLGASMIX was able to further include non-ideal solution phases. A Gibbs energy function $G(Y)$ is first defined using Eqs (9.1) and (9.15) with a trial set of composition inputs $y_1^g, y_2^g, \ldots, y_m^g; y_1^c, y_2^c, \ldots, y_s^c$, where m is the number of substances in the gas phases, s the number of condensed substances in the equilibrium and where the various values of y are subject to the mass balance constraints described by Eq. (9.16). A quadratic approximation of $G(Y)$ is then expanded using Taylor's approximation and minimised utilising Lagrange multipliers to incorporate the mass balance constraint. The relevant set of linear equations are solved to produce interim values for x_i^g and x_i^c which are then used as estimates for y_i^g and y_i^c in the next iteration. Convergence is considered to have been reached when the difference in Gibbs energy between the two steps is below a predefined limit of accuracy. A short additional procedure eliminates the possibility of obtaining negative values of x_i and guarantees convergence; all mole numbers are screened to remain positive and the directional derivatives has to remain negative, i.e., the minimum point is not passed (White *et al.* 1958).

The minimisation methods used by later programmes such as ChemSage, the successor to SOLGASMIX, (Eriksson and Hack 1990), F*A*C*T (Thompson *et al.* 1988), Thermo-Calc (Sundman 1991), MTDATA (Davies *et al.* 1991) and PMLFKT (Lukas *et al.* 1982, Kattner *et al.* 1996) are, in the broadest sense, similar in principle to that described above, although there are clear differences in their actual operation. Thermodynamic models are now more complex, which makes it necessary to consider further degrees of freedom but, essentially, constraints are made to the system such that the Gibbs energy may be calculated as a function of extensive variables, such as the amount of each phase present in the system. Initial guesses are made for the Gibbs energy as a function of phases present, their amounts and composition, etc. The Gibbs energy is then calculated and some numerical method is used, whether it be through Lagrangian multipliers (Jansson 1984a) or a Newton–Raphson method (Hodson 1989), so that new values which will cause the Gibbs energy to be decreased can be estimated. When the difference in calculated Gibbs energy between the iterative steps reaches some small enough value, the calculation is taken to have converged. Recently, Bogdanoff (1995) described the development of new code which provides an interesting insight to some of the computational problems which can arise during Gibbs energy minimisation.

9.2.4 Stepping and mapping

The Gibbs energy minimisation process described in Sections 9.2.2 and 9.2.3 is specific in that it gives a calculated phase equilibrium at a defined composition, temperature and pressure. This can be modified so that phase equilibria are calculated as one particular property is changed, for example, temperature. This would be considered 'stepping' as only one variable is being considered. Plotting of various calculated properties such as the amount of a phase (Fig. 9.16), composition of phases (Fig. 9.17) and element distribution (Fig. 9.18) can then be made. Stepping is probably the most valuable form of calculation for multi-component alloys.

Mapping is the process by which phase diagrams are plotted and is the most common way for binary and ternary systems to be represented. Usually two degrees of freedom are considered and diagrams are created by finding boundaries where the phases in equilibrium differ on either side. The simplest forms are binary temperature vs composition diagrams and ternary isothermal sections which are easy to map. Figures 9.15 and 9.19 show, respectively, a ternary isothermal section involving three phases and a simple binary eutectic diagram. In both of these cases the tie-lines are in the plane of the figure and any calculation of a two- or three-phase field defines the positions of these boundaries on the diagram. Essentially, all that is required to map a phase boundary is a code which intelligently steps either the composition or temperature of an alloy whose composition lies in a two-phase field. It is then necessary to recognise when a new phase will appear and define the position of the three-phase field.

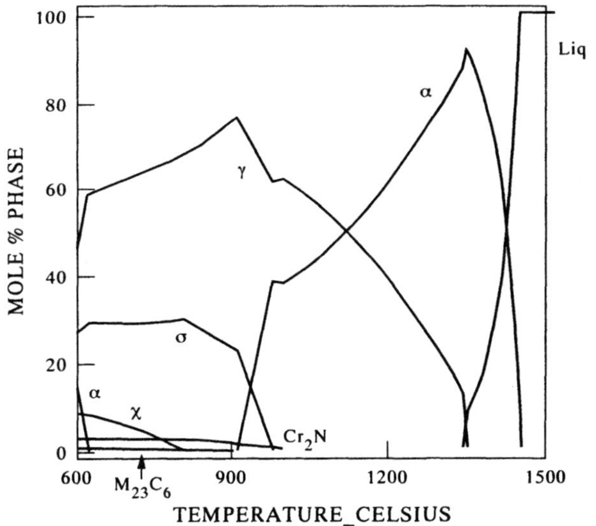

Figure 9.16. Mole % phase vs temperature plot for a ZERON 100 duplex stainless steel.

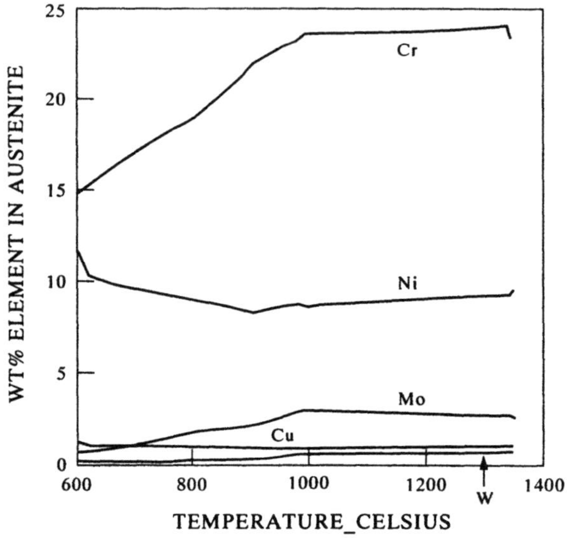

Figure 9.17. Composition of γ-phase in ZERON 100 duplex stainless steel as a function of temperature.

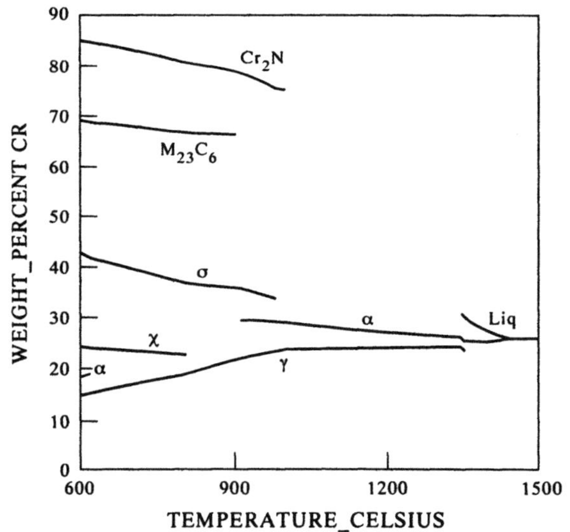

Figure 9.18. Distribution of Cr in the various phases of a ZERON 100 duplex stainless steel as a function of temperature.

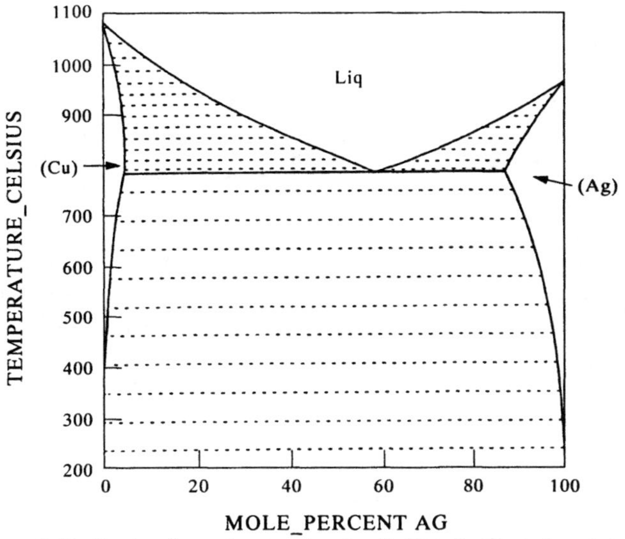

Figure 9.19. Cu–Ag phase diagram showing tie-lines in the various two-phase fields.

In a binary diagram the position of the three-phase line can be calculated utilising a method whereby the step size is changed when a phase boundary is reached. For example, the calculation begins with an alloy in the $(\alpha + \beta)$ phase field. The temperature is increased by 10°C steps and its composition maintained so that it exists in the $(\alpha + \beta)$ phase field. At each new step the stability of the liquid is checked. Once the liquid becomes stable the previous temperature is used as a start point and the temperature step is decreased. This process is repeated with subsequent decreased step sizes until a the temperature is defined within a critical step size. This method is cumbersome and more intelligent searching routines can be used. But in the end the temperature will be defined within a critical step size. Alternatively, the temperature where the activity/chemical potential of A and B in the three phases is equal can be explicitly calculated.

In a ternary isothermal section a similar procedure is used where an alloy is stepped such that its composition remains in a two-phase field. The three-phase field is now exactly defined by the composition of the phases in equilibrium and this also provides the limiting binary tie-lines which can used as start points for calculating the next two-phase equilibrium.

In vertical sections through ternary and higher-order systems or isothermal sections for quaternary and higher systems, the position becomes more complex as tie-lines do not lie in the plane of the diagram. They therefore cannot be used to define the positions of phase boundaries and the procedures described above become inoperative. New concepts are required, such as viewing the diagram in a

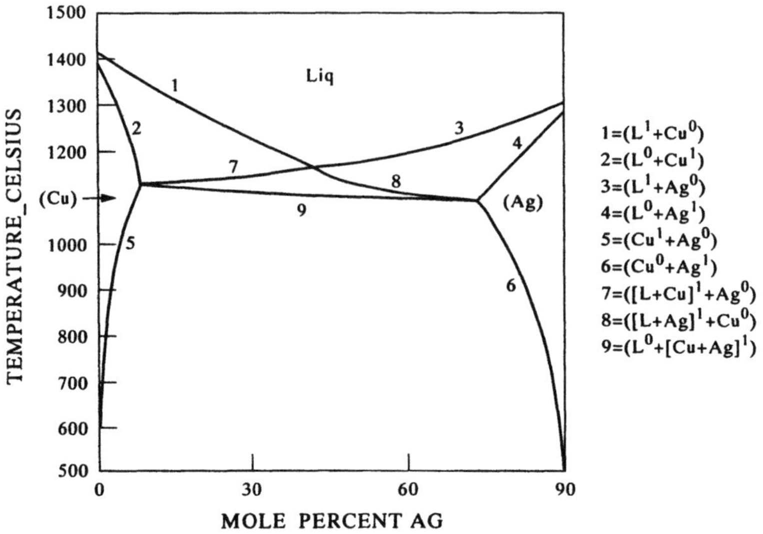

Figure 9.20. Vertical section through Cu–Ni–Ag, at a constant 10at%Ni, showing definition of phase-boundary lines.

topological way and considering what generally defines the lines/boundary positions. A vertical section through a ternary system will be used as an example (Fig. 9.20). The definition of the lines in this diagram is quite simple as they delineate the positions where more than one phase is in equilibrium and the mole fraction of one of the phases is zero. This definition is a general one and applicable to phase diagrams of higher-order systems. Invariant reactions have their own unique properties and occur at a fixed temperature. Mapping then requires that these positions are found, which requires a further step in the computing routines.

The discussion above centred around diagrams where the axes were composition or temperature. It is quite possible to use other variables in mapping routines, for example, activity/chemical potential and pressure. Further, it is possible to consider mapping other features, for example, the liquid invariant lines in a ternary system (Fig. 9.21). In such cases the positions of the lines are defined by other criteria than described above and new search routines are required.

A great strength of the CALPHAD approach is that various other properties are automatically obtained during the calculation procedure and these can also be plotted. Figure 9.22(a) shows the chemical potential of Mg (with respect to Mg liquid) as a function of temperature in Cu–Mg and Fig. 9.22(b) shows the corresponding T/x diagram (Coughnanowr *et al.* 1991). The chemical potential diagram is interesting because there are no two-phase fields since $\bar{G}_i$ is constant in such a region and, therefore, only varies in the single-phase fields. In a ternary system the extra degree of freedom means that $\bar{G}_i$ varies in the two-phase region but is constant in the

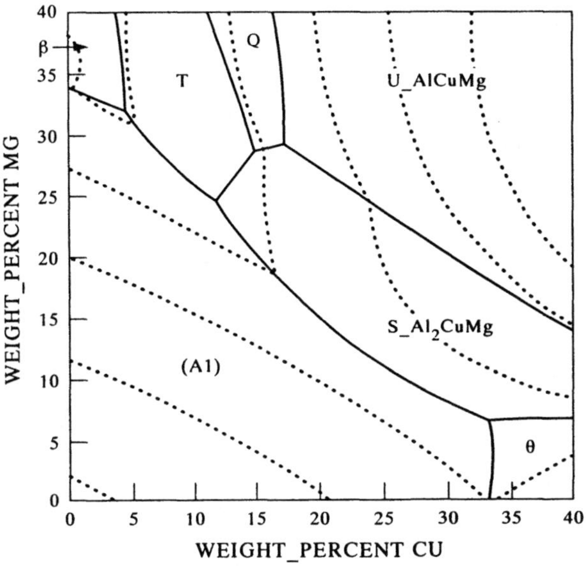

Figure 9.21. Calculated liquidus surface for Al–Cu–Mg.

three-phase region. Figure 9.23 shows a diagram for Fe–Cr–N from Frisk (1990) with the chemical potential axes converted to activity for the purpose of variety. Although there is potential for plotting all types of axes, there is also considerable danger that the diagram may become difficult to interpret. Hillert (1996) addressed this problem and produced the following definition of what is a true phase diagram.

> *"A phase diagram may be defined as any diagram showing the range of existence of individual phases and of combinations of phases. Each such range is a phase field and it can have the same dimensionality as the diagram itself or a lower dimensionality. By selecting conditions corresponding to a point in the diagram, one can directly read what phases could be present at equilibrium. A diagram should not be regarded as a true phase diagram if it does not give a unique answer."*

Examples of diagrams which failed the above criteria were shown and a number of conjugate pairs of variables which would satisfy the above definition were also given. It is hoped that in future all software will have algorithms recognising which axes will provide true equilibrium diagrams and warn users if this is not the case.

9.2.5 Robustness and speed of calculation

First a clear distinction must be made between robustness and speed of calculation, although in certain cases they are clearly linked. Speed may be altered by straightforwardly changing factors such as the processing power of the computer, using a

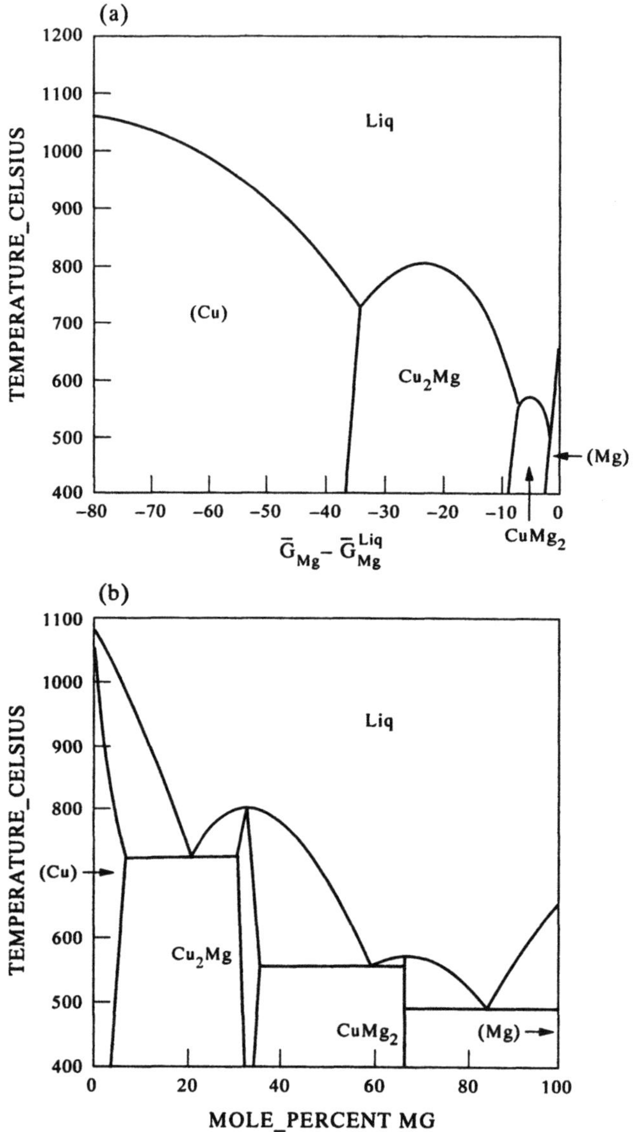

Figure 9.22. (a) Chemical potential vs temperature diagram for Cu–Mg and (b) phase diagram for Cu–Mg (Coughnanowr *et al.* 1991).

more efficient compiler, changing the operating system, etc. These are extrinsic factors which, while making the calculation faster, do not ensure a more robust calculation; a robust code being defined here as one which will ensure that *true*

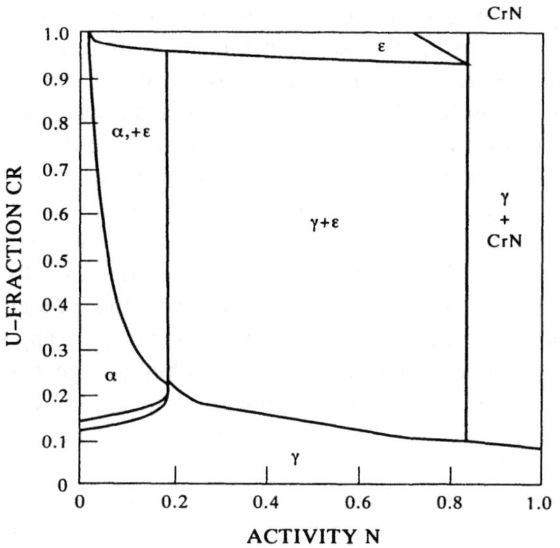

Figure 9.23. Isothermal section for Fe–Cr–N using activity and U-fractions from Frisk (1990).

equilibrium is obtained. By speeding up the calculation, and using a code which is not robust, the wrong answer is only achieved faster, or the programme crashes more quickly. Robustness and speed are linked, however, by choice of start points which are the initial guesses for the amount and composition of each phase given to the minimisation code. Clearly the more accurate the start points, the less work needs to be done by the Gibbs energy minimisation sub-routine and the more quickly the calculation is performed. Further, there is less likelihood of local minima affecting the calculation.

If codes are based on local minimisation methods, such as steepest descent or Newton–Raphson, problems will always arise due to the existence of unknown local minima in Gibbs energy space. More robust codes based on global optimisation methods can be designed, but these must sample a larger portion of Gibbs energy space so that all of the local minima are recognised. Unfortunately, computational times can then be greatly increased. If speed is required (which is usually the case) it is necessary to use local minimisation methods and it is, therefore, necessary to ensure that good guesses are provided for the phases which are present and their compositions. This was one of the main problem areas for the first attempts at producing sound code. Most programmes now address this reasonably well, but it is the authors' experience that it is still possible for the minimisation routines to get caught in local minima, rather than find true equilibrium. Therefore, the user should always be aware of this possibility and, where necessary, apply some form of checking. Some programmes allow this by providing a facility for the user to input alternative start points before calculations.

One of the main reasons for the occurrence of local minima is the presence of miscibility gaps in various phases. In binary and ternary systems such effects are well established and manifest themselves in phase separation. The minimisation routines must then recognise their existence and account for them. It is possible to have a datafile which contains two identical sets of parameters with phases called FCC#1 and FCC#2; alternatively the software produces a 'virtual' FCC#2 phase based on the pre-existing entry for FCC. The programme must then provide different start points for FCC#1 and FCC#2 and ensure that that convergence accounts for both sets of potential minima. However, this pre-supposes that the potential occurrence of a miscibility gap is known. Such an approach can work satisfactorily for a well-understood class of material but, as a generalised method, it would fail if unknown miscibility gaps were present. In this circumstance it is possible to look at the derivatives of the Gibbs energy as a function of composition to check for potential local minima.

In some thermodynamic models there are also potential minima associated with different site occupations, even though the composition may not vary, e.g., a phase with an order/disorder transformation. This must be handled in a somewhat different fashion and the variation in Gibbs energy as a function of site fraction occupation must be examined. Although this is not, perhaps, traditionally recognised as a miscibility gap, there are a number of similarities in dealing with the problem. In this case, however, it is the occupation of sites which govern the local minima and not the overall composition, *per se*.

There may be problems if the Gibbs energy of two phases converge either above the temperature of the miscibility gap or above the ordering temperature. The minimisation code will then see two phases with identical Gibbs energy, which can cause considerable confusion. The problem is now one of strategy and programming. Miscibility gaps are handled in all of the major software codes, but in different ways. There is no one ideal solution to this problem whereby true robustness is guaranteed. Searching derivatives of Gibbs energy vs composition for all phases is potentially a robust solution, especially if a completely generalised method is required. However, the time penalty to check all phases in a real, multi-component material is huge and, in reality, impractical. Some knowledge of potential ordering or miscibility gap formation is, therefore, always necessary to complete the calculation in a reasonable time.

9.3. THERMODYNAMIC OPTIMISATION OF PHASE DIAGRAMS

9.3.1 Introduction

Thermodynamic optimisation is probably the largest single area of reported work in the journal CALPHAD. It invariably concerns binary and ternary systems and can

References are listed on pp. 294–296.

be defined as a fitting process whereby the adjustable coefficients in the Gibbs energy equations are altered such that the best representation of *both* the experimentally measured phase diagram and thermodynamic properties are obtained. The accuracy of this representation can be defined mathematically through some form of least-squares algorithm, and this forms the basis for optimisation software such as the Lukas programme (Lukas *et al.* 1977), PARROT (Jansson 1984b), FITBIN (Thompson *et al.* 1988) and ChemOpt (Königsberger and Eriksson 1995).

In earlier times, thermodynamic and phase diagram optimisation were often separate exercises. For example, the compilations of binary diagrams by Hansen and Anderko (1958), Elliott (1965) and Shunk (1969) did not normally take into account thermodynamic information. The compilations of Hultgren *et al.* (1973) concentrated on thermodynamic measurements, but the phase diagram was also an important factor in their assessment procedure. However, this was mainly for phase identification purposes rather than for making a self-consistent assessment of both phase diagram and thermodynamic measurement.

Essentially, any optimisation process requires good judgement. This is basically because the errors which are incurred during measurement are not completely random. There are systematic errors which can arise from, for example, faulty calibration or unsuitability of a particular measurement technique. It is worth, therefore, considering a question and answer session which succeeded a paper presented by one of the present authors at the Royal Society (Saunders 1995):

"Q. You hinted at a rather remarkable claim. When there are several distinct measurements of a phase equilibrium feature, you suggest you can pick out a faulty measurement by comparison with a CALPHAD calculation. But surely the precision of such a calculation is linked to the precision of the thermochemical measurements which contribute the input. Shouldn't a curve calculated by CALPHAD actually show upper and lower confidence limits depending on your judgement of the reliability of the input measurements?

"A. The answer to your question is yes and no. In an ideal world, where all experimentation is reliable and of good quality, I would agree that the curve calculated by a CALPHAD calculation could have such confidence limits imposed. Unfortunately, this is not often the case and, for example, sets of measured datapoints for a liquidus in a simple binary system can vary considerably. In some high-melting-point systems it is not unusual for these differences to be of the order of 100°C or higher. Sometimes a simple comparison of the original datapoints can show that one particular set of results is obviously inconsistent with other reported values and this forms the basis of experimental phase diagram assessment work by people such as Hansen. Unfortunately in others cases the position is not so clear cut. The CALPHAD calculation does impose *a self-consistency between the underlying thermodynamics and the phase diagram* and therefore can be used as an arbiter between conflicting values."

In essence this summarises the position of optimisation. Hopefully, systematic errors are exposed but this does rely on there being enough experimental information available so that they are clearly shown. The general practice of optimisation at the present time is performed through a process of trial-and-error combinations of the adjustable parameters. Optimising codes such as the Lukas programme or PARROT reduce the error using mathematical algorithms while a 'manual' procedure means that the user changes these parameters according to personal judgement. Although a mathematical least-squares-type approach might be considered as ideal, a great deal of judgement is still required in the process, especially when there is little experimental information available.

A good example of an optimisation in practice is the assessment of Ga–Te (Oh and Lee 1992). The liquid phase was modelled using the ionic two-sublattice model (see Section 9.5.3) which readily models systems with negative mixing enthalpies but which also have high-temperature miscibility gaps. Figures 9.24–9.27 show comparisons between calculated and experimental thermodynamic properties in the liquid. It is interesting to note that a concentration structure factor could also be

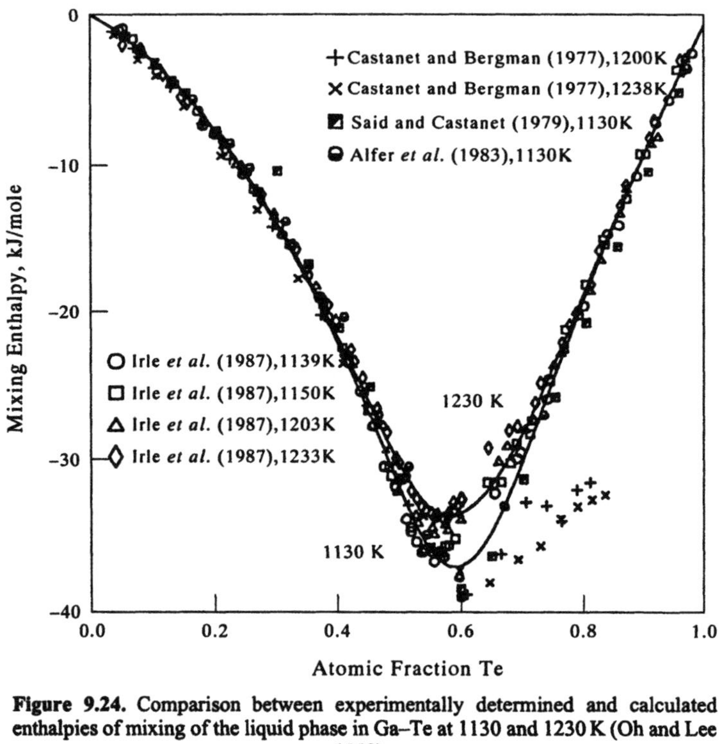

Figure 9.24. Comparison between experimentally determined and calculated enthalpies of mixing of the liquid phase in Ga–Te at 1130 and 1230 K (Oh and Lee 1992).

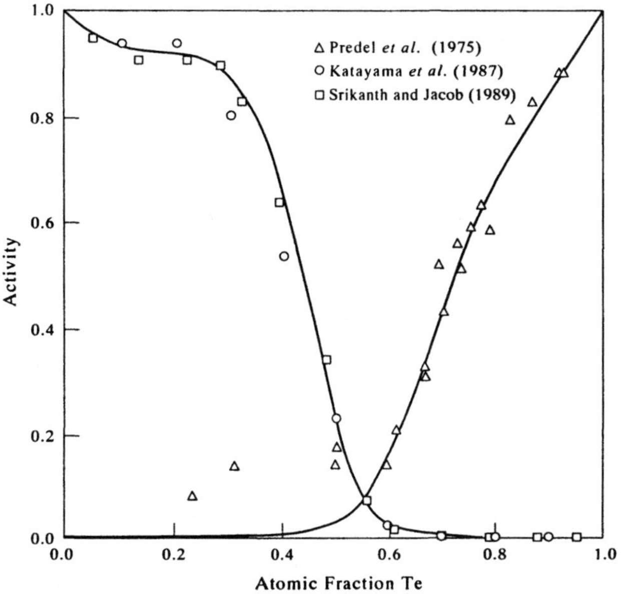

Figure 9.25. Comparison between experimentally determined and calculated activities of Ga and Te in the liquid phase in Ga–Te at 1120 K (Oh and Lee 1992).

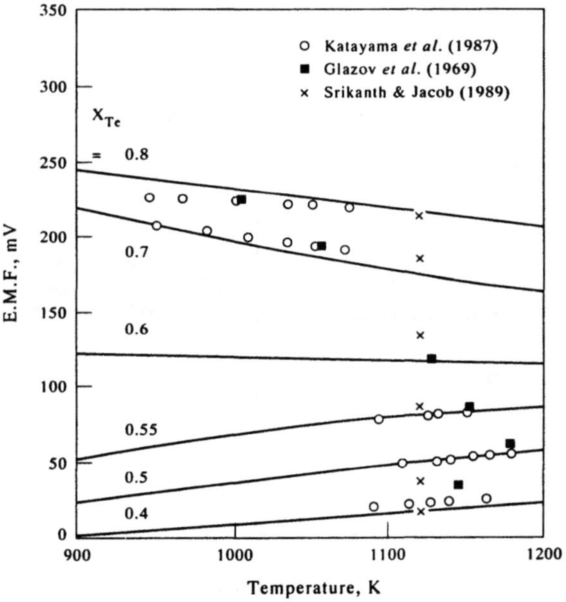

Figure 9.26. Comparison between experimentally determined and calculated EMF values as a function of temperature for various liquid alloys in Ga–Te (Oh and Lee 1992).

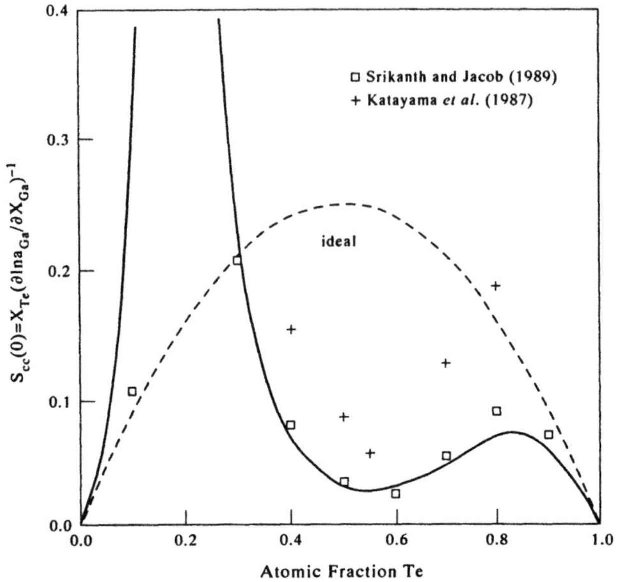

Figure 9.27. Comparison between experimentally determined and calculated concentration–concentration structure factors at 1120 K in liquid phase of Ga–Te at 1120 K (Oh and Lee 1992).

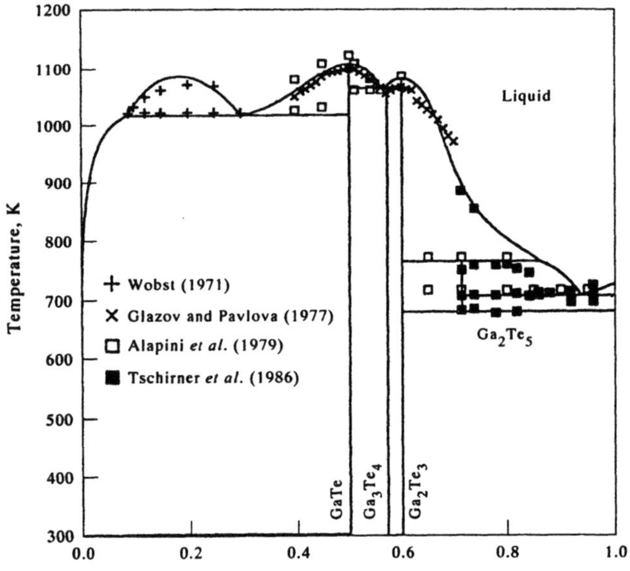

Figure 9.28. Comparison between experimentally determined and calculated Ga–Te phase diagram (Oh and Lee 1992).

Table 9.1. Invariant equilibria in the Ga–Te binary system from Oh and Lee (1992).

Phases	Concentration (X_{Te}^{liq})	Temperature (K)	Reference
Ga/liquid/GaTe	—	302.9	Klemm and Vogel (1934)
	5.8 10^{-11}	302.9	This study
Liquid/liquid/GaTe	—/0.33	1013	Klemm and Vogel (1934)
	—/0.30	1019	Newman *et al.* (1961)
	—	1023	Alapini *et al.* (1979)
	—	1019	Blachnik and Irle (1985)
	0.09/0.30	1018	Wobst (1971)
	0.088/0.302	1015	This study
GaTe/liquid/Ga_3Te_4	—	1049	Blachnik and Irle (1985)
	0.561	1061	This study
Liquid/Ga_3Te_4/Ga_2Te_3	—	1057	Blachnik and Irle (1985)
	0.570	1063	This study
GaTe/Ga_2Te_5/liquid	—	768	Alapini *et al.* (1979)
	—	761	Blachnik and Irle (1985)
	0.780	755	Tschirner *et al.* (1986)
	0.862	761	This study
Ga_2Te_5/liquid/Te	0.9	713	Alapini *et al.* (1979)
	—	709	Blachnik and Irle (1985)
	0.9	703	Tschirner *et al.* (1986)
	0.937	703	This study
GaTe/Ga_2Te_5/Te	—	673	Alapini *et al.* (1979)
	—	684	Tschirner *et al.* (1986)
	—	677	This study
Critical point	0.20	1069	Wobst (1971)
	0.183	1083	This study

plotted and compared with results from two separate studies. Table 9.1 shows the comparison between various calculated and experimental quantities for the compounds in the system. Finally the calculated phase diagram is shown in Fig. 9.28 with detailed comparison between calculated and experimentally measured invariant reactions shown in Table 9.2.

Firstly, the optimisation shows how well the various thermodynamic quantities are matched and the excellent agreement with the experimentally observed phase diagram. It also shows a clear discrepancy between one set of experimental results and the optimised values for the mixing enthalpy in the liquid, emphasising the point that the combined thermodynamic and phase-diagram optimisation has been able to differentiate between conflicting experiments.

The assessment of Oh and Lee (1992) utilised the PARROT programme (Jansson 1984b) and it is now worth discussing this programme and the one developed by Lukas *et al.* (1977) as these are the main codes presently used for optimisation.

Table 9.2. Thermodynamic quantities for GaTe and Ga_2Te_3 phases (J/g-at.) from Oh and Lee (1992).

Quantity	Value	Method	Reference
		*GaTe	
ΔH^f_{298}	-39,400	Calorimetry	Said and Castanet (1977)
	-39,000	Calorimetry	Said and Castanet (1977)
	-59,800	Calorimetry	Hahn and Burow (1956)
	-62,800	Potentiometry	Abbasov *et al.* (1964)
	-39,205	Calculated	This study
S^0_{298}	42.7	Calorimetry	Kerimov *et al.* (1971)
	41.3	Calculated	This study
ΔH^{fus}	19,100	Calorimetry	Said and Castanet (1977)
	17,991	DTA	Kuliev *et al.* (1976)
	18,041	Calculated	This study
T_m	1108	DTA	Newman *et al.* (1961)
	1123	DTA	Alapini *et al.* (1979)
	1121	DTA	Blachnik and Irle (1985)
	1099	DTA	Glazov and Pavlova (1977)
	1107	Enthalpimetry	Said and Castanet (1977)
	1107	Calculated	This study
		*Ga_2Te_3	
ΔH^f_{298}	-39,800	Calorimetry	Said and Castanet (1977)
	-38,600	Calorimetry	Said and Castanet (1977)
	-54,400	Calorimetry	Hahn and Burow (1956)
	-55,000	Potentiometry	Abbasov *et al.* (1964)
	-37,879	Calculated	This study
S^0_{298}	42.7	Calculated	Mills (1974)
	43.1	Calculated	This study
ΔH^{fus}	15,800	Calorimetry	Said and Castanet (1977)
	12,385	DTA	Babanly and Kuliev (1976)
	13,477	Calculated	This study
T_m	1065	DTA	Newman *et al.* (1961)
	1083	DTA	Alapini *et al.* (1979)
	1085	DTA	Blachnik and Irle (1985)
	1063	DTA	Glazov and Pavlova (1977)
	1071	Enthalpimetry	Said and Castanet (1977)
	1079	Calculated	This study

9.3.2 The Lukas programme

The Lukas programme (Lukas *et al.* 1977) was one of the first dedicated optimising programmes for use by CALPHAD practitioners. It is freely available, on request, and forms the basis for a substantial amount of work published in CALPHAD. Two main forms of the Lukas programme are now available, BINGSS and TERGSS, which are applicable to binary and ternary systems respectively. The original code used a Gaussian least-squares method (Zimmerman 1976) but later versions also include a method suggested by Marquardt (1963). The programme separates the

References are listed on pp. 294–296.

experimental measurements into three distinct types. (1) Calorimetric measurements, which are usually enthalpies; (2) EMF and vapour pressure measurements, which provide partial Gibbs energy; and (3) phase-diagram information. Each of these types of measurements is given a different equation of error.

For calorimetric measurements these equations are governed by the number of phases involved and the temperature of the measurement. Three types of equations are defined. If one phase only is programmed, the measured value describes the heat of formation of a phase from its pure components. As the enthalpy of the pure components are defined before the optimisation takes place, there is only one unknown. The equation of error for two phases applies to enthalpies of melting and transformation. In this case there are two unknowns; which are the enthalpies of the phases involved in the transformation. Enthalpic changes involving three phases can also be considered, for example the enthalpy change on mixing two liquids of different compositions. In this case the enthalpy of the two original liquids and the final liquid are needed.

For phase diagrams, a phase boundary is one end of a tie-line and, therefore, is dependent on the phase which exists at the other end of the tie-line. In a binary system, two independent measurements are therefore needed to define the tie-line; in the case of a liquid/solid phase boundary this would be x_B^{sol} and x_B^{liq} at temperature T. Ideally it would be desirable to have these two compositions as independent variables giving rise to two independent equations of error. The Lukas programme does this by making two equations but where the dependence of error on one of the measurements is weak. This is important if the two concentrations have different accuracies. For some types of experimental values newer versions of the Lukas programme offer different kinds of equations of error (Lukas and Fries 1992).

EMF and vapour pressure measurements are dependent on the temperature, the number of phases involved and, importantly, the reference state of the component in question. The problem with the reference state is important as experimentally stated values of partial Gibbs energies will be dependent on this value. The standard states are fixed before optimisation and may actually have values different from those used by the original author. Therefore, as far as possible like should be compared with like.

Once the equations of error are defined, 'weighting factors', in terms of estimated accuracy of the experiment, are included. These are either taken as the accuracy of the corresponding measurement as given by the original author or estimated by the person who is doing the optimisation. The error is then divided by these weighting factors to provide a dimensionless relative error for all types of experimental measurement. In addition to this, the sensitivity to the measured value of changes in temperature and composition are considered.

If the Lukas programme is run with all experimental data, including reasonable estimates for accuracy, it performs its fitting operation by assuming that errors arise from random effects rather than systematic inaccuracies. Systematic errors can be taken into account in at least two ways. Firstly, it is possible to ignore the particular experiments completely, and there is a command in the programme which allows this to be done. Secondly, the estimated experimental error can be increased by increasing the corresponding value in the data file which contains the experimental information. Both routes emphasise the fact that a purely mathematical approach to optimisation is usually not possible and some form of personal judgement is required.

9.3.3 The PARROT programme

The PARROT programme (Jansson 1984b) is integrated into the Thermo-Calc software package. It relies on principles different from those applied in the Lukas programme although it has the same aim of reducing the error between calculated and experi-mental quantities. Primary governing principles in PARROT involve establishing a criterion for the best fit, separating data into sets of different accuracies and making a distinction between independent and dependent variables.

The criterion for best fit is based on the maximum likelihood principle (Fisher 1922) where the best estimates of the model parameters should maximise the likelihood function, L, for the observation of N different experimental observations,

$$L = \prod_{i=1}^{N} F_i(\bar{z}_i^{\text{o}}, \bar{w}_i) \tag{9.19}$$

where F_i is the multi-variable density function for the distribution of the measured values in experiment i, $\bar{z}_i^{\text{o}}$, are measured experimental values which might differ significantly from their true value and w_i is used to denote the statistical parameters in the probability density function concerning experiment i. Equation (9.19) requires that the joint probability density function of all experimental data is known. However, the available data rarely allow the determination of probability functions. Therefore, in order to apply Eq. (9.19) it is assumed that the statistical density functions of the experiments is Gaussian in form. The normal distribution of experimental data (the probability of observing the values $\bar{z}_i^{\text{o}}$) given that their true values are $\bar{\mu}_i$ is then expressed by

$$F_i = 2\pi - \frac{nz_i}{2}(\det V_i)^{1/2}\exp\left(-\frac{1}{2}(\bar{z}_i^{\text{o}} - \bar{\mu}_i)^T V_i^{-1}(\bar{z}_i^{\text{o}} - \bar{\mu}_i)\right) \tag{9.20}$$

where V_i is the variance–covariance matrix of the measured variable in experiment i. The maximisation of Eq. (9.19) combined with Eq. (9.20) is complex and the problem was simplified by assuming that the maximum of L coincides with the minimum of the exponential factors in Eq. (9.20). Their product, which is the estimator of the goodness of fit, is then represented by exponent S where

$$S = \sum_{i=1}^{N}\left((\bar{z}_i^{\text{o}} - \bar{\mu}_i)^T V_i^{-1}(\bar{z}_i^{\text{o}} - \bar{\mu}_i)\right). \tag{9.21}$$

References are listed on pp. 294–296.

Further simplifications can be made by assuming the off-diagonal elements of V_i, which represent the covariance between the measured quantities, are zero; this effectively means that there is no coupling between the different experimental determinations in the experimental procedure, which is reasonable.

PARROT allows for the fact that, in addition to the experimental quantities $\bar{z}_i^{\text{o}}$ which exhibit significant deviations from their true value, there is another set of quantities $\bar{u}_i^{\text{o}}$ which are considered free from significant inaccuracy. Also, *independent* and *dependent* variables are defined, where independent variables are measured values which can be considered to define the conditions for the equilibrium state while dependent variables are regarded as the responses of the system to the these prescribed conditions. Independent variables should come primarily from the $\bar{u}_i^{\text{o}}$ set, but if recourse has to be made to the $\bar{z}_i^{\text{o}}$ set the latter can be further divided into two sets, one which contains independent variables $\bar{x}_i^{\text{o}}$ and the other containing dependent variables $\bar{y}_i^{\text{o}}$. Whether a measured independent variable should be included in the $\bar{u}_i$ or $\bar{x}_i$ set depends on the sensitivity of the $\bar{y}_i$ quantities to fluctuations in that value. If the fluctuations in the $\bar{y}_i$ quantities are less than the accuracy of the experimental measurement when the $\bar{u}_i$ or $\bar{x}_i$ value is varied within its own accuracy limit, then it should be included in the $\bar{u}_i$ set.

Taking these additional features into account, the estimator of the goodness of fit then becomes the weighted sum of the squared deviation of the observed values from their calculated values.

$$S = \sum_{i=1}^{N} \left\{ \sum_{j=1}^{nx_i} \left(\frac{x_{i,j}^{\text{o}} - x_{i,j}^{*}}{\sigma x_{i,j}} \right)^2 + \sum_{j=1}^{ny_i} \left(\frac{y_{i,j}^{\text{o}} - y_{i,j}^{c}(\bar{x}_i^{*}, \bar{\mathbf{A}}^{*})}{\sigma y_{i,j}} \right)^2 \right\} \tag{9.22}$$

where * denotes an estimated value, $\bar{\mathbf{A}}$ is a vector with its elements equal to the model parameters to be estimated, $\sigma x_{i,j}$ and $\sigma y_{i,j}$ are respectively the standard deviations of the *j*th independent and dependent variable in experiment i and $\bar{y}_{i,j}^{c}$ is the calculated dependent variable. The minimisation of Eq. (9.22) in PARROT utilises neither the Gaussian method or the Marquardt modification, which are the normal methods of the Lukas programme. Instead a finite difference approximation method after Powell (1965) is implemented via a subroutine VA050 from the Harwell Subroutine Library (Hopper).

The PARROT programme uses the Poly-3 subroutine in Thermo-Calc to calculate Gibbs energies of the various phases and find the equilibrium state. In such equilibrium calculations the temperature, pressure and chemical potentials are treated as independent variables, and preselected state variables are used to define the conditions for an equilibrium calculation. The dependent state variables, i.e., the responses to the system, can then be given as a function of the independent state variable and the model parameters. It is thus possible to use almost any type of experimental information in the evaluation of the model parameters.

9.3.4 Summary

It must always be remembered that optimisation is not an exact science and, therefore, it is sometimes difficult to define confidence limits in the final optimised values for the coefficients used in the thermodynamic models. The final outcome is at least dependent on the number of experimental measurements, their accuracy and the ability to differentiate between random and systematic errors. Concepts of quality can, therefore, be difficult to define. It is the author's experience that it is quite possible to have at least two versions of an optimised diagram with quite different underlying thermodynamic properties. This may be because only experimental enthalpy data were available and different entropy functions were chosen for the different phases. Also one of the versions may have rejected certain experimental measurements which the other version accepted. This emphasises the fact that judgement plays a vital role in the optimisation process and the use of optimising codes as 'black boxes' is dangerous.

REFERENCES

Abbasov, A. S., Nikolskaia, A. V., Gerasimov, Y. I. and Vasilev, V. P. (1964) *Dokl. Akad. Nauk, SSSR*, **156,** 1140.

Alapini, F., Flahaut, J., Guittarad, M., Jaulmes, S. and Julien-Pouzol, M. (1979) *J. Solid State Chem.*, **28,** 309.

Al'fer, S. A., Mechkovski, L. A. and Vecher, A. A. (1983) *Thermochimica Acta*, **88,** 493.

Ansara, I. (1979) *Int. Met. Reviews*, **22,** 20.

Babanly, M. B. and Kuliev, A. A. (1976) *Izv. Akad. Nauk, AzSSR, Ser. Fiz.-Tekh. Mat. Nauk*, **4,** 145.

Blachnik, R. and Irle, E. (1985) *J. Less Common Metals*, **113,** L1.

Bogdanoff, P. (1995) M.Sc. Thesis, University of Cambridge, UK.

Brebrick, R. F. (1976) *Met. Trans. A*, **7A,** 1609.

Brebrick, R. F. (1977) *Met. Trans. A*, **8A,** 403.

Brinkley, S. R. (1946) *J. Phys. Chem.*, **14,** 563.

Brinkley, S. R. (1947) *J. Phys. Chem.*, **15,** 107.

Castanet, R. and Bergman, C. (1977) **9,** 1127.

Coughnanowr, C. A., Ansara, I., Luoma, R., Hamalainen, M. and Lukas, H. L. (1991) *Z. Metallkde.*, **82,** 574.

Dantzig, G. B., Johnson, S. M. and White, W. B. (1957) The RAND Corporation, Paper P-1060, April 15.

Davies, R. H., Dinsdale, A. T., Hodson, S. M., Gisby, J. A., Pugh, N. J., Barry, T. L. and Chart, T. G. (1991) in *User Aspects of Phase Diagrams*, ed. Hayes, F. H. (Inst. Metals, London), p. 140.

Elliott, R. P. (1965) *Constitution of Binary Alloys, 1st Supplement* (McGraw-Hill, New York).

Eriksson, G. (1971) *Acta Chem. Scand.*, **25,** 2651.

Eriksson, G. (1975) *Chemica Scripta*, **8,** 100.

Erikkson, G. and Rosen, E. (1973) *Chemica Scripta*, **4,** 193.
Eriksson, G. and Hack, K. (1990) *Met. Trans. B*, **21B,** 1013.
Fisher, R. A. (1922) *Phil. Trans. A*, **A222**, 309.
Frisk, K. (1990) *Met. Trans. A*, **21A,** 2477.
Glazov, V. M. and Pavlova, L. M. (1977) *Inorg. Mater.*, **13,** 184.
Glazov, V. M., Chizhevskaya, S. N. and Glagoleva, N. N. (1969) "Liquid Semiconductors" (Plenum, New York).
Hahn, H. and Burow, F. (1956) *Angew. Chem.*, **68,** 382.
Hansen, M. and Anderko, K. (1958) *Constitution of Binary Alloys* (McGraw-Hill, New York).
Hillert, M. (1996) "How to select axes for a phase diagram", presented at CALPHAD XXXV, Erice, Sicily, May 26–31.
Hodson, S. M. (1989) "MTDATA handbook: multiphase theory", NPL report, 3 February.
Hopper, M. J. Harwell Subroutine Library, A.E.R.E., Harwell, Didcot, UK.
Hultgren, R., Desai, P., Hawkins, D. T., Gleiser, M. and Kelley, K. K. (1973) *Selected Values of the Thermodynamic Properties of Binary Alloys* (ASM, Metals Park, OH).
Irle, E., Gather, B., Blachnik, R., Kattner, U., Lukas, H. L. and Petzow, G. (1987) *Z. Metallkde.*, **78,** 535.
Jansson, B. (1984a) "A general method for calculating phase equilibria under different types of conditions", TRITA-MAC-0233, Division of Physical Metallurgy, Royal Institute of Technology, Stockholm, Sweden.
Jansson, B. (1984b) "Evaluation of parameters in thermochemical models using different types of experimental data simultaneously", TRITA-MAC-0234, Division of Physical Metallurgy, Royal Institute of Technology, Stockholm, Sweden.
Kandiner, H. J. and Brinkley, S. R. (1950) *Ind. Eng. Chem.*, **42,** 850.
Katayama, I., Nakayama, J.-I., Nakai, T. and Kozuka, Z. (1987) *Trans JIM*, **28,** 129.
Kattner, U. R., Boettinger, W. J. and Coriell, S. R. (1996) *Z. Metallkde.*, **87,** 522.
Kaufman, L. and Bernstein, H. (1970) *Computer Calculation of Phase Diagrams* (Academic Press, New York).
Kerimov, I. G., Mamedov, K. K., Mekhtiev, M. I. and Kostryukov, V. N. (1971) *J. Phys. Chem.*, **45,** 1118.
Klemm, W. and Vogel, H. U. V. (1934) *Z. Anorg. Chem.*, **219,** 45.
Königsberger, E and Eriksson, G. (1995) *CALPHAD*, **19,** 207.
Krieger, F. G. and White, W. B. (1948) *J. Chem. Phys.*, **16,** 358.
Kuliev, A. A., Babanly, M. B. and Kagrammanian, S. G. (1976) *Azerb. Khim. Zh.*, **2,** 113.
Lukas, H. L., Henig, E. Th. and Zimmerman, B. (1977) *CALPHAD*, **1,** 225.
Lukas, H. L., Weiss, J. and Henig, E.-Th. (1982) *CALPHAD*, **6,** 229.
Lukas, H. L. and Fries, S. G. (1992) *J. Phase Equilibria*, **13,** 532.
Marquardt, D. W. (1963) *SIAM J.*, **11,** 431.
Meijering, J. L. (1957) *Acta Met.*, **5,** 257.
Mills, K. C. (1974) "Thermodynamic Data for Inorganic Sulfides, Selenides and Tellurides" (Butterworth, London).
Newman, P. C., Brice, J. C. and Wright, H. C. (1961) *Philips Res. Rept.*, **16,** 41.
Oh, C.-S. and Lee, D. N. (1992) *CALPHAD*, **16,** 317.
Powell, M. J. D. (1965) *Comput. J.*, **7,** 303.
Predel, B., Piehl, J. and Pool, M. J. (1975) *Z. Metallkde.*, **66,** 268.

Said, H. and Castanet, R. (1979) *J. Less Common Metals*, **68,** 213.
Saunders, N. (1995) *Phil. Trans. R. Soc. (Lond.) A*, **351,** 543.
Shunk, F. A. (1969) *Constitution of Binary Alloys, 2nd Supplement* (McGraw-Hill, New York).
Srikanth, S. and Jacob, K. T. (1989) *Thermochimica Acta*, **153,** 27.
Sundman, B. (1991) in *User Aspects of Phase Diagrams*, ed. Hayes, F. H. (Inst. Metals, London), p. 130.
Thompson, W. T., Eriksson, G., Pelton, A. D. and Bale, C. W. (1988) *Proc. Met. Soc. CIM*, **11,** 87.
Tschirner, K.-U., Garlipp, B. and Rentzsch, R. (1986) *Z. Metallkde.*, **77,** 811.
White, W. B., Johnson, S. M. and White, W. B. (1958) *J. Chem. Phys.*, **28,** 751.
Wobst, M. (1971) *Scripta Met.*, **5,** 583.
Zimmerman, B. (1976) Ph.D. Thesis, University of Stuttgart, Germany.

Chapter 10

The Application of CALPHAD Methods

Chapter 10
The Application of CALPHAD Methods

10.1. INTRODUCTION

The purpose of this chapter is to describe the current state of the art for what can be achieved in terms of CALPHAD calculations. It will start with a brief history on how CALPHAD calculations have been used in the past and then concentrate on the complex multi-component materials that can be treated by the software packages described in the previous chapter on computational methods. It will also concentrate on calculations for 'real' materials, which are mainly multi-component and, where possible, comparisons will be made with experimental observations. Little time will, therefore, be spent on binary and ternary calculations, although these can be quite complex and have their own points of interest. There are now so many such calculations that it would be impossible to cover the full range here. Also they must now be considered rather as building blocks for use in calculations which are multi-component in nature. For more extended information the reader is therefore pointed to the 20-year index of the CALPHAD journal (*CALPHAD* 1997) where calculations of binary and ternary systems are regularly published. Some examples of calculated binary and ternary diagrams are shown in Figs 10.1 and 10.2. They show the formation of non-stoichiometric binary and ternary compounds, order–disorder transformations (β/B2 in Ti–Al–Nb), ionic liquids with miscibility gaps, multiple sub-lattice phases, spinel phases, etc., and help demonstrate the state of the art which can now be achieved.

It is possible to separate calculations made under the CALPHAD umbrella into a number of strands. One major area is concerned with substance databases where, with the general exception of the gas phase, phases are modelled as stoichiometric. The problems associated with Gibbs energy minimisation are simpler for such calculations, but they can often involve much larger numbers of phases than would be considered in a multi-component alloy. Alloys probably still account for the majority of CALPHAD publications, as a review of the CALPHAD journal shows a preponderance of papers in this area. This relates back to the development of solution databases, which are more complex in nature than substance databases, and also to the intrinsic early problems of CALPHAD associated with factors such as lattice stabilities and development of new models for metallic systems. The development of models for ionic systems is also another significant, and growing area, sharing much in common with the principles driving the development of alloy

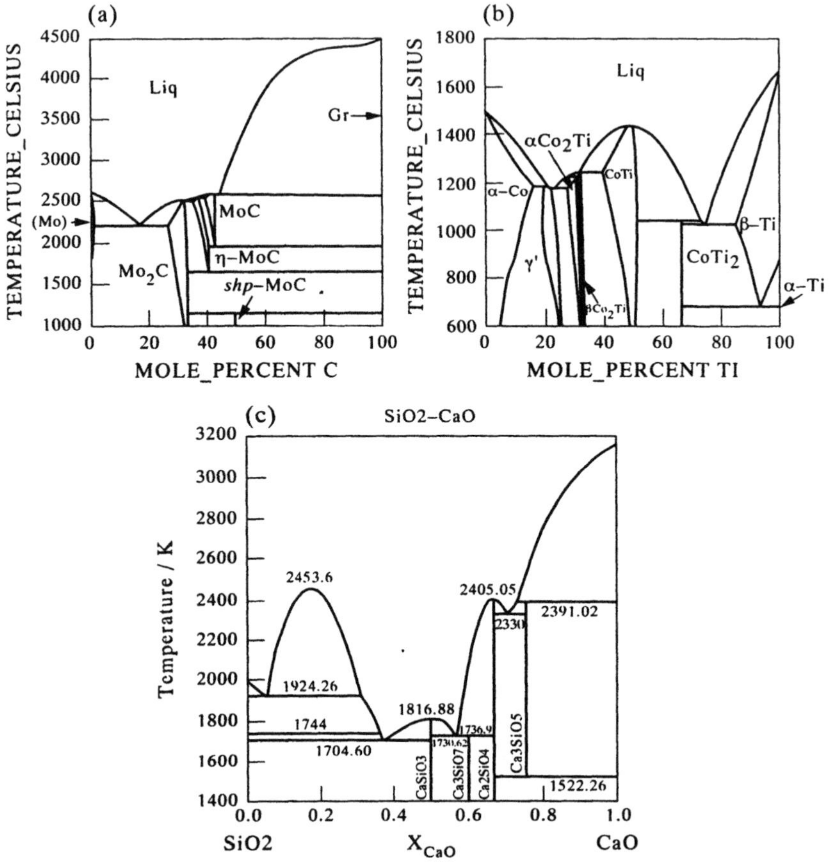

Figure 10.1 Calculated binary phase diagrams for (a) Mo–C (Andersson 1988), (b) Co–Ti (Saunders 1996c) and (c) CaO–SiO_2 (Taylor and Dinsdale 1990).

calculations. In ionic systems the aim is to develop models which represent both the mixing properties of ionic substances as well as allowing these to be combined with substance and alloy databases. At present most CALPHAD software packages include models which deal with aqueous solutions, but their general usage has not been great, which is also true for organic systems.

10.2. EARLY CALPHAD APPLICATIONS

Broadly speaking, the first application of CALPHAD methods was intrinsically coupled to experimental thermodynamic or phase-diagram measurements. For

References are listed on pp. 402–408.

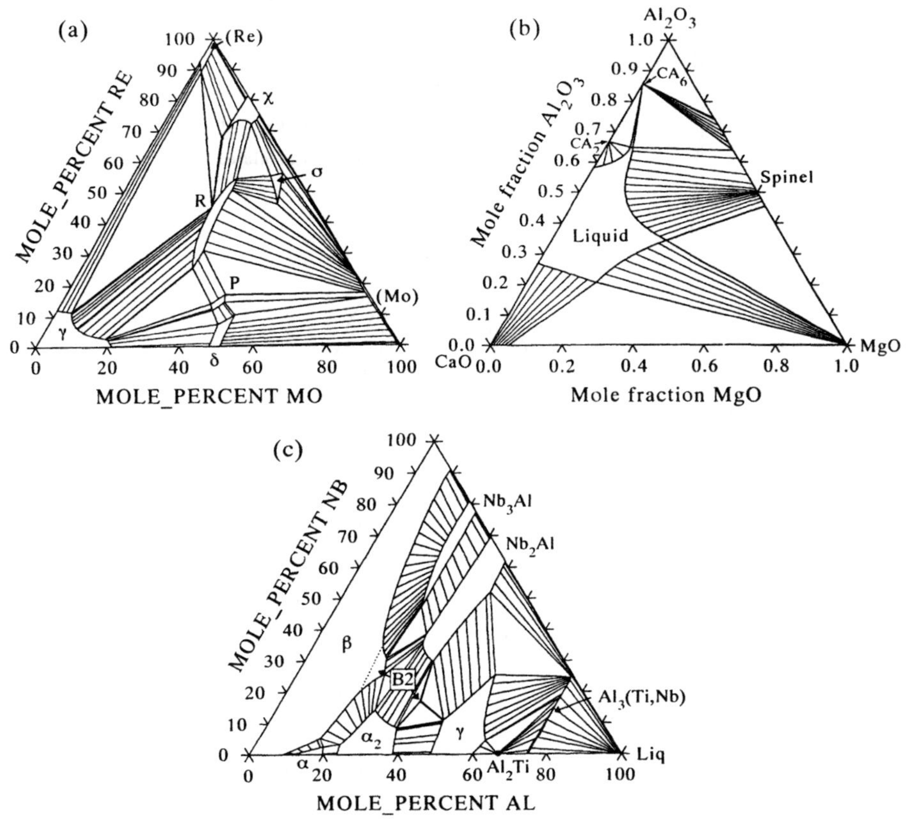

Figure 10.2 Calculated ternary isothermal sections for (a) Ni–Mo–Re at 1000°C (Saunders 1997c), (b) CaO–MgO–Al2O3 at 2000 K (Hallstedt 1992) and (c) isothermal section of Ti–Al–Nb at 1000°C (Saunders 1996a).

example, early work by Kaufman and Clougherty (1964) on Ti–O both explored aspects of modelling as well as coupling of their calculations to experimentally determined O activities. In Europe a number of groups, particularly some of the founder-members of the Scientific Group Thermodata Europe (SGTE), were strongly interested in systems of interest to Fe-based alloys, and work by Counsell *et al.* (1972) on Fe–Ni–(Cu, Cr), by Harvig *et al.* (1972) on Fe–X–C systems and Ansara and Rand (1980) on Fe–Cr–Ni–C clearly demonstrates the concept of coupling of a CALPHAD calculation with experimental determination of phase boundaries and thermodynamics.

An example of an early paper on a binary system is the work of Spencer and Putland (1973) on Fe–V. This combined a review of the thermodynamics and phase diagram of the Fe–V system with new, selective experimental thermodynamic

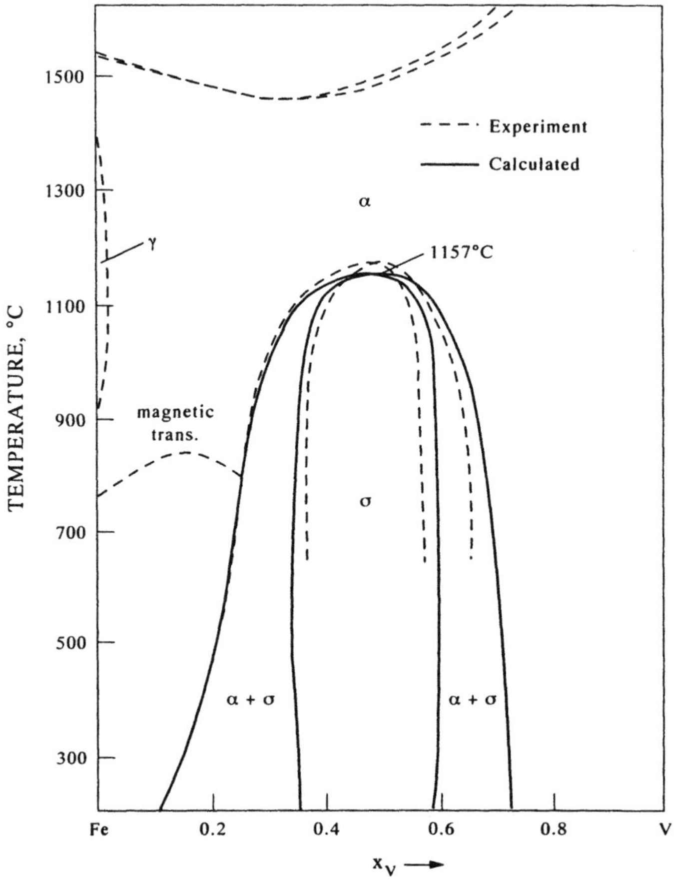

Figure 10.3 Calculated $\alpha + \sigma$ phase boundary for Fe–V with experimental boundaries superimposed.

measurements in order to provide a thermodynamic assessment of the α/σ-phase boundary in this system. Figure 10.3 shows a diagram calculated with their parameters and compares this with the experimentally determined phase diagram. Since then, new assessments have been made for this system, but apart from changes in models and the inclusion of the liquid and γ phases (which are in themselves important developments) the form of the calculated diagram in the region of interest to Spencer and Putland (1973), as well as the calculated thermodynamic properties, have not essentially changed.

In ternary alloys, coupling thermodynamic calculation with experimental work is often essential and a good example of such work is found in the assessment of Cr–Fe–W by Gustafsson (1988). Figure 10.4 shows a calculated isothermal section

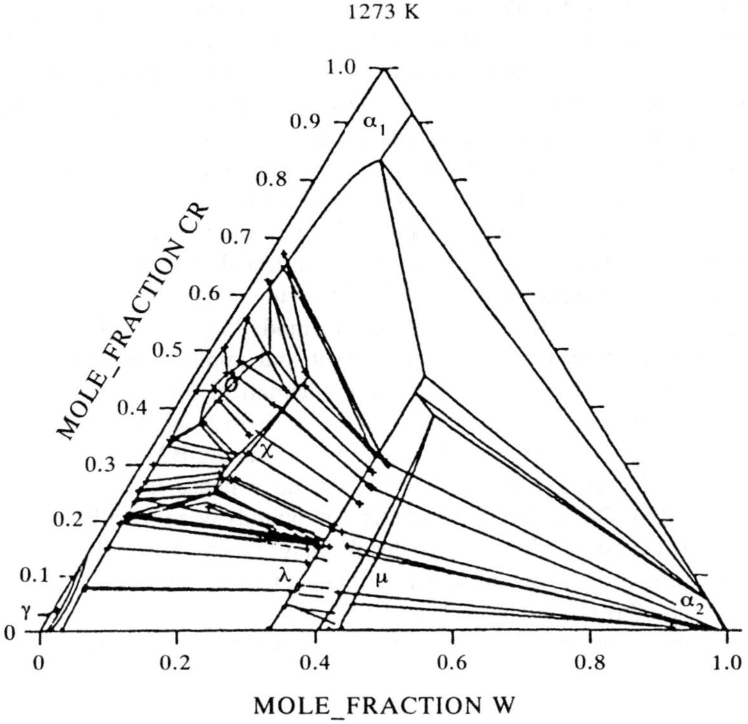

Figure 10.4 Calculated isothermal section of the Cr–Fe–W system with experimental tie-lines superimposed (Gustafson 1988).

at 1273 K which is compared with wide-ranging tie-line experiments carried out by the same author. Ideally, it is helpful to have new experimental backing for assessment work but enough information may exist so that a good assessment can be made with data already available in the literature. Increasingly, the latter situation is becoming the more usual route to providing ternary assessments but, in the authors' opinion, if high-quality results are needed, and there are discrepancies between various literature references, highly accurate experimentation on a small scale is a worthwhile undertaking. An example of such an approach is that undertaken by Yang *et al.* (1992a, b) for the Ni–Al–Ti system, where a few alloys in the $\beta/\beta'/\gamma'$ ternary phase field were all that was necessary to accurately fix this particular phase equilibrium, which was crucial to the development of a new type of intermetallic alloy. The concept of a detailed optimisation of thermodynamic parameters, although initially slower in generating large quantities of data, has eventually led to the ability to make highly accurate calculations in higher-order systems.

The development of models and concepts was also of major importance in the development of CALPHAD techniques, particularly when dealing with the so-called 'lattice stability' concept and intermetallic phases. Kaufman (1967) and Kaufman and Bernstein (1970) constructed many different phase diagrams using ideal or regular solution theory and assessed values for the energy differences between the f.c.c., b.c.c. and c.p.h. phases for many elements. The work clearly demonstrated that the correct forms of phase diagrams could be matched with simple concepts and models. Although these diagrams cannot now be considered accurate enough for detailed application, there is little doubt that correct trends were predicted, and this more than anything confirmed the validity of a number of the fundamental concepts of the CALPHAD route.

Probably the most fundamental change in modelling, particularly in the solid state, has occurred through the development and application of the sub-lattice model (see Chapter 5). In early years the mathematical formalisms for describing substitutional solid solutions had been well established and it was these types of models which were first employed for phases such as the B2, NiAl phase in Ni–Al (Kaufman and Nesor 1978b) (Fig. 10.5) and the σ phase in Cr–Fe (Chart *et al.* 1980) (Fig. 10.6). Otherwise intermetallic compounds were usually treated as line compounds (Figs 10.5 and 10.7). The disadvantages of both models are substantial. Firstly, in multi-component space it is difficult to model properties of compounds which have preferred stoichiometries and sub-lattice occupation, using a model which essentially describes random mixing on a single sub-lattice. Secondly, in

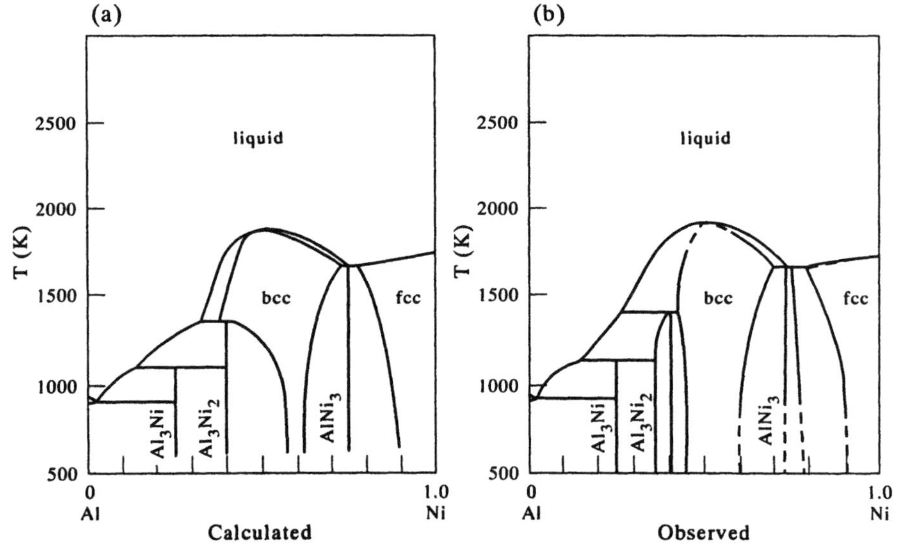

Figure 10.5 (a) Calculated and (b) observed Ni–Al phase diagram (from Kaufman and Nesor 1978b).

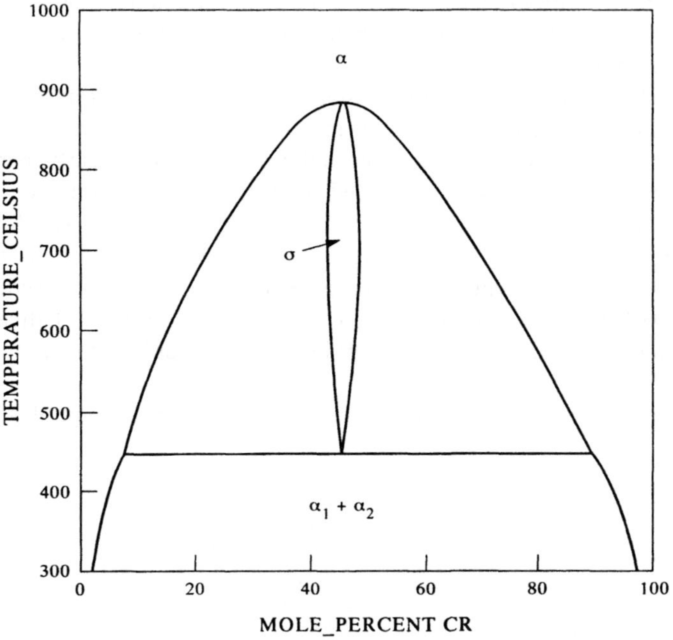

Figure 10.6 Calculated low-temperature region of the Cr–Fe system (Chart *et al.* 1980).

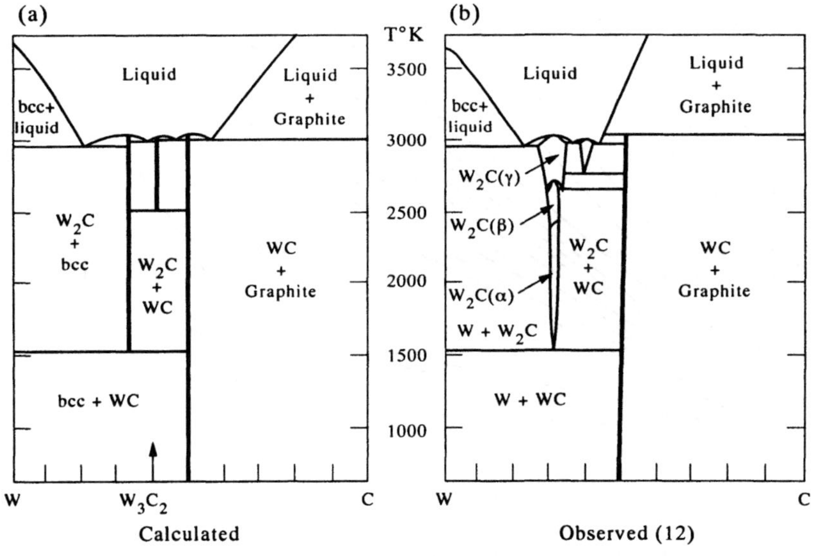

Figure 10.7 (a) Calculated and (b) observed W–C phase diagram (from Kaufman and Nesor 1978a).

systems where solubility ranges are very important, for example the γ' phase in Ni-based superalloys, the stoichiometric compound approach can never achieve high levels of accuracy.

The sub-lattice model is now the predominant model used in most CALPHAD calculations, whether it be to model an interstitial solid solution, an intermetallic compound such as γ-TiAl or an ionic solution. Numerous early papers, often centred around Fe–X–C systems, showed how the Hillert–Staffansson sub-lattice formalism (Hillert and Staffansson 1970) could be applied (see for example Lundberg *et al.* (1977) on Fe–Cr–C (Fig. 10.8) and Chatfield and Hillert (1977) on Fe–Mo–C (Fig. 10.9)). Later work on systems such as Cr–Fe (Andersson and Sundman 1987) (Fig. 10.10) showed how a more generalised sub-lattice treatment developed by Sundman and Ågren (1981) could be applied to multi-sub-lattice phases such as σ.

Alongside calculations for alloy systems, complex chemical equilibria were also

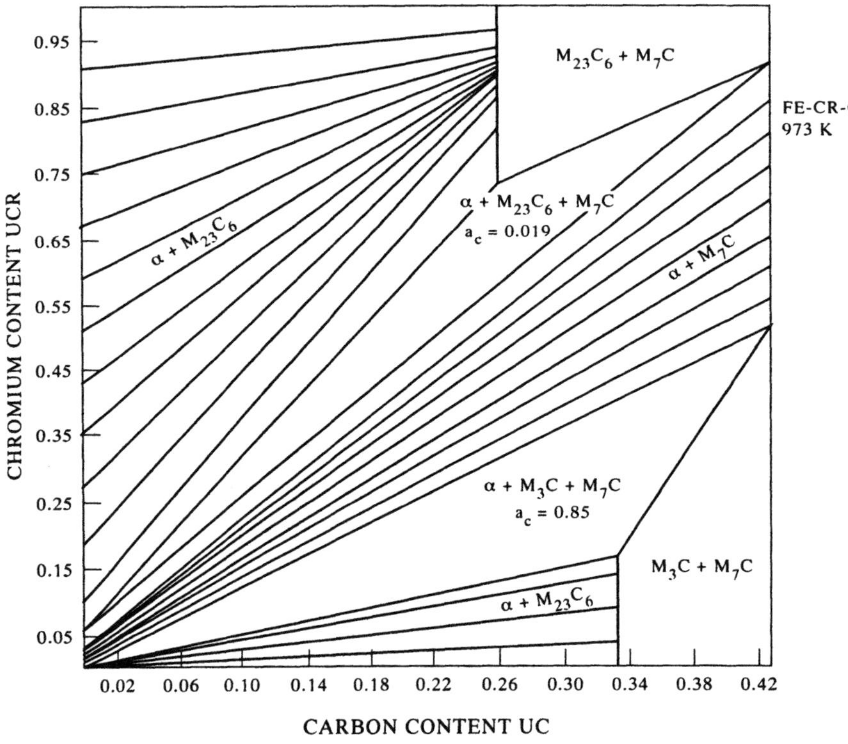

Figure 10.8 Calculated isothermal section for Fe–Cr–C at 973 K (Lundberg *et al.* 1977). Axes are in U-fractions where $UCR = x_{Cr}/(1-x_C)$ and $UC = x_C/(1-x_C)$.

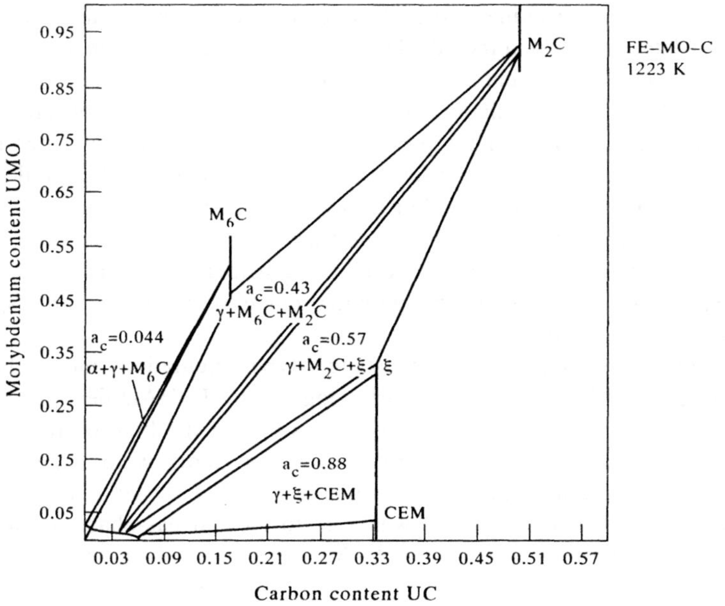

Figure 10.9 Calculated isothermal section for Fe–Mo–C at 1223 K (Chatfield and Hillert 1977). Axes are in U-fractions where $UMO = x_{Mo}/(1-x_C)$ and $UC = x_C/(1-x_C)$.

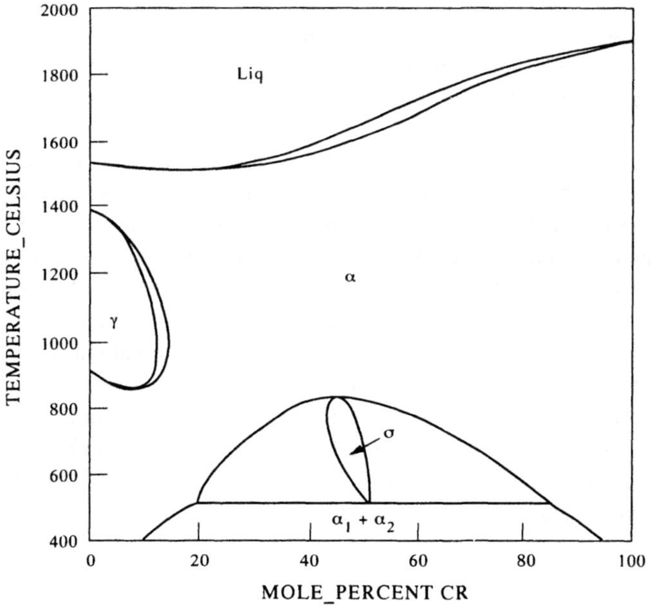

Figure 10.10 Calculated Cr–Fe phase diagram (Andersson and Sundman 1987).

being dealt with on a routine basis. These calculations centred around so-called 'substance' databases, which include data for condensed, stoichiometric substances and an ideal gas phase containing many different species. In essence, the computational aspects of such calculations are less severe than for alloy or ionic systems which are dominated by more complex models. However, they have a complexity of their own as the input of only a few elements gives rise to substantial numbers of substances and gaseous species. For example, in a gas-phase reaction involving C–H–N–O there could be some 60 various gaseous species. Furthermore, in fields such as environmental studies, it is important to consider low levels of activity, so it is not always possible to simplify the potential reactions by 'suspending' species if they exist below a certain critical limit.

The simplicity of modelling in 'substance'-type calculations meant that complex systems could be studied at an early stage of CALPHAD development. The SolGasMix programme of Eriksson (1975) was one of the first to provide such a capability, and within a decade very complex systems were being routinely handled; two good examples are given in the CALPHAD volume dedicated to Eriksson in 1983. Lindemer (1983) was able to look at gas-phase reactions in a fission reaction involving U–Cs–I–Ba–Zr–Mo considering 18 gaseous species and 20 condensed substances. Another interesting example of a calculation based mainly on a substance database is given in the work of Lorenz *et al.* (1983) who were interested in the stability of SiC-based ceramics containing ZrO_2 and other oxides. This involved the quaternary system $SiC–ZrO_2–Al_2O_3–SiO_2$ reacting in inert argon gas, and gave rise to 36 gaseous species and 22 stoichiometric, solid substances. Calculations of the reaction for various composite mixtures of SiC and oxides were then made as functions of temperature and pressure. Calculations involving so many phases were not being routinely handled at this time for systems such as steels, where non-ideal mixing in most of the solid phases is the norm.

There now exists a thriving community of researchers who regularly publish thermodynamic assessments for binary and ternary systems. Many are extremely accurate and, due to a formalisation of unary data proposed by SGTE (Dinsdale 1991), these can be combined with existing or future work in the establishment of multi-component solution databases, where the promise of CALPHAD methods is greatest. In many ways an accurate thermodynamic assessment of a binary or ternary system only proves again the validity of the fundamental concepts laid down in earlier years. The models have improved and, therefore, levels of accuracy level have improved, but most commonly used materials are multi-component in nature and the degree to which binary and ternary assessments can be applied *per se* is fundamentally limited. The key issue is how to extend the CALPHAD route to 'real' rather than 'ideal' materials, and the next section will concentrate on this issue.

References are listed on pp. 402–408.

10.3. GENERAL BACKGROUND TO MULTI-COMPONENT CALCULATIONS

10.3.1 Introduction

From their earliest days CALPHAD methods have always promised to be extendable to complex materials. Certainly, the necessary mathematical formulations to handle multi-component systems have existed for some time and have been programmed into the various software packages for calculation of phase equilibria. However, it is interesting to note that, until quite recently (with the exception of steels), there has been little actual application to complex systems which exist in technological or industrial practice, other than through calculations using simple stoichiometric substances, ideal gas reactions and dilute solution models. The latter have been used for some time, as it is not intensive in computational terms, and some industrially important materials, although containing many elements, are actually low in total alloy or impurity content, e.g., HSLA steels. Examples in this area can be found in Kirkaldy *et al.* (1978), Bhadeshia and Edmond (1980), Hack and Spencer (1985) and Kroupa and Kirkaldy (1993). The limitations of dilute solution models have been discussed earlier in Chapter 5 and, although useful for certain limited applications, they are not applicable in a generalised way and could not begin to handle, with any accuracy, complex alloy types such as stainless steels or Ni-based superalloys. Substance calculations, while containing large numbers of species and condensed phases, are, in many ways, even more limited in their application to alloys as they do not consider interactions in phases involving substantial mixing of the components.

The main areas of application for more generalised models have, until recently, been restricted to binary and ternary systems or limited to 'ideal industrial materials' where only major elements were included. The key to general application of CALPHAD methods in multi-component systems is the development of sound, validated thermodynamic databases which can be accessed by the computing software and, until recently, there has been a dearth of such databases.

The notable exception to this trend was steels and, in particular, stainless and high-speed steels where alloy contents can rise to well above 20wt%. For such alloys a concentrated solution database (Fe-base) has existed since 1978, based on work done at the Royal Institute of Technology (KTH), Stockholm, Sweden. However, although it is far more generalised than dilute solution databases, its range of applicability is limited in temperature to between 700° and 1200°C. Work since 1978, mainly by the Royal Institute of Technology, has seen the development of a new steel database, TCFe, for use in Thermo-Calc. This work now forms the basis for steel calculations in the SGTE solution database. More recently, TCFe has been extended and improved by Saunders and Sundman (1996). These newer databases have a number of distinct advantages over the old Fe-base, not least in that liquid-phase equilibria is now taken into account.

The lack of similar databases for other material types presented severe problems for CALPHAD calculations with any of the other commonly used materials and led to a concentration of application to steels. However, in the past four years further multi-component databases have been developed for use with Al-, Ni-, Ti- and TiAl-based alloys (Saunders 1996a–c, 1997a,b). These databases have been created mainly for use with industrial, complex alloys and the accuracy of computed results has been validated to an extent not previously attempted. Simple, statistical analysis of average deviation of calculated result from experimental measurement in 'real', highly alloyed, multi-component alloys has demonstrated that CALPHAD methods can provide predictions for phase equilibria whose accuracy lies close to that of experimental measurements.

The importance of validation of computed results cannot be stressed too highly. We are now in a position where computational speed has allowed the development of modelling in many related areas. These models often rely on input data which can be time-consuming to measure but can be readily predicted via CALPHAD and related methods. Therefore, CALPHAD results may be used as input for other models, for example, in the manufacture of a steel starting from the initial stages in a blast furnace, through the refinement stages to a casting shop, followed by heat treatment and thermomechanical processing to the final product form. All of these stages can be modelled and all use input data which can be provided either directly or indirectly from CALPHAD calculations. Such a future total modelling capability will never materialise properly until confidence can be placed in the predictions of each of the building blocks. In the case of CALPHAD methods, the key to this is the availability of high-quality databases and the rest of this section will concentrate on databases and discuss some of the strategies in their construction.

10.3.2 Databases

10.3.2.1 'Substance' databases. Basically substance databases have little complexity as they are assemblages of assessed data for stoichiometric condensed phases and gaseous species. They have none of the difficulties associated with non-ideal mixing of substances, which is the case for a 'solution' database. However, an internal self-consistency must still be maintained. For example, thermodynamic data for $C_{(s)}$, $O_{2<g>}$ and $CO_{2<g>}$ are held as individual entries, which provide their requisite properties as a function of temperature and pressure. However, when put together in a calculation, they must combine to give the correct Gibbs energy change for the reaction $C_{(s)}+O_{2<g>} \rightleftharpoons CO_{2<g>}$. This is a simple example, but substance databases can contain more than 10,000 different substances and, therefore, it is a major task to ensure internal self-consistency so that all experimentally known Gibbs energies of reaction are well represented. Examples of substance databases of this type can be found in the review of Bale and Eriksson (1990).

10.3.2.2 'Solution' databases. Solution databases, unlike substance databases, consider thermodynamic descriptions for phases which have potentially wide ranges of existence both in terms of temperature and composition. For example, the liquid-phase can usually extend across the whole of the compositional space encompassed by complete mixing of all of the elements. Unlike an ideal solution, the shape of the Gibbs energy space which arises from non-ideal interactions can become extremely complex, especially if non-regular terms are used. Although it may seem an obvious statement, it is nevertheless important to remember that thermodynamic calculations for complex systems are multi-dimensional in nature. This means that it becomes impossible to visualise the types of Gibbs energy curves illustrated in earlier chapters which lead to the easy conceptualisation of miscibility gaps, invariant reactions, etc. In multi-component space such things are often difficult to understand, l*et al*one conceptualise. Miscibility gaps can appear in ternary and higher-order systems even though no miscibility gap exists in the lower-order systems. The Gibbs phase rule becomes vitally important in understanding reaction sequences, but often one has to accept the computer predictions which can be surprising at times. This emphasises the need to validate the database for multi-component systems and leads inexorably to the concept of two types of database.

10.3.3 The database as a collection of lower order assessments

Essentially this is the basic concept of any database, but an unthinking application of this concept is dangerous. It can be easily demonstrated that in multi-component calculations the properties of some substances, or lower-order interactions in solution phases, are ineffective in modifying phase equilibria, while in other cases some are extremely critical. This may be because the total energy of the system is very exothermic and a particular Gibbs energy term is close to ideal. In this case a change of a few hundred percent in a binary value actually alters things very little. Other reasons may exist for the precise value of an interaction being non-critical. For example, the equilibrium solubility of elements in a particular phase may be small and the $\sum_i \sum_j x_i x_j$ solubility product subsequently produces small changes in total energy, even if interaction coefficients are heavily modified. This leads to a number of important questions and concepts. The first and most important question is how many of the constituent substances or lower-order interactions must be accurately represented before a successful calculation can be guaranteed.

Let us take the example of a simple commercial alloy such as Ti–6Al–4V. This is the most popular structural Ti alloy used worldwide. Essentially one would need to consider Ti–Al, Ti–V and Al–V binary interactions and Ti–Al–V ternary interactions. Unfortunately, although called Ti–6Al–4V, this alloy also contains small amounts of O, C, N and Fe and it therefore exists in the multi-component space within the Ti–Al–V–O–C–N–Fe system. There are then 21 potential binary interactions and 35 possible ternary interactions to consider. The number of thermodynamic assessments necessary to obtain all of these parameters is obviously massive, and the inclusion of an additional element to the alloy means a further 7 binary and 21 ternary assessments would potentially need to be made. This would make a total of 28 binary and 56 ternary assessments. The effort to do all of this is so large that it is much easier (and cheaper) to consider an almost exclusively experimental route to determining phase equilibria in the commercial Ti–6Al–4V alloy. This can be enhanced with regression analysis techniques to specify the effect of various elements on critical features such as the temperature when the alloy becomes fully β.

Fortunately, it is not necessary to perform all of these thermodynamic assessments. In essence one should ensure that all of the binary systems are completed, but the levels of C and N are so low that it is possible to effectively ignore interaction parameters between these two elements, even if they were possible to determine. The percentage of ternary assessments which is necessary to provide an accurate calculation is, in reality, small, mainly because the ternary $\sum_i \sum_j \sum_k x_i x_j x_k$ solubility product can be small for the minor elements. For example, the effect of including ternary parameters for Fe–C–N, which are basically impurity elements, is negligible and little effect is found from the Fe–V–N system even though it contains one of the major elements.

It can therefore be seen that if we wish to consider making calculations for the Ti–6Al–4V alloy, the actual amount of work is much reduced from the theoretical number of permutations. However, this is a particularly simple Ti-alloy and another type such as IMI 834 typically contains Ti, Al, Sn, Zr, Nb, Si, C, O, N and Fe, where seven elements have a significant effect on phase equilibria. The general problem, therefore, remains as to how to judge the number of the potential interactions which must be included, so that a successful multi-component calculation can be made. This cannot be answered in a simple fashion and the position is considerably exacerbated if one wishes to make a generalised database applicable for many material types. If reliable and accurate multi-component calculations are to be made, new paradigms are required and it is no longer possible to consider using databases which are basically constructed as collections of assessed binary and ternary systems which might be available at the time. *The database itself must be assessed.*

10.3.4 Assessed databases

By definition 'assessed databases' are focused, usually on material types. The recent Al-, Ni- and Ti-databases (Saunders 1996a–c) and, to a large degree, the Fe-databases produced by KTH in Stockholm are good examples. They contain up to 15 elements and have been designed for use within the *composition space* associated with the different material types. All, or most, of the *critical* binary and ternary systems have been assessed and calculated results have been *validated* as being successful. The words italicised in the previous sentence hold the key paradigms which need to be employed when designing an assessed database.

Composition space. It is firstly of critical importance that a well-understood and properly circumscribed composition space is defined. This is best done by considering databases for use with particular material types, for example steels, conventional Ti alloys, etc. This firstly limits the number of elements which need to be considered and also helps to define concentration limits for these elements.

Critical systems. It is impossible just by looking at a list of elements to decide which are the critical binary and ternary systems that must be critically assessed. However, some clear pointers can be gained by looking at the composition space in which the database is to be used. For example, B levels in Ni-based superalloys as a rule do not exceed 0.1wt% and, therefore, assessment of B-rich ternary and higher-order alloys is unnecessary. There are, however, critical B-containing ternary systems which must be assessed to understand the thermodynamics of the M_3B_2 phase which can appear. Likewise, the thermodynamics of the MC carbide must be well defined and this includes a large number of carbon-containing ternary systems. On the other hand, although Ti, Ta and Nb may appear in the alloy in much larger amounts than B and C, a ternary assessment of Ti–Ta–Nb is not critical as the magnitude of the thermodynamic interactions are small which, combined with the $\sum_i \sum_j \sum_k x_i x_j x_k$ solubility product term, makes for small Gibbs energy changes. The understanding of the critical systems in an assessed database is in the hands of the developer of the database and can often only be understood after a series of multi-component calculations are made.

Validation of the database. This is the final part in producing an assessed database and must be undertaken systematically. There are certain critical features such as melting points which are well documented for complex industrial alloys. In steels, volume fractions of austenite and ferrite in duplex stainless steels are also well documented, as are γ' solvus temperatures (γ'_s) in Ni-based superalloys. These must be well matched and preferably some form of statistics for the accuracy of calculated results should be given.

Only after at least these three steps are taken can a database then be considered as an assessed database and used with confidence in an application involving a complex multi-component alloy.

10.4. STEP-BY-STEP EXAMPLES OF MULTI-COMPONENT CALCULATIONS

This section will take three commercial alloys and analyse how the final calculated result took form, starting with some of the important constituent binary and ternary diagrams, and seeing how the various features of the final alloy are controlled by these underlying systems.

10.4.1 A high-strength versatile Ti-alloy (Ti–6Al–4V)

Ti–6Al–4V is probably the most widely used Ti alloy in the world. It is an alloy with a duplex structure containing solid solutions based on the α, c.p.h._A3 and β, b.c.c._A2 allotropes of Ti. In its final heat-treated form it consists predominantly of α and its high strength is partly derived from its final microstructure which is manipulated by a series of thermomechanical treatments that include hot isothermal forging just below its β transus temperature (T^{β}). The interest is, in the first place, to predict T^{β} and how the amounts of α and β vary with temperature.

There are empirical relationships which relate alloy content to T^{β}, but these are not usually applicable to all types of Ti alloy and can suffer from a lack of accuracy. Significantly, there are no such relationships which can be generally used for predicting the amount of α and β in commercial alloys as a function of temperature and composition and little work has been undertaken to quantitatively understand the partitioning of elements between the α and β phases.

The alloy combines the features of two different binary systems, Ti–Al and Ti–V (Figs 10.11 and 10.12). Al is an 'α-stabiliser' while V is a strong 'β-stabiliser'. It is the combination of these two types of diagram which produces a wide two-phase $\alpha+\beta$ region, and Fig. 10.13 shows the behaviour of the basic ternary alloy as a function of temperature. Although more heavily alloyed in Al, the alloy never

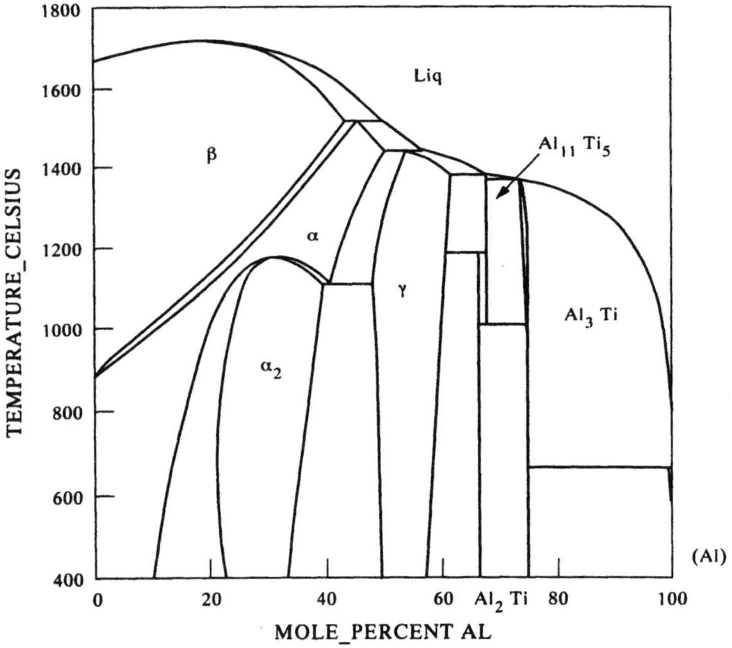

Figure 10.11 Calculated Ti–Al phase diagram.

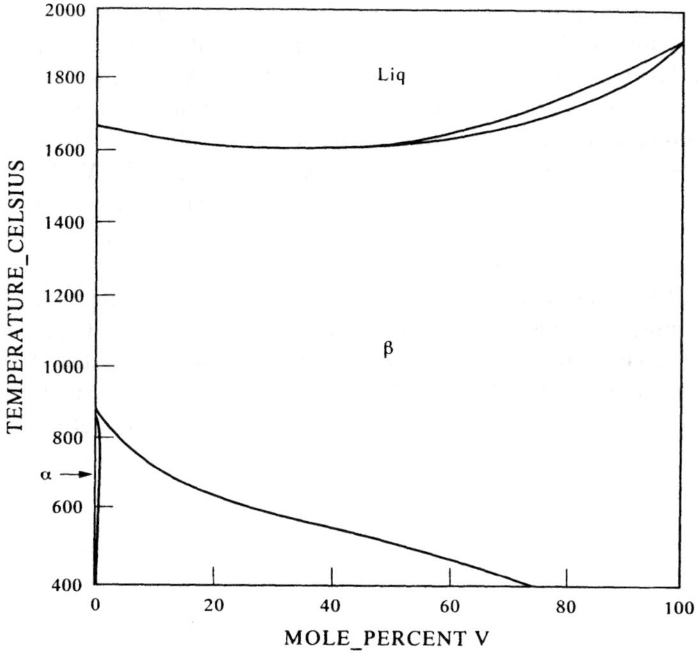

Figure 10.12 Calculated Ti–V phase diagram.

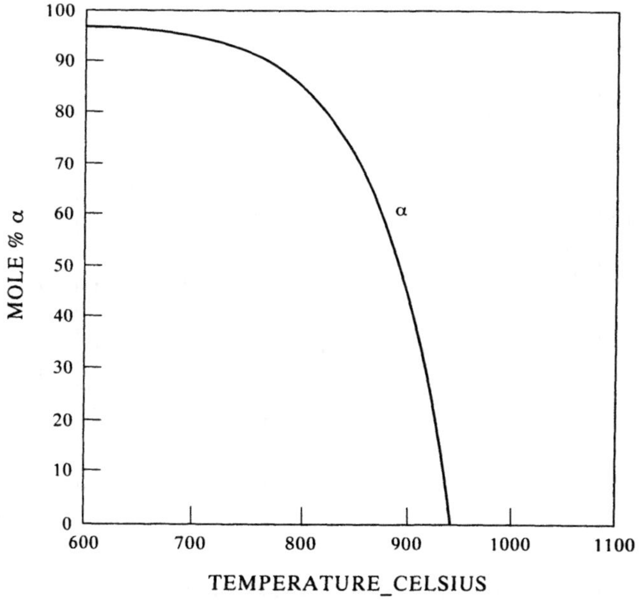

Figure 10.13 Calculated mole % α vs temperature plot for a Ti–6Al–4V ternary alloy.

becomes fully α, as just enough V is added to retain some β. The basic behaviour of the alloy is thus defined by the Ti–Al–V ternary but Ti alloys always contain significant levels of O and N, and in the case of Ti–6Al–4V usually some Fe. At high levels (>5000 ppm) O generally acts as an embrittling agent, but at lower levels (2000 ppm) it can be used to enhance strength (Jaffee 1958) and is therefore a deliberate addition to 'conventional' alloys such as Ti–6Al–4V. Apart from the physical effects of O, this also produces significant phase-boundary shifts even at low levels of uptake, and it is therefore necessary to include the effects of at least this element in the calculations.

Figure 10.14 shows the calculated Ti–O diagram from Lee and Saunders (1997). In the composition range of interest it is of a simple type but O has a powerful effect on T^{β}. This effect is carried over to the critical ternary system Ti–Al–O and Figs 10.15 and 10.16 show how the Ti–Al system changes as O is added (Lee and Saunders 1997). No such information is available for Ti–V–O but it is interesting to note the predicted effect of O on Ti–V. Figure 10.17 shows a section through

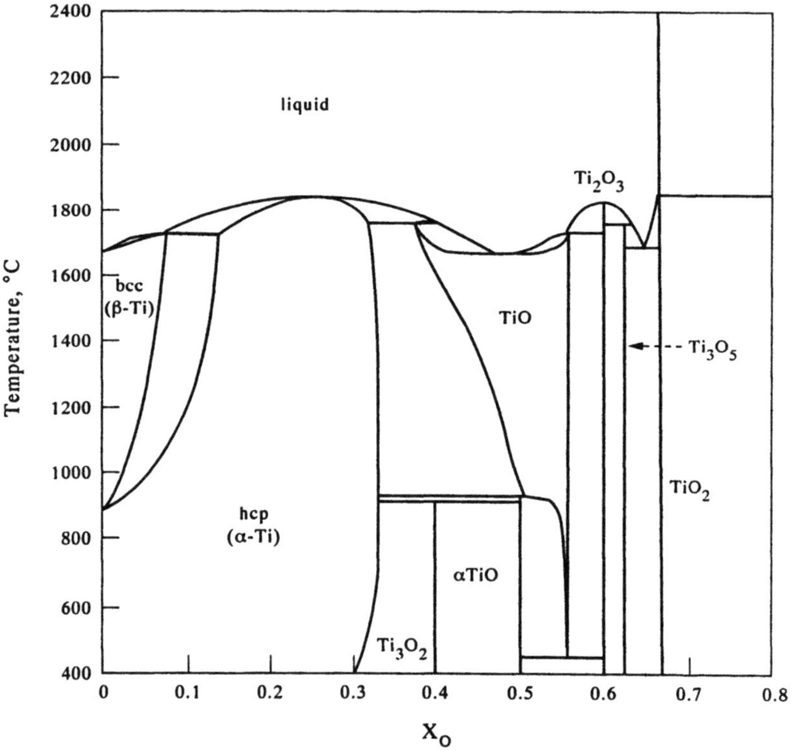

Figure 10.14 Calculated Ti–O phase diagram (from Lee and Saunders 1997).

References are listed on pp. 402–408.

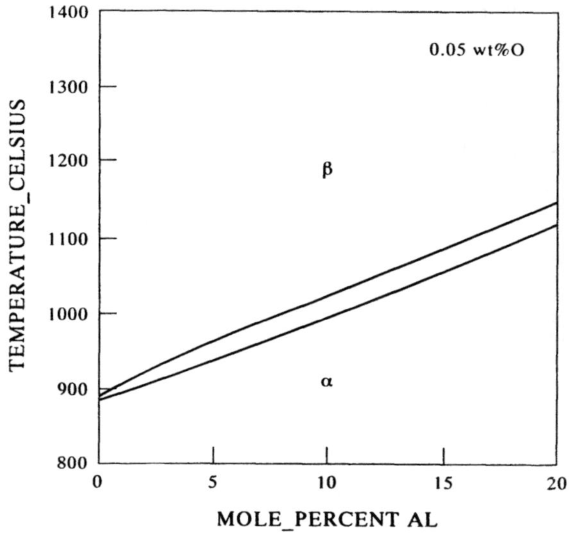

Figure 10.15 Calculated vertical section through Ti–Al–O at 0.05wt%O.

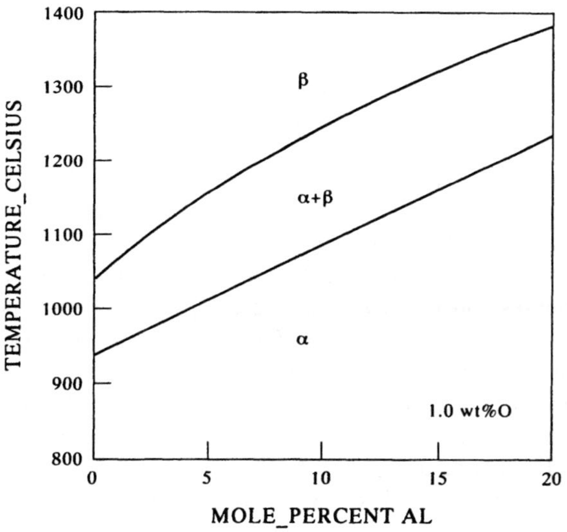

Figure 10.16 Calculated vertical section through Ti–Al–O at 1.0wt%O.

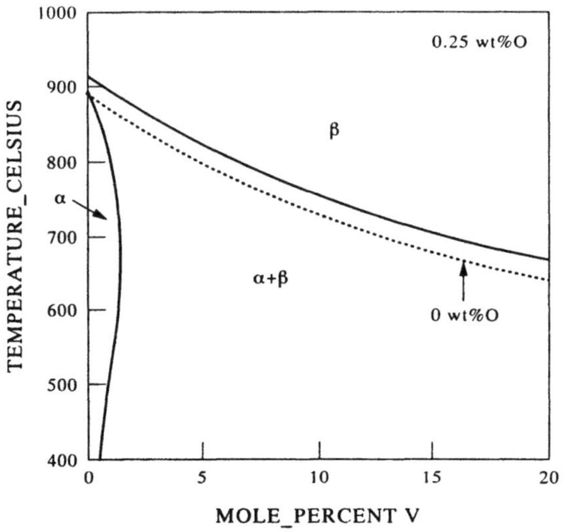

Figure 10.17 Calculated vertical section through Ti–V–O at 0.25wt%O.

Ti–V–O at a constant 2500 ppm of O and the effect of O on T^β is significant, while the effect on the position of the α-phase boundary is minimal.

Taking the necessary binary assessments for the inclusion of C, N and Fe and the assessments for Ti–Al–O and Ti–V–O, the effect of O on the T^β of Ti–6Al–4V with typical C, N and Fe impurity levels, was calculated and compared with experiment. The agreement between the calculations and experimental results of Kahveci and Welsch (1986) is good (Fig. 10.18). Figure 10.19 further shows some calculated phase % vs temperature plots for three Ti–6Al–4V commercial alloys and compares these with experiment. The advantage of the CALPHAD route becomes increasingly apparent because, as well predicting T^β, the calculations have also given good results for the amounts of α and β.

Furthermore, it is now possible to look at the partitioning of the various elements to the α and β phases and Fig. 10.20 shows comparisons with experiment for the various metallic elements. One of the V results for the β phase has an arrow indicating that the true experimental result was considered to be higher than that shown in Fig. 10.20 (Lasalmonie and Loubradou 1979). This is because the β grains were so small that some overlap with the α matrix occurred during measurement by EPMA, resulting in a V reading that is almost certainly too low.

As previously stated, the levels of the light elements such as O are important in determining physical properties of Ti and it is also possible to look at the partitioning of O between the α and β phases (Fig. 10.21). It can be seen that the level of O in α just below T^β is extremely high. This has significant consequences for thermomechanical processing for Ti–6Al–4V at these temperatures as yield

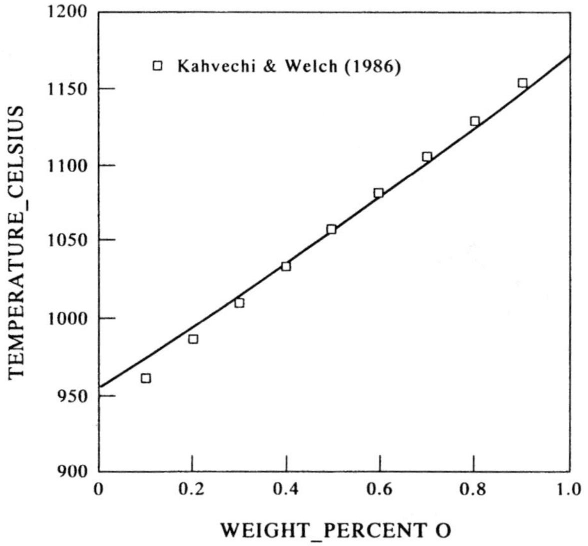

Figure 10.18 Comparison between calculated and observed β-transus for Ti–6Al–4V alloys as a function of O concentration.

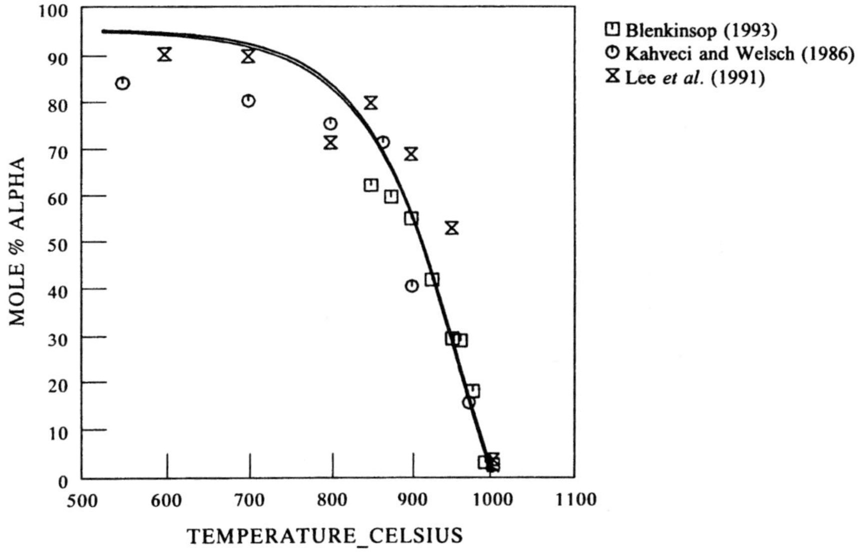

Figure 10.19 Calculated mole % phase vs temperature plots for three Ti–6Al–4V alloys with experimental data superimposed.

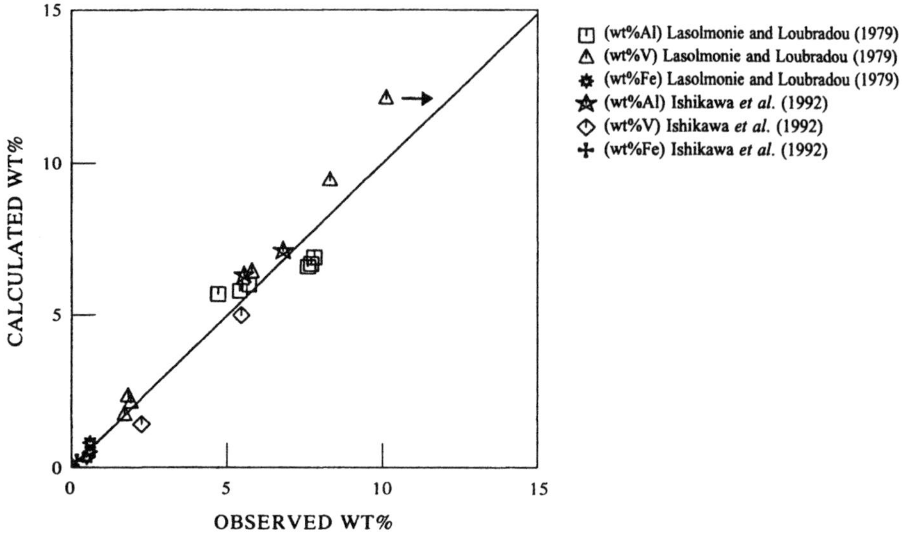

Figure 10.20 Comparison between calculated and experimental values for the concentration of Al, V and Fe in the α and β phase in Ti–6Al–4V alloys.

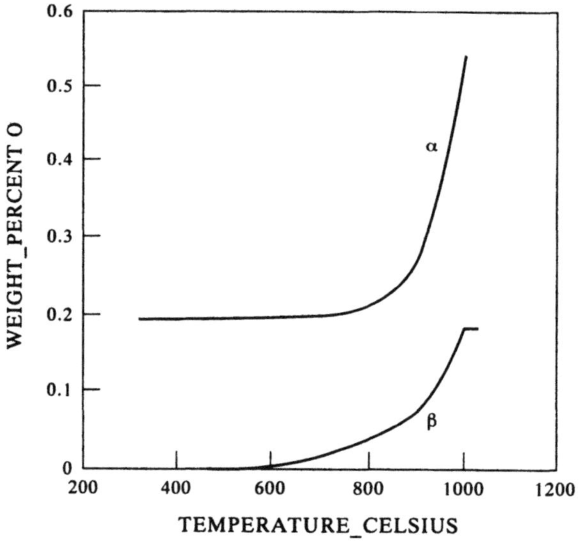

Figure 10.21 Calculated concentration of O in the α and β phase for a Ti–6Al–4V alloy.

strengths in α will be higher and ductility levels lower than would be expected by just taking total levels of impurity as a guide.

10.4.2 A high-tonnage Al casting alloy (AA3004)

Aluminium alloys form one of the most widely used groups of materials in existence. They make products which are often cheap and can be applied to many different areas. Extensive work has been done on the experimental determination of binary and ternary phase diagrams, mainly during the mid-part of this century, and researchers such as Phillips (1961) and Mondolfo (1976) have produced detailed reviews of the literature which provide industry standard publications. However, although some important Al-alloys are based on ternary systems, such as the LM25/356 casting alloy based on Al–Mg–Si, in practice they inevitably include small amounts of Cu, Mn, Fe, Ti etc., all of which can significantly modify the castability and properties of the final product. The situation is further exacerbated by the use of scrap material. It is therefore useful to be able to predict phase equilibria in multi-component alloys.

The modelling issues for Al alloys turn out to be reasonably straightforward. Unlike superalloys or steels there are few intermetallic phases with wide regions of stoichiometry. A large number of the compounds tend to be stoichiometric in nature, for example, Mg_2Si and Al_2CuMg. Where there is substantial solubility, such as in the Al_6Mn and α-AlFeMnSi phases, the transition metals basically mix on one sub-lattice while Si mixes on the Al sub-lattice. The phases can be then treated as conventional line compounds and complexities of modelling associated with phases such as σ and γ', where many elements may mix on more than one sub-lattice, do not arise. There is also limited solubility in the Al solid solution for most elements which are usually added to Al alloys which means that, for this phase, the effect of most ternary interactions is completely negligible. Nevertheless, Al alloys more than make up for this simplicity in modelling by exhibiting reaction schemes which can be far more complex than usually found in systems involving more complex models. Because of their inherent simplicity in modelling terms, Al systems offer a good example of how a database can be constructed and the AA3004 alloy, which is based on the Al–Fe–Mn–Si system, will now be discussed in more detail.

The Al–Si diagram and Al-rich regions of the Al–Mn and Al–Fe diagrams are shown in Figs 10.22–10.24. The Al–Mn and Al–Fe systems are modifications based on the work of Jansson (1992) and Saunders and Rivlin (1987). Unless stated, all other diagrams are from Saunders (1996b). The Al–Mn and Al–Fe diagrams are complex but in terms of Al alloys only the Al_6Mn, Al_4Mn and Al_3Fe phases are of importance. This leads to a large degree of simplification in considering the ternary modelling.

Figure 10.25 shows the liquidus projection for the Al–Fe–Mn system which is characterised by a substantial extension of the Al_6Mn phase into the ternary and

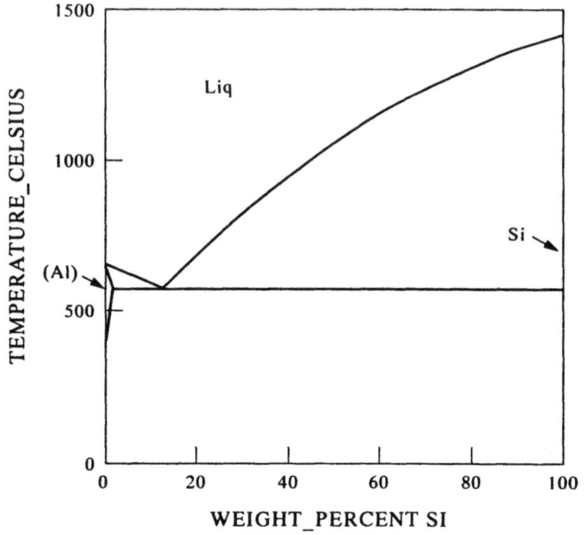

Figure 10.22 Calculated Al–Si phase diagram.

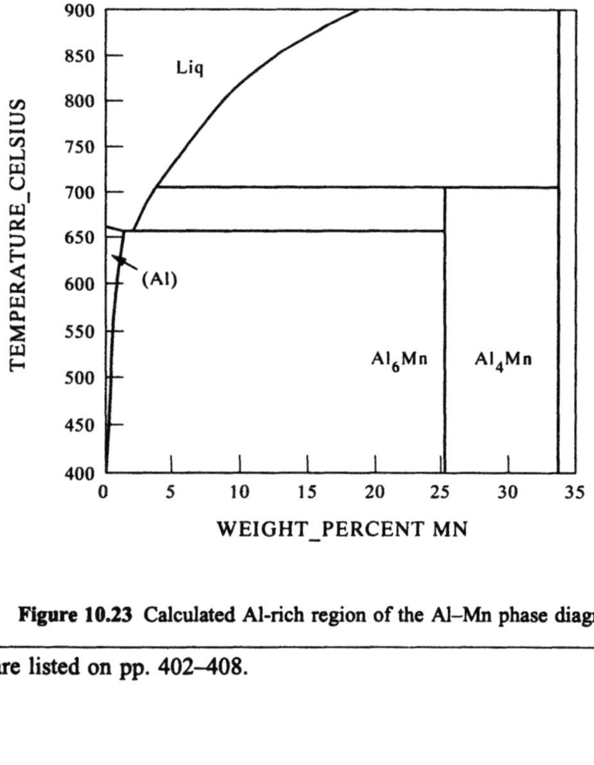

Figure 10.23 Calculated Al-rich region of the Al–Mn phase diagram.

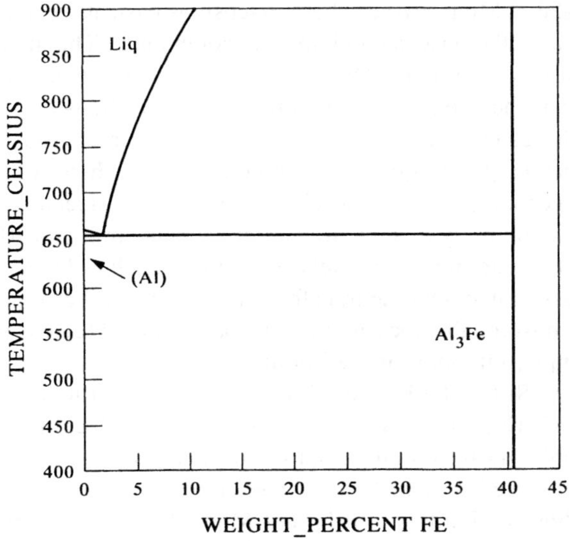

Figure 10.24 Calculated Al-rich region of the Al–Fe phase diagram.

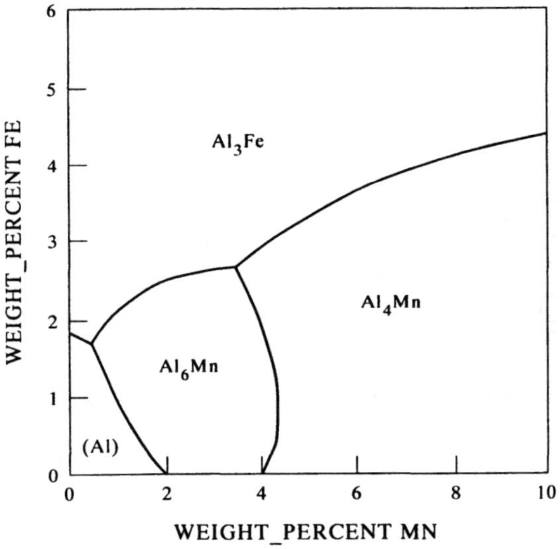

Figure 10.25 Calculated liquidus projection for Al–Fe–Mn.

some slight uptake of Mn in Al_3Fe. These extensions basically comprise Mn and Fe substitution in an otherwise stoichiometric compound. The diagram has been reviewed by Phillips (1961) and Mondolfo (1976) where there is good agreement on features such as the eutectic temperature and compositions of phases. The main problem with the assessment of Al systems is in interpretation of experimental information concerning the liquidus of the compounds, where results can be scattered. This is a general problem and relates partly to the steepness of the liquidus slope and the low solubility of some transition elements in liquid Al. Such liquidus lines are difficult to determine accurately by methods such as DTA, as there is often very little heat associated with the transformation, and significant undercooling can also occur. The most reliable measurement techniques tend to be isothermal in their nature, for example, sampling of the liquid.

The addition of Si to Al–Mn and Al–Fe leads to the formation of a number of ternary compounds (Figs 10.26 and 10.27). There is significant mixing between Al and Si in the cubic α-AlMnSi phase but the other phases are treated accurately as compounds with fixed stoichiometry. The agreement between the calculated diagrams and those of Phillips (1961) and Mondolfo (1976) is again good, with compositions of invariant reactions typically being to within 1 wt% of each element and 5°C in temperature.

The joining of the ternary systems into Al–Fe–Mn–Si basically involves considering the α-AlMnSi compound which extends almost completely to Al–Fe–Si. In fact, in early work on Al–Fe–Si alloys, which contained minor levels of Mn, the

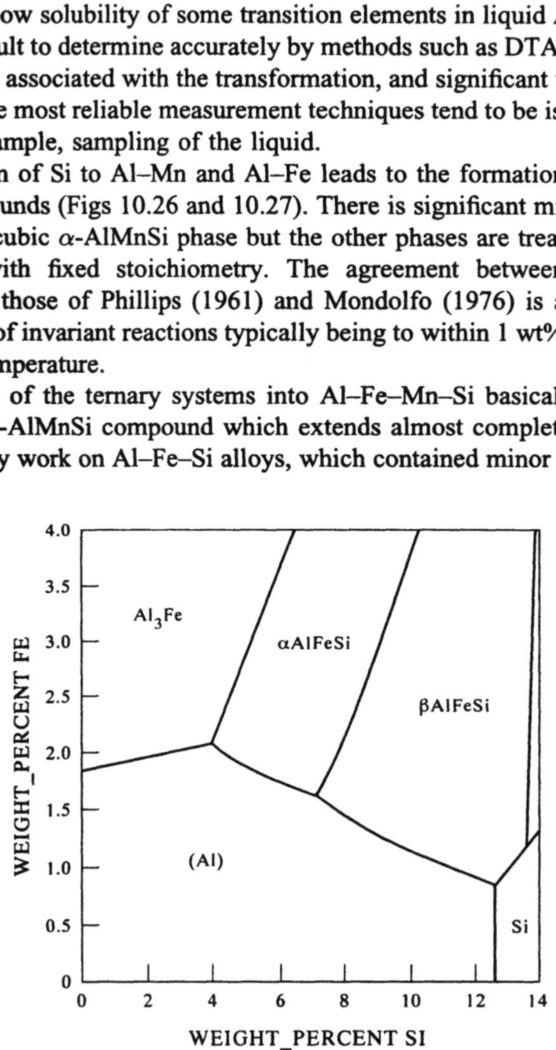

Figure 10.26 Calculated liquidus projection for Al–Fe–Si.

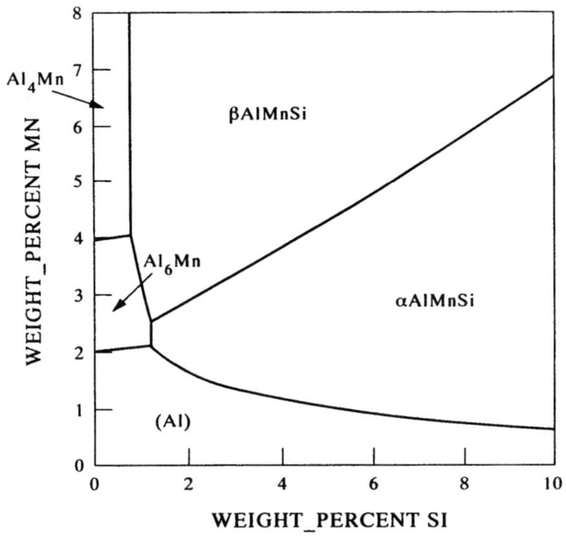

Figure 10.27 Calculated liquidus projection for Al–Mn–Si.

α-AlFeSi phase was considered to be isomorphous with α-AlMnSi. Later work showed this not to be the case and its stable structure is hexagonal (Munson 1967, Mondolfo 1976, Rivlin and Raynor 1988). The addition of Fe to the α-AlMnSi phase is simply achieved by making the Gibbs energy of the cubic AlFeSi compound only just metastable. Other elements such as Cr and V also partition to the cubic α phase and this can also be taken into account.

Combining this quaternary with Mg, and in particular Al–Mg–Si, it is now possible to consider a reasonably pure AA3004 alloy which is used extensively for thin-walled containers such as drink cans. Figures 10.28(a,b) show phase % vs temperature plots for an alloy Al–1Mn–1.2Mg–0.5Fe–0.2Si (in wt%). On solidification the primary phase is Al, with Al_6Mn appearing soon afterwards. There is a subsequent peritectic reaction involving α-AlFeMnSi (which will now just be called α) which partly consumes the Al_6Mn phase. The amount of α increases as the alloy is cooled below its solidus and it becomes the dominant solid-state intermetallic just below 600°C. However, it disappears around 400°C, as Si is taken up by the formation of Mg_2Si which acts as a precipitation hardening phase. The interplay between the α and Al_6Mn is critical, as the surface finish during fabrication of cans is much improved if α particles, rather than Al_6Mn, predominate in 3XXX alloys of this type (Anyalebechi 1992, Marshall 1996). It is therefore now possible for CALPHAD methods to be used as a tool in helping to model and control this reaction.

Cama *et al.* (1997) studied an alloy with the same composition as used in the last

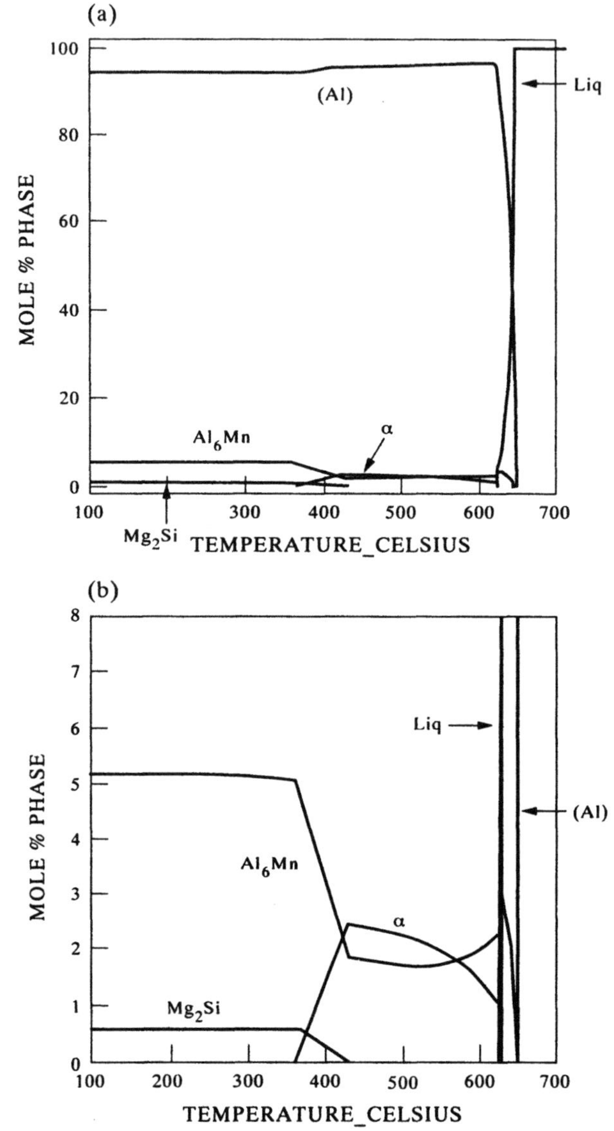

Figure 10.28 (a) Calculated mole % phase vs temperature plot for a AA3004 alloy. (b) Expanded region of Fig. 10.26(a).

calculation, but with 0.2wt%Cu added, and performed long-term anneals between 550° and 630°C. They measured the relative levels of Al_6Mn and α and reported results as a percentage of α observed. Calculations were therefore made for their alloy so that computed results could be compared with experiment. Experimental results varied between 61 and 39% while the calculations predict values of 53–33%. The calculations suggest some small amount of liquid would be present at 630°C and the lower value is quoted at 620°C. The results, while underestimating the measured values, are still in very reasonable agreement and the temperature dependency of the conversion of Al_6Mn to α is almost exactly predicted. Furthermore, Marshall (1996) reported results for the closely related AA3104 alloy where the transition from Mg_2Si to α is observed somewhere between 350 and 400°C. Although Marshall (1996) did not provide a composition for the alloys which were used, calculations for an ideal 3104 composition, following Sigli *et al.* (1996), suggest this transition occurs between 360° and 440°C, in good agreement with observation.

In alloys such as AA3004 some of the major issues concern solidification and therefore it is interesting to look at this in detail. However, as solidification in Al alloys rarely occurs under equilibrium conditions, a more detailed examination of this issue will be found in the next chapter.

10.4.3 A versatile corrosion-resistant duplex stainless steel (SAF2205)

Duplex stainless steels are a highly formable, strong, yet highly corrosion-resistant series of alloys. The 'ideal' duplex structure is aimed to be a 50/50 mixture of austenite (γ) and ferrite (α). The microstructure can be manipulated by thermo-mechanical processing to produce an alloy with high strength. They also have a high Pitting Resistance Equivalent (PRE), where PRE can be related to the levels of Cr, Mo, W and N by the empirical formula (Hertzman 1995)

$$\mathrm{PRE} = \mathrm{wt\%Cr} + 3.3(\mathrm{wt\%Mo} + \mathrm{wt\%W}) + 16\mathrm{wt\%N}.$$

A popular alloy of this type is SAF 2205. The alloy is predominantly an Fe–Cr–Ni alloy with significant additions of Mo, Mn, Si, C and N. The composition may typically be Fe–22Cr–5.5Ni–3Mo–1.7Mn–0.4Si–0.14N–0.024C (in wt%). Figure 10.29 shows an isothermal section for Fe–Cr–Ni at 1000°C and Fig. 10.30 shows a phase% vs temperature plot for a Fe–22Cr–5.5Ni alloy. It has a narrow liquid+solid region and it is already duplex below 1216°C, reaching a 50/50 $\gamma + \alpha$ mixture at 1015°C, close to the final annealing temperature of the full composition alloy. The σ phase forms below 730°C at the expense of α, but this is low compared to the temperature where it is observed in real SAF2205 (Thorvaldsson *et al.* 1985). The PRE number for this ternary alloy is only 22 and values around 30–40 are necessary for adequate corrosion resistance.

The addition of 3%Mo improves its pitting resistance equivalent (PRE) but causes substantial changes (Fig. 10.31). The level of austenite is substantially decreased, only forming below 1134°C and never reaching more than 40% in the

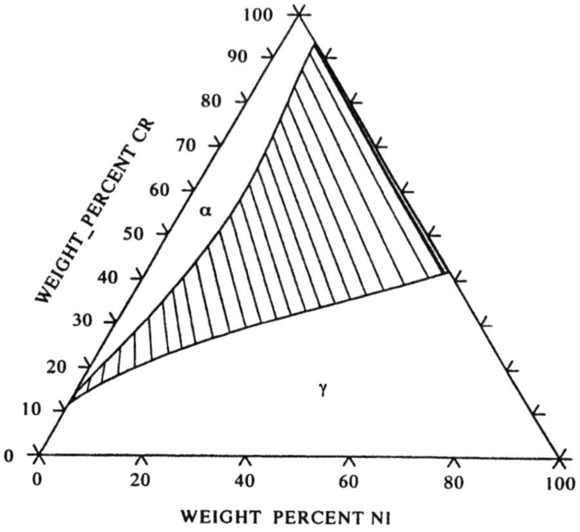

Figure 10.29 Calculated isothermal section for Fe–Cr–Ni at 1000°C.

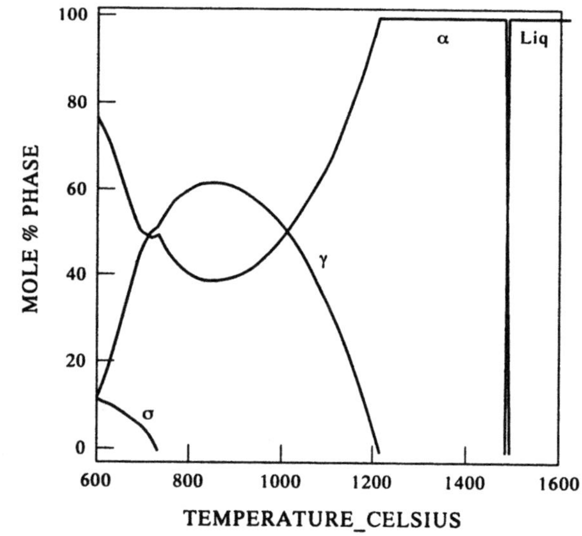

Figure 10.30 Calculated mole % phase vs temperature plots for a Fe–22Cr–5.5Ni alloy.

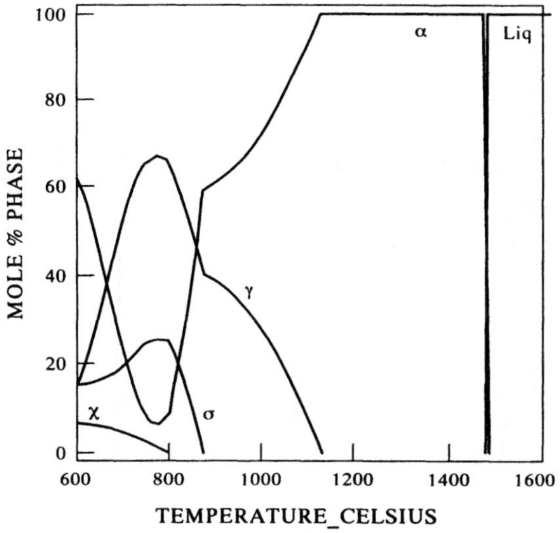

Figure 10.31 Calculated mole % phase vs temperature plots for a Fe–22Cr–5.5Ni–3Mo alloy.

duplex region. The stability of σ is markedly increased: it now forms below 875°C and some χ forms below 800°C. Both σ and χ are actually seen in SAF2205. Figure 10.32 shows an isothermal section for Fe–Cr–Mo which shows the expansive region of σ and the formation of a ternary χ phase.

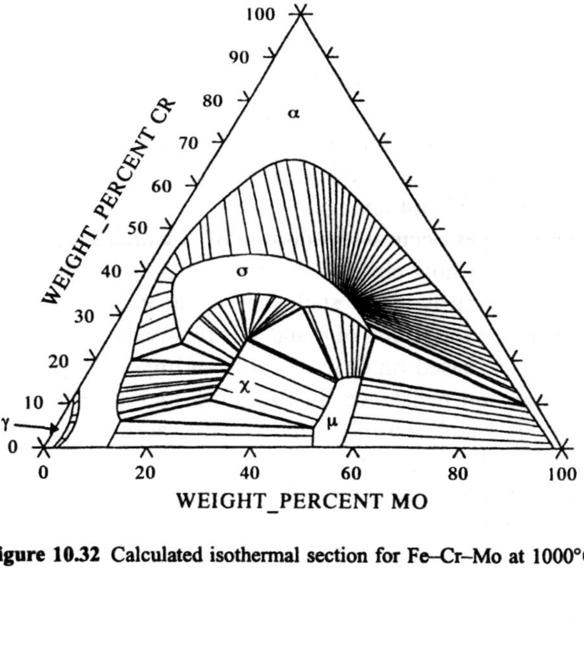

Figure 10.32 Calculated isothermal section for Fe–Cr–Mo at 1000°C.

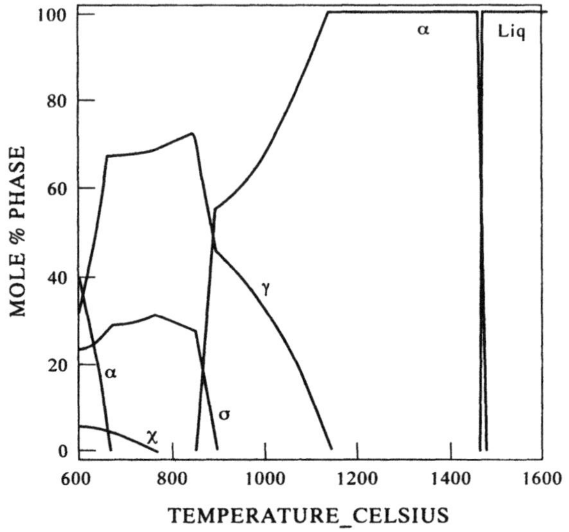

Figure 10.33 Calculated mole % phase vs temperature plots for a Fe–22Cr–5.5Ni–3Mo–1.7Mn alloy.

The addition of 1.7%Mn does not make such a large difference as the Mo addition (Fig. 10.33). It is, however, noticeable that all of the α has disappeared between 670–850°C by the reaction to form $(\gamma + \sigma)$. This is because Mn is both a γ stabiliser and it enhances the formation of σ which competes with α for α-stabilising elements such as Cr and Mo.

The addition of 0.4wt%Si does not alter the general behaviour of the alloy significantly (Fig. 10.34). It is known that a ternary σ phase forms in Fe–Cr–Si but Si is also a powerful α stabiliser. It is noticeable that there has been sufficient α stabilisation to delay the onset of the $\alpha \rightarrow (\gamma + \sigma)$ transformation and α is stable to lower temperatures than previously but, in the end, the level of addition of Si is insufficient to make significant changes.

The next major changes occur with addition of N which substantially stabilises γ (Fig. 10.35). The alloy is now close to its final composition and it can be seen that the N has stabilised γ sufficiently such that it becomes the predominant phase below 1050°C while the formation of σ and χ is relatively unchanged. A new phase is now observed, M_2N, based on Cr_2N. This is an important phase as it can cause sensitisation to corrosion resistance. In SAF2205 its temperature of formation is close to that of σ. The addition of C additionally causes the formation of $M_{23}C_6$ below 900°C (Fig. 10.36) and slightly lowers the solidus.

The final predicted behaviour for SAF2205 is close to that found in practice. The amount of γ in the alloy as a function of temperature is in excellent agreement with experimental results (Hayes 1985) and the behaviour of the minor phases is also

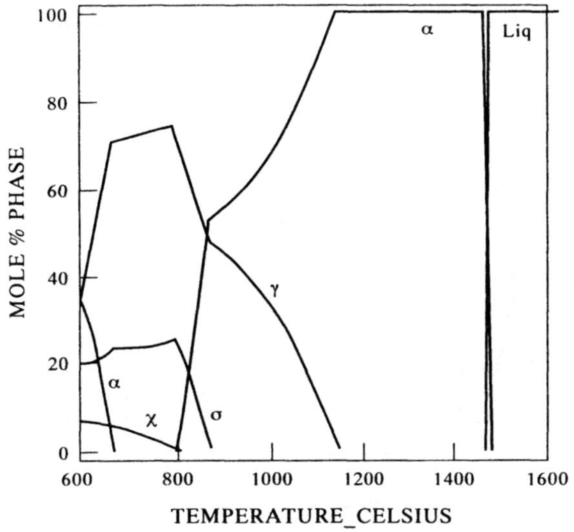

Figure 10.34 Calculated mole % phase vs temperature plots for a Fe–22Cr–5.5Ni–3Mo–1.7Mn–0.4Si alloy.

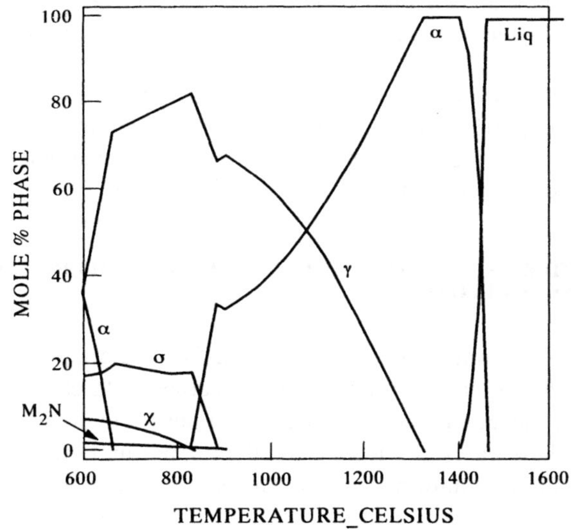

Figure 10.35 Calculated mole % phase vs temperature plots for a Fe–22Cr–5.5Ni–3Mo–1.7Mn–0.4Si–0.14N alloy.

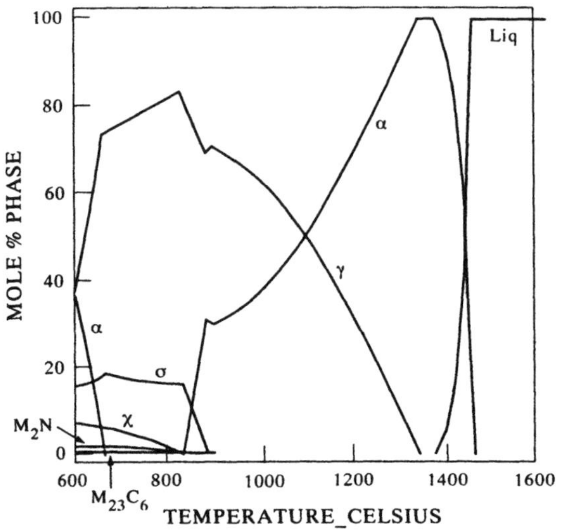

Figure 10.36 Calculated mole % phase vs temperature plots for a SAF2205 alloy with composition Fe–22Cr–5.5Ni–3Mo–1.7Mn–0.4Si–0.14N–0.24C.

well predicted. The temperature of σ and M_2N formation is close to that observed in practice and the $M_{23}C_6$ and χ phases are predicted to form as observed (Thorvaldsson *et al.* 1985). The behaviour of the χ phase is interesting as it is commonly seen as one of the first minor phases to form in practice. Thorvaldsson *et al.* (1985) showed that in SAF2205 the sequence of phase formation at 850°C would be χ followed by σ, with σ finally being the stable phase and χ disappearing after long anneals. The predicted solvus temperature for χ is close to that of σ but at 850°C it is not yet stable when σ forms. This would be quite consistent with the observed behaviour of SAF2205.

10.5. QUANTITATIVE VERIFICATION OF CALCULATED EQUILIBRIA IN MULTI-COMPONENT ALLOYS

This section will give examples of how CALPHAD calculations have been used for materials which are in practical use and is concerned with calculations of critical temperatures and the amount and composition of phases in duplex and multi-phase types of alloy. These cases provide an excellent opportunity to compare predicted calculations of phase equilibria against an extensive literature of experimental measurements. This can be used to show that the CALPHAD route provides results whose accuracy lies close to what would be expected from experimental measurements. The ability to statistically validate databases is a key factor in seeing the CALPHAD methodology become increasingly used in practical applications.

10.5.1 Calculations of critical temperatures

In terms of practical use, one of the most important features of phase equilibria can often be the effect of composition on some critical temperature. This can be a liquidus or solidus or a solid-state transformation temperature, such as the β-transus temperature, (T^{β}), in a Ti alloy. The solidus value can be critical, as solution heat-treatment windows may be limited by incipient melting. In some materials a solid-state transformation temperature may be of prime importance. For example, in Ti alloys it may be specified that thermomechanical processing is performed at some well-defined temperature below the β-transus temperature. The CALPHAD route provides a method where such temperatures can be quickly and reliably calculated.

10.5.1.1 Steels. One of the most striking successes of the CALPHAD technique has been in the highly accurate calculation of liquidus and solidus temperatures. Because of their inherent importance in material processing, there are numerous reported measurements of these values, which can be used to judge how well CALPHAD calculations perform in practice. For example, detailed measurements of liquidus and solidus values for steels of all types have been made by Jernkontoret (1977). The values were obtained on cooling at three different cooling rates, 0.5, 1 and 5°C sec^{-1}. The effect of cooling rate is seen to be marginal on the liquidus but could be profound on the solidus due to the effects of non-equilibrium segregation during the liquid→solid transformation. Calculations for the liquidus and solidus were made for these alloys using the Fe-DATA database (Saunders and Sundman 1996) and compared with the results obtained at the lowest cooling rate. Figure 10.37 shows the results of this comparison and the accuracy of the predictions is impressive, particularly for the liquidus values which exhibit an average deviation from experiment ($\bar{d}$) of only 6°C. It is also pleasing to note how well the solidus values are predicted with an average deviation of just under 10°C. Three solidus values are not matched so well and are highlighted. In these alloys low-melting eutectics were observed, but not predicted, and it is uncertain if the difference is due to an inherent inaccuracy in the prediction or to the persistence of non-equilibrium segregation during solidification.

10.5.1.2 Ti alloys. In Ti alloys there are numerous measurements of the T^{β} as this is a very critical temperature for these alloys. Figure 10.38 shows the comparison between predicted and measured values for Ti alloys of all types, ranging from β-type alloys such as Ti–10V–2Al–3Fe through to the α types such as IMI834. The results exhibit an average deviation from experiment of less than 15°C which is very good for the measurements of a solid-state transformation such as T^{β}.

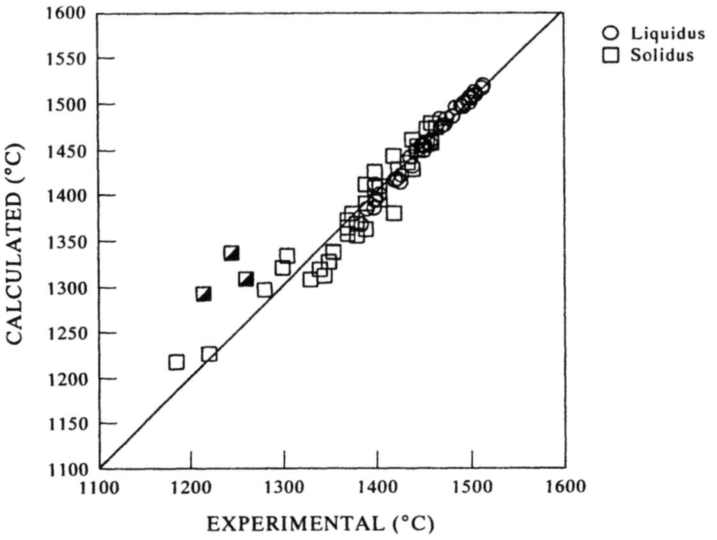

Figure 10.37 Comparison between calculated and experimental (Jernkonteret 1977) solidus and liquidus values of a range of steels.

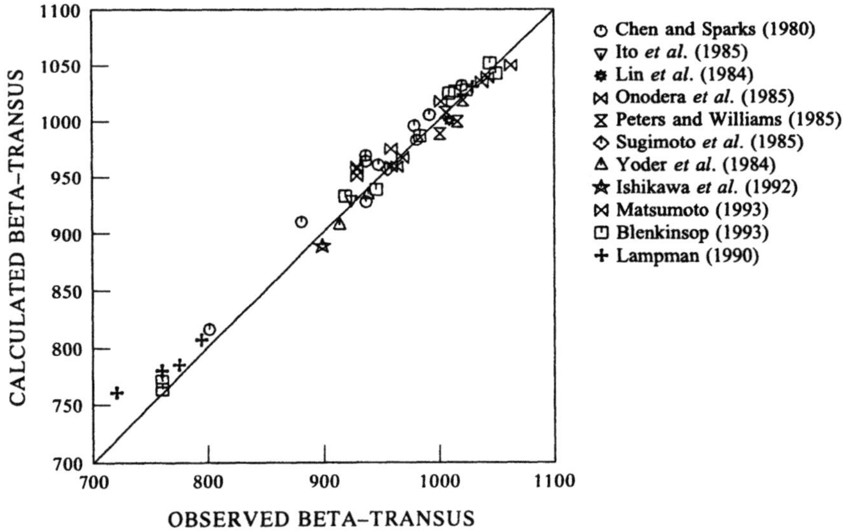

Figure 10.38 Comparison between calculated and experimental β-transus temperatures in Ti–alloys (from Saunders 1996a).

10.5.1.3 Ni-based superalloys. In Ni-based superalloys, containing high volumes of γ', the temperature window where an alloy can be heat treated in the fully γ state is a critical feature both in alloy design and practical usage. This heat treatment window is controlled both by γ'_s and the solidus and there have therefore been numerous experimental measurements of these properties. A further key experimental feature for cast alloys is the liquidus and, similarly, numerous measurements have also been made for this temperature. Figure 10.39 shows a comparison plot for γ'_s, liquidus and solidus for wide variety of Ni-base superalloys and average deviations from experiment are typically the same as for steels and Ti alloys, with $\bar{d}$ for liquidus and solidus being 6°C and 10°C respectively while $\bar{d}$ for the γ'_s is less than 15°C.

10.5.2 Calculations for duplex and multi-phase materials

10.5.2.1 Duplex stainless steels. Duplex stainless steels have provided a fruitful area for CALPHAD calculations and have been an example of where high levels of success have been achieved for practical materials. An early study (Hayes 1985) was able to demonstrate that reasonable predictions for amounts of austenite could be obtained for a variety of different duplex stainless steels, demonstrating the

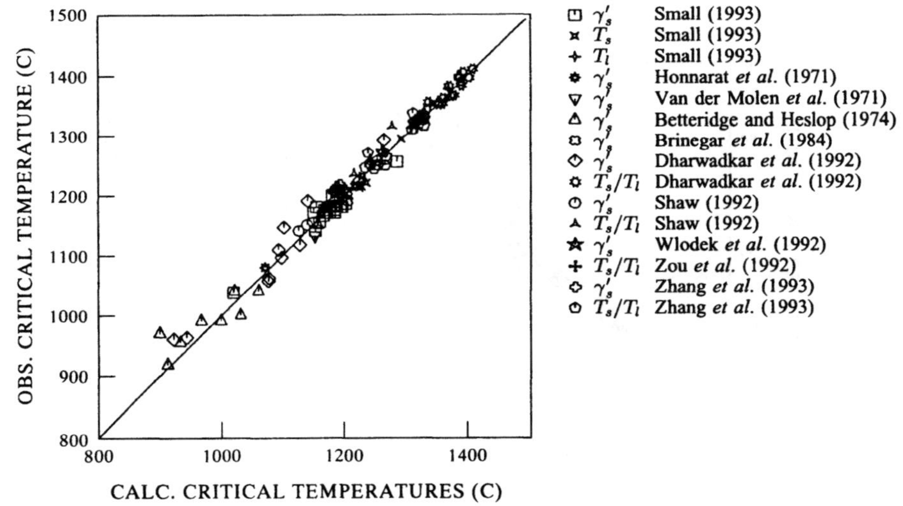

Figure 10.39 Comparison between calculated and experimental critical temperatures for Ni-based superalloys (from Saunders 1996c).

general applicability of CALPHAD calculations to these materials. The later work of Longbottom and Hayes (1994) shows how a combination of CALPHAD calculation and experiment can provide accurate formulae for the variation in austenite and ferrite as a function of composition and heat-treatment temperature in Zeron 100 stainless steels. These formulae can then be used during production of the material to help define temperatures for thermomechanical processing. The steel database (TCFe) developed at KTH has been used in a number of publications, notably by Nilsson (1992) and Hertzman (1995).

More recently, Fe-DATA (Saunders and Sundman 1996) was used in calculations for a wide variety of duplex stainless steels, and detailed comparisons were made for amounts of austenite, as a function of temperature, and the partition coefficients of various elements in austenite and ferrite. The results of these comparisons are shown in Figs 10.40 and 10.41. In Fig. 10.40, experimental results which have been given as volume fractions have been compared with mole% predictions, which is reasonable as molar volumes of the two phases are very similar. $\bar{d}$ for the amount of austenite is less than 4%, of the same order as would be expected for experimental accuracy, and the comparison of elemental partition coefficients is good. C and N levels, which are difficult to measure in practice, are automatically calculated. Where such measurements have been made the comparison is good and the advantage of using a calculation route is further emphasised.

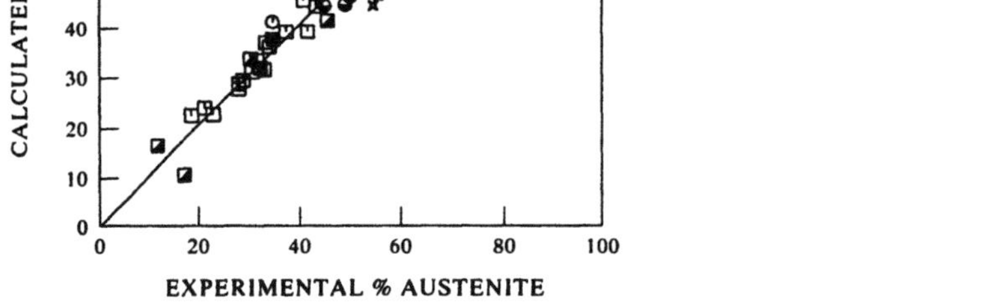

Figure 10.40 Comparison between calculated and experimentally observed % of austenite in duplex stainless steels. (Data from Longbottom and Hayes (1991) represent dual phase steels.).

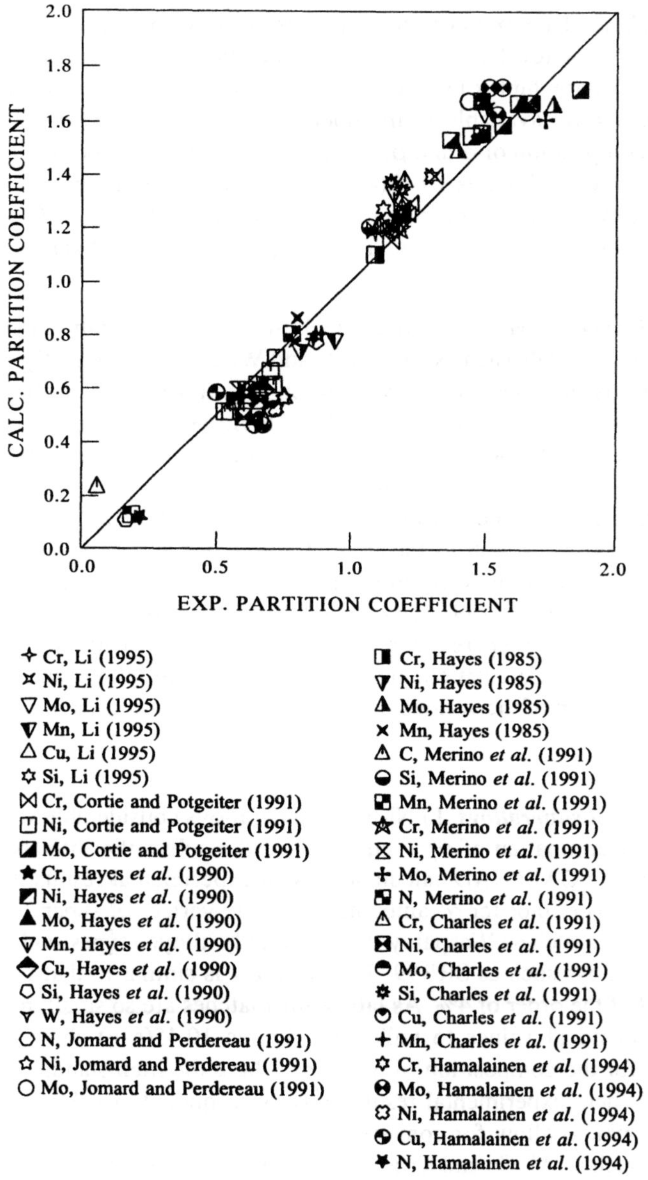

Figure 10.41 Comparison between calculated and experimentally observed partition coefficients between austenite and ferrite in duplex stainless steels.

10.5.2.2 Ti alloys. Duplex microstructures are usually formed in Ti alloys which are classed using the level of α-Ti or β-Ti in the alloy. There are a few fully α and β alloys but most are duplex in nature. Much work has been done in measuring T^{β} but fewer results are available in the open literature for the variation of volume fraction and composition of α and β. Fewer experiments are therefore available for comparison purposes. Figures 10.42(a–c) show calculated phase% vs temperature plots for three types of commercial alloys, an α-type, IMI834, an (α/β)-type, SP 700 and a β-type Ti–10V–2Al–3Fe. The agreement is very satisfactory.

10.5.2.3 High-speed steels. High-speed steels are alloys with high C levels and containing refractory metals such as Cr, Mo, V and W. They subsequently contain high levels of various carbides, typically between 5 and 15 vol%, with some alloys having more than 20 vol% (Riedl *et al.* 1987, Hoyle 1988, Wisell 1991, Rong *et al.* 1992). They are useful practical alloys for cutting tools where their wear resistance, toughness and resistance to tempering during machining is controlled by the types, amounts and shapes of the various primary carbides M_6C, M_2C and MC. The composition of the steel controls the amounts and types of the various carbides and, therefore, knowledge of phase equilibria in these alloys is important. Detailed work has been done by Wisell (1991) characterising the various carbides in tool steels. Fifty-six alloys were prepared and compositions of the carbides measured by microprobe analysis. Calculations have been made for all alloys and comparisons with the composition of the various M_6C, M_2C and MC carbides are shown in Figs. 10.43(a–c).

10.5.2.4 Ni-based superalloys. In Ni-based superalloys considerable work has been done on the determination of γ/γ' equilibria and a substantial literature exists by which to compare CALPHAD calculations with experimental results. Figure 10.44 shows a comparison plot for γ' amounts in a wide variety of superalloys, ranging from low γ' types such as Waspaloy through highly alloyed types such as IN939 to single crystal alloys such as SRR99. The accuracy is similar to that for the duplex steels, with $\bar{d}$ of the order of 4%. As lattice mismatches are so small, mole% values give almost identical values to vol% and figures 10.45(a–e) show some of the comparisons for the composition of γ and γ' where the high standard of results is maintained. Where experimental results have been quoted in wt% they have been converted to at% to allow for consistency of comparison. The average difference for elements such as Al, Co and Cr is close to 1at% while for Mo, Ta, Ti and W this value is close to 0.5at%. Too few experimental values for Hf and Nb were found to be statistically meaningful but where possible these were compared and results for average differences were found to be slightly better than obtained for Mo, Ta, Ti and W.

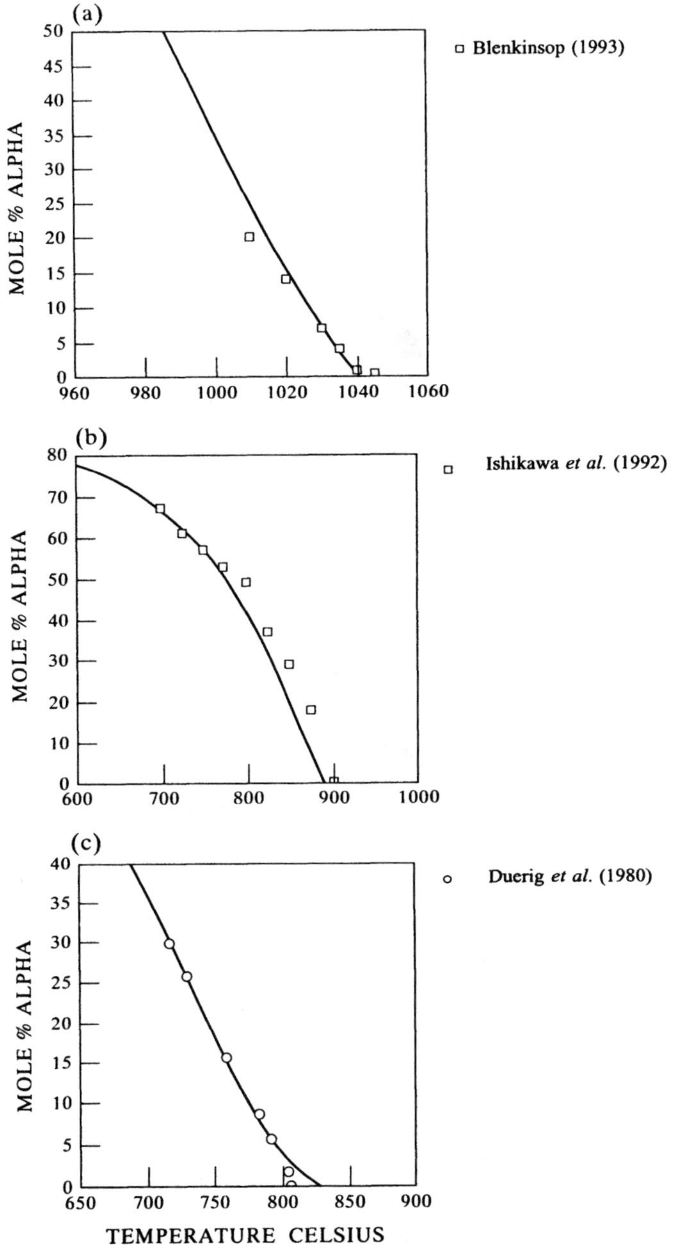

Figure 10.42 Calculated mole % phase vs temperature plots for three types of commercial alloys: (a) an α-type, IMI 834, (b) an (α/β)-type, SP 700 and (c) a β-type Ti–10V–2Al–3Fe (from Saunders 1996a).

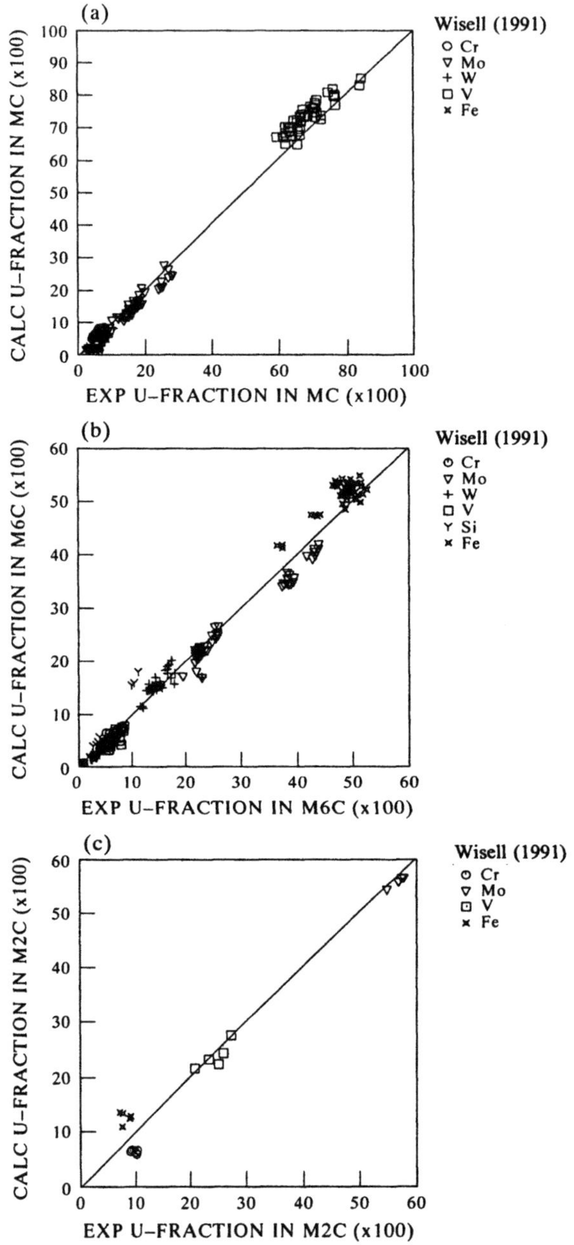

Figure 10.43 Comparison between calculated and experimental values (Wisell 1991) for the concen-tration of Cr, Mo, W, V and Fe in (a) the MC, (b) the M_6C and (c) M_2C phases of high-speed steels.

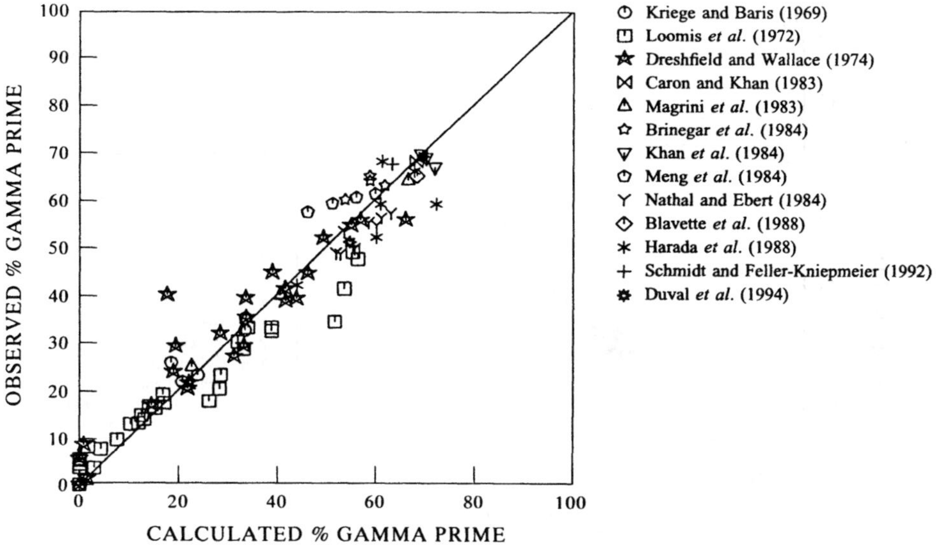

Figure 10.44 Comparison between observed and calculated amounts of γ' in Ni-based superalloys (from Saunders 1996c).

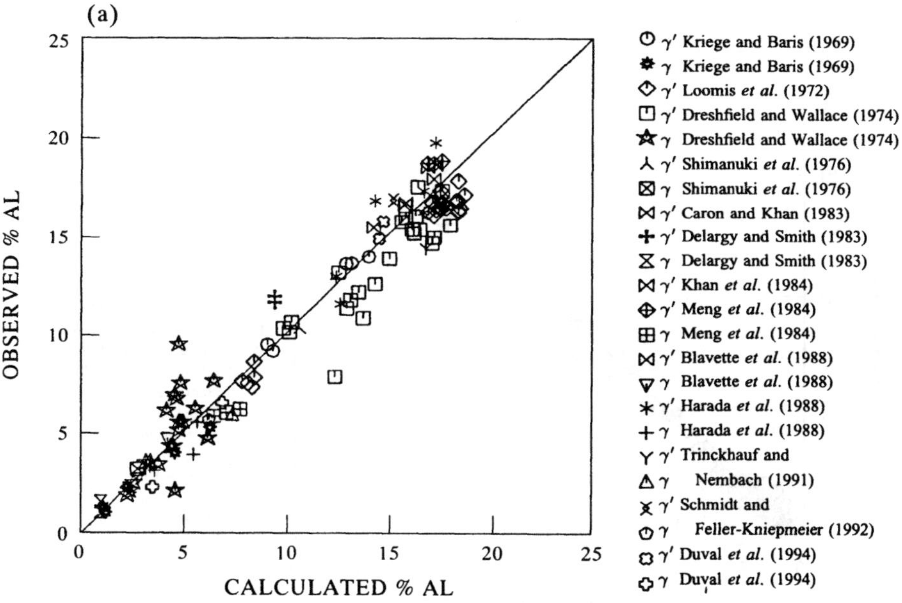

Figure 10.45 Comparison between calculated and observed compositions of γ and γ' in Ni-based superalloys: (a) Al, (b) Co, (c) Cr, (d) Mo and (e) W. (at %) (from Saunders 1996c)

(b)

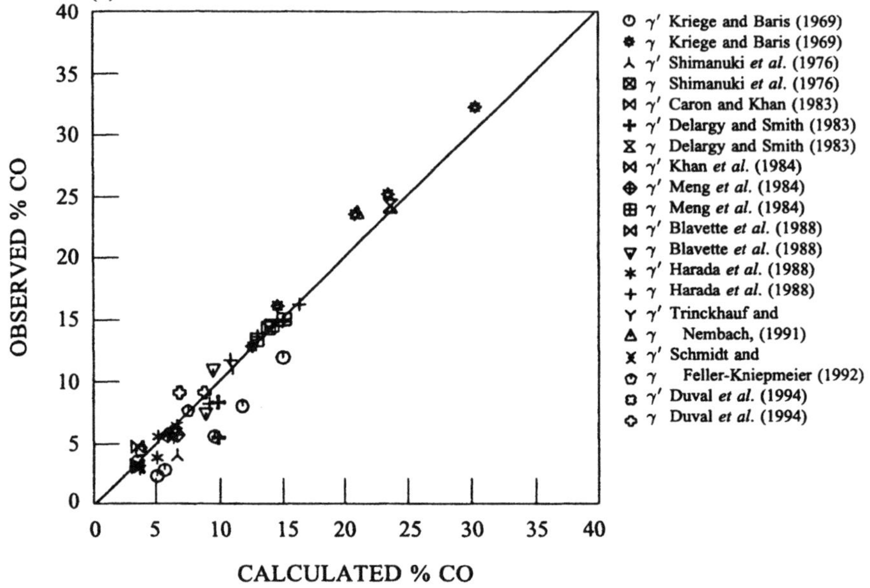

(c)

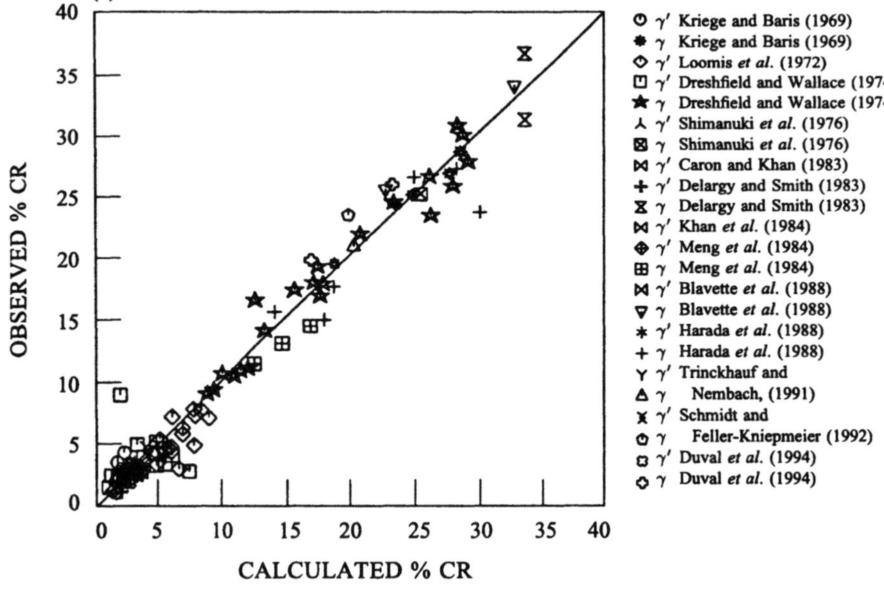

Figure 10.45 (b) and (c).

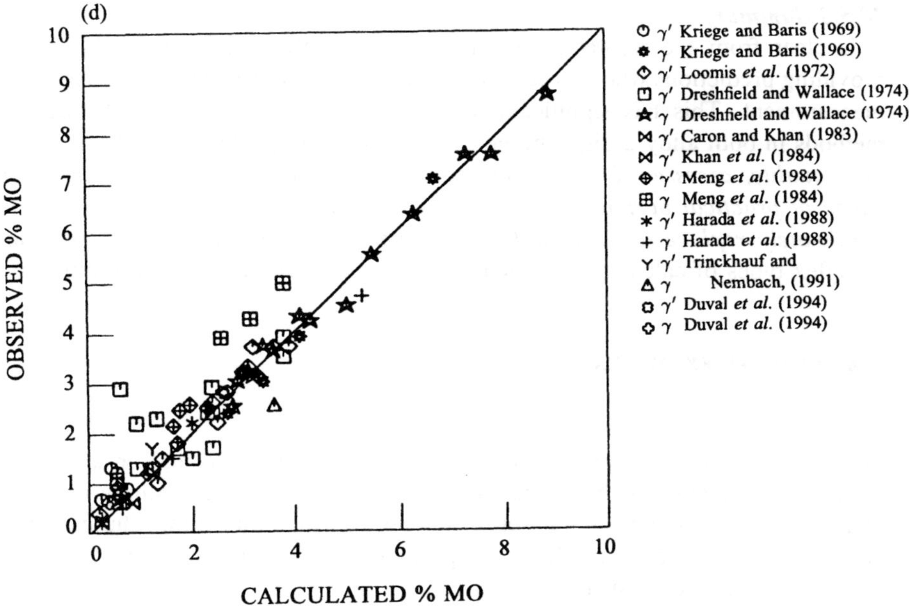

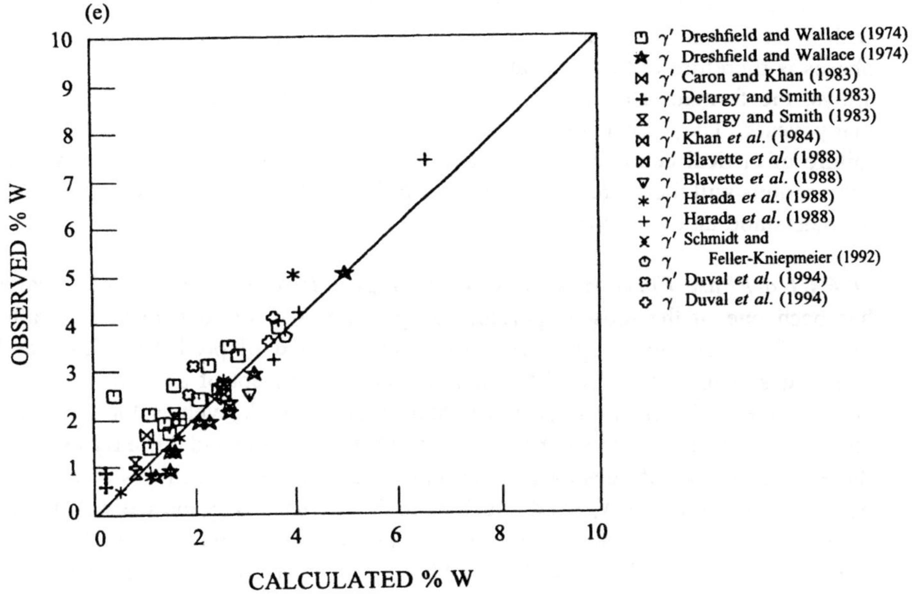

Figure 10.45 (d) and (e).

10.5.3 Summary

It is clear from the results shown in this section that the CALPHAD route is providing predictions whose accuracy lies close to that expected from experimental measurement. This has significant consequences when considering CALPHAD methods in both alloy design and general everyday usage as *the combination of a high quality, assessed database and suitable software package can, for a wide range of practical purposes, be considered as an information source which can legitimately replace experimental measurement.* The next sections discuss more complex types of calculations which are geared to specific, practical problems.

10.6. SELECTED EXAMPLES

10.6.1 Formation of deleterious phases

Formation of secondary phases is a feature of many materials and, in the context used here, is defined as the formation of phases other than the primary hardening phases or the predominant phases in duplex alloys. Embrittling phases can be carbides, borides, topologically close-packed (TCP) phases such as σ or μ, or 'insoluble' compounds such as Al_7Cu_2Fe in Al alloys. They can also be beneficial, providing secondary hardening reactions as for example in the low-temperature precipitation of η phase in the γ'-hardened alloy IN939 (Delargy and Smith 1983). But, more often, they produce a degradation in mechanical properties as is found with σ formation in stainless steels. The understanding of the formation of these phases is therefore critical in material design and processing. For Ni-based superalloys, the formation of σ and related phases has concerned alloy designers for many years. They are major materials in aerospace gas turbine engines where failure of critical components can have catastrophic consequences. The next two sub-sections will therefore show examples of how CALPHAD methods can be used to understand and help control TCP phases.

10.6.1.1 σ-Phase Formation in Ni-based Superalloys. The concept of 'σ-safety' has been one of the most important design criteria in the design of superalloys (Sims 1987), and in the past the most usual method of predicting this was by techniques such as PHACOMP which rely on the concept of an average electron hole number, $\bar{N}_v$, made up of a weighted average of $\bar{N}_v$ values for the various elements. In itself the concept behind PHACOMP is simple and it is easy to use, but there are a number of questions concerning its use and theoretical justification. For example, the values of N_v used to calculate N_v are usually empirically adjusted to fit experience and the model fails to explain why σ appears in the Ni–Cr–Mo ternary but is not observed in binary Ni–Cr or Ni–Mo. Furthermore, although it supposedly correlates with the phase boundary of γ and σ, it gives no information on the temperature range where σ may be stable, nor does it provide any information on the interaction of this boundary with the γ/μ or γ/Laves boundaries.

Using the CALPHAD route an actual σ-solvus temperature can now be calculated which defines the temperature below which σ will form and can be unambiguously used to help define 'σ-safety'. A good example of this concept is in Udimet® 720 (U720) whose composition is given below (Keefe *et al.* 1992).

	C	Cr	Co	Mo	W	Ti	Al	B	Zr
U720	0.035	18.0	14.7	3.0	1.25	5.0	2.5	0.033	0.03
U720LI	0.01	16.0	14.7	3.0	1.25	5.0	2.5	0.015	0.03

This alloy was first used in land-based gas turbine engines and for long-term use at up to 900°C (Keefe *et al.* 1992), but its excellent all-round properties suggested that it could also be used as an aerospace disc alloy. Unfortunately, while long-term exposure at high temperatures produced only minor susceptibility to σ, its use at 750°C quickly led to large amounts of σ being formed (Keefe *et al.* 1992). Clearly the alloy was either close to, or just below, its σ-solvus at the higher temperature and it was found necessary to reduce Cr levels to destabilise σ at lower temperatures. This led to the development of U720LI with 2wt% less Cr than for U720. C and B levels were also lowered to reduce formation of borides and carbides which acted as nucleation sites for σ formation.

Figure 10.46 shows a calculated phase % vs temperature plot for U720 and it can be seen that its σ-solvus is indeed close to 900°C and at 750°C the alloy would

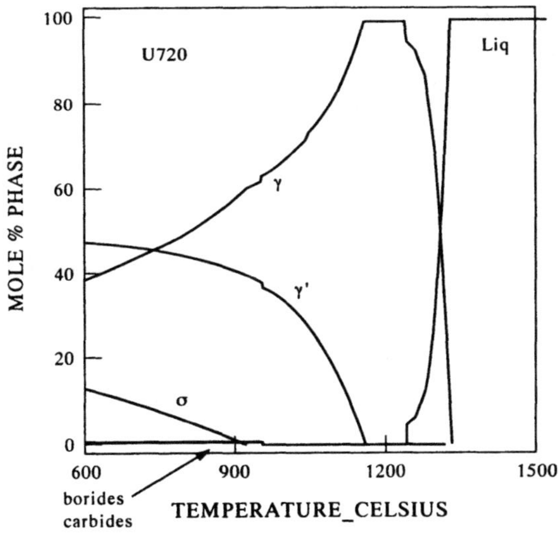

Figure 10.46 Calculated mole % phase vs temperature plot for U720 (from Saunders 1996c).

contain substantial levels of σ, in excellent accord with experimental observation. Keefe *et al.* (1992) further determined TTT diagrams for both U720 and U720LI, which are shown in Figure 10.47. Decreasing the Cr levels must decrease the σ-solvus and, as the high-temperature part of the TTT diagram asymptotes to the σ-solvus temperature, the two TTT diagrams should have distinct and separate curves. Taking the σ-solvus calculated for U720 and U720LI, it was proposed (Saunders 1995, 1996c) that the TTT diagrams should have the form as shown in Fig. 10.48.

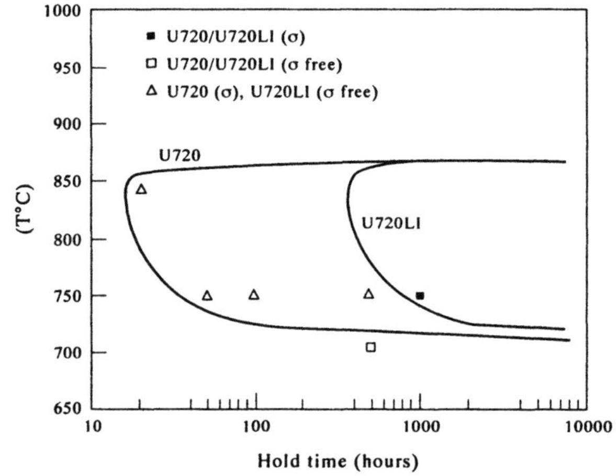

Figure 10.47 TTT diagrams for σ formation for U720 and U720LI after Keefe *et al.* (1992).

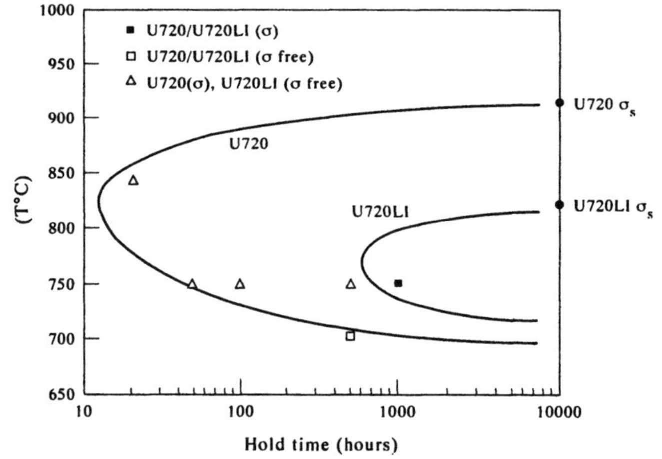

Figure 10.48 TTT diagrams for U720 and U720LI based on calculated σ-solvus temperatures.

10.6.1.2 The effect of Re on TCP formation in Ni-based superalloys. As gas turbine engines are designed and manufactured to work at higher and higher thrust-to-weight ratios, the temperature capability of most components has had to be increased. This has applied particularly to the high-pressure (HP) turbine blades where the development of new alloys has followed lines associated both with the development of microstructure and new chemistries. HP turbine blades are now commonly single-crystal materials where the grain boundaries have been removed by novel casting techniques. HP blades have also seen the development of alloy variants which typically contain Re at levels of between 2 and 6wt%. Because Re is such a heavy element, this relates to a small addition in atomic terms, but the effect on properties such as creep and strength is pronounced. Re also has a profound effect on the temperature at which TCPs are observed, raising this substantially (Darolia *et al.* 1988, Erickson *et al.* 1985). The reason for this is not readily understood in terms of a PHACOMP approach (Darolia *et al.* 1988), and it therefore is interesting to look at how CALPHAD methods can deal with an element like Re.

CMSX-4 is a second generation single crystal superalloy and typically has the composition Ni–6.3Cr–9Co–0.6Mo–6W–6.5Ta–3Re–5.6Al–1Ti–0.1Hf (in wt%). Figure 10.49 shows a calculated phase % vs temperature plot for this alloy and most aspects of its phase behaviour appear well matched. The high-temperature TCP phase is the *R* phase, in excellent agreement with the work of Proctor (1992) and its predicted composition is also in good agreement with Proctor (1992), with W and Re in almost equal proportions (~35wt% each). It is clear that the partitioning of Re

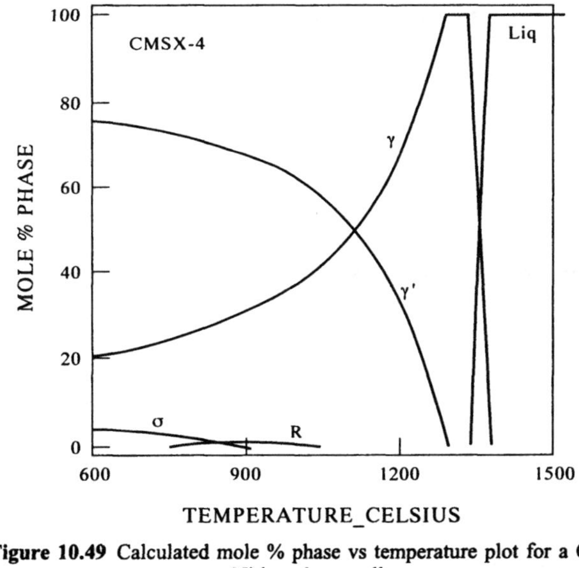

Figure 10.49 Calculated mole % phase vs temperature plot for a CMSX-4 Ni-based superalloy.

to the R phase is particularly strong and even stronger in σ, where levels can reach 45wt%. Although the formation of TCP phases is clearly enhanced because they are so rich in Re, their amounts in the alloy are restricted by the total level of Re, in this case only 3wt%. Therefore, the amounts of σ and R never rise to catastrophic levels as was evident in the previous example of U720. This has the corollary that Re-containing single-crystal alloys, such as CMSX-4, may be relatively tolerant to TCP formation even though their temperatures of formation can be high.

The sensitivity of TCP formation to Re can be further emphasised by examining what happens to CMSX-4 when the Re level is reduced by ½ to 1.5wt% accompanied by a corresponding increase in W level. For this case the solvus temperature falls by some 50°C. A similar exercise of replacing Re with W was performed by Darolia *et al.* (1988) and their observed variations in the start temperature of TCP formation are in reasonable accord with this.

It is also evident that there is a complex interplay between the three main TCP phases, σ, R and P, with μ occasionally being observed (Darolia *et al.* 1988, Proctor 1992, Walston *et al.* 1996). To examine this more closely, calculations were made for CMSX-4 by alternatively suspending two of the three main TCP phases and calculating the behaviour of the alloy with just one of the phases allowed to form at any one time (Fig. 10.50). The solvus temperatures for R, P and σ respectively were calculated as being close to each other at 1045°, 1034° and 1026° respectively, and the R and P phases are particularly close in stability over the whole temperature range. It is clear, therefore, that the interplay between the various TCP phases will be sensitive to the alloy composition and indeed changes in Re, W and Cr levels will cause either σ, R or P to become the dominant TCP phase.

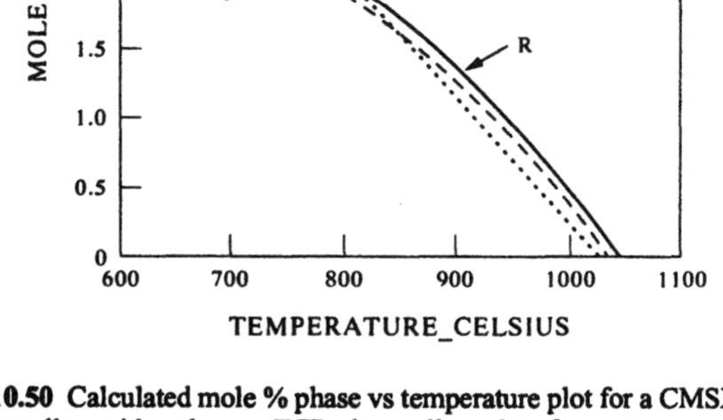

Figure 10.50 Calculated mole % phase vs temperature plot for a CMSX-4 Ni-based superalloy with only one TCP phase allowed to form at any one time.

10.6.2 Complex precipitation sequences

10.6.2.1 7000 series Al-alloys. Probably the most complex type of Al alloys are the 7000 or 7XXX series. These are based on the Al–Cu–Mg–Zn quaternary system and hardening reactions are based on one or more of three phases: (1) η, which is based on the binary Mg–Zn Laves compound but also exists in Al–Cu–Mg; (2) *T*_AlCuMgZn, which exists in both the Al–Cu–Mg and Al–Mg–Zn systems; and (3) *S*_Al_2CuMg (as in an AA2024 alloy). In these alloys it is the Zn/Mg ratio which is considered most critical in deciding the type of precipitation reaction which takes place. It should be noted that in practice an alloy such as AA7075 achieves maximum hardness due to the precipitation of the metastable η' phase which is structurally related to η. Also, both the *T*_AlCuMgZn and *S*_Al_2CuMg phases can form similar metastable hardening phases. At present, thermodynamic descriptions for these phases do not exist but, because of the inherent relationship between the metastable and stable forms of the compounds, it is reasonable to expect that the metastable precipitate which forms will be closely related to the respective stable precipitate which predominates in the alloy. Also, a number of important questions connected with processing are directly related to the stable forms; for example, solution temperatures and intermediate heat-treatment temperatures and their formation during non-equilibrium solidification. It is therefore instructive to look at a series of calculations for 7XXX alloys to observe how the calculated diagrams vary as the composition of the alloys is changed.

Figures 10.51a–d show the precipitation of the three major hardening phases, η, *T* and *S*, as well as ε-AlCrMnMg and Mg_2Si, for a series of high-strength 7000 series alloys with Zn:Mg ratios as given in Table 10.1. For simplicity, other phases which may appear, such as Al_6Mn, α, etc., have not been included. AA7049 represents the highest Zn:Mg ratio and also gives one of the highest values for the total percentage (Zn+Mg) total. In this case the hardening phase is almost completely η and its level rises to a maximum of 6% in molar units. As the Zn:Mg ratio is reduced from 3.1 to 2.7 in the AA7050 alloy the competition between *T* and η is still won easily by η but, due to the high levels of Cu in this alloy, it will also be significantly prone to the formation of *S*_Al_2CuMg. The 7075 alloy has a lower Zn:Mg ratio of 2.2 and although still predominantly hardened by η, some *T* phase

Table 10.1 Composition and Zn/Mg ratio of various 7XXX alloys (after Polmear 1989). Values for Si, Fe and Mn relate to maximum values

Alloy	Si	Fe	Cu	Mn	Mg	Zn	Cr	Zn/Mg
AA7049	0.25	0.35	1.5	0.2	2.5	7.7	0.15	3.1
AA7050	0.12	0.15	2.3	0.1	2.3	6.2	0.04	2.7
AA7075	0.4	0.5	1.6	0.3	2.5	5.6	0.23	2.2
AA7079	0.3	0.4	0.6	0.2	3.3	4.3	0.18	1.3

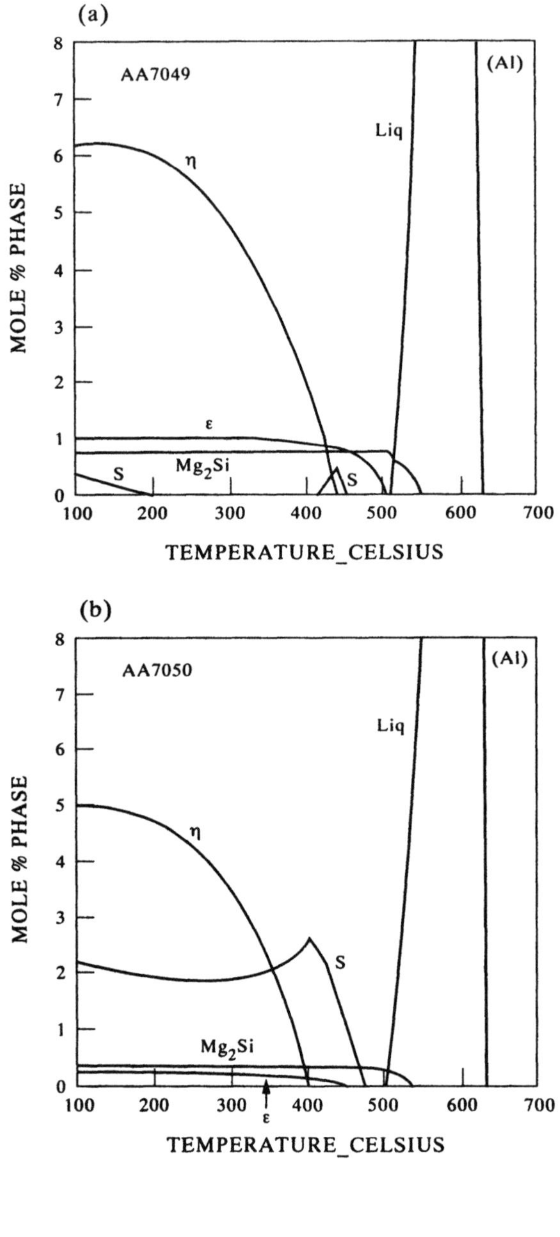

Figure 10.51 Calculated mole % phase vs temperature plots for 7XXX series Al-alloys. (a) AA7049, (b) AA7050, (c) AA7075 and (d) AA7079.

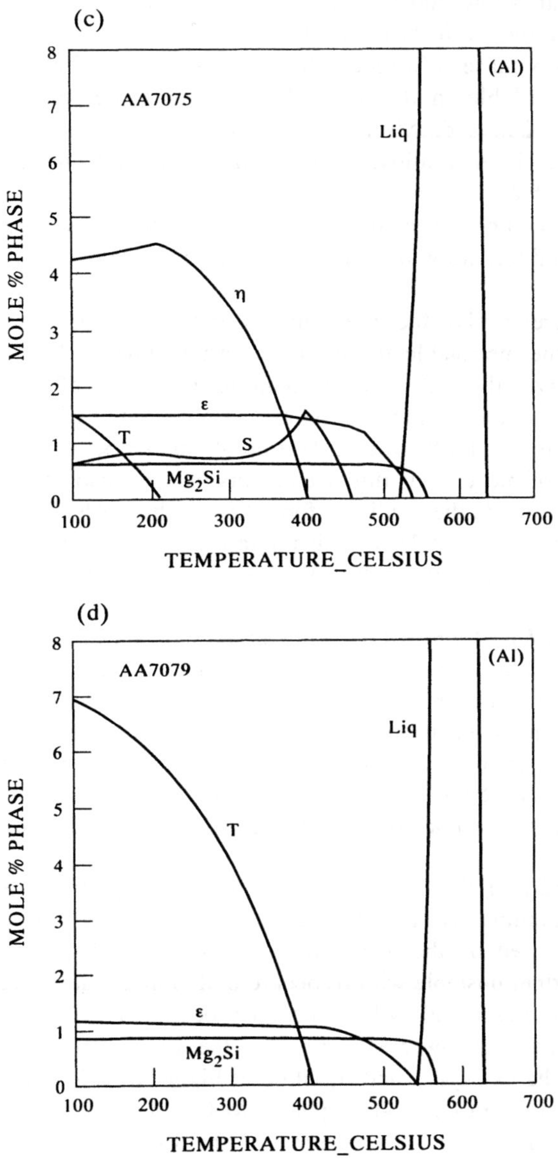

Figure 10.51 (c) and (d).

begins to appear below 200°C. It is also noteworthy that the level of S phase is reduced. Both of these effects are due to the reduction of Cu levels in AA7075. This means that less S phase is formed, which releases Mg, and effectively means that more Mg is available in the alloy. We are therefore forced to consider the concept of an *effective* Zn:Mg ratio which is potentially affected by the level of Cu in the alloy. The final alloy, AA7079, has the lowest Zn:Mg ratio of 1.3, well below a value of 2 which is considered the critical point at which the T phase is favoured. As would be expected, the main precipitation now occurs by the T phase with no predicted formation of η and no S phase because of the effective absence of Cu.

As stated previously, the predominant hardening in these alloys may be controlled by the metastable forms of the various phases, but it is clear that the calculations have allowed a quick scan to be made of the major types of equilibrium precipitates and this must have a significant bearing on the formation of the metastable forms. Furthermore, the role of Cu can be better understood as can also the effect of the various minor elements on the 'insoluble' compounds. It is also interesting to note the predicted formation of the ε-AlCrMnMg phase with the addition of Cr. In AA7075 it helps retard recrystallisation during the high-temperature heat treatment for this alloy and is, therefore, beneficial.

10.6.2.2 (Ni, Fe)-based superalloys. Ni,Fe-based superalloys, such as 718, can behave in a complex fashion, which is associated with the formation of various carbides and the interplay between three major precipitated phases; δ based on Ni_3Nb, γ' based on Ni_3Al and a metastable phase γ'' which is related to the δ phase. Inconel® 625 (IN625) was the prototype for the Nb-hardened NiFe-type superalloys and it is instructive to look at the complex precipitation phenomena which occur in this alloy which has the composition Ni–21.5Cr–9Mo–3.6Nb–5Fe–0.2Al–0.2Ti–0.05C (in wt%).

Figures 10.52(a,b) show phase % vs temperature plots for a standard IN625 alloy. It is predominantly strengthened at low temperatures by the formation of δ and is characterised by the formation of three types of carbide, MC, M_6C and $M_{23}C_6$. It is further unstable with respect to both σ and μ at low temperatures. This behaviour shows that the alloy lies close to a 'boundary' which is controlled by the Mo/Cr ratio where, at lower Mo/Cr ratios, the alloy would contain MC and the more Cr rich phases $M_{23}C_6$ and σ, while at higher Mo/Cr ratios the alloy would contain MC and the more Mo-rich phases M_6C and μ.

The temperature range of formation of the various carbides and the solvus temperatures of δ are well established (Ferrer *et al.* 1991, Vernot-Loier and Cortial 1991) and the calculated diagram is in excellent agreement with this experimental information. However, during processing it is usually the γ'' phase which is formed instead of δ. This is due to much faster transformation kinetics of this phase and hardening Ni,Fe-based superalloys is usually due to a combination of γ' and γ''.

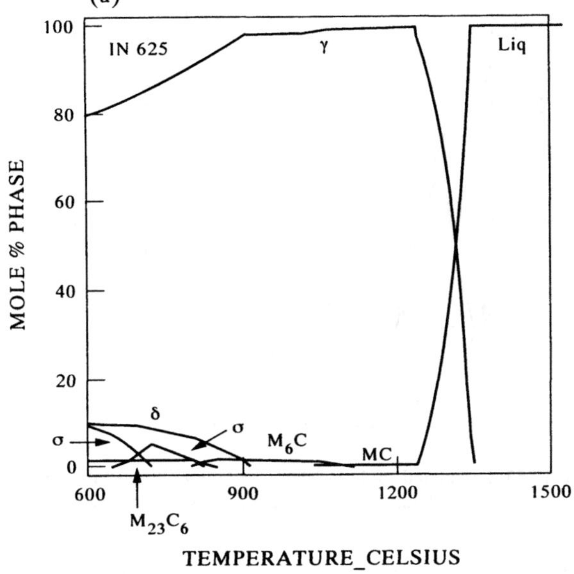

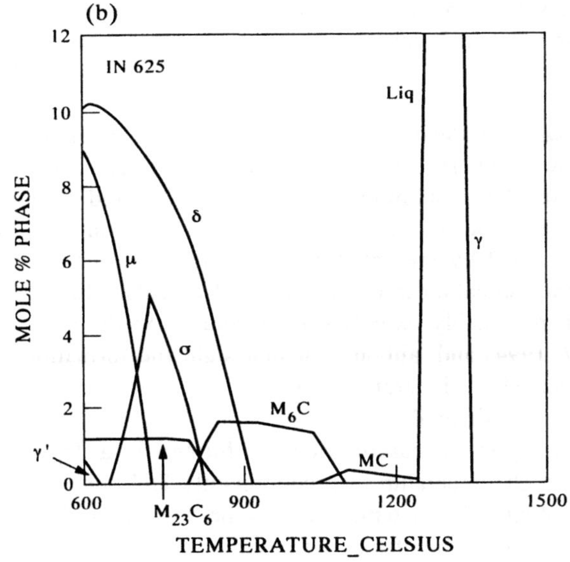

Figure 10.52 (a) Calculated mole % phase vs temperature plots for an IN625 Ni-based superalloy. (b) Expanded region of Fig. 10.50(a).

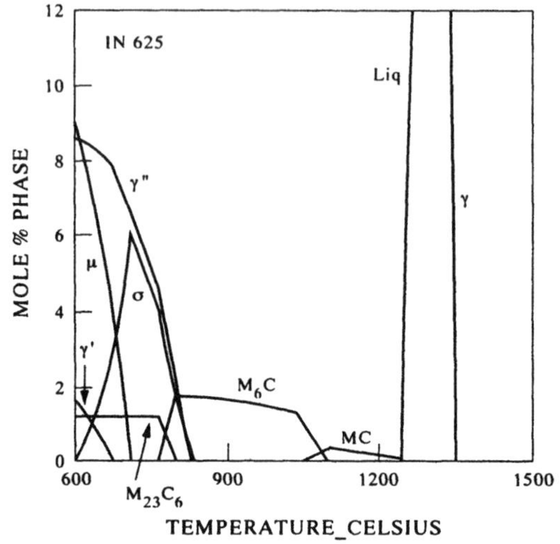

Figure 10.53 Calculated mole % phase vs temperature plots for an IN625 Ni-based superalloy with the δ phase suppressed.

This can be modelled by suspending the δ phase from the calculations and Figure 10.53 shows the subsequently calculated phase % plot. Again, the solvus temperature for γ'' is well matched with a consequent increase in the γ' solvus temperature of some 50°C.

10.6.2.3 Micro-alloyed steels. In the past 20 years the strength and toughness of high-strength low-alloy (HSLA) steels has improved considerably through micro-alloying additions. These additions, typically Ti, Nb and V, form stable carbonitrides which help both by reducing grain growth during the high-temperature annealing process and by precipitation hardening at lower temperatures. Similar effects occur with so-called interstitial-free (IF) steels. Modelling work on such alloys has been reasonably extensive (Zou and Kirkaldy 1992, Houghton 1993, Akamatsu *et al.* 1994) and, although at first sight the formation of such carbonitrides may appear relatively straightforward, there is an internal complexity which should not be underestimated.

In essence two types of carbonitride are formed in a Ti,Nb-hardened micro-alloyed steel. At high temperatures a predominantly TiN-rich carbonitride is formed. However, on cooling to lower temperatures a predominantly NbC-rich carbonitride also precipitates. Both carbonitrides are based on the NaCl structure and form part of a continuum usually described by a formula such as $(Ti_xNb_{1-x})(C_zN_{1-z})$. This can be expanded to include elements such as V and Ta, so the formula becomes $(Ta_xTi_yNb_zV_{1-x-y-z})(C_zN_{1-z})$. The formation of two types of carbonitride can be considered due to 'phase separation' and Fig. 10.54 shows a projected miscibility gap for the simple case of $(Ti_xNb_{1-x})(C_zN_{1-z})$ from Akamatsu *et al.* (1994). If the composition of the carbonitride lies outside the miscibility gap, simple single-phase carbonitrides form. But as the temperature is lowered the miscibility gap extends over much of the composition space and breakdown to TiN-rich and NbC-rich phases is possible.

In reality, the modelling issues are even more interesting as the NaCl carbonitride structure is formed by the filling of the octahedral sites of austenite and there is actually a continuum between austenite and the carbonitride phase. The duplex precipitation is therefore occurring by a breakdown of an Fe-rich phase to first form a TiN-rich carbonitride which subsequently undergoes further separation. Providing the f.c.c._A1 phase is modelled correctly, the calculation can be made by considering only two phases, the ferritic b.c.c._A2 and the f.c.c._A1 phase which then separates to form the austenite, the TiN-rich carbonitride and the NbC-rich carbonitride phases. This has the advantage that non-stoichiometric effects in the carbonitride can also be considered and the formula becomes $(Fe_wTi_xNb_{1-w-x})(C_yN_zVa_{1-y-z})$. Figure 10.55 shows a calculation for a micro-alloyed steel with a composition after Zou and Kirkaldy (1992) and the formation of the two types of carbonitride is predicted to form, as is observed in practice.

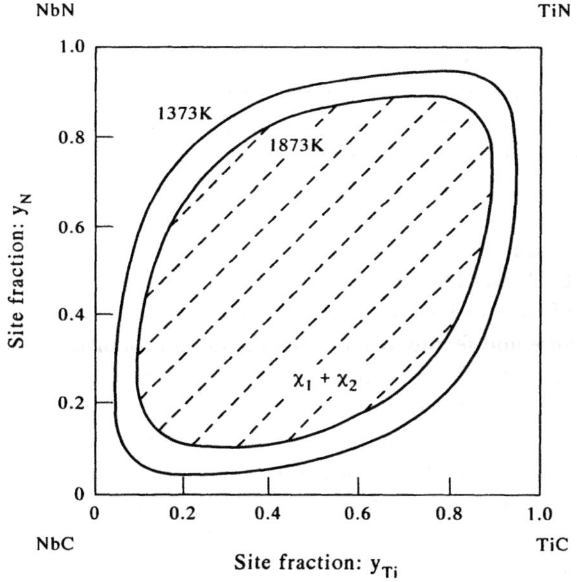

Figure 10.54 Projected miscibility gap for the case of $(Ti_xNb_{1-x})(C_zN_{1-z})$ from Akamatsu *et al.* (1994).

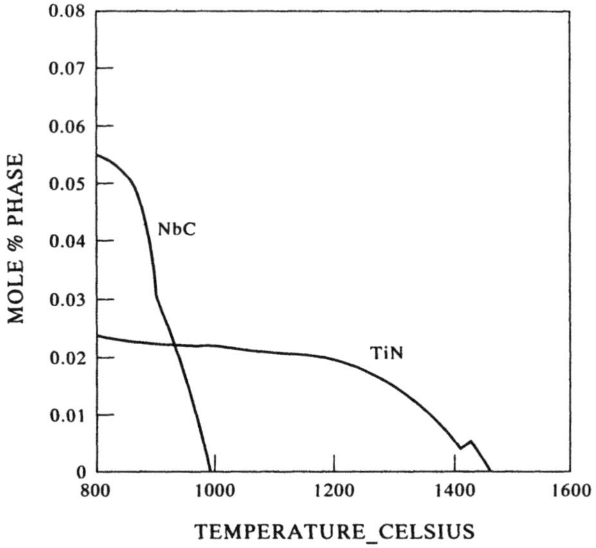

Figure 10.55 Calculated mole % vs temperature plot for a micro-alloyed steel showing precipitation of TiN- and NbC-based carbides.

10.6.3 Sensitivity factor analysis

An interesting and important corollary of CALPHAD calculations is to predict sensitivity factors for various important material properties, such as liquidus and solidus temperature, solid state transus temperatures, heat-treatment windows, etc. These can then be utilised both in alloy design and in production of alloys or components.

10.6.3.1 Heat treatment of duplex stainless steels. An excellent example of the application of CALPHAD methods to sensitivity analysis is in the work of Longbottom and Hayes (1994) on duplex stainless steels. These materials contain a mixture of austenite and ferrite, usually with a 50:50 proportion of each phase, and are produced by suitable heat treatment, typically in the range 1050–1150°C. The amount of austenite and ferrite is critically controlled by the composition of the alloy which dictates the temperature where there is the ideal 50:50 ratio. The particular alloy of interest in their study was a Zeron 100 alloy with the nominal composition Fe–25Cr–3.5Mo–6.5Ni–1Mn–1Si–0.75Cu–0.75W–0.25N–0.15C (in wt%). Using original experimentation, Longbottom and Hayes (1994) confirmed that the correct austenite levels, as a function of temperature, were calculated for a set of four Zeron 100 variants. Figure 10.56 shows the agreement between their experimental work and calculations and, except for three cases, the calculations agree within the error bar of the measurements.

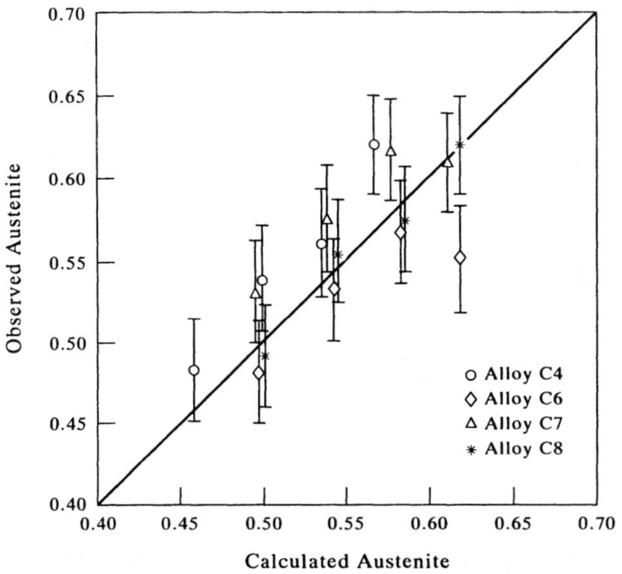

Figure 10.56 Comparison between thermodynamically calculated and experimentally observed amounts of austenite in Zeron 100 duplex stainless steels (Longbottom and Hayes 1994).

Rather than utilise a full thermodynamic calculation for their final sensitivity analysis, they fitted their calculated results to a formula which could be used within the composition specification range of Zeron 100 alloys, such that the fraction of austenite, f^γ, was given by

$$
\begin{aligned}
f^\gamma = &+ 2.2 - 1.39 \times 10^{-3} T \\
&+ (1.35 \times 10^{-3} T - 0.78)\%\text{C} + (1.30 \times 10^{-5} T - 0.037)\%\text{Cr} \\
&- (2.44 \times 10^{-6} T - 0.099)\%\text{Cu} - (5 \times 10^{-5} T - 0.061)\%\text{Mn} \\
&- (3.08 \times 10^{-6}, T + 0.0081)\%\text{Mo} + (1.56 \times 10^{-3} T - 1.40)\%\text{N} \\
&- (3.97 \times 10^{-5} T - 0.11)\%\text{Ni} + (3.32 \times 10^{-4} - 0.61)\%\text{Si} \\
&+ (2.11 \times 10^{-5} - 0.038)\%\text{W}.
\end{aligned} \tag{10.1}
$$

Figure 10.57 compares the results obtained using this formula with the thermodynamic calculations, while Fig. 10.58 shows the comparison with experimental volume fractions. This formula can now be written in simple software code to provide an almost instant answer to the temperature at which heat treatment will give the ideal 50:50 ratio of austenite to ferrite.

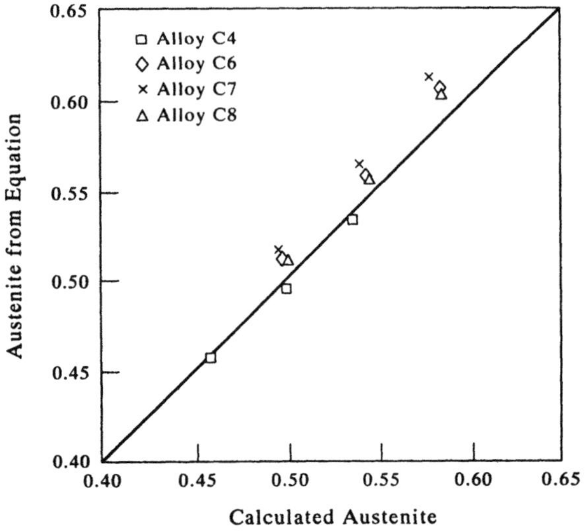

Figure 10.57 Comparison between austenite amounts calculated thermodynamically and calculated using Eq. (10.1) (Longbottom and Hayes 1994).

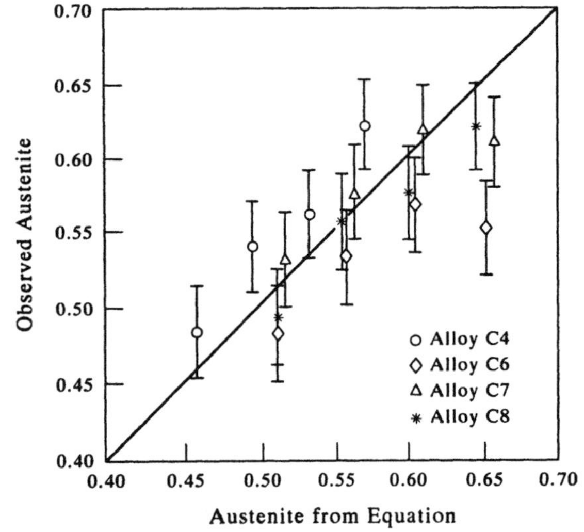

Figure 10.58 Comparison between experimental austenite amounts and those calculated using Eq. (10.1).

10.6.3.2 σ phase in Ni-based superalloys. Section 10.6.1.1 showed how CALPHAD calculations could be used to predict σ formation in Udimet 720 and 720LI alloys. A corollary is to calculate how the σ-solvus varies as the contents of the different elements are altered, and Fig. 10.59 shows such variations in σ-solvus temperature for U720LI as each element is changed within its nominal specified composition limits. It is interesting to note that the greatest sensitivity is to Al, with a Ti sensitivity similar to that of Cr. This is because additions of Al and Ti increase the levels of γ' in the alloy and reject σ-forming elements such as Cr, Mo and W into the γ phase. As the amount of γ decreases there is a significant increase in concentration of these elements in γ, which leads to higher susceptibility to σ formation. Taking the information in Fig. 10.59 it is now possible to define σ-sensitivity factors for each of the elements with a simple mathematical formula, as in the previous section. This can then be used to monitor σ susceptibility of different heats during alloy production, replacing PHACOMP methods. Such sensitivity factors can be defined for all types of production features where phase equilibria are important for Ni-based superalloys, for example γ' heat-treatment windows, levels of γ' at heat-treatment or forging temperatures, solidus and liquidus temperatures, etc.

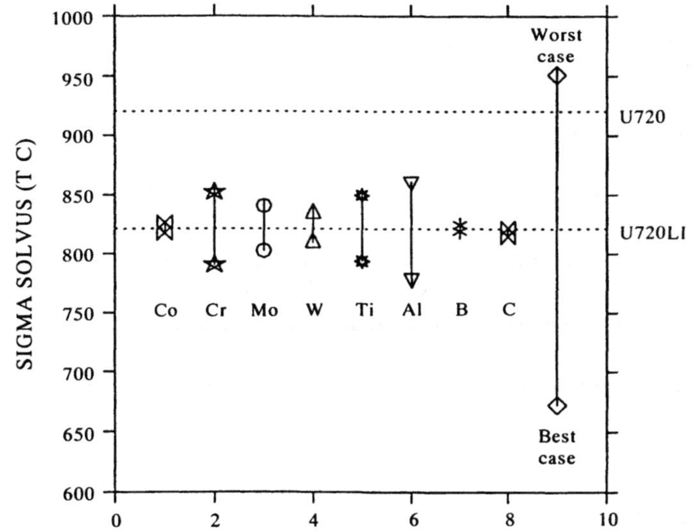

Figure 10.59 Variation in calculated σ-solvus temperature as elements in U720LI change between maximum and minimum specified limits.

10.6.3.3 Liquid-phase sintering of high-speed M2 steels. High-speed steels, which typically contain at least seven elements, can be produced in a variety of ways, for example by casting, by water or gas atomisation with subsequent hot compaction of the atomised powder, and by liquid-phase sintering. The latter provides a production route whereby the steel can be produced directly to the correct shape. The process relies on there being sufficient liquid to infiltrate the pores of a powder compact and produce densification, but not too much, as this would cause the compact to lose structural strength and distort. A reasonable level for liquid-phase sintering is approximately 10% of the volume. It is also critical that the volume of liquid is fairly insensitive to fluctuations in temperature in the furnace, i.e., dV_{liq}/dT is as small as possible, otherwise the sensitivity of the process to fluctuations in temperature control, or the intrinsic variations in temperature within the furnace itself, makes the process uncontrollable.

Figure 10.60 shows a phase % vs temperature plot for an M2 steel with a composition (in wt%) of Fe–6.2W–5.1Mo–4.2Cr–1.85V–0.3Si–0.8C. The alloy is characterised by the formation of two types of carbide, M_6C, which is the predominant form, and MC. The calculated melting temperature is close to that observed for such alloys and the amount of M_6C and MC is also in good agreement with experimental observation (Hoyle 1988). Chandrasekaran and Miodownik (1989, 1990) and Miodownik (1989) examined M2 steels, with particular emphasis on the effect of increasing V levels to increase the formation of the V-rich MC carbide. Figure 10.61 shows the change in behaviour of M2 when the V level is increased from 1.85 to 2.5wt%, calculated using Fe-DATA (Saunders and Sundman 1996). The MC carbide is stabilised as expected, but α is also stabilised and becomes involved in a fairly complex reaction scheme with the liquid, γ, M_6C and MC. Both of these effects are qualitatively expected, but now the CALPHAD route enables calculations to be made for the amount of liquid as a function of temperature. Figure 10.62 shows such calculations as V levels are raised from 1.75% to 2.5% and the increase in solidus temperature with increasing V addition reflects well the experimental measurements quoted by Chandrasekaran and Miodownik (1989). It can now be seen that the slope of the liquid curve is flat for the lower levels of V, i.e., dV_{liq}/dT is small, but as the V levels reach 2.5wt%V, the slope becomes steep. Clearly there will be a limit around this composition at which the alloy will become difficult to fabricate by liquid-phase sintering.

10.6.4 Intermetallic alloys

Intermetallic alloys (compounds) are becoming of increasing interest as materials which possess significantly enhanced high temperature capabilities compared to many conventional metallic alloys. However, they suffer intrinsic problems associated with low-temperature ductility and fracture toughness. Two types of

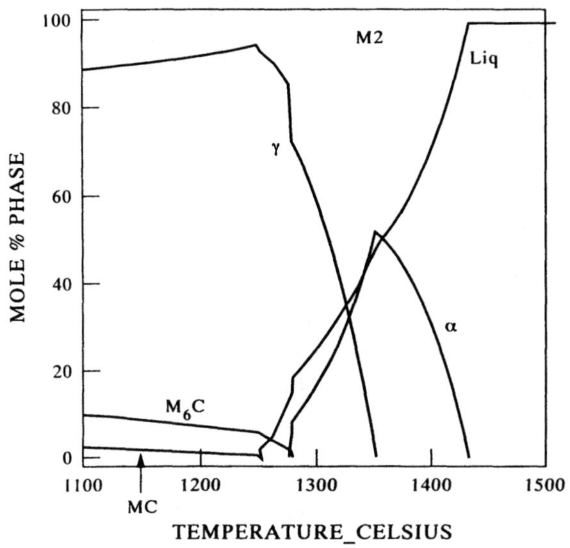

Figure 10.60 Calculated mole % phase vs temperature plot for an M2 steel with normal V level.

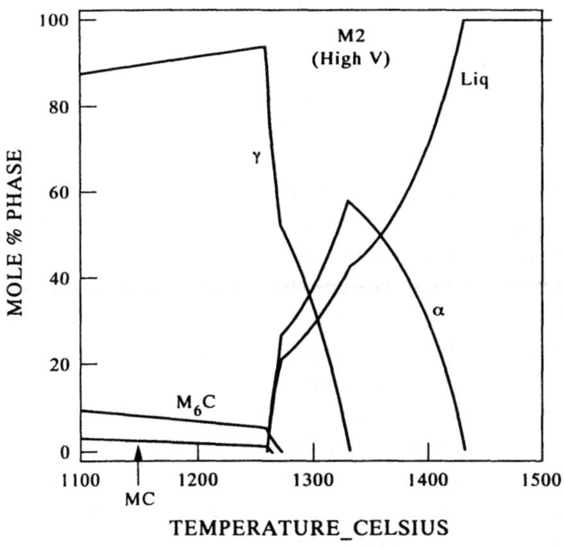

Figure 10.61 Calculated mole % phase vs temperature plot for an M2 steel with high V level.

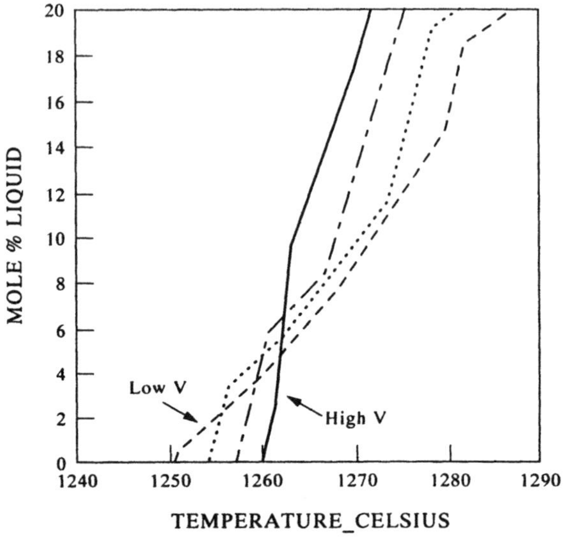

Figure 10.62 Calculate mole % liquid vs temperature plots for an M2 steel as V levels are changed between 1.75 and 2.5wt%.

intermetallic alloys have attracted wide attention, those based on NiAl and those on TiAl. The next two sections describe work on both types of alloy.

10.6.4.1 NiAl-based intermetallic alloys. Phase-diagram studies have been used extensively in NiAl-based alloys as a method of designing microstructures. NiAl is difficult to prepare as a binary alloy with reasonable levels of ductility and fracture toughness. Therefore considerable effort has been placed in designing microstructures which might reduce this inherent brittleness. An approach has been to make alloy additions to NiAl so that it forms a two-phase structure with a more ductile phase, such as the f.c.c. Ni–Fe-based solid solution, γ, (Ishida *et al.* 1991), or even γ' based on Ni_3Al (Kainuma *et al.* 1996). This has been shown to produce high levels of ductility for such materials but, unfortunately, the process of ductilisation by a softer second phase does lead to a degradation of intrinsic strength and high-temperature creep capability. Another approach to designing NiAl alloys has been to produce multi-phase structures containing NiAl + other intermetallic compounds such as the Heusler-type phase Ni_2AlTi and γ' (Yang *et al.* 1992a, b). These alloys are less ductile but have significantly higher yield strengths and potentially enhanced creep properties.

The approach of Ishida *et al.* (1991) and Kainuma *et al.* (1996) utilised both experimentally determined and calculated diagrams for Ni–Al–Fe (Fig. 10.63) to

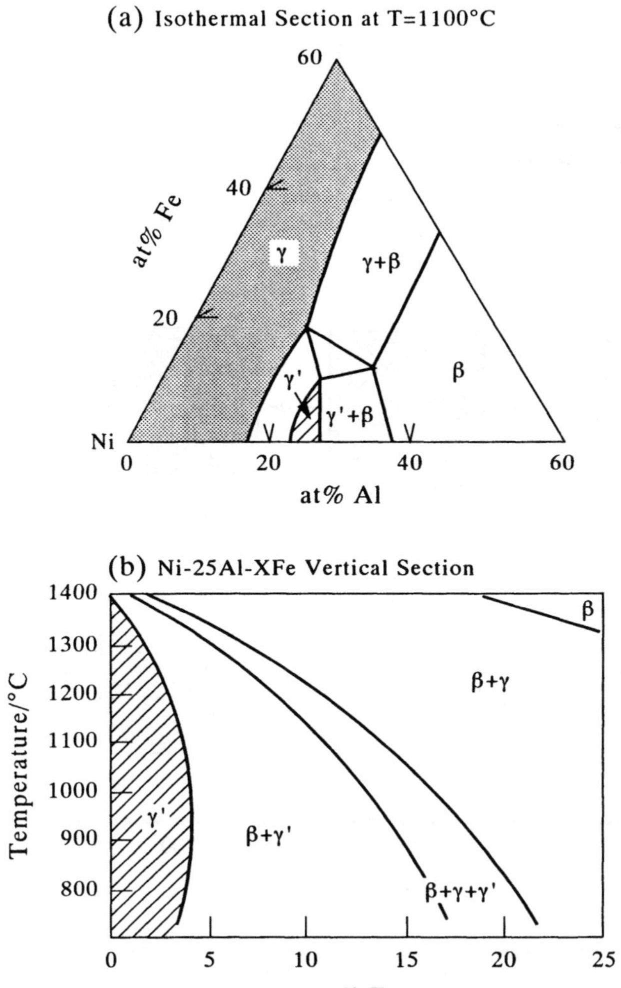

Figure 10.63 Phase equilibria in the Ni–Al–Fe system from Kainuma *et al.* (1996).

define compositions whereby various mixtures of NiAl and γ/γ' could be produced. These authors optimised their alloys on the basis of the amounts of the various phases as well as their microstructure. In their case, the initial microstructure was an $L1_0$ martensite formed by transformation of the β phase on cooling. By using a series of heat treatments and varying the composition of the alloy they were able to design three distinct types of microstructure (Fig. 10.64) and successfully produce

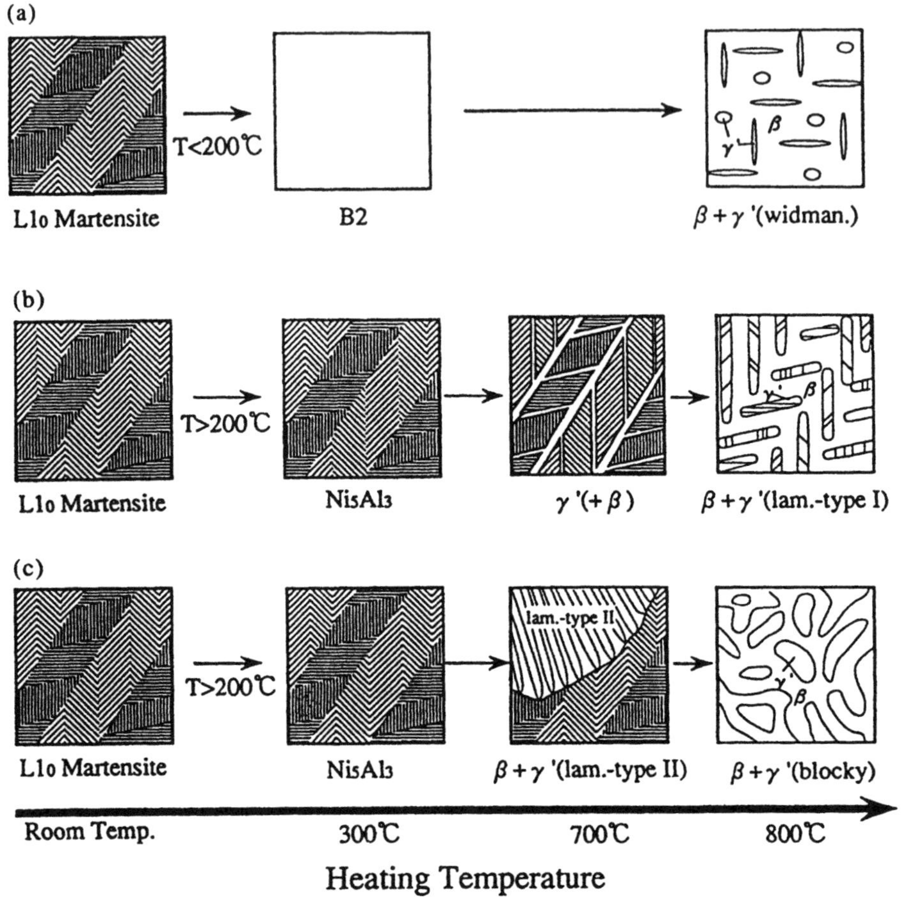

Figure 10.64 Schematic illustrations showing the microstructural evolution of three kinds of $\beta+\gamma'$ microstructures; widmanstätten, type 1 lamellar and blocky type 2 lamellar structures in Ni–25Al–(a)18Fe, (b)15Fe and (c)13Fe alloys.

NiAl-based alloys with ductilities as high as 10% in tension and strengths in the region of 750–1000 MPa.

Yang *et al.* (1992a, b) also utilised a combination of experiment and calculation to critically determine the phase region for the β-NiAl, γ'-Ni_3Al and β'-Ni_2AlTi phases. The philosophy of their approach was to produce an alloy with high levels of β and β', as mixtures of these phases had been shown to have enhanced creep resistance in comparison to the monolithic phases themselves (Polvani *et al.* 1976). The combination of experiment and calculated phase % vs temperature plots (Figs.

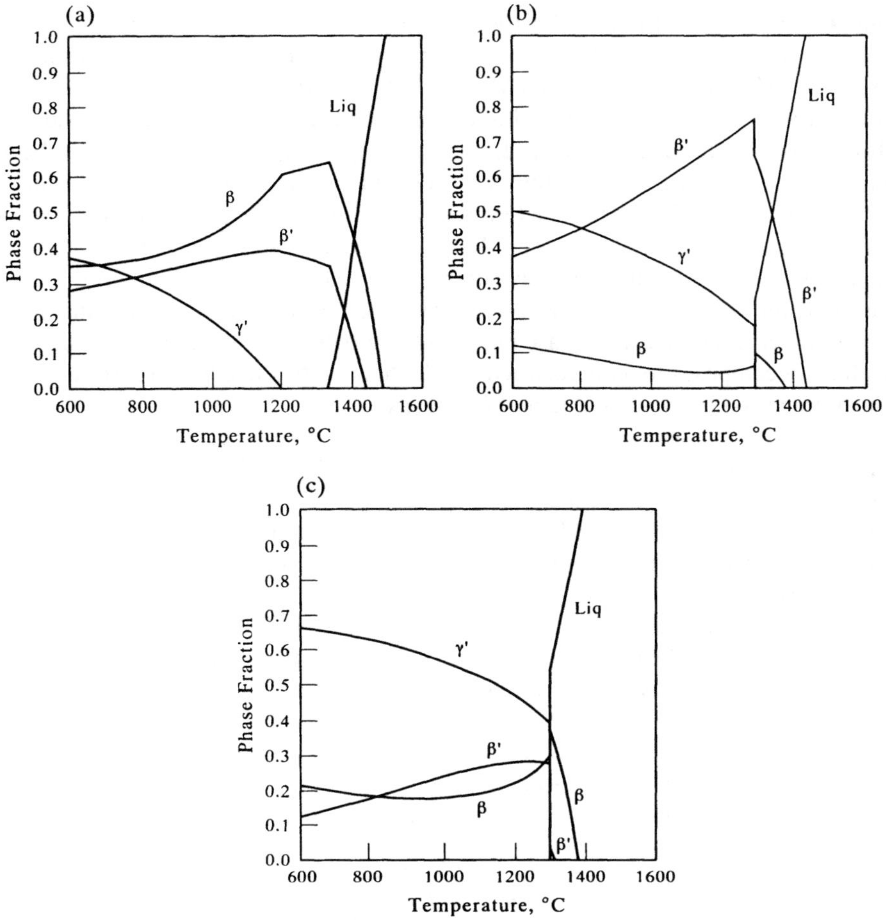

Figure 10.65 Calculated mole % phase vs temperature plots for Ni–Al–Ti intermetallic alloys. (a) Ni–28Al–12Ti, (b) Ni–22Al–15Ti and (c) Ni–20Al–13Ti.

10.65(a–c)) allowed precise microstructures to be defined and subsequently produced in candidate alloys. In these alloys the microstructure was as much controlled by the solidification mechanism as by any solid state transformation and, as such, the calculations were invaluable as they could be used to predict solidification patterns. Alloys produced as part of this programme had high compressive yield stresses, in the range of 1000–2000 MPa, and, although the alloys still had quite low tensile ductility, there is a significant enhancement in ductility due to the generation of dislocations at interphase boundaries.

10.6.4.2 TiAl-based intermetallic alloys. Alloys based on the γ-TiAl compound are still very much in the early stages of development. Commonly used variants such as Ti–48Al–2Mn–2Cr and Ti–48Al–2Mn–Nb (at%) are quaternary in nature and it is only recently that alloys of a multi-component nature have been developed. As alloy development has progressed, increasingly complex behaviour has been observed and changes in transformation behaviour can now be tracked using a recently developed TiAl database (Saunders 1997b). All alloys are based on the γ-TiAl compound of Ti–Al and Fig. 10.66 shows the calculated diagram for the binary system Ti–Al. Alloys usually contain between 43–50at%Al and, for the most part, contain α_2-Ti_3Al as well as γ-TiAl. They are usually heat treated in the high-temperature, single-phase α-phase region, or just below it, before cooling when the α phase decomposes to a microstructure which exhibits a fine lamellar structure of transformed γ-TiAl with fine laths of α_2-Ti_3Al.

Figure 10.67 shows a mole% phase vs temperature plot for a Ti–48Al–2Mn–2Nb alloy of nominal composition; variations of this alloy with Al in the range 45–48at% are commonly used. It is a straightforward α_2/γ alloy which can be heat treated in the α condition before transforming on cooling to $(\alpha_2 + \gamma)$ and very much exhibits the prototype microstructure for the early type of alloys.

The replacement of Mn by Cr in the Ti–48Al–2Cr–2Nb alloy (Fig. 10.68) causes the predicted onset of instability with respect to a Cr-rich B2 phase. This in good accord with experimental observations in alloys of this type (Fuchs 1995, Kelly and Austin 1996) and at low temperatures there is also a potential for the formation of

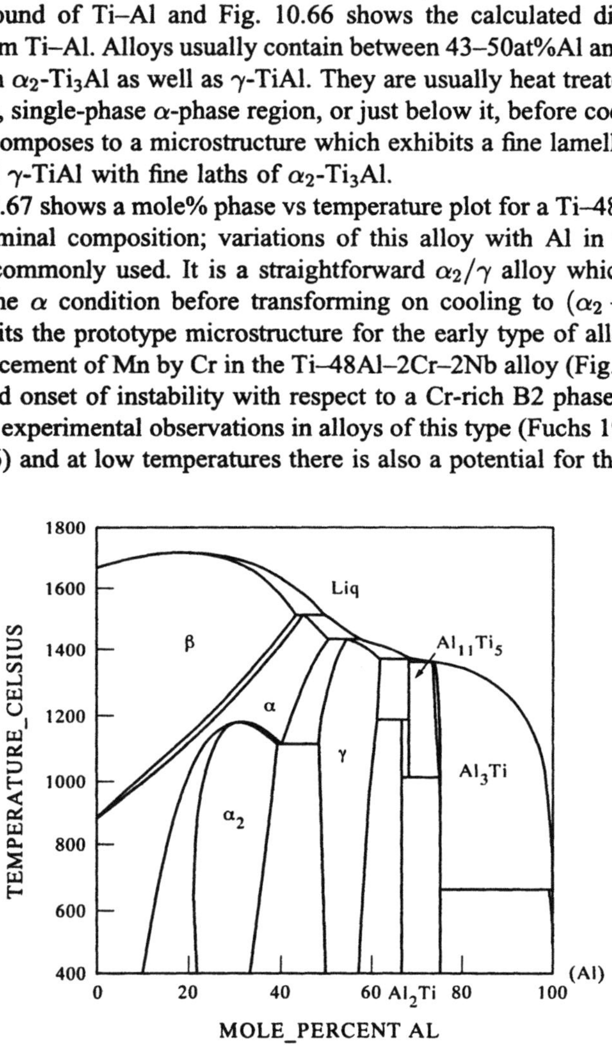

Figure 10.66 Calculated Ti–Al phase diagram.

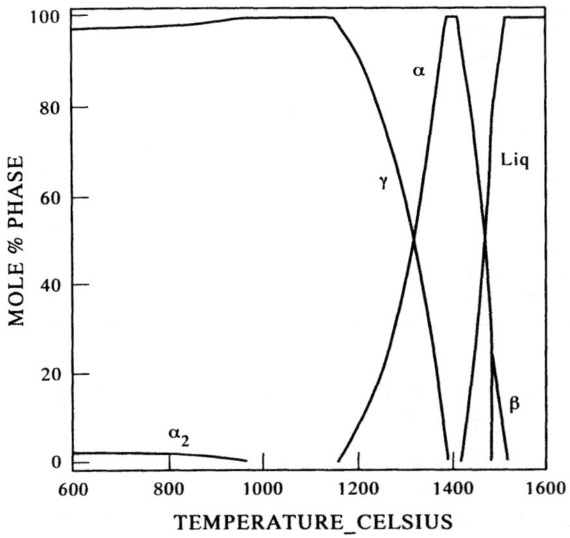

Figure 10.67 Calculated mole % phase vs temperature plot for a Ti–48Al–2Mn–2Nb alloy.

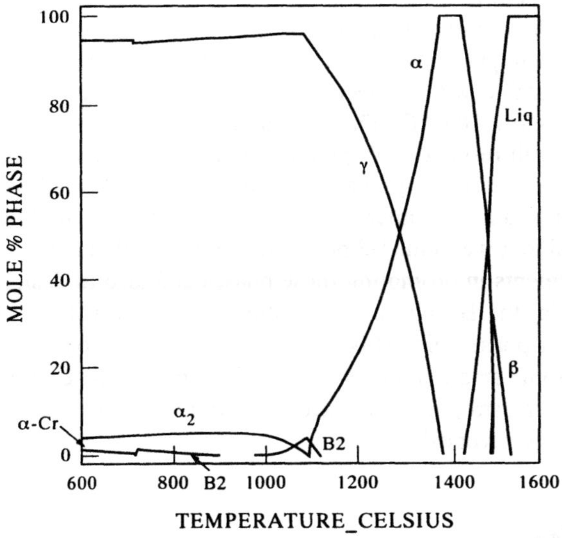

Figure 10.68 Calculated mole % phase vs temperature plot for a Ti–48Al–2Cr–2Nb alloy.

Table 10.2. Some recent γ-TiAl-based alloys with comparison between experimentally observed and calculated α-transus temperatures

Alloy	α-transus (exp.)	α-transus (calc.)	Reference
Ti–48Al–2Cr–2Nb	1364	1357	Fuchs (1993)
Ti–49.5Al–1.1Mn–2.5Nb	1436	1425	Lombard *et al.* (1992)
Ti–47Al–2.5Nb–1Cr–1V	1350	1357	Kim (1992)
Ti–46.7Al–3Nb–1W	1350	1353	Bhowal *et al.* (1995)
Ti–48Al–2Nb–2Cr–1W	1358	1371	Fuchs (1995)
Ti–47Al–1.5Cr–0.5Mn–2.6Nb–0.15B	1365	1367	Kim (1995)
Ti–47Al–2Nb–1Mn–0.5W–0.5Mo–0.2Si	—	1366	Seo *et al.* (1995)

almost pure Cr. Newer alloys have centred around the substitution and addition of elements to the 48–2–2-type alloys and have compositions of the type shown in Table 10.2, which also gives comparisons for experimentally observed and calculated α-transus temperatures.

It is noticeable that heavy elements such as W and Ta are now being increasingly used and that alloys are subsequently becoming more prone to formation of the B2 phase. This is seen for example in alloys such as Ti–48Al–2Nb–2Cr–1W, and their variants (Fuchs 1995). One of the most complex alloys developed so far is Ti–47Al–2Nb–1Mn–0.5W–0.5Mo–0.2Si. However, although truly multi-component in nature, the total refractory metal addition to a basic Ti–Al alloy is still only 4at%. The microstructure of this alloy consists of second-phase particles enriched in Mo and W and there is also evidence for silicides (Seo *et al.* 1995). Figure 10.69 shows the calculated phase % vs temperature plot for this alloy; the alloy is predicted to be unstable with respect to a B2 phase enriched in Mo and W, and Ti_5Si_3 will form below 1020°C. The calculation also predicts that the alloy may become unstable with respect to W precipitation at low temperatures.

It is clear that as these alloys have become more highly alloyed, the simple prototype concept of an α_2/γ microstructure is having to be modified to take into account borides, silicides and the B2 phase. It is now possible to understand the role of the various elements in promoting these phases and also to design with them in mind. For example, the B2 phase can inhibit grain growth if it is present in the high-temperature α-phase field. This appears somewhat analogous to conventional Ti alloys, where some small amounts of α in β grain boundaries during processing and help prevent catastrophic grain growth during forging, but the role of the phases is completely reversed.

10.6.5 Alloy design

Most of the previous examples show that CALPHAD calculations can enhance knowledge of the general phase behaviour of alloys when a suitable thermodynamic

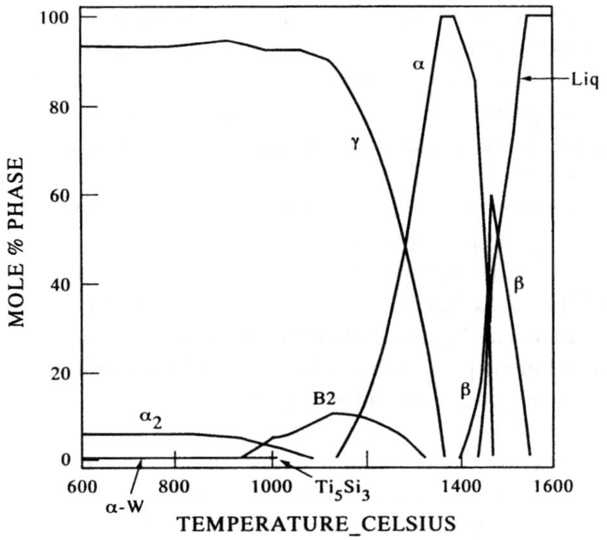

Figure 10.69 Calculated mole % phase vs temperature plot for a Ti–47Al–2Nb–1Mn–0.5W–0.5Mo–0.2Si alloy.

database is available. However, it is also interesting to look at some CALPHAD-based work where the main aim has been to help design or optimise some specific feature, or combinations of features, which are affected by phase behaviour. Such features may be associated with a physical property such as modulus and magnetism, a mechanical property such as strength, a chemical property such as corrosion resistance or some more complex combination which might effect a property such as weldability. The best examples centre around Fe-based materials, which probably relates to historical reasons as Fe-based systems were one of the first types of alloy to be thermodynamically characterised in any great detail. However, it is clear that once good thermodynamic characterisations are available the use of CALPHAD methods need not be limited to any particular material type (Miodownik 1993).

10.6.5.1 Magnetic materials. The value of CALPHAD calculations for magnetic alloys has been demonstrated for Fe–Cr–Co-based alloys, where alloys with improved magnetic properties were made by designing their composition and heat-treatment temperature to lie in a miscibility caused by magnetic forces (Nishizawa *et al.* 1979a, Homma *et al.* 1981, Ishida and Nishizawa 1991). Inden (1987) has described in detail the background to the production of such hard magnets which can be summarised as follows.

A good design for a hard magnetic material is to have small, elongated particles, aligned and magnetised along their axis, embedded in a paramagnetic matrix. One of the easiest ways to produce such a microstructure is by some form of precipitation reaction, where the size of the particles can be controlled by heat treatment. This then leaves the problem of producing the requisite shape and alignment of the particles which leads to a further prerequisite that the precipitate/matrix interface must have a low energy. It is not always possible to produce 'ideal' rod shapes, but precipitates with a more plate-like morphology can also be aligned by applying a magnetic field during the precipitation process.

For such materials it is desirable for there to be a miscibility gap due to magnetism (Fig. 10.70(a, b)) as described in Chapter 8. In these circumstances, spinodal decomposition can occur by the formation of paramagnetic and ferromagnetic phases. The best magnetic properties can be produced if annealing is performed just below the temperature where instability begins.

Figure 10.71 shows a calculated isothermal section for Fe–Cr–Co at 1200 K, showing how the miscibility gap in the binary systems is expanded in the ternary. It should be noted that these sections are metastable, as the σ phase would precipitate out in equilibrium. However, if heat treatments are done for sufficiently short times, spinodal decomposition occurs without precipitation of σ. The effect of the expansion of the miscibility gap in the ternary means that higher heat-treatment temperatures can be used, allowing a material with a higher Curie temperature and subsequently higher saturation magnetisation to be produced. Heat treatments for these alloys can be sensitive to small changes in composition. Figure 10.72 shows a vertical section in Fe–Co–Cr at a ratio of $x_{Fe}/x_{Co} = 83/17$ and it can be seen that only a few at% differentiates alloys with a paramagnetic or a ferromagnetic matrix.

Al–Ni and Al–Ni–Co magnets work on identical principles to the Fe–Cr–Co magnets but, instead of considering a miscibility gap between paramagnetic and ferromagnetic phases which are disordered, the paramagnetic phase is, in this case, ordered. Figures 10.73(a, b) show an isothermal and a vertical section through the Fe–Ni–Al ternary system produced by a combination of experiment and calculation (Hao *et al.* 1984). Additions of Co were made to this basic system and, in accord with thermodynamic calculations (Nishizawa *et al.* 1983), both the peak temperature of the miscibility gap and the region of immiscibility in the quaternary are increased.

Good magnetic properties in these alloys are obtained under the following conditions: (1) The ferromagnetic α_1 phase is present as isolated particles embedded in a weakly magnetic or non-magnetic matrix of α_2; (2) the particles are uniaxially aligned; and (3) the volume fraction of α_1 is in the range 0.4–0.6

Unfortunately, Al–Ni and Al–Ni–Co magnets suffer from the disadvantage of being hard and brittle, leading to problems with machining which becomes expensive and difficult. Using a mainly experimental phase-diagram approach, Ishida *et al.* (1991) were able to design an alloy which contained a small amount of ductile f.c.c. γ phase distributed along the grain boundaries of the alloy. This

References are listed on pp. 402–408.

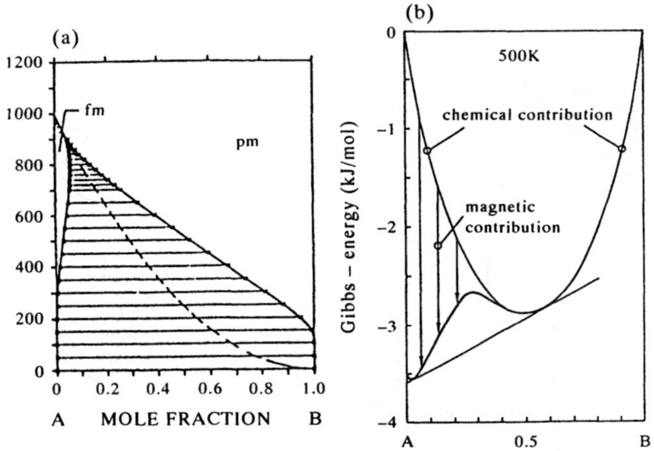

Figure 10.70 (a) Phase diagram for an A–B alloy system exhibiting phase separation due to magnetism and (b) underlying Gibbs energy vs composition curve showing the contribution from chemical and magnetic energies (from Inden 1987).

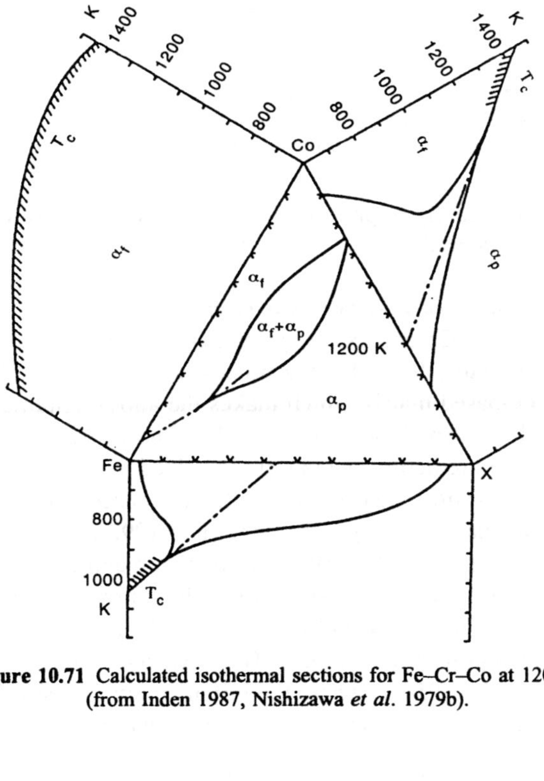

Figure 10.71 Calculated isothermal sections for Fe–Cr–Co at 1200 K (from Inden 1987, Nishizawa *et al.* 1979b).

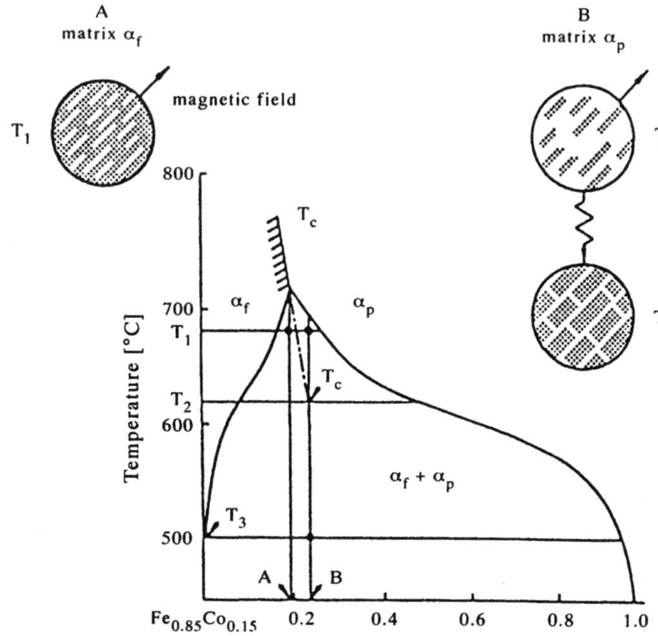

Figure 10.72 Calculated vertical section through the Fe–Cr–Co system at a constant ratio x_{Fe}:x_{Co} of 83:17 (from Inden 1987, Homma *et al.* 1981). T_c is the Curie temperature.

microstructure significantly improves hot workability and leads to the possibility of producing ductile material.

10.6.5.2 Rapidly solidified in-situ metal matrix composites. A design project for alloys based on the Fe–Cr–Mo–Ni–B system, and produced by rapid solidification, was undertaken by Pan (1992). During processing a mixture of borides is formed inside a ductile Fe-based matrix which makes the alloys extremely hard with high moduli. These alloys provide a good example of how phase-diagram calculations were able to provide predictions which firstly helped to identify unexpected boride formation (Saunders *et al.* 1992) and were ultimately used in the optimisation of the modulus of a shaft material for gas turbines (Pan 1992).

The alloys are first produced by rapid solidification and are amorphous in nature. They are either directly fabricated as powders, by a process such as high-pressure gas atomisation (HPGA), or by melt-spinning of ribbons, which are subsequently pulverised to form a powder (<150 μm). The powders are then consolidated by hot extrusion between 950–1050°C where the initial amorphous structure breaks down and forms a fine dispersion of stable borides in a ductile Fe-based matrix.

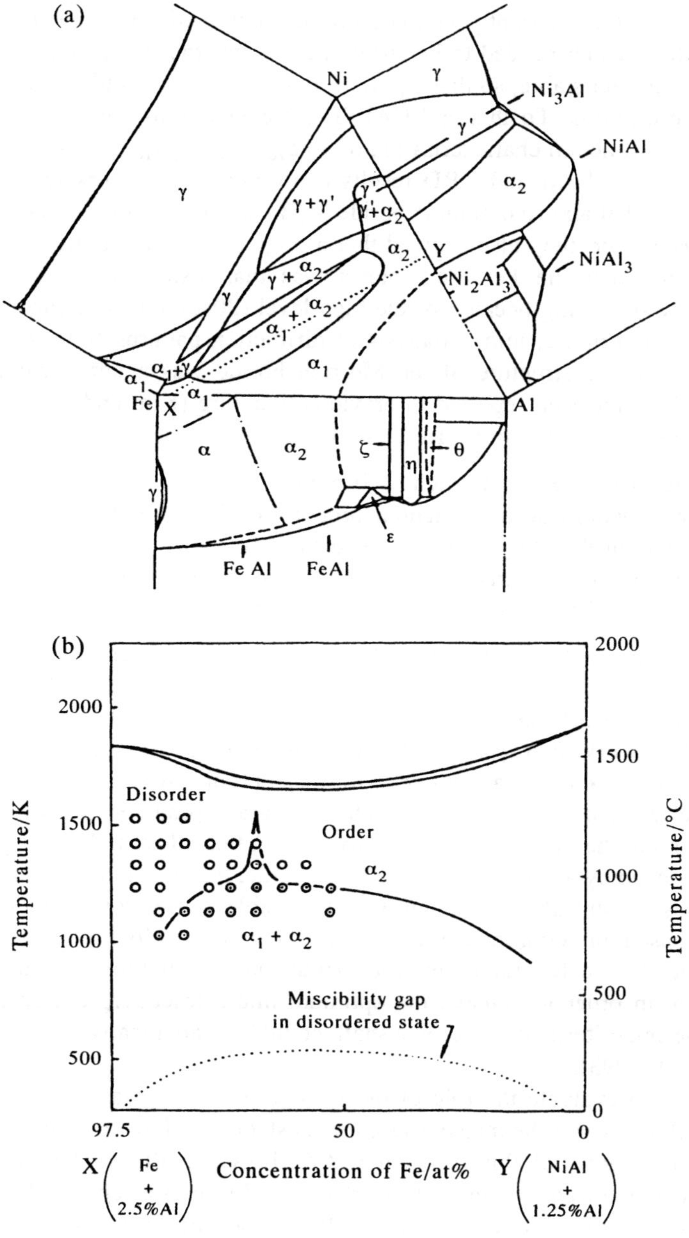

Figure 10.73 (a) Isothermal and (b) vertical sections from the Ni–Fe–Al phase diagram (Ishida and Nishizawa 1991).

Because of the inherently non-equilibrium nature of the production route, the first question which needed to be answered was whether the phases present in the alloy were in fact stable, so that equilibrium calculations could actually be used to design these alloys. To this end CALPHAD calculations were combined with a detailed experimental characterisation of a $Fe_{70}Cr_{18}Mo_2B_{10}$ alloy (Kim *et al.* 1990, Pan 1992). The TEM and XRD results confirmed earlier work (Xu *et al.* 1985) which stated that an orthorhombic boride M_2B was present and its composition was Cr-rich. However, they also showed that a proportion of the borides (~10%) were Mo-rich and that the Fe-based matrix was martensitic. The latter result was particularly surprising because of the high level (20at%) of α-ferrite stabilisers Cr and Mo. Furthermore, initial analysis of diffraction patterns from the TEM work indicated that the structure of the Mo-rich boride was a tetragonal type whose structure had not been reported in previous literature (Kim and Cantor 1988).

Calculated phase equilibria for this alloy produced the following predictions:

(1) The existence of a Cr-rich M_2B phase.
(2) There would be a substantial proportion (7.3%) of the U_3Si_2-type, M_3B_2 boride in the alloy with high levels of Mo.
(3) The Fe-based matrix would be austenitic at 1000°C but transform to ferrite below 857°C, thus giving a reason for the observed martensitic structure.

Based on the prediction for the M_3B_2 boride, and its observed lattice parameter variation with Fe, Cr and Mo levels, the structure of the Mo-rich phase was re-evaluated and clearly shown to be primitive tetragonal of the U_3Si_2-type (Kim *et al.* 1990). This work also gave results for ratio of Fe:Cr:Mo in the various phases which were in excellent agreement with those predicted at 1000°C.

It is clear that the input of the phase-diagram predictions greatly helped in understanding the evolution of microstructure in this alloy, and although the alloys were produced by a highly non-equilibrium route, the calculations also showed that the phases present after extrusion were the stable phases for the alloy, so design criteria based on equilibrium calculations could therefore be used. A further advantage of the calculation route was that the number of alloys which needed to be examined, in order to achieve the optimum microstructure/property combination for the design criteria of the turbine shaft, could be dramatically reduced (Pan 1992, Miodownik 1993).

In order to evaluate the role of the various phases, it was necessary to design distinct alloys where the matrix was either austenitic or ferritic. Figure 10.74 shows the principle by which this was achieved for Fe–Cr–B alloys where it can be seen that only a small change in Cr level puts an alloy into a region where it is fully ferritic at consolidation temperatures. Alloys from the Fe–Cr–Mo–B system were then designed so that the matrix was fully ferritic and a further set of alloys from the Fe–Cr–Ni–B system which were fully austenitic. Although producing stable microstructures, significant drops in strength and hardness were found in alloys

References are listed on pp. 402–408.

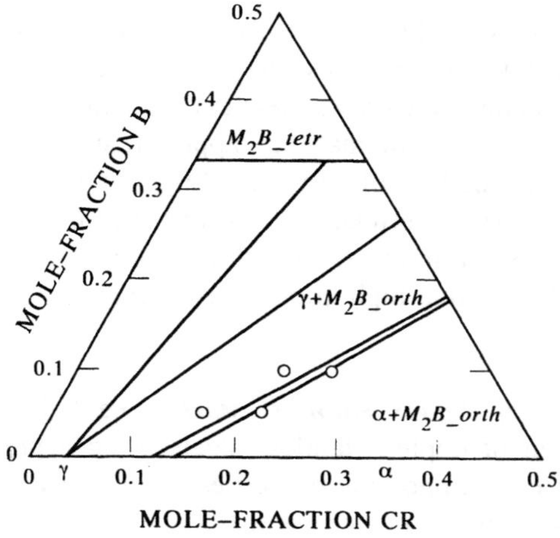

Figure 10.74 Calculated isothermal for Fe–Cr–B at 1000°C showing placement of high-B alloys which have either α or γ matrices at processing temperature (Pan 1992).

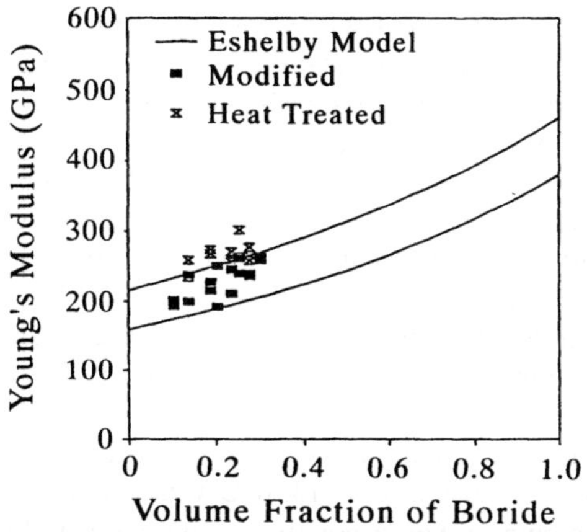

Figure 10.75 Comparison of calculated Eshelby curves for modulus with experimentally determined moduli for high-B ferrous base alloys (Pan 1992).

whose matrix was either fully ferritic or austenitic. It also became clear that small amounts of C were present in all alloys which was adding to the hardness of the martensite, hence enhancing its properties.

The next step was to optimise the ratio of M_3B_2 to MB_2 and various alloys were designed and tested to improve factors such as strength and modulus (Pan 1992). Eshelby analysis (Eshelby 1957, Withers *et al.* 1989) was used to predict modulus and good agreement was found with experimental measurement (Figure 10.75). It was clear that modulus was significantly enhanced by increasing boride level but owing to morphological considerations (the borides were not aligned and had low aspect ratios) the modulus increase was less than was hoped for. In the end, alloys were designed to have martensitic matrices, approximately equal volumes of M_3B_2 and MB_2 and predicted modulii of between 245–255 GPa. All of these properties were achieved in the final experiments . One of the limiting factors in looking at higher amounts of M_3B_2, which were more finely dispersed and, hence, potentially gave greater strength, was that high liquidus temperatures made it difficult to melt such alloys.

10.6.5.3 The design of duplex stainless steels. Lee (1995) reported on work relating to the design of duplex stainless steels using a predominantly thermodynamic approach. Four important properties were chosen as being critical to performance, and these were: (1) strength, (2) toughness, (3) weldability and (4) corrosion resistance. Of these, strength was considered to be the least critical in the alloy design process, as this was heavily controlled by factors such as work hardening and grain size, and would be more effectively controlled by thermomechanical processing rather than design of composition. Lee (1995) then identified critical factors associated with phase equilibria which would affect the other three properties. These were the stability of austenite at high temperatures for weldability, sigma formation for toughness and the composition of austenite in local equilibrium with $M_{23}C_6$ for corrosion resistance. These factors could then be combined to give an overall performance criterion. The rationale for the choice of favourable thermodynamic properties was based on the following arguments.

Weldability. It had been previously reported (Norstrom *et al.* 1981) that an increase in austenite amount was beneficial to weldability. Further, Cao and Hertzman (1991) had reported that loss of impact toughness because of ferrite formation during welding could be recovered by re-formation of austenite at lower temperatures. Lee (1995) subsequently concluded that increased stability of austenite above the heat-treatment temperature would be beneficial to weldability. Therefore, the amount of austenite at 1350°C was taken as a criterion for alloy design, higher values being given a positive rating compared to lower values.

Sigma formation. While σ-phase formation is known to cause embrittlement in duplex stainless steels, Lee (1995) argued that the important factor in the process was the driving force for precipitation rather than the potential amount of σ which might be formed. This was because only small amounts are actually necessary to cause embrittlement (Norstrom *et al.* 1981) and, therefore, the early stage of precipitation should be considered more important than the final equilibrium state. While nucleation is a complex process, Lee (1995) suggested that, for the purpose of alloy design, nucleation parameters for σ such as interfacial energies, misfit strain energies, etc., would be approximately equal for alloys which lay close in composition to the original alloy. This would mean that the driving force for

Table 10.3. Composition of candidate alloys and results of calculations on toughness, weldability and corrosion resistance. Calculation was conducted assuming an annealing temperature of 1050°C, 0.8%Mn, 0.7%Si and 0.024%C (wt%). PRE value in the heading of the table indicates value for SAF2205 (from Lee 1995)

Alloy composition (wt%)				Toughness	Weldability	Corrosion resistance	
N	Ni	Mo	Cr	ΔG for σ	High temp. austenite %	Cr content in austenite	PRE (37.28)
0.14	7.5	0	27.15	X	—	—	
0.14	7.5	1	26.21	X	—	—	
0.14	7.5	2	25.27	X	—	—	
0.14	7.5	3	24.33	X	—	—	
0.14	7.5	4	23.40	X	—	—	
0.14	6.5	0	25.93	X	—	—	
0.14	6.5	1	24.95	X	—	—	
0.14	6.5	2	23.97	X	—	—	
0.14	6.5	3	22.99	X	—	—	
0.14	6.5	4	22.01	√	√ (3.2)	X	41.55
0.14	5.5	0	24.69	X	—	—	
0.14	5.5	1	23.67	√	X (0.3)	√	32.80
0.14	5.5	2	22.64	√	X (0.8)	√	35.08
0.14	5.5	3	21.62	√	X (1.4)	X	37.36
0.14	5.5	4	20.60	√	X (1.9)	X	39.63
0.14	4.5	0	23.43	√	XXX (0.0)	√√√	28.76
0.14	4.5	1	22.37	√	XX (0.0)	√√	31.00
0.14	4.5	2	21.29	√	X (0.0)	√	33.23
0.14	4.5	3	20.21	√√	X (0.2)	√	35.45
0.14	4.5	4	19.13	√√√	X (1.0)	X	37.67
0.20	7.5	0	28.47	X	—	—	
0.20	7.5	1	27.54	X	—	—	
0.20	7.5	2	26.61	X	—	—	
0.20	7.5	3	25.69	X	—	—	
0.20	7.5	4	24.77	X	—	—	
0.20	6.5	0	27.24	X	—	—	
0.20	6.5	1	26.27	X	—	—	
0.20	6.5	2	25.31	X	—	—	
0.20	6.5	3	24.34	X	—	—	
0.20	6.5	4	23.38	X	—	—	
0.20	5.5	0	25.99	X	—	—	
0.20	5.5	1	24.99	X	—	—	
0.20	5.5	2	23.98	√	√ (6.5)	√	38.22
0.20	5.5	3	22.97	√	√√ (7.2)	√	40.51
0.20	5.5	4	21.97	√	√√√ (7.8)	√	42.80
0.20	4.5	0	24.73	√	√ (3.6)	√√√	31.87
0.20	4.5	1	23.68	√	√ (4.6)	√√	34.12
0.20	4.5	2	22.63	√	√ (5.4)	√	36.37
0.20	4.5	3	21.57	√	√ (6.3)	√	38.61
0.20	4.5	4	20.51	√√	√√ (7.0)	√	40.85

nucleation would effectively be the dominating criteria for the initial precipitation.

Corrosion resistance. The main criterion here was judged to be the level of Cr in the matrix phase and the avoidance of Cr depletion in the austenite which would be controlled by the precipitation of $M_{23}C_6$ in the early stages. A criterion was therefore placed on alloy design where high levels of Cr in the austenite phase in 'local equilibrium' with $M_{23}C_6$ would be given a positive rating.

It is also desirable for the alloy to have as high a Pitting Resistance Equivalent (PRE) value as possible and Lee (1995) calculated this number using the formula

$$\text{PRE} = \text{wt\%Cr} + 3.3\text{wt\%Mo} + 30\text{wt\%N} + 0.5\text{wtNi} - 0.5\text{wt\%Mn} - 30\text{wt\%C}.$$

Lee (1995) then constructed a matrix of 40 candidate alloys, close to the SAF2205 composition of Fe–22Cr–5.5Ni–3Mo–1.7Mn–0.4Si–0.14N–0.025C (in wt%) which are given in Table 10.3. Calculations were then made to establish the magnitude of the various criteria described previously and a tick ($\surd$) is shown where improvement was predicted over SAF2205, (X) where there would be a deterioration in the property and (—) denotes that this property was not calculated. On the basis of this approach it is clear that five alloys with high N would give an improvement in all three criteria and also give a higher PRE number.

10.6.5.4 Design of high-strength Co–Ni steels. Grujicic *et al.* (1987) and Grujicic (1991) have developed a procedure to design ultra-high-strength alloys based on M_2C-hardened Co–Ni steels. They set a design criterion for producing an alloy with good fracture toughness (K_{IC}~70–130 MPa$\sqrt{\text{m}}$) and an ultimate tensile strength (UTS) level between 2100 and 2400 MPa. They began with a commercial alloy AF1410 which, although falling far short of their ultimate goal, served as a starting point for their design programme. The composition of AF1410 is Fe–14Co–10Ni–2Cr–1Mo–0.15C (in wt%). It is primarily strengthened by M_2C, where M=Cr and Mo, and peak hardness is found on tempering at 510°C for 5 hrs.

Their initial aims were to assess the role played by Co on the carbide phases and to optimise the Cr, Mo and W levels subject to the requirements of a balanced composition where $x_{\text{Mo}} + x_{\text{Cr}} + x_{\text{W}} = 2x_{\text{C}}$. Further, constraints were to limit the C level at 0.25wt% and the nickel level at 10wt%.

The first part of the work was to calculate the driving force for the precipitation of M_2C from their alloys, as this is the factor which would govern its precipitation behaviour in the initial stages. They did this for an alloy with the fixed-base composition Fe–14Co–10Ni–0.25C, at the proposed heat-treatment temperature of 783 K, and varied the W, Mo and Cr levels using the formula given in the previous paragraph. Results were plotted in a ternary format (Fig. 10.76(a)) and show how the greatest driving force is found for Mo-rich alloys with the relative effect of various elements on ΔG being in the order Mo>Cr>W. As carbides other than M_2C would also form in their alloys, they plotted driving forces to form $M_{23}C_6$ and M_6C, which are shown in Figs. 10.76(b, c). It can be seen that the lowest driving forces to form these carbides were in Mo-rich alloys and, hence, such alloys would maximise the yield of M_2C.

Grujicic (1991) then made a series of calculations to estimate the M_2C coarsening rate in these types of alloys. This is important because metastable cementite is often a predecessor precipitate to M_2C and deleterious to mechanical properties. The cementite can be removed by prolonging the tempering treatment, but this causes subsequent coarsening of the desirable M_2C. Using simple coarsening theory, they were able to determine that W-rich alloys would have the greatest resistance to

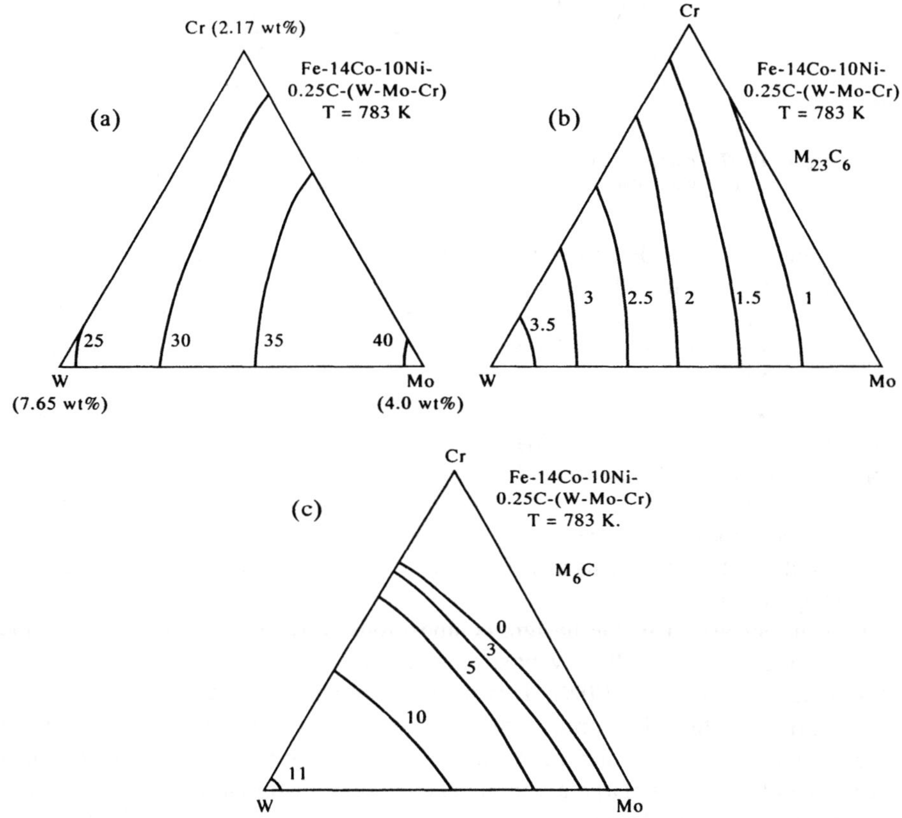

Figure 10.76 Variation of the driving force (kJ mol^{-1}) for precipitation of (a) M_2C, (b) $M_{23}C_6$ and (c) M_6C from ferrite at 783 K in a Fe–14Co–10Ni–0.25C–(W–Mo–Cr) alloys with balanced alloy additions of Mo, Cr and W (Grujicic *et al.* 1987).

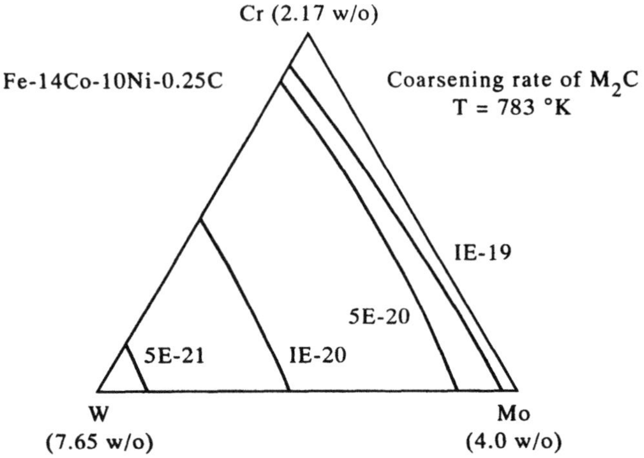

Figure 10.77 Coarsening rate of M_2C at 783 K in a Fe–14Co–10Ni–0.25C–(W–Mo–Cr) alloys with balanced alloy additions of Mo, Cr and W (Grujicic 1991).

coarsening; their calculated results are shown in Fig. 10.77, again using ternary axes. The design of the alloy now becomes a subtle balance of increasing W levels to enhance the coarsening resistance of M_2C, but keeping Cr and Mo levels sufficiently high to minimise the potential for precipitation of the other carbides, $M_{23}C_6$ and M_6C.

The general design process for these alloys was extended by Grujicic and Olson (1988) to consider how ferrite/M_2C equilibria could be adjusted by considering coherency strains at the matrix/carbide interface during the early stages of precipitation. Their calculations indicated that Fe would substitute for Cr in the case of coherent equilibrium and that the C stoichiometry would be shifted significantly from an ideal M_2C ratio. This affects coarsening resistance as well as driving force for precipitation.

In a further aspect of the design, compositions were adjusted to maximise the fracture toughness of the alloy by maximising its transformation toughening behaviour (Haidemenopoulos *et al.* 1989). This toughening occurs because of deformation-induced martensite which forms at the crack tip as it moves through the dispersion of austenite which is retained in the alloy. The design criteria used here were that there should be both as large a volume change on transformation to martensite as possible and also that the driving force to form martensite should be as high as possible in the retained austenite. Magnetic Gibbs energy contributions, as well as those due to composition, were taken into account and it was concluded that deformation-induced transformation toughness would be maximised by moving from the 'ideal' AF1410 composition to an alloy with significantly higher levels of Ni.

10.6.6 Slag and slag–metal equilibria

10.6.6.1 Matte–slag–gas reactions in Cu–Fe–Ni sulphide ores. Sulphide ores are a major source of Cu, Ni and precious metals. A basic principle of the extraction processes is to blow air into the molten sulphide in order to oxidise (1) S, which forms a gas and (2) Fe, which forms predominantly FeO and then partitions to a slag phase which covers the matte. A key element in the recovery of the metals is the solidification of the matte which separates into a sulphur-rich matte and metal-rich liquid. This process may occur under non-equilibrium conditions with precious metals concentrating in the last metallic liquid.

To obtain a better understanding of the process, calculations were performed by Dinsdale *et al.* (1988) and Taylor and Dinsdale (1990) for a pre-fused matte of Ni_3S_2, Cu_2S and FeS, heated to around 1270°C with an equivalent amount of oxide slag, and with O being blown into the matte. Calculations from the model system Cu–Fe–Ni–S–O are presented in Fig. 10.78 which shows the comparison between calculated and experimental values of the Fe/S partitioning in the matte as a function of SO_2 levels.

Calculations were then made for the amount of Cu and Ni levels in the slag as a function both of Fe in the matte and SO_2 levels. This showed that partitioning was

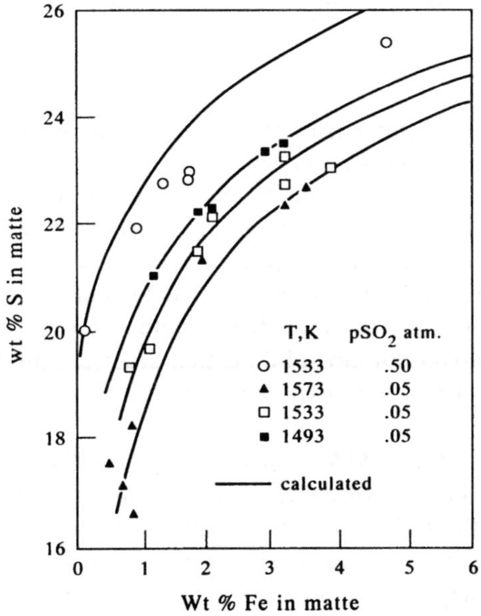

Figure 10.78 Comparison between calculated and experimental matte compositions (Ni:Cu wt ratio 2:1) at different temperatures and SO_2 partial pressures (from Taylor and Dinsdale 1990).

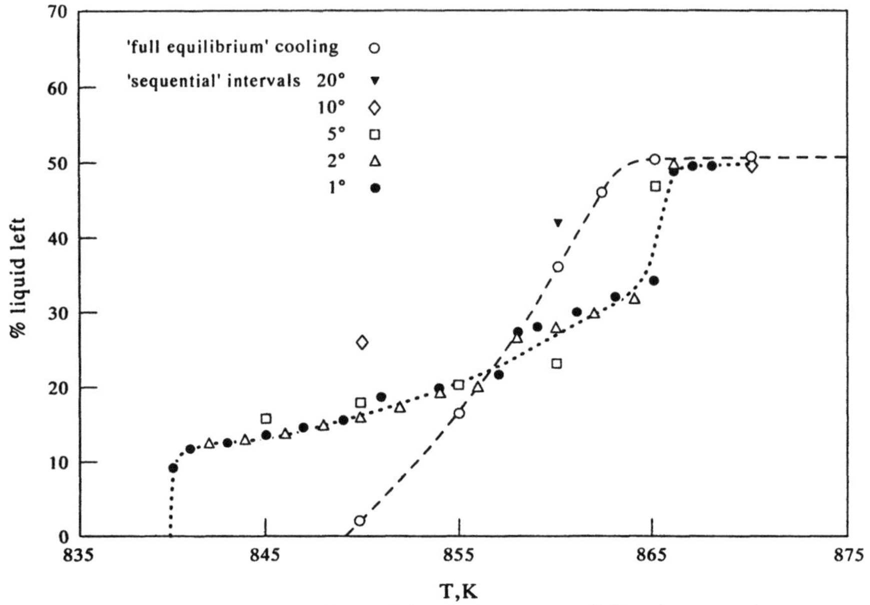

Figure 10.79 Percent liquid remaining during matte solidification under both equilibrium and 'Scheil–Gulliver' solidification conditions (from Taylor and Dinsdale 1990).

insensitive to SO_2 levels and it was only necessary to maintain significant levels of Fe in the matte to ensure that Cu and Ni uptake in the slag was minimised. The solidification of the matte was then modelled both when solidification occurred by an equilibrium process and alternatively in a non-equilibrium fashion, diffusion in the solid being considered negligible (the so-called 'Scheil–Gulliver' conditions, see Chapter 11). Figure 10.79 shows the calculated results and it is evident that the liquid is retained to lower temperatures than would occur by equilibrium freezing, and the model for non-equilibrium solidification predicted the formation of a new phase, which was consistent with experimental information.

10.6.6.2 Calculation of sulphide capacities of multi-component slags. The thermo-dynamics and kinetics of desulphurisation are of great importance to iron and steel making and the ability to predict the behaviour of sulphur associated with multi-component slags is, therefore, very desirable. To this end Pelton *et al.* (1993) recently described an approach to predict the sulphide-removing capacity in multi-component oxide slags. While a comprehensive database for oxides was already available, the inclusion of S had not yet been undertaken. Therefore it was not possible to calculate the necessary S activity and solubility directly through a

CALPHAD calculation. To overcome this difficulty they combined a semi-empirical model from Reddy and Blander (1987, 1989) with an equilibrium CALPHAD calculation for the multi-component oxide system SiO_2–Al_2O_3–TiO_2–CaO–MgO–MnO–FeO. The approach can be summarised as follows.

The exchange of O and S between an oxide slag and other phases can be written in general terms as

$$A_nO_{(\mathrm{liq})} + \frac{1}{2}S_{2\,(\mathrm{gas})} = A_nS_{(\mathrm{liq})} + \frac{1}{2}O_{2\,(\mathrm{gas})} \tag{10.2}$$

where A_nO and A_nS are components of a slag, with A being a particular cation. An equilibrium constant can be defined for this reaction as

$$K_A = \frac{a_{A_nS}}{a_{A_nO}}\left(\frac{P_{\mathrm{O_2}}}{P_{\mathrm{S_2}}}\right)^{1/2} = \exp\left(\frac{-G_A^\circ}{RT}\right) \tag{10.3}$$

where a_{A_nO} and a_{A_nS} are the activities in the slag and G_A° is the standard Gibbs energy of reaction. As the amount of S in solution is small, it was assumed that Henry's law could be applied such that a_{A_nS} varies directly as the amount of dissolved S. Further, it was assumed that for a given oxide slag composition, a_{A_nO} would be nearly equal to its value in the sulphur- free slag. From this the sulphide capacity of a slag (C_S) can be derived (Fincham and Richardson 1964) as

$$C_S = (\mathrm{wt\%S})\left(\frac{P_{\mathrm{O_2}}}{P_{\mathrm{S_2}}}\right)^{1/2} \tag{10.4}$$

where (wt%S) is the amount of dissolved sulphur. The sulphide capacity for any given slag will be constant as long as the dissolved S is relatively low and, therefore, the higher the sulphide capacity, the higher the S content of the slag.

Equation (10.4) relies on a knowledge of both the sulphide and oxide activities. However, S was not yet included in their database for oxides. They therefore utilised the approach of Reddy and Blander (1987, 1989) to relate a_{A_nS} to the S content and sulphide capacities could then be predicted through calculation of a_{A_nO} alone. The approach yields the following expressions for sulphide capacities for basic and acid slags:

Basic slags $x_{\mathrm{SiO_2}} \leq 1/3$

$$\left(\frac{C_S}{C_S^\circ}\right) = a_{A_nO}\frac{(1 - 2x_{\mathrm{SiO_2}})}{(x_{A_nO}M_{A_nO} + x_{\mathrm{SiO_2}}M_{\mathrm{SiO_2}})} \tag{10.5}$$

Acid slags $x_{\mathrm{SiO_2}} \geq 1/3$

$$\left(\frac{C_S}{C_S^\circ}\right) = a_{A_nO}\frac{x_{\mathrm{SiO_2}}}{(x_{A_nO}M_{A_nO} + x_{\mathrm{SiO_2}}M_{\mathrm{SiO_2}})}\cdot\left(\frac{1 - x_{\mathrm{SiO_2}}}{2x_{\mathrm{SiO_2}}}\right)^2 \tag{10.6}$$

and

$$C_S^o = \frac{1000\ M_S K_A}{\gamma^o_{A_nS}} \tag{10.7}$$

where x and M denote, respectively, the moles and molecular weights of the various components. The Henrian activity coefficient of A_nS, $\gamma^o_{A_nS}$, is then assumed to be equal to 1, and if K_A for each component is known, the sulphide capacity can be calculated for each oxide component. Values of K_A were obtained from assessed thermodynamic values. The various values of C_S for each oxide component were then obtained from the following equation:

$$\log C_S = y_A \log C_{S(A)} + y_B \log C_{S(B)} + y_C \log C_{S(C)} + \cdots \tag{10.8}$$

where $\log C_{A(S)}$ is the value of $\log C_S$ in the binary A_nO–SiO_2 at the same value of x_{SiO_2} as in the multi-component system and

$$y_A = \frac{x_{A_nO}}{(x_{A_nO} + x_{B_nO} + x_{C_nO} + \cdots)}. \tag{10.9}$$

The approach yielded excellent results and Figs 10.80–10.83 show comparison of calculated and experimentally determined sulphide capacities in various multi-component slags.

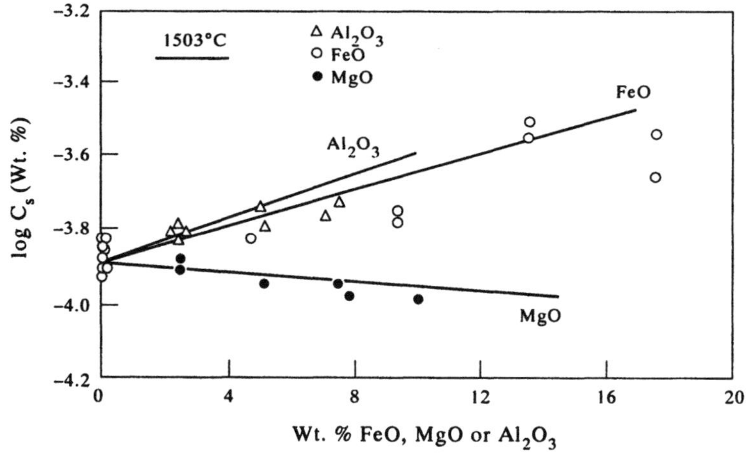

Figure 10.80 Comparison between calculated and experimental (Bronson and St-Pierre 1981) effects of Al_2O_3, FeO and MgO additions upon sulphide capacities of CaO–SiO_2 slags at a constant $(CaO+AO)/(SiO_2+Al_2O_3)$ ratio, where A = Fe or Mg.

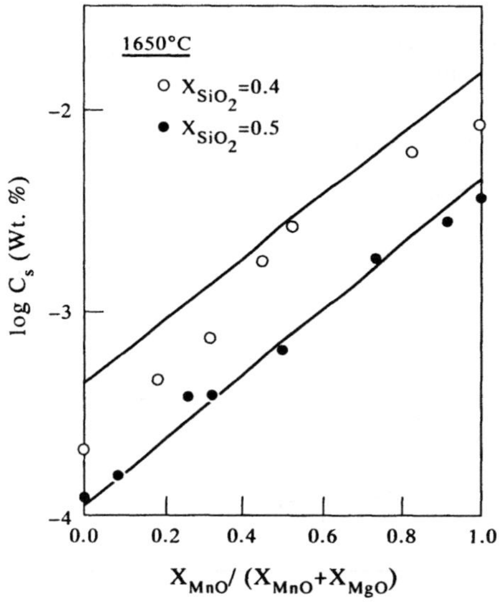

Figure 10.81 Comparison between calculated and experimental (Sharma and Richardson 1965) sulphide capacities at constant SiO_2 fraction in MnO–MgO–SiO_2 slags.

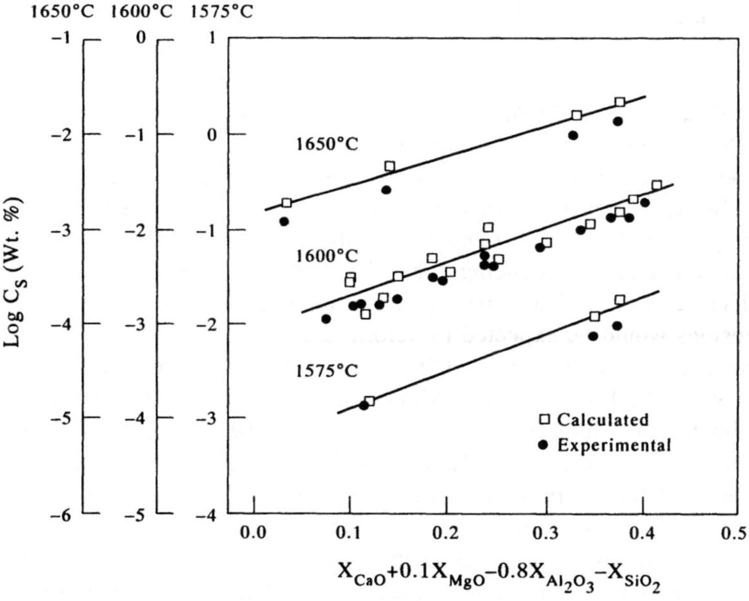

Figure 10.82 Comparison between calculated and experimental (Bronson and St-Pierre 1981) sulphide capacities in SiO_2–Al_2O_3–MgO–CaO–FeO slags.

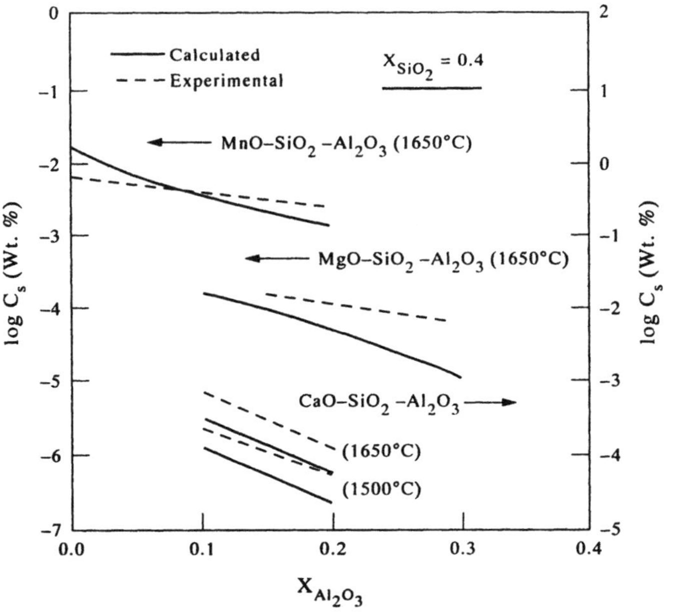

Figure 10.83 Comparison between calculated and experimental (Fincham and Richardson 1954, 1964, Sharma and Richardson 1965) sulphide capacities in $AO–SiO_2–Al_2O_3$ slags (where A = Mn, Mg, Ca) at a constant SiO_2 fraction of 0.4.

10.6.6.3 Estimation of liquidus and solidus temperatures of oxide inclusions in steels. The deformation of inclusions in steels has significant consequences on the hot workability of steels as well as for the mechanical properties of the final product. In order to increase their deformability there are at least three strategies (Matsumiya *et al.* 1987): (1), Reduction of their melting point; (2), deceleration of crystallisation; and (3), reducing their flow stress. If the melting point can be reduced sufficiently so that some liquid is present at the hot-working temperature, the inclusions would be expected to deform easily.

To this end, Matsumiya *et al.* (1987) made calculations for inclusions from the quinary system $SiO_2–Al_2O_3–CaO–MgO–MnO$, based on assessments of the component binary and ternary sub-systems. Figures 10.84(a, b) show two of the calculated ternary diagrams and Table 10.4 shows comparisons between calculated and experimentally determined liquidus temperatures for five oxide inclusions.

Figures 10.85(a, b) show phase fraction plots for inclusions 1 and 2 in Table 10.4. These are plotted in such a way as to show the cumulative amount of all phases as well as their individual amounts. A quasi-ternary diagram was then plotted for an 'ideal' inclusion with a fixed level of $Al_2O_3 = 20.4$wt% and $MgO = 8.2$wt% (Fig. 10.86). From this it can be seen that a slight increase in SiO_2 reduces the liquidus

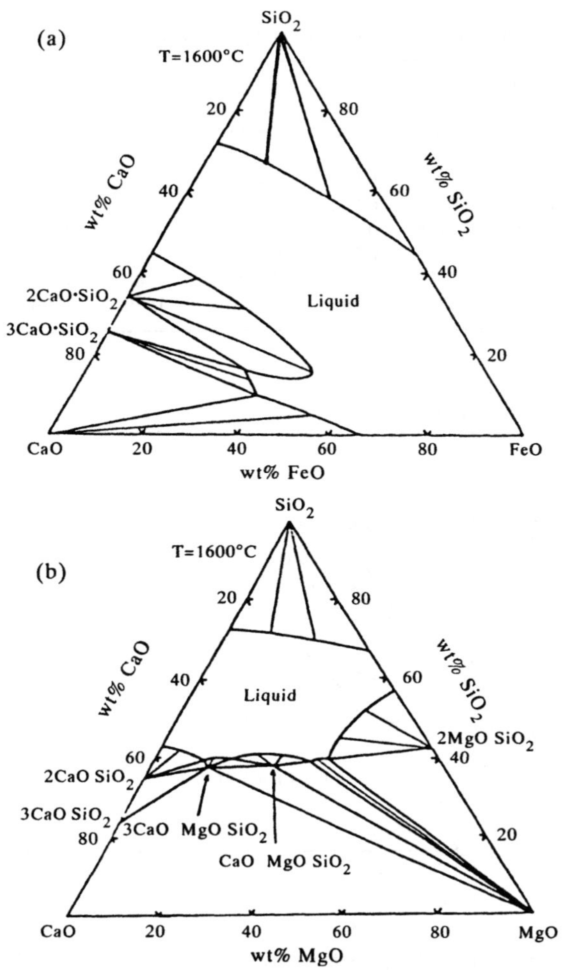

Figure 10.84 Calculated isothermal sections at 1600°C for (a) CaO–FeO–SiO_2 and (b) CaO–MgO–SiO_2 (Matsumiya *et al.* 1987).

temperature of inclusion 1 from Table 10.4. Taking this into account in a new 'ideal' inclusion, they made further calculations by varying the various levels of the component oxides with fixed SiO_2 content. From these calculations they were able to conclude that the liquidus temperature was sensitive to changes in the component oxides with sensitivity factors in the following order: MnO>Al_2O_3>MgO>CaO. To realise the potential reductions in liquidus temperature of the inclusions, changes would be needed during the steelmaking process in the de-oxidiser, refining slags and refractory materials.

Table 10.4. Compositions, liquidus temperatures and primary phases of oxide inclusions

	Oxide composition (weight fractions)					Liquidus (°C)		
No.	SiO_2	Al_2O_3	CaO	MgO	MnO	Calc.	Exp.	Primary phase
1	0.459	0.204	0.153	0.082	0.102	1290	1300	$MgO.Al_2O_3$
2	0.500	0.050	0.020	0.230	0.200	1390	1404	$2MgO.SiO_2$
3	0.4208	0.2956	0.2108	0.0441	0.0287	1450	1460	$CaO.2SiO_2.Al_2O_3$
4	0.5081	0.3559	0.0833	0.0437	0.0090	1550	>1500	$CaO.2SiO_2.3Al_2O_3$
5	0.7749	0.1830	0.0012	0.0054	0.0355	1280	>1650	$2SiO_2.3Al_2O_3$
6	0.5647	0.2712	0.0270	0.0010	0.1361	1470	>1500	$2SiO_2.3Al_2O_3$

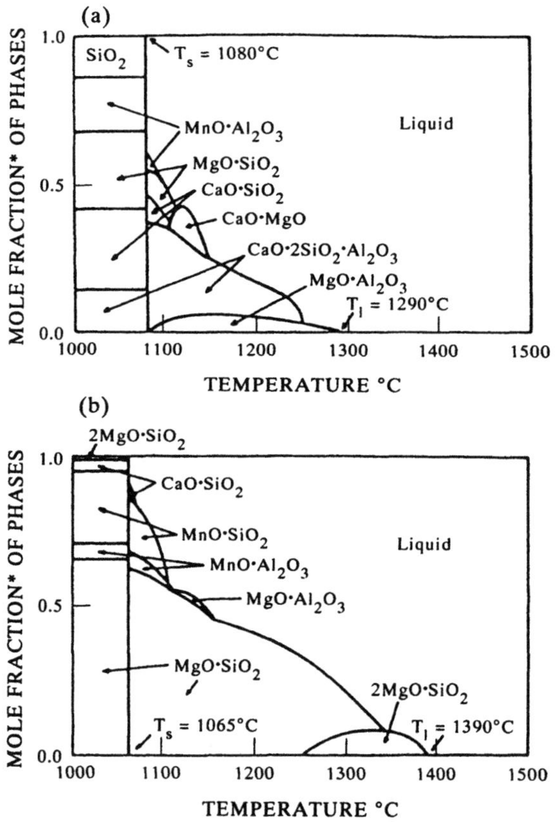

Figure 10.85 Calculated phase fraction vs temperature plots for oxide inclusions nos (a) 1, (b) 2 and (c) 3 from Table 10.4.

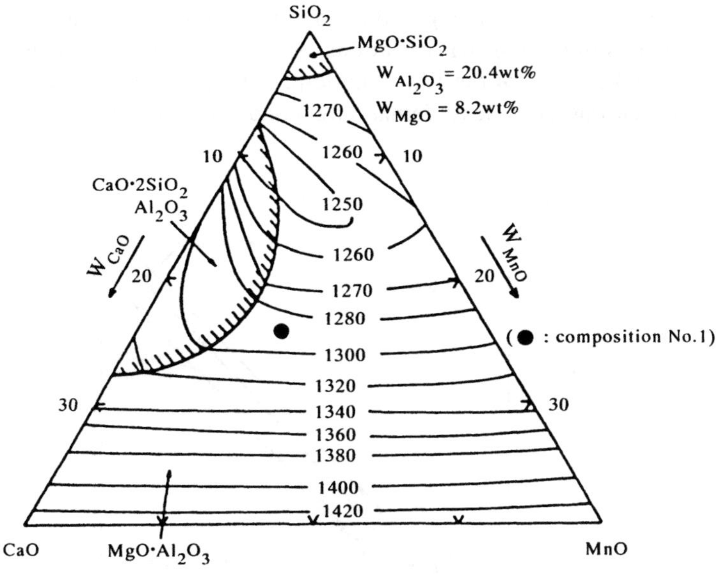

Figure 10.86 Calculated liquidus surface of the five-component oxide system with fixed values of 20.4wt%Al_2O_3 and 8.2 wt%MgO.

10.6.7 Complex chemical equilibria

10.6.7.1 CVD processing. CVD (chemical vapour deposition) processing is an area where calculations using mainly a substance database have proved valuable in understanding the process conditions for deposition of silicides, borides, carbides, etc. A good example of such calculations is provided by Vahlas *et al.* (1996) who looked at the deposition of WSi_2 from various gaseous environments. They utilised a substance database for the elements Si–W–Cl–H–O–Ar where 46 species in the gas phase and 20 stoichiometric condensed phases were considered in the calculation.

They first calculated a so-called 'CVD diagram' for the system WF_6–SiH_4–H_2–Ar, representing the incoming gas, which shows the phases which will be deposited as a function of partial pressure of WF_6 and SiH_4 (Fig. 10.87). The lines separating the various deposition regimes have 'bands of uncertainty' superimposed to reflect the uncertainty arising from the assessment of the thermodynamic data for the total system. It is clearly seen that the deposition of *pure* WSi_2 will be difficult using these input gases. In order to find a less sensitive system the case of WCl_4–SiH_4–H_2–Ar was considered, whose CVD diagram is shown in Fig. 10.88. It can now be seen that the single-phase region for WSi_2 has significantly expanded which means that the CVD process will be less sensitive to composition fluctuations in the input gas itself, or fluctuations in the gas due to transport phenomena within the CVD deposition chamber. Further variations occur when SiH_4 is substituted for by $SiCl_2H_2$ and the width of the WSi_2 single-phase region expands even further (Fig. 10.89), providing even less demanding conditions for the deposition of pure WSi_2.

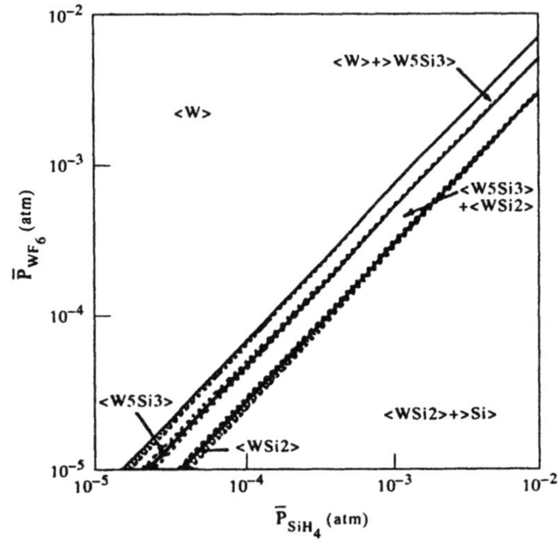

Figure 10.87 Calculated CVD phase diagram for the WF_6–SiH_4–H_2–Ar system at $T = 1000$ K, $P_{total} = 1$ atm and $P_{ar} = 0.9$ atm (from Vahlas *et al.* 1996).

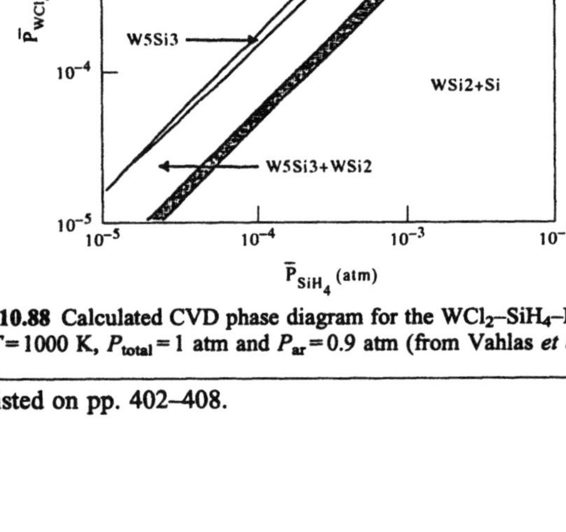

Figure 10.88 Calculated CVD phase diagram for the WCl_2–SiH_4–H_2–Ar system at $T = 1000$ K, $P_{total} = 1$ atm and $P_{ar} = 0.9$ atm (from Vahlas *et al.* 1996).

The substitution of $TiCl_4$ for WCl_4 yields the CVD diagram as shown in Fig. 10.90 (Vahlas *et al.* 1996) and it can clearly be seen that the CVD diagram is totally different, demonstrating both the system specific nature of such diagrams and the importance of the underlying thermodynamics in determining how the process will occur in practice.

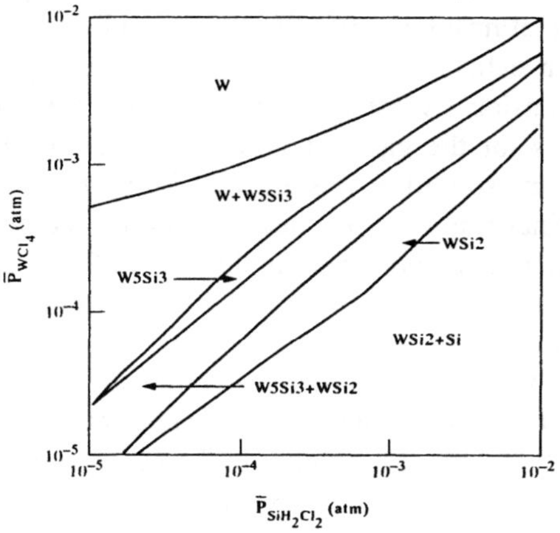

Figure 10.89 Calculated CVD phase diagram for the WCl_4–$SiCl_2H_2$–H_2–Ar system at $T = 1000$ K, $P_{total} = 1$ atm and $P_{ar} = 0.9$ atm (from Vahlas *et al.* 1996).

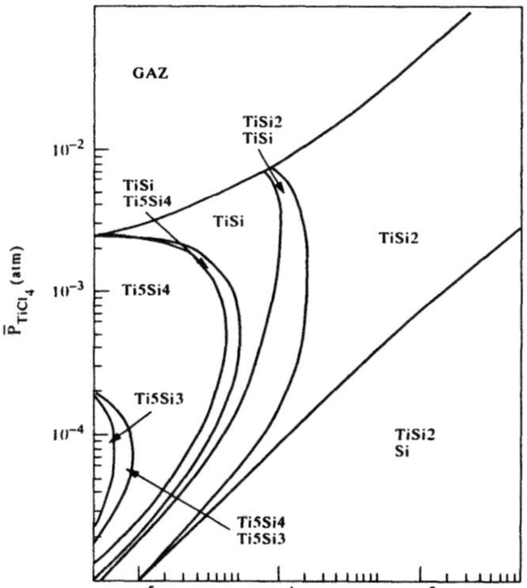

Figure 10.90 Calculated CVD phase diagram for the $TiCl_4$–SiH_4–H_2–Ar system at $T = 1000$ K, $P_{total} = 1$ atm and $P_{ar} = 0.9$ atm (from Vahlas *et al.* 1996).

10.6.7.2 Hot-salt corrosion in gas turbines. In a wide variety of cases, Ni-based super-alloys are protected from corrosive attack in gas turbine engines by the formation of a surface layer of Cr_2O_3. However, in marine environments, NaCl can become concentrated in the incoming gas and lead to reaction with sulphur-containing combustion products to form Na_2SO_4. This then condenses onto the Cr_2O_3 layer which leads to the dissolution of Cr_2O_3, exposing the alloy itself to corrosive attack from the atmosphere, which is both oxidising and sulphidising. This type of attack is often called 'hot-salt corrosion' and Barry and Dinsdale (1996) examined this process using a substance database as the basis for their calculations.

The required data is for the gas and solid phases in the system NaCl–NaOH–Na_2CrO_4–Na_2SO_4. Some solubility was allowed for in the condensed solid phases, mainly in terms of binary interactions between the respective components which form simple phase diagrams (see for example Fig. 10.91). Calculations were then made relating to a gas turbine operating at a fuel/air ratio of 50:1, at 15 bar and 750°C. The fuel was taken to have a composition approximating to the formula $CH_{1.8}$. On the basis of these assumptions, the combustion products were calculated to be predominantly N_2, O_2, CO_2 and H_2O, their respective partial pressures being, 11.6, 2.216, 0.61 and 0.55 bars. 1% sulphur in the fuel would then cause the sum of the pressures of SO_2 and SO_3 to be 0.0026 bar. If only 1 ppm of NaCl were present this would cause the partial pressure of HCl in the gas to be 0.0046 bar, the residual sodium forming mainly as Na_2SO_4 on any exposed surfaces.

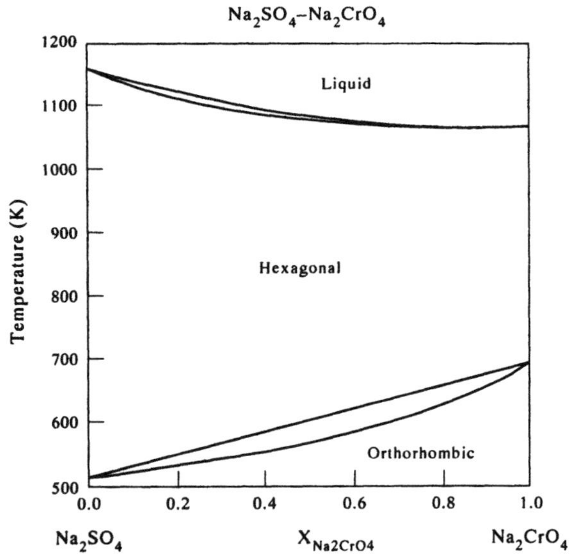

Figure 10.91 Calculated phase diagram for the Na_2SO_4–Na_2CrO_4 binary system (Barry and Dinsdale 1996).

10.6.7.3 Production of silicon in an electric arc furnace. Although not particularly complex in terms of chemical equilibrium, the production of Si in an electric arc furnace is a neat example of how equilibrium calculations can be made which take into account material flow during a dynamic process (Eriksson and Hack 1990). In practice, the furnace works continuously, with raw starting material supplied at the top of the furnace while gases circulate in such a way that there is a flow upwards with a counter-current flow of solid material which falls downwards. In order to simulate the reactions taking place in the furnace, it is necessary to take into account the fact that substances move in a temperature gradient during the process and that the temperature is controlled as much by the heat exchange and enthalpy of reactions as by external heating.

Eriksson and Hack (1990) developed a module for the ChemSage software code which would help examine cases where there is material and heat flow as well as chemical reactions. This is achieved by conceptually separating up the furnace into a number of separate parts or 'stages' where local chemical equilibrium can be assumed. Flow is then modelled by including distribution coefficients between the stages. So, for example, in a reactor with two stages, material which reacts in one stage can move to another stage, dependent on the flow direction. The accuracy of the programme in dealing with dynamic flow, which is non-equilibrium in nature, lies in how many stages are used. The use of many stages will allow smooth changes to be considered but will cause a substantial increase in computational time. For the Si arc furnace four stages were considered and these are shown schematically below.

	GAS ↑			
Stage 1	↑	*ΔH/kJ = 0 T/K = 1784 P/bar = 1.0	↓	*Input* C SiO_2 (quartz)
Stage 2	↑	*ΔH/kJ = 0 T/K = 2059 P/bar = 1.0	↓	
Stage 3	↑	*ΔH/kJ = 0 T/K = 2079 P/bar = 1.0	↓	
Stage 4	↑	*ΔH/kJ = 875 T/K = 2355 P/bar = 1.0	↓	
		*regulated quantity	↓ Condensed phases	

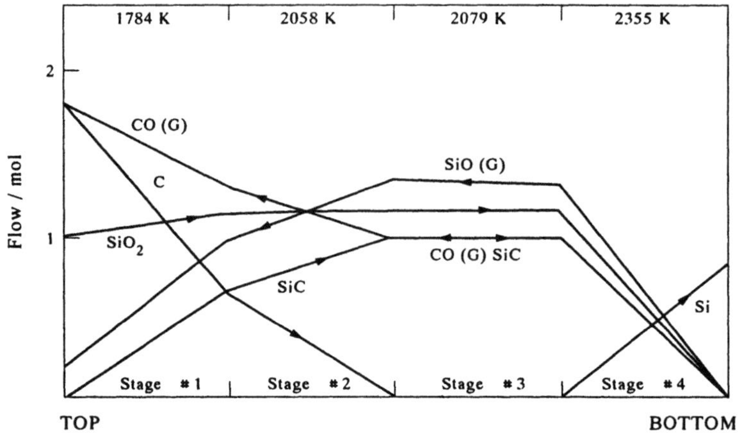

Figure 10.92 Flow scheme for various phases in a Si electric arc furnace (from Eriksson and Hack 1990).

The raw materials, 1 mole of SiO_2 and 1.8 moles of C are introduced into the top stage where the temperature is 1784 K, while the energy from the arc, 875 kJ per mole of SiO_2, was taken to be released only in the bottom stage of the reactor. The reactor assumes each stage has fixed temperature and therefore the enthalpy input to the other three stages arises solely from heat exchange and reaction. The flow scheme (Fig. 10.92), distribution coefficients, values for the heat balances in each stage, input substances and initial temperatures are actually obtained by an optimisation process and relate to a particular furnace rather than generically to all Si arc furnaces. The example above is from Eriksson and Hack (1990) but later calculations by the same authors (Eriksson and Hack 1996) give somewhat different results. However, the latter relate to a specific Si arc furnace at KemaNord, Ljungaverk, in Sweden.

10.6.8 Nuclear applications

Nuclear applications constituted a strong area in the early stages of CALPHAD calculations, not only to obtain a better understanding of alloying in U- and Pu-based systems but also for handling complex gas reactions. The interest in applying CALPHAD to nuclear problems is well demonstrated by the papers of Potter and Rand (1980, 1983) which reviewed work on a variety of problems, including simple calculations for the U–O and U–Pt–O systems, and more complex calculations for irradiated fuels and coolant reactions. Recently Ball *et al.* (1989, 1996) have looked at the application of phase-equilibrium calculations to cladding failure in irradiated pins for water-cooled thermal and liquid-metal-cooled fast-breeder reactors and the analysis of accidents in nuclear reactors.

10.6.8.1 Cladding failure in oxide fuel pins of nuclear reactors. The long-term operational performance of nuclear fuel pins is critically governed by the reactions that occur in the gap between the fuel and its cladding. Ball *et al.* (1989) examined this for the cases of (1) Zircaloy-clad pellets of UO_{2+x} in a pressurised water reactor (PWR) and (2) stainless-steel-clad pellets of $(U, P)O_{2+x}$ in a liquid-metal-cooled fast-breeder reactor (LMFBR). In particular they were interested in the influence of O potential on Cs, I, Te and Mo and the effects of irradiation on the gaseous species within the fuel-clad gaps.

In the case of PWRs, which operate at relatively low centre temperatures (~1500 K), it was considered that little diffusion of fission product elements would occur to the gap, the majority of material in the gap arising from athermal processes such as fission-fragment recoil. The oxygen potential was taken to increase with increasing irradiation and a series of calculations were made at 650 K to predict the pressure of the predominant gas species as a function of oxygen potential. Two cases were considered: (1), Mo and Zr being zero, i.e., no presence of the fission-product Mo or interaction with the Zircaloy (Fig. 10.93(a)), and (2) with Mo and Zr included (Fig. 10.93(b)). Although qualitatively similar, the calculations show that the inclusion of Mo increased the I pressure in the gas due to the reaction of CsI with Mo-containing species to form Cs_2MoO_4.

In LMBFRs, operating temperatures at the centre are nearer 2300 K, far higher than in PWRs, and the fuel is a mixture of U and P oxide $(U, P)O_{2+x}$. Any reactions must now include Cr from the oxide film of the stainless-steel cladding which has replaced Zircaloy. Figure 10.94(a) then show the calculated ratios of Te:Cs, I:Cs and Te:I for a reaction at 850 K, assuming that no fuel was involved in the reaction, which is then predominantly controlled by Mo and Cr, while Fig. 10.94(b) shows what happens when fuel is included, assuming similar levels of Mo and Cr. Again, the calculations show that predictions are qualitatively similar in both cases, but there are significant differences in the amounts of reaction.

On the basis of the above calculations, Ball *et al.* (1989) concluded that, for PWRs, the potential of I in the gas from equilibrium reactions would be insufficient to account for stress corrosion cracking (SCC) of the Zircaloy. The inclusion of Mo, which can be present in the gap as a fission product, raised the iodine potential but, again, not to a level sufficient to account for any SCC. However, the concentrations of the elemental gaseous species could increase more significantly due to irradiation by fission fragments, and this may be sufficient to produce SCC of the Zircaloy. For the case of LMBFRs, I levels were also increased due to increased O potential. More importantly, Cs and Te levels would also be increased and, with the ratio Te:Cs>0.5, the calculations showed that these elements may deposit as solids onto the cladding surface leading to corrosion of the stainless-steel cladding.

10.6.8.2 Accident analysis during melt-down of a nuclear reactor. The problem to be considered here is the erosion of concrete by liquid material during a 'melt-

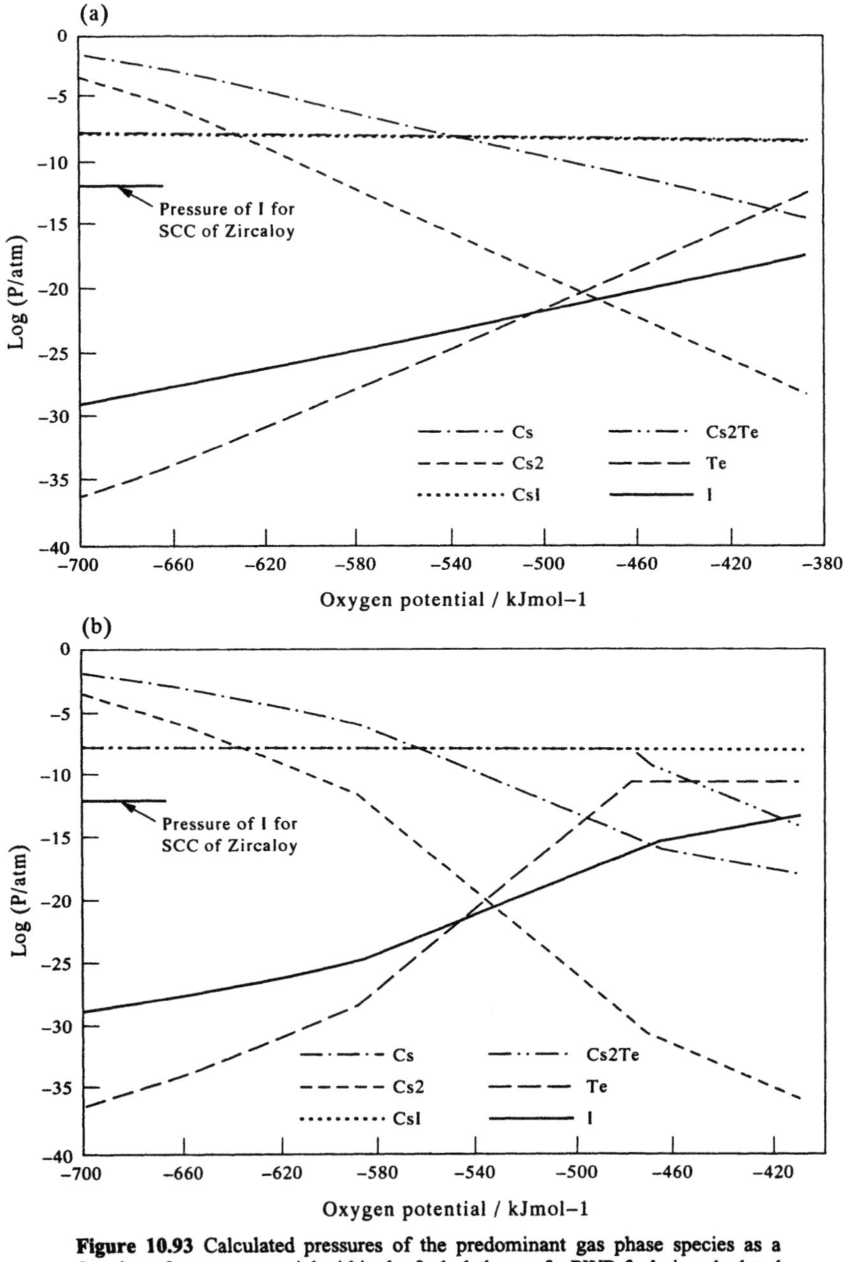

Figure 10.93 Calculated pressures of the predominant gas phase species as a function of oxygen potential within the fuel-clad gap of a PWR fuel pin calculated for the case (a) where Mo and Zr are absent and (b) where Mo and Zr are included (Ball *et al.* 1989).

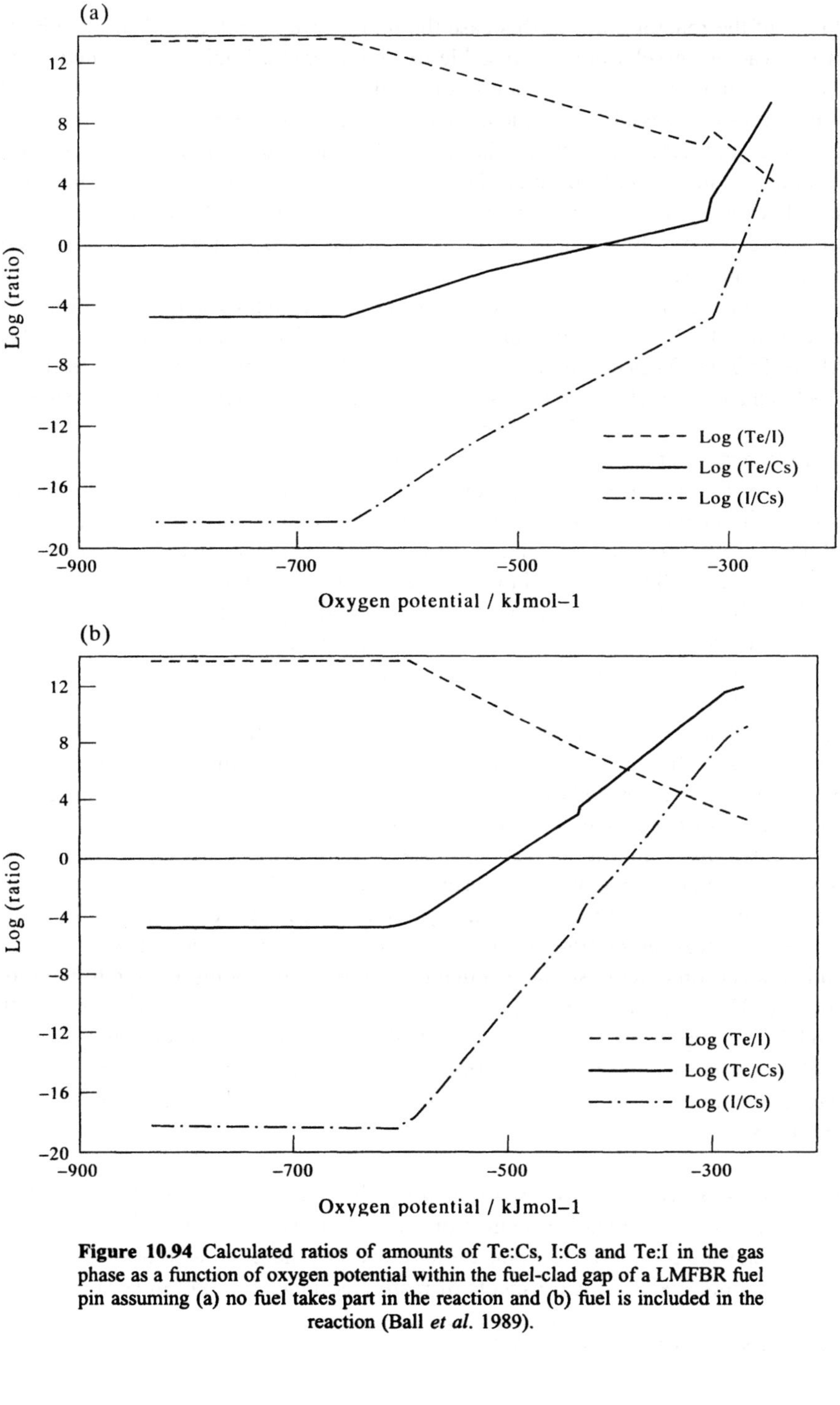

Figure 10.94 Calculated ratios of amounts of Te:Cs, I:Cs and Te:I in the gas phase as a function of oxygen potential within the fuel-clad gap of a LMFBR fuel pin assuming (a) no fuel takes part in the reaction and (b) fuel is included in the reaction (Ball *et al.* 1989).

down' of the reactor core. In this case the liquid core is 'relocated' into the bottom of the reactor vessel where immiscible oxide and metal liquids are formed. This is initially contained within a crucible of condensed material but, if the wall of the reactor pressure vessel is breached, the molten 'corium', comprising components of the core and structural materials, is ejected onto a concrete basemat in the secondary containment building. This concrete basemat is both thermally ablated and attacked by the molten 'corium–concrete' interaction (MCCI). In addition there are gas reactions which can potentially lead to the release of dangerous radioactive material into the atmosphere.

To examine this problem more closely it was necessary to develop (1) a model for the nine-component oxide system UO_2–ZrO_2–SiO_2–CaO–MgO–Al_2O_3–SrO–BaO–La_2O_3 to account for the MCCI (Chevalier 1992, Ball *et al.* 1993) and (2) develop a database for the gas-phase reactions in the oxide sub-system UO_2–ZrO_2–SiO_2–CaO–MgO–Al_2O_3. The final oxide database included four solution phases and 70 condensed stoichiometric phases.

Figures 10.95(a, b) show isopleths calculated between (a) corium and siliceous concrete and (b) corium and limestone concrete. Comparison between experimental (Roche *et al.* 1993) and calculated values for the solidus are in reasonable agreement, but two of the calculated liquidus values are substantially different. However, as the solidus temperature is more critical in the process, the calculations can clearly provide quite good-quality data for use in subsequent process simulations. Solidus values are critical factors in controlling the extent of crust formation between the melt-concrete and melt-atmosphere interface, which can lead to thermal insulation and so produce higher melt temperatures. Also the solidus, and proportions of liquid and solid as a function of temperature, are important input parameters into other software codes which model thermal hydraulic progression and viscosity of the melt (Cole *et al.* 1984).

Calculations including the vapour phase were then made to determine the extent of release of various components during the reaction. Two types of calculation were made, one where ideal mixing in the solution phases was considered and the other where non-ideal interactions were taken into account. For elements such as Ba, U and, to a certain extent, Si, the calculations were relatively insensitive to the model adopted. However, the amount of Sr in the gas was 24 times higher in the 'full model' in comparison to the ideal model. This led to the conclusion that sensitivity analysis was necessary to determine the extent to which accuracy of the thermodynamic parameters used in the model affected the final outcome of the predictions.

10.6.8.3 The effect of radiation on the precipitation of silicides in Ni alloys. While chemical reactions in nuclear generators have dominated how CALPHAD methods have been used in practice for nuclear applications, there has also been a significant interest in the metallurgical aspects of materials under irradiation (Kaufman *et al.*

References are listed on pp. 402–408.

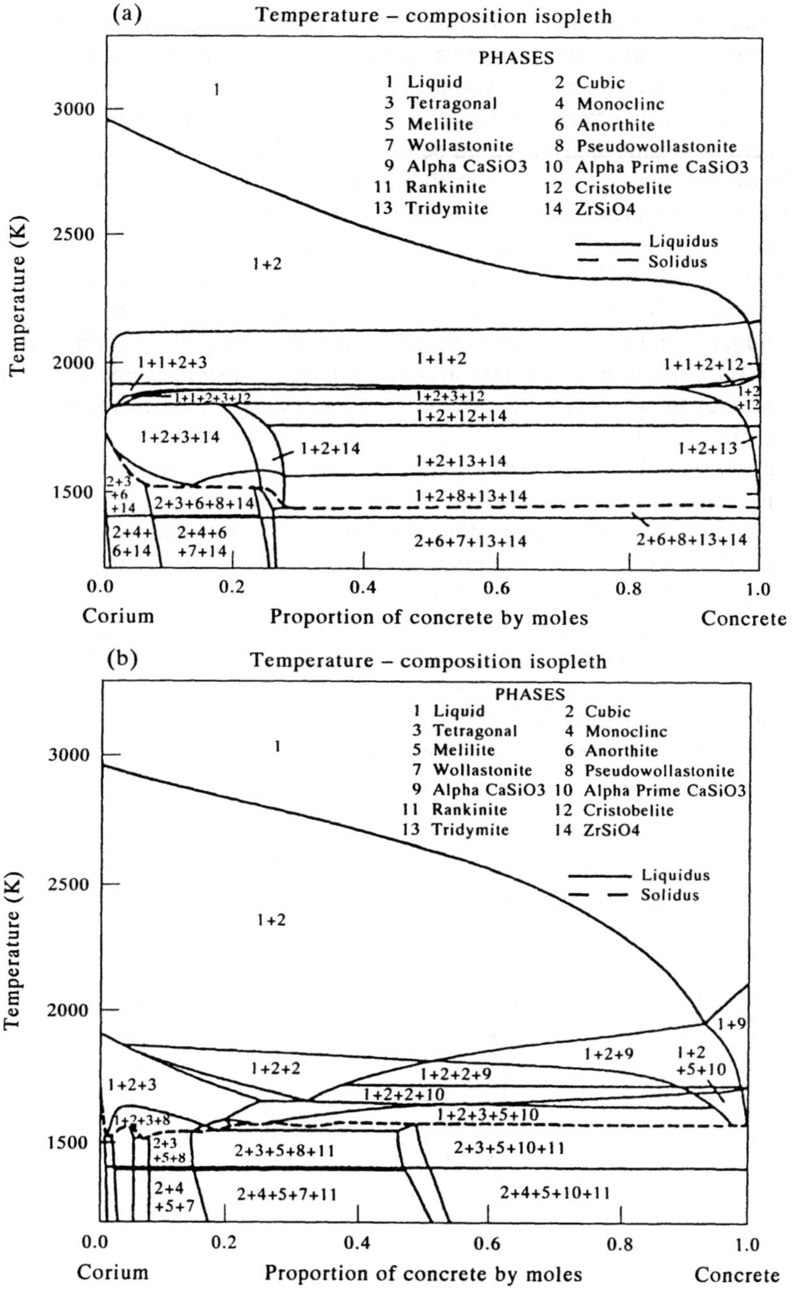

Figure 10.95 Calculated isopleths from 100% corium to 100% concrete. (a) For sicliceous concrete and (b) limestone concrete (Ball *et al.* 1996).

1978, Miodownik *et al.* 1979, Watkin 1979). Irradiation can cause void-swelling, suppression of σ formation in stainless steels and non-equilibrium precipitation of silicides. These phenomena are complex and occur by a combination of thermodynamic and kinetic effects. However, it was shown by Miodownik *et al.* (1979) that a thermodynamic analysis could be used to good effect to rationalise the effect of radiation on silicide formation. Although the work was done for a simple alloy system, it demonstrates how thermodynamics can be used in unusual circumstances.

Barbu and Martin (1977) observed that increasing Ni^+ dose rate enhanced the formation of Ni_3Si in Ni–Si alloys (Fig. 10.96), causing it to form in 'sub-saturated' alloys whose Si levels were below the solvus composition for the formation of equilibrium Ni_3Si. From a purely thermodynamic point of view an additional Gibbs energy (ΔG^*) has to be taken into account which is shown schematically in Fig. 10.97. Miodownik *et al.* (1979) correlated this additional Gibbs energy to dose rate for sub-saturated Ni–Si alloys with 2 and 6at%Si (Fig. 10.98), and generated a diagram which showed how the formation of Ni_3Si would occur in these alloys as a function of temperature and dose rate (Fig. 10.99). The diagram is analogous to a time–temperature–transformation diagram with the dose rate taking the place of the time axis. Rotating the axes also gives a form of diagram similar to that of Barbu and Martin (1977) shown in Fig. 10.96.

Although the work was in the form of pseudo-equilibrium in the presence of irradiation, and did not take into account the effect of kinetics, it nevertheless gives a clear indication of an additional energy term due to irradiation. Similar studies were done by Kaufman *et al.* (1978) to show the magnitude of the Gibbs energy necessary to prevent σ formation in Fe–Cr–Ni alloys.

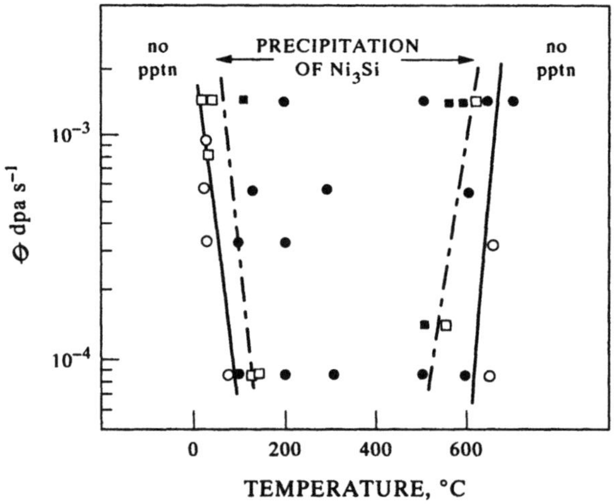

Figure 10.96 Combinations of Ni^+ dose-rate and temperature leading to precipitation of Ni_3Si (from Barbu and Martin 1977).

References are listed on pp. 402–408.

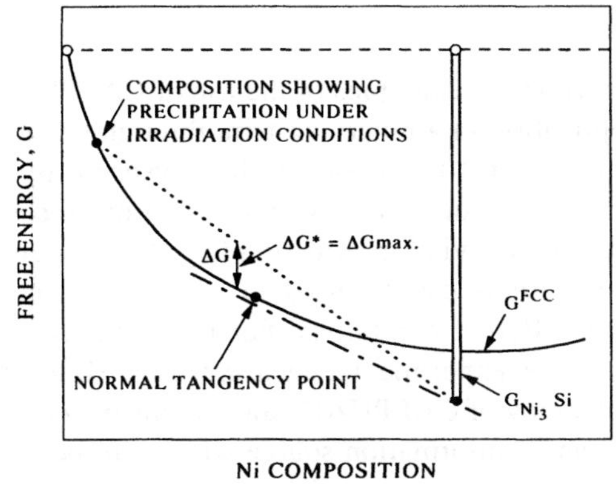

Figure 10.97 Definition of the Gibbs energy (ΔG^*) required in order to precipitate Ni_3Si in subsaturated solid solutions (from Miodownik *et al.* 1979).

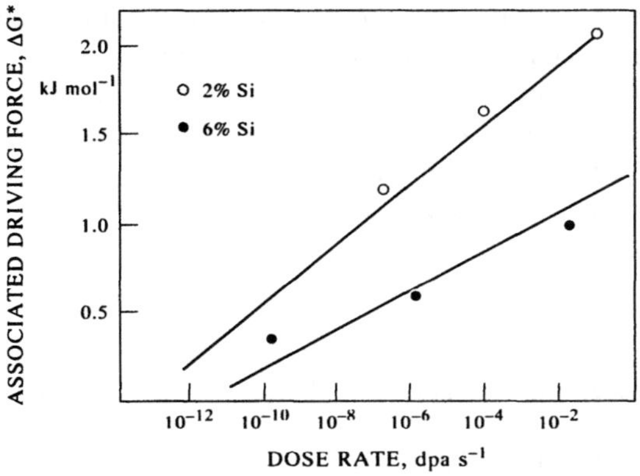

Figure 10.98 Correlation of ΔG^* with dose rate for 2at% and 6at%Si alloys (from Miodownik *et al.* 1979).

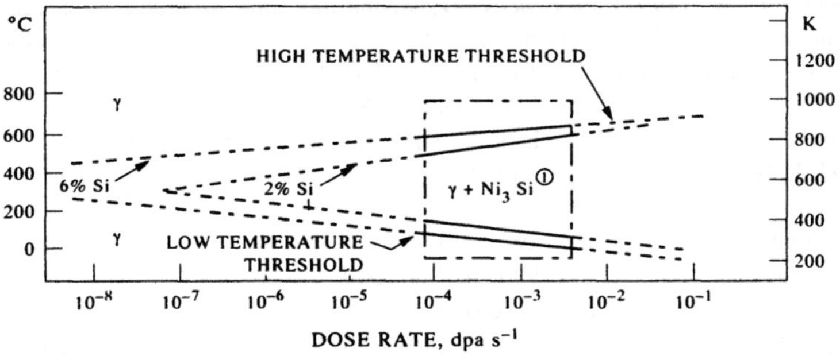

Figure 10.99 Calculated amount of Ni_3Si precipitated in subsaturated Ni–Si alloys in the presence of Ni^+ irradiation (from Miodownik *et al.* 1979).

10.7. SUMMARY

This chapter has shown many examples of the use of CALPHAD methods, ranging from an unusual application in a binary system, through complex equilibrium calculations to calculations for 10-component alloy systems. In all cases the use of CALPHAD methods has enhanced the understanding of processes, clearly defined alloy behaviour and provided vital information for other models, etc. It is also clear that *equilibrium* calculations can be used in many different areas and under a surprising number of different conditions. For numerous reasons, modelling will never completely replace experimental measurement. However, the quantitative verification of the accuracy of CALPHAD calculations now means that they can be seriously considered as an information source which can be used as an alternative to experimental measurement in a number of areas and can also enhance interpretation of experimental results.

For a number of applications, particularly those associated with conditions of continuous cooling or heating, equilibrium is clearly never approached and calculations must be modified to take kinetic factors into account. For example, solidification rarely occurs via equilibrium, amorphous phases are formed by a variety of non-equilibrium processing routes and in solid-state transformations in low-alloy steels much work is done to understand time–temperature–transformation diagrams which are non-equilibrium in nature. The next chapter shows how CALPHAD methods can be extended to such cases.

REFERENCES

Akamatsu, S., Hasebe, M., Senuma, T., Matsumura, Y. and Akisue, O. (1994) *ISIJ International*, **34**, 9.

Andersson, J.-O. (1988) *CALPHAD*, **12**, 1.

Andersson, J.-O. and Sundman, B. (1987) *CALPHAD*, **11**, 83.

Ansara, I. and Rand, M. H. (1980) *The Industrial Use of Thermochemical Data*, ed. Barry, T. I. (Chemical Society, London), p. 149.

Anyalebechi, P. N. (1992) in *Processing, Properties and Application of Metallic and Ceramic Materials*, eds Loretto, M. H. and Beevers, C. J. (MCE publications, UK), p. 219.

Bale, C. W. and Eriksson, G. (1990) *Canadian Metallurgical Quarterly*, **289**, 105.

Ball, R. G. J., Mason, P. K. and Mignanelli, M. A. (1996) in *The SGTE Casebook—Thermodynamics at Work*, ed. Hack, K. (Inst. Materials, London), p. 135.

Ball, R. G. J., Mignanelli, M. A., Barry, T. L. and Gisby, J. A. (1993) *J. Nucl. Materials*, **201**, 238.

Barbu, A. and Martin, G. (1977) *Scripta Met.*, **11**, 771.

Barry, T. I. and Dinsdale, A. T. (1996) in *The SGTE Casebook—Thermodynamics at Work*, ed. Hack, K. (Inst. Materials, London), p. 56.

Betteridge, W. and Heslop, J. (1974) in *The NIMONIC Alloys and other Ni-Based High Temperature Alloys: 2nd Edition*, (Edward Arnold Ltd, 1974).

Bhadeshia, H. K. D. B and Edmond, H. V. (1980) *Acta Met.*, **28,** 1265.

Bhowal, G. E., Konkel, W. A. and Merrick, H. F. (1995) in *Gamma Titanium Aluminides*, eds Kim, Y.-W. *et al.* (TMS, Warrendale, OH), p. 787.

Blavette, B., Caron, P. and Khan, T. (1988) in *Superalloys 1988*, eds Reichman, S. *et al.* (TMS, Warrendale), p. 305.

Blenkinsop, P. (1993) "IRC in materials for high performance applications", University of Birmingham, U.K., private communication.

Boniardi, M., Iacoviello, F. and La Vecchia, G. M. (1994) in *Proc. Conf. Duplex Stainless Steels '94* (Welding Institute, Cambridge), Paper 89.

Brinegar, J. R., Mihalisin, J. R. and Van der Sluis, J. (1984) in *Superalloys 1984*, eds Gell, M. *et al.* (Met. Soc. AIME, Warrendale), p. 53.

Bronson, A. and St-Pierre, G. R. (1981) *Met. Trans. B*, **12B,** 729.

Cama, H., Worth, J., Evans, P. V., Bosland, A. and Brown, J. M. (1997) in *Solidification Processing 1997*, eds Beech, J. and Jones, H. (University of Sheffield, UK), p. 555.

Cao, H.-L. and Hertzmann, S. (1991) in *Proc. Conf. Duplex Stainless Steels '91* (Les Editions Physique, Les Ulis, France), p. 1273.

Caron, P. and Khan, T. (1983) *Mat. Sci. Eng.*, **61,** 173.

Chandrasekaran, L. and Miodownik, A. P. (1989) "Phase equilibria relevant to liquid phase sintering in Fe-base alloys", Report to SERC under contract GR/D 99935 (University of Surrey, Guildford, January).

Chandrasekaran, L. and Miodownik, A. P. (1990) in *PM into the 90s; Proc. Int. Conf. PM90* (Inst. Metals, London), p. 398.

Charles, J., Dupoiron, F., Soulignac, P. and Gagnepain, J. C. (1991) in *Proc. Conf. Duplex Stainless Steels '91* (Les Editions Physique, Les Ulis, France), p. 1273.

Chart, T. G., Putland, F. and Dinsdale, A. T. (1980) *CALPHAD*, **4,** 27.

Chatfield, C. and Hillert, M. (1977) *CALPHAD*, **1,** 201.

Chen, C. C. and Sparks, R. B. (1980) in *Titanium Science and Technology*, eds Kimura, H. and Izumi, O. (Met. Soc. AIME, Warrendale), p. 2929.

Chevalier, P. Y. (1992) *J. Nucl. Materials*, **186,** 212.

Cole, R. K., Kelley, D. P. and Ellis, M. A. (1984) "CORCON-Mod 2: A computer programme for analysis of molten core–concrete interactions", NUREG/CR-3920, August.

Counsell, J. F., Lees, E. B and Spencer, P. J. (1972) in *Metallurgical Thermochemistry* (NPL–HMSO, London), p. 451.

Cortie, P. and Potgeiter, J. H. (1991) *Met. Trans. A*, **22A,** 2173.

Darolia, R., Lahrman, D. F. and Field, R. D. (1988) in *Superalloys 1988*, eds Reichmann, S. *et al.* (TMS, Warrendale), p. 255.

Delargy, K. M. and Smith, G. D. W. (1983) *Met. Trans. A*, **14A,** 1771.

Dharwadkar, S. R., Hilpert, K., Schubert, F. and Venugopal, V. (1992) *Z. Metallkde.*, **83,** 744.

Dinsdale, A. T. (1991) *CALPHAD*, **15,** 319.

Dinsdale, A. T., Hodson, S. M., Barry, T. I. and Taylor, J. R. (1988) *Computer Software in Chemical & Extractive Metallurgy*, Proc. Met. Soc. CIM, Montreal '88, **11**.

Dreshfield, R. L. and Wallace, J. F. (1974) *Met. Trans*, **5,** 71.

Duerig, T. W., Terlinde, G. T. and Williams, J. C. (1980) *Met. Trans. A*, **11A,** 1987.

Duval, S., Chambreland, S., Caron, P. and Blavette, D. (1994) *Acta Met. Mat.*, **42,** 185.

Eriksson, G. (1975) *Chemica Scripta*, **8**, 100.

Eriksson, G. and Hack, K. (1990) *Met. Trans. B*, **21B**, 1013.

Eriksson, G. and Hack, K. (1996) *The SGTE Casebook, Thermodynamics at Work*, ed. Hack, K. (Inst. Materials, London), p. 200.

Erickson, G. L., Harris, K. and Schwer, R. E. (1985) "The development of CMSX-5 a 3rd generation high strength single crystal alloy", presented at the TMS Annual Meeting, New York.

Eshelby, J. D. (1957) *Proc. Royal Soc., London*, **241A**, 376.

Ferrer, L., Pieraggi, B. and Uginet, J. F. (1991) in *Superalloys 718, 625 and Various Derivatives*, ed. Loria, A. (TMS, Warrendale, PA), p. 217.

Fincham, C. J. B. and Richardson, F. D. (1954) *J. Iron & Steel Inst.*, **178**, 4.

Fincham, C. J. B. and Richardson, F. D. (1964) *Proc. Roy. Soc. (London)*, **223A**, 40.

Fuchs, G. E. (1993) in *Structural Intermetallics*, eds Darolia, R. *et al.* (TMS, Warrendale, OH), p. 195.

Fuchs, G. E. (1995) *Mat. Sci. Eng. A*, **A192/193**, 707.

Grujicic, M. (1991) *CALPHAD*, **15**, 179.

Grujicic, M. and Olson, G. B. (1988) *CALPHAD*, **12**, 405.

Grujicic, M., Lee, H.-M. and Allen, S. M. (1987) in *User Applications of Alloy Phase Diagrams*, ed. Kaufman, L. (ASM, Metals Park, OH), p. 195.

Gustafson, P. (1988) *Met. Trans. A*, **19A**, 2547.

Gysel, W. and Schenk, R. (1991) in *Proc. Conf. Duplex Stainless Steels '91* (Les Editions Physique, Les Ulis, France), p. 331.

Hack, K. and Spencer, P. J. (1985) *Steel Res.*, **56**, 1.

Haidemenopoulos, G. N., Grujicic, M., Olson, G. B. (1989) *CALPHAD*, **13**, 215.

Hallstedt, B. (1992) Ph.D.Thesis, Royal Institute of Technology, Stockholm.

Hamalainen, E., Laitinen, A., Hanninen, H. and Liimatainen, J. (1994) in *Proc. Conf. Duplex Stainless Steels '94*, (Welding Institute, Cambridge), Paper 122.

Hao, S. M., Talayama, T., Ishida, K. and Nishizawa, T. (1984) *Met. Trans. A*, **15A**, 1819.

Harada, H., Ohno, K., Yamagata, T, Yokokawa, T. and Yamazaki, M. (1988) in *Superalloys 1988*, eds Reichman, S. *et al.* (TMS, Warrendale), p. 733.

Harvig, H., Nishizawa, T. and Uhrenius, B. (1972) in *Metallurgical Thermochemistry* (NPL–HMSO, London). p. 431.

Hayes, F. H. (1985) *J. Less Common Metals*, **114**, 89.

Hayes, F. H., Hetherington, M. G. and Longbottom, R. D. (1990) *Mat. Sci. Tech.*, **6**, 263.

Hertzman, S. (1995) *Scand. J. Met.*, **24**, 140.

Hillert, M. and Staffansson, L.-I. (1970) *Acta Chem. Scand.*, **24**, 3618.

Homma, M., Okada, M., Minowa, T. and Horikoshi, E. (1981) *IEEE Trans. Magn.*, Mag-17, 3473.

Honnarat, Y., Davidson, J. and Duffaut, F. (1971) *Mem. Sci. Rev.*, **68**, 105.

Houghton, D. C. (1993) *Acta Met. Mater.*, **41**, 2993.

Hoyle, G. (1988) *High Speed Steels* (Butterworths, London).

Inden, G. (1987) in *User Applications of Alloy Phase Diagrams*, ed. Kaufman, L. (ASM, Metals Park, OH), p. 25.

Ishida, K. and Nishizawa, T. (1991) in *User Aspects of Phase Diagrams*, ed. Hayes, F. H. (Inst. Materials, London), p. 185.

Ishida, K., Kainuma, R., Ueno, U. and Nishizawa, T. (1991) *Met. Trans. A*, **22A**, 441.

Ishikawa, M., Kuboyana, O., Niikura, M. and Ouchi, C. (1992) "Microstructure and mechanical properties relationship of β-rich, $\alpha - \beta$ Ti-alloy; SP-700", NKK Corporation, Kawasaki 210, Japan.

Ito, Y., Moriguchi, Y., Nishimura, T. and Nagai, N. (1985) in *Titanium Science and Technology*, eds Lutjering, G. *et al.* (Deutsche. Gess. Metallkunde E.V., Oberursal), p. 1643.

Jaffee, R. I. (1958) in *Progress in Metal Physics: Vol. 7*, eds Chalmers, B. and King, R. (Pergamon, London), p. 65.

Jannson, Å. (1991) *Met. Trans. A*, **23A,** 2953.

Jernkontoret (1977) *A Guide to Solidification of Steels* (Jernkontoret, Stockholm).

Jomard, F. and Perdereau, M. (1991) in *Proc. Conf. Duplex Stainless Steels '91* (Les Editions Physique, Les Ulis, France), p. 719.

Kahveci, A. I. and Welsch, G. E. (1986) *Scripta Met.*, **20,** 1287.

Kainuma, R., Imano, S., Ohtani, H. and Ishida, K. (1996) *Intermetallics*, **4,** 37.

Kaufman, L. (1967) in *Phase Stability in Metals and Alloys*, (McGraw-Hill, New York), p. 125.

Kaufman, L. and Bernstein, H. (1970) in *Computer Calculation of Phase Diagrams* (Academic Press, New York).

Kaufman, L. and Clougherty, E. V. (1964) in *Metallurgy at High Pressures and High Temperatures: Met. Soc. AIME Symp. Vol. 22*, eds Schneider, K.A. *et al.* (Gordon & Breach Science, New York), p. 322.

Kaufman, L. and Nesor, H. (1978a) *CALPHAD*, **2,** 295.

Kaufman, L. and Nesor, H. (1978b) *CALPHAD*, **2,** 325.

Kaufman, L., Watkin, J. S., Gittus, J. H. and Miodownik, A. P. (1978) in *Applications of Phase Diagrams in Metalllugy and Ceramics* (NBS Special Publication 496, Gaithesburg), p. 1065.

Keefe, P. W., Mancuso, S. O. and Maurer, G. E. (1992) in *Superalloys 1992*, eds Antolovich, S. D. *et al.* (TMS, Warrendale), p. 487.

Kelly, T. J. and Austin, C. M. (1996) in *Titanium '95: Science and Technology*, eds Blenkinsop, P. *et al.* (Inst. Materials, London), p. 192.

Khan, T. Caron, P. and Duret, C. (1984) in *Superalloys 1984*, eds Gell, M. *et al.* (Met. Soc. AIME, Warrendale), p. 145.

Kim, W. T. and Cantor, B. (1988) University of Oxford, Oxford, UK, private communication.

Kim, W. T., Cantor, B., Clay, K. and Small, C. (1990) in *Fundamental Relationships Between Microstructure and Mechanical Properties of Metal Matrix Composites*, eds Liaw, P. K. and Gungor, M. N. (TMS, Warrendale), p. 89.

Kim, Y.-W., (1992) *Acta Met. Mat.*, **40,** 1121.

Kim, Y.-W., (1995) *Mat. Sci. Eng. A*, **A192/193,** 519.

Kirkaldy, J. S., Thomson, B. A. and Baganis, E. A. (1978) in *Hardenability Concepts with Applications to Steel*, eds Kirkaldy, J. S. and Doane, D. V. (TMS, Warrendale), p. 82.

Kriege, O. H. and Baris, J. M. (1969) *Trans. ASM*, **62,** 195.

Kroupa, A. and Kirkaldy, J. S. (1993) *J. Phase Equilibria*, **14,** 150.

Lampman, S. (1990) *Metals Handbook*, 10th Edition, Vol. 2 (ASM International, Materials Park, OH), p. 592.

Lasalmonie, A. and Loubradou, M. (1979) *J. Mat. Sci.*, **14,** 2589.

Lee, B. J. (1995) in *Applications of Thermodynamics in the Synthesis and Processing of Materials*, eds Nash, P. and Sundman, B. (TMS, Warrendale), p. 215.
Lee, B. J. and Saunders, N. (1997) *Z. Metallkde.*, **88**, 152.
Lee, Y. T., Peters, M. and Welsch, G. (1991) *Met. Trans. A*, **22A**, 709.
Lemire, R. J., Paquette, J., Torgeson, D. F., Wren, D. J. and Fletcher, J. W. (1981) *Assessment of Iodine Behaviour in Reactor Containment Buildings from a Chemical Perspective*, Atomic Energy of Canada Ltd., AECL-6812.
Li, X. (1995) M.Sc. Thesis, Univ. of Birmingham, Edgbaston, UK.
Lin, F. S., Sarke Jr., E. A., Chakraborty, S. B. and Gybor, A. (1984) *Met. Trans. A*, **15A**, 1229.
Lindemer, T. B. (1983) *CALPHAD*, **7**, 125.
Lombard, C. M., Nekkanti, R. M. and Seetharaman, V. (1992) *Scripta Met.*, **26**, 1559.
Longbottom, R. D. and Hayes, F. H. (1991) *User Aspects of Phase Diagrams* (Inst. Metals, London), p. 32.
Longbottom, R. D. and Hayes, F. H. (1994) in *Proc. Conf. Duplex Stainless Steels '94*, (Welding Institute, Cambridge), Paper 124.
Loomis, W. T., Freeman, J. W. and Sponseller, D. L. (1972) *Met. Trans.*, **3**, 989.
Lorenz, J., Lukas, H. L., Hucke, E. E. and Gaukler, L. J. (1983) *CALPHAD*, **7**, 125.
Lundberg, R., Waldenström, M. and Uhrenius, B. (1977) *CALPHAD*, **1**, 159.
Maehara, Y., Ohmori, Y., Murayama, J., Fujino, N. and Kunitake, T. (1983) *Metal Science*, **17**, 541.
Magrini, M., Badan, B. and Ramous, E. (1983) *Z. Metallkde.*, **74**, 314.
Marshall, G. J. (1996), *Materials Science Forum*, **217–222**, 19.
Matsumoto, T. (1993) *Kobe Steel Engineering Reports*, **43**, 103.
Matsumya, T., Yamada, W. and Ohashi, T. (1987) in *User Applications of Alloy Phase Diagrams*, ed. Kaufman, L. (ASM, Metals Park, OH), p. 137.
Meng, Z.-Y. Sun, G.-C. and Li, M.-L. (1984) in *Superalloys 1984*, eds Gell, M. *et al.* (Met. Soc. AIME, Warrendale), p. 563.
Merino, P., Novoa, X. R., Pena, G., Porto, E. and Espada, L. (1991) in *Proc. Conf. Duplex Stainless Steels '91* (Les Editions Physique, Les Ulis, France), p. 719.
Miodownik, A. P. (1989) *Powder Metallurgy*, **32**, 269.
Miodownik, A. P. (1993) in *Proc. Conf. Advances in Materials & Processes*, ed. Ramakrishnan, P. (Oxford & IBBH, New Delhi), p. 87.
Miodownik, A. P., Watkin, J. S. and Gittus, J. H. (1979) "Calculation of the driving force for the radiation induced precipitation of Ni_3Si in nickel silicon alloys", UKAEA Report ND-R-283(S), February.
Mondolfo, L. F. (1976) *Aluminium Alloys: Structure and Properties* (Butterworths, London).
Munson, D. (1967) *J. Inst. Metals*, **95**, 217.
Nathal, M. V. and Ebert, L. J. (1984) in *Superalloys 1984*, eds Gell, M. *et al.* (Met. Soc. AIME, Warrendale), p. 125.
Nilsson, J.-O. (1992) *Mat. Sci. Tech.*, **8**, 685.
Nishizawa, T., Hasebe, M. and Ko, M. (1979a) *Acta Met.*, **27**, 817.
Nishizawa, T., Hasebe, M. and Ko, M. (1979b) Proc. CALPHAD VIII, Stockholm, Sweden, 113.
Nishizawa, T., Hao, S. M., Hasebe, M. and Ishida, K. (1983) *Acta Met.*, **31**, 1403.
Norstrom, L.-A., Pettersson, S. and Nordin, S. (1981) *Z. Werkstofftech.*, **12**, 229.

Nystrom, M. and Karlsson, B. (1994) in *Proc. Conf. Duplex Stainless Steels '94*, (Welding Institute, Cambridge), Paper 104.
Onodera, H., Ro, Y., Yamagata, T. and Yamazaki, M. (1985) *Titanium Science and Technology*, eds Lutjering, G. *et al.* (Deutsche. Gess. fur Metallkunde E.V., Oberursal), p. 1883.
Pan, L.-M. (1992) Ph.D. Thesis, University of Surrey, Guildford, UK.
Pelton, A. D., Eriksson, G. and Romero-Serrano, A. (1993) *Met. Trans. B*, **24B,** 817.
Peters, M. and Williams, J. C. (1985) *Titanium Science and Technology*, eds Lutjering, G., Zwicker, U. and Bunk, W. (Deutsche. Gess. fur Metallkunde E.V., Oberursal), p. 1843.
Phillips, H. W. L. (1961) *Equilibrium Diagrams of Aluminium Alloy Systems* (The Aluminium Development Assoc., London, UK, 1961).
Polmear, I. J. (1989) *Light Alloys: Metallurgy of the Light Metals, 2nd Edition* (Edward Arnold, London).
Polvani, R. S., Tzeng, W. S. and Strutt, P. R. (1976) *Met. Trans. A*, **7A,** 23, 31.
Potter, P. E. and Rand, M. H. (1980) in *The Industrial Use of Thermochemical Data*, ed. Barry, T .I. (Chemical Society, London), p. 149.
Potter, P. E. and Rand, M. H. (1983) *CALPHAD*, **7,** 165.
Proctor, C. S. (1992) "Characterisation of a needl morphology phase observed in CMSX-4", Report to Rolls-Royce, University of Cambridge, May 1992.
Rand, M. H. (1980) in *The Industrial Use of Thermochemical Data* ed. Barry, T. I. (Chemical Society, London), p. 105.
Reddy, R. G. and Blander, M. (1987) *Met. Trans. B*, **18B,** 591.
Reddy, R. G. and Blander, M. (1989) *Met. Trans. B*, **20B,** 137.
Riedl, R., Karagöz, S., Fischmeister, H. and Jeglitsch, F. (1987) *Steel Research 58*, No. 8, 339.
Rivlin, V. G. and Raynor G. V. (1988) *Phase Equilibria in Iron Ternary Alloys* (Inst. Metals, London).
Roche, M. F., Liebowitz, L., Fink, J. K. and Baker, Jr., L. (1993) NUREG report CR-6032, June.
Rong, W., Andrén, H.-O., Wisell, H. and Dunlop, G. L. (1992) *Acta Met. Mater.*, **40,** 1727.
Saunders, N. (1995) *Phil. Trans. A*, **351,** 543.
Saunders, N. (1996a) in *Titanium '95: Science and Technology*, eds Blenckinsop, P. *et al.* (Inst. Materials, London), p. 2167.
Saunders, N. (1996b) *Materials Science Forum*, **217–222,** 667.
Saunders, N. (1996c) In *Superalloys 1996*, eds Kissinger, R. D. *et al.* (TMS, Warrendale), p. 101.
Saunders, N. (1997a) in *Light Metals*, ed. Huglen, R. (TMS, Warrendale, PA), p. 911.
Saunders, N. (1997b) "Phase diagram modelling of TiAl alloys", presented at the Symp. Fundamentals of γ Titanium Aluminides, TMS Annual Meeting, Orlando, Florida, 10–13 February 1997.
Saunders, N. (1997c) unpublished research
Saunders, N. and Rivlin, V. G. (1987) *Z. Metallkde.*, **78,** 795.
Saunders, N. and Sundman, B. (1996) Fe-DATA, a database for calculation of phase equilibria in Fe-based alloys.
Saunders, N., Pan, L.-M., Clay, K., Small, C. and Miodownik, A. P. (1992) in *User Aspects of Phase Diagrams*, ed. Hayes, F. (Inst. Metals, London), p. 64.
Schmidt, R. and Feller-Kniepmeier, M. (1992) *Scripta Met. Mat.*, **26,** 1919.

Seo, D. Y., An, S. U., Bieler, T. R., Larsen, D. E., Bhowal, P. and Merrick, H. (1995) in *Gamma Titanium Aluminides*, eds Kim, Y.-W. *et al.* (TMS, Warrendale), p. 745.

Sharma, R. A. and Richardson, F. D. (1965) *Trans. AIME*, **233**, 1586.

Shaw, S. K. (1992) University of Birmingham, Edgbaston, Birmingham, UK, private communication.

Shimanuki, Y., Masui, M. and Doi, H. (1976) *Scripta Met.*, **10**, 805.

Sigli, C., Vichery, H. and Grange, V. (1996) *Materials Science Forum*, **217–222**, 391.

Sims, C. T. (1987) in *Superalloys II*, eds Sims, T. C. *et al.* (Wiley Interscience, New York), p. 217.

Small, C. (1993) Rolls-Royce plc, Derby DE24 8BJ, UK, private communication.

Spencer, P. J. and Putland, F. H., (1973) *J. Iron & Steel Inst.*, **211**, 293.

Sugimoto, T., Kamei, K., Komatsu, S. and Sugimoto, K. (1985) *Titanium Science and Technology*, eds Lutjering, G. *et al.* (Deutsche. Gess. fur Metallkunde E.V., Oberursal), p. 1583.

Sundman, B. and Ågren, J. (1981) *J. Phys. Chem. Solids*, **42**, 297.

Taylor, J. R. and Dinsdale, A. T. (1990) *CALPHAD*, **14**, 71.

Thorvaldsson, T., Eriksson, H., Kutka, J. and Salwen, A. (1985) in *Proc. Conf. Stainless Steels, Göteburg, Sweden, 1984* (Inst. Metals, London), p. 101.

Trinckhauf, K. and Nembach, E. (1991) *Acta Met. Mater.*, **39**, 3057.

Vahlas, C., Bernard, C. and Madar, R. (1996) in *The SGTE Casebook—Thermodynamics at Work*, ed. Hack, K. (Inst. Materials, London), p. 108.

Van der Molen, E. H., Oblak, J. M. and Kriege, O. H. (1971) *Met. Trans.*, **2**, 1627.

Vernot-Loier, C. and Cortial, F. (1991) in *Superalloys 718, 625 and Various Derivatives*, ed. Loria, A. (TMS, Warrendale, PA), p. 409.

Walston, W. S., Schaeffer, J. C. and Murphy, W. H. (1996) in *Superalloys 1996*, eds Kissinger *et al.* (TMS, Warrendale), p. 9.

Watkin, J. S. (1979) "A review of suggested mechanisms for the phase stability of alloys under irradiation", in Proc. CALPHAD VIII, Stockholm, Sweden, 21–25 May.

Wisell, H. (1991) *Met. Trans. A*, **22A**, 1391.

Withers, P. J., Stobbs, W. M. and Pedersen, O. B. (1989) *Acta Met.*, **37**, 3061.

Wlodek, S. T., Kellu, M. and Alden, D. (1992) in *Superalloys 1992*, eds Antolovich, S. D. *et al.* (TMS, Warrendale), p. 165.

Xu, J., Lin, Y., Zhang, J., You, Y., Zhang, X., Huang, W., Zhang, W. and Ke, J. (1985) in *Rapidly Solidified Materials*, eds Lee, P. W. and Carbonara, R. J. (ASM, Metals Park), p. 283.

Yang, R, Leake, J. A. and Cahn, R. W. (1992a) *Mater. Sci. Eng.*, **A152**, 227.

Yang, R., Saunders, N., Leake, J. A. and Cahn, R. W. (1992b) *Acta Met. Mater.*, **40**, 1553.

Yoder, G. R., Froes, F. H. and Eylon, D. (1984) *Met. Trans. A*, **15A**, 183.

Zou, H. and Kirkaldy, J. S. (1992) *Met. Trans. A*, **23A**, 651.

Zou, J., Wang, H. P., Doherty, R. and Perry, E. M. (1992) in *Superalloys 1992*, eds Antolovich, S. D. *et al.* (TMS, Warrendale), p. 165.

Chapter 11
Combining Thermodynamics and Kinetics

Chapter 11
Combining Thermodynamics and Kinetics

11.1. INTRODUCTION

Knowledge of phase equilibria in its own right is a powerful tool for the understanding of materials behaviour even in situations where true equilibrium is never reached. Knowledge of critical temperatures, pressure and composition ranges for the formation of phases is very important in defining limiting parameters for usage or processing. However, certain processes develop microstructures which are well away from the equilibrium state, for example rapid solidification, mechanical alloying, vapour deposition, etc., and these cannot be directly interpreted using equilibrium phase diagrams. Phase equilibria are also not directly applicable to predicting the development of microstructure and, in this case, some further knowledge and application of phase transformation theory is necessary.

These limitations to a 'traditional' CALPHAD approach have long been understood and the combination of thermodynamics and kinetics is seen as a logical extension of the CALPHAD methodology. While phase transformation theory is often seen as a somewhat separate area, certain critical input parameters can only be obtained from the understanding of phase equilibria. For example, the simplest case of transformation in steels requires, at the very least, knowledge of the temperature below which austenite becomes unstable to the precipitation of ferrite ($T^{\gamma\rightarrow\alpha}$). If one wishes to use some form of growth equation, the driving force for the growth of ferrite in austenite is required. This driving force is often approximated as some function of the undercooling below $T^{\gamma\rightarrow\alpha}$ and hidden within a series of adjustable parameters which take into account other effects. It is, therefore, not always explicitly recognised as a thermodynamic input parameter.

The aim of this chapter is to introduce the reader to some of the ways in which the CALPHAD approach has been combined with kinetics to predict the formation of phases and/or microstructures under conditions which are not considered to be in equilibrium. Broadly speaking, the combination of thermodynamics and kinetics can be broken down into at least two separate approaches: (1) the calculation of metastable equilibria and (2) the direct coupling of thermodynamic and kinetic modelling.

In its simplest form, the calculation of metastable equilibria merely requires the 'suspension' of an equilibrium phase, for example, the calculation of the Fe–C phase diagram with or without graphite to give respectively the Fe + graphite phase diagram or the Fe–Fe_3C phase diagram. This is often done empirically; for example the Fe-graphite phase diagram is used for cast irons while the Fe–Fe_3C diagram is used for steels. Rapid solidification processing is another area where kinetic limitations are often arbitrarily placed on the system in order to use conventional G/x diagrams as a tool to help interpret phase formation.

Such an approach should not be confused with the direct coupling of thermodynamic and kinetic parameters. Broadly speaking, this can achieved either

1. By the direct coupling of thermodynamic and kinetic models in a single software package, where the kinetic model can call the thermodynamic calculation part as a subroutine for the calculation of critical input parameters.

or

2. Thermodynamic calculations are made separately and calculated data are input into a kinetic model, perhaps from a datafile.

Direct coupling has a number of significant advantages as 'local' effects can be directly calculated rather than globally approximated under a single external input. The present chapter will present detailed examples of the above approaches and discuss their advantages/disadvantages and areas of applicability.

11.2. THE CALCULATION OF METASTABLE EQUILIBRIA

11.2.1 General concepts and sample calculations

An advantage of the CALPHAD method over experimental measurement of phase diagrams is that it is possible to make predictions for the extension of various phase fields outside their range of equilibrium stability. For example, Ansara (1979) was able to straightforwardly show the variety of metastable phase diagrams that could exist for a system such as Ni–Ti (Fig. 11.1). Some of the diagrams are for regions which would be highly metastable and, therefore, unlikely to be accessible, but the figure clearly shows how a phase diagram could be modified if suppression of various phases were to occur.

Figure 11.2 shows the calculated stable phase diagram for Fe–C from Gustafson (1985) which is characterised by a eutectic between γ-Fe and graphite. This eutectic occurs for many cast irons and graphite forms from the liquid state. However, if the C level is low, as in steels, solidification can occur directly to δ- or γ-Fe. On cooling, the steel becomes unstable to the formation of graphite and there is a driving force for its precipitation. However, the volume change on forming graphite from the solid state is high and the nucleation of graphite becomes difficult. A variety of conditions then lead to the formation of the metastable phase Fe_3C which has transus temperatures only marginally below those found for equilibrium with graphite. It is difficult to overemphasise the importance of this effect as many

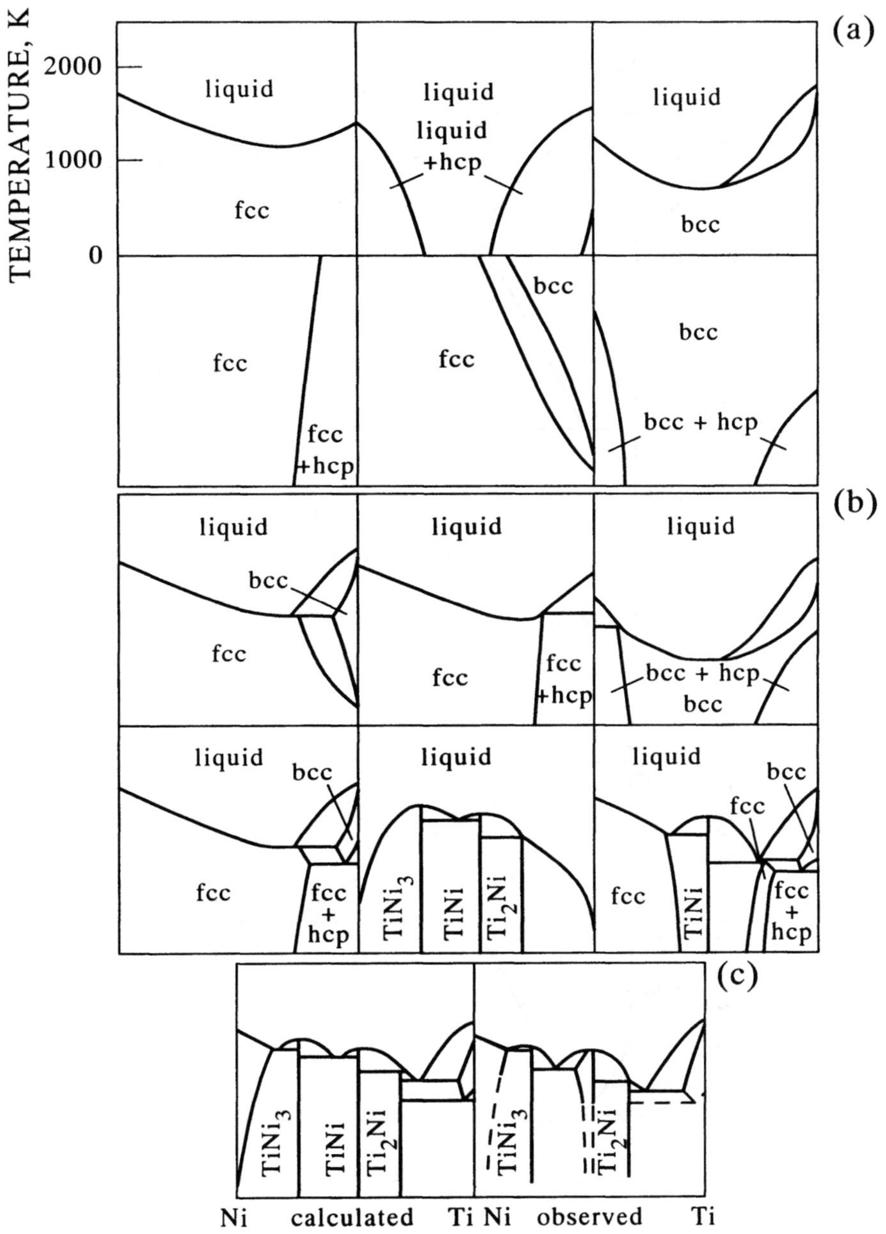

Figure 11.1. Calculated (a) metastable phase diagrams in the Ni–Ti system, (b) some combinations of these digrams and (c) comparison of calculated and observed stable diagram (Ansara 1979).

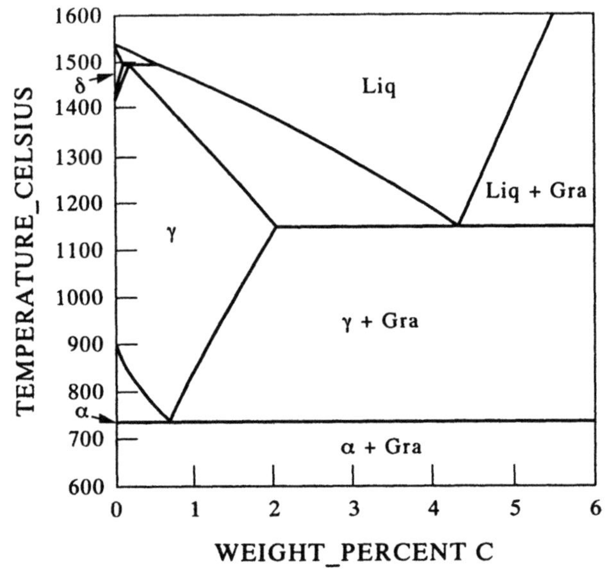

Figure 11.2. Calculated Fe–C phase diagram.

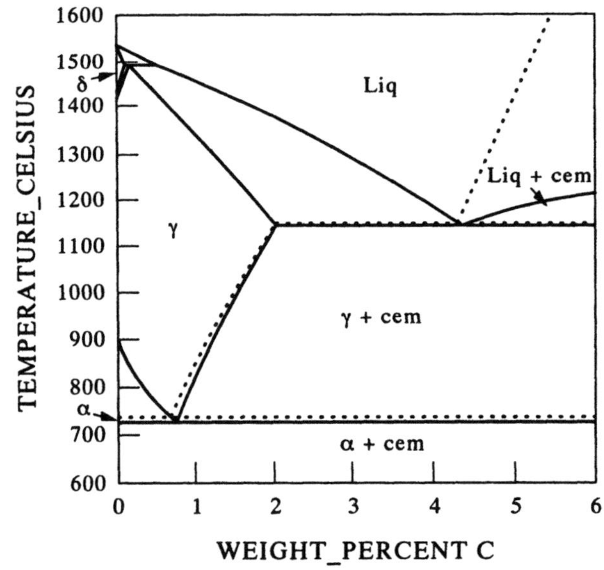

Figure 11.3. Calculated metastable Fe–C phase diagram. Stable diagram with graphite is shown by dotted lines (Gustafson 1985).

common steels are metastable with respect to graphite. It is only because the kinetics of transformation to graphite is so slow that the breakdown of the desirable carbide structure to a much weaker graphite-containing alloy is prevented. This effect has been known empirically for many years and CALPHAD calculations can be made for the metastable Fe–Fe_3C phase diagram (Fig. 11.3). This is done by 'suspending' graphite from the calculations but otherwise keeping the same model parameters for the liquid, f.c.c. and b.c.c. phases.

Another such example occurs in Ti–Si alloys. The stable phase diagram of Ti–Si indicates that, below 1170°C, β- and α-Ti become unstable with respect to Ti_3Si. However, in practice, Ti_3Si only forms after lengthy anneals at high temperature and Ti_5Si_3 is formed instead. Calculations for commercial Ti-alloys (Saunders 1996a) are then routinely made considering only Ti_5Si_3 and its phase boundaries are extrapolated by the calculation below 1170°C.

Another use of metastable diagrams is in the processing of some Nimonic superalloys which are unstable to η phase formation (Betteridge and Heslop 1974). Usually, alloys of this type rely on the precipitation-hardening of γ, the Ni-based matrix phase, by γ', an ordered intermetallic based on Ni_3Al. However, when the ratio of Ti/Al becomes too high, a η-phase based on Ni_3Ti can form as well. As the kinetics of γ' precipitation is faster than η, most processing can be done so that γ' forms instead, and at some service temperatures transformation kinetics may be sluggish enough so that the metastable γ' does not then transform to η. It is therefore of value to be able to predict both the stable and metastable solvus temperatures for η and γ' and understand the behaviour of the alloy with either both phases appearing or with only γ' taking part. A Nimonic 263 alloy with a composition (in wt%) Ni–0.5Al–20Co–20Cr–5.9Mo–2.1Ti–0.001B–0.06C can be used as an example. Figures 11.4 and 11.5 show phase % plots for this alloy with η respectively included and excluded. It is interesting to note that the γ' is only just metastable and its amount when η is excluded is similar to the amount of equilibrium η itself.

While the approach outlined above has proved useful in predicting phase equilibria for metastable alloys, it is rather limited as it does not allow any predictions to be made for the conditions under which metastability can be achieved. This means that the use of the CALPHAD route has to be tempered by the user's judgement and past experience. Any quantification is not straightforward when phase competition is close, for example, in the case of some cast irons where the Fe_3C and graphite phase boundaries are very similar. In this case some cooling conditions give rise to graphite while others give rise to Fe_3C. However, when processing conditions are far from equilibrium it is possible to be more certain about the conditions where metastable equilibria are likely to be observed. An example of such an area is rapid-solidification processing.

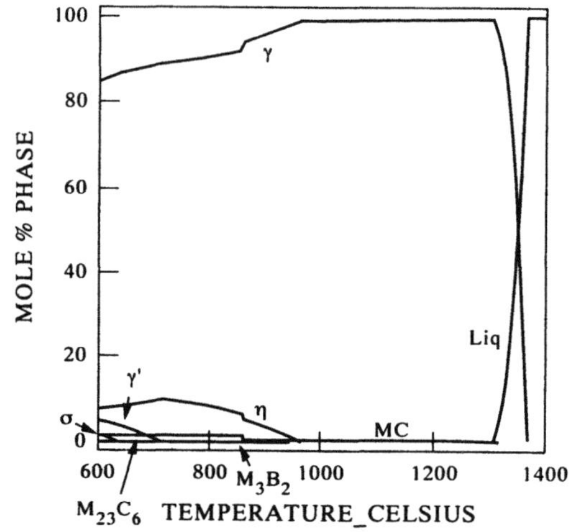

Figure 11.4. Calculated phase % vs temperature plot for a Nimonic 263 Ni-based superalloy.

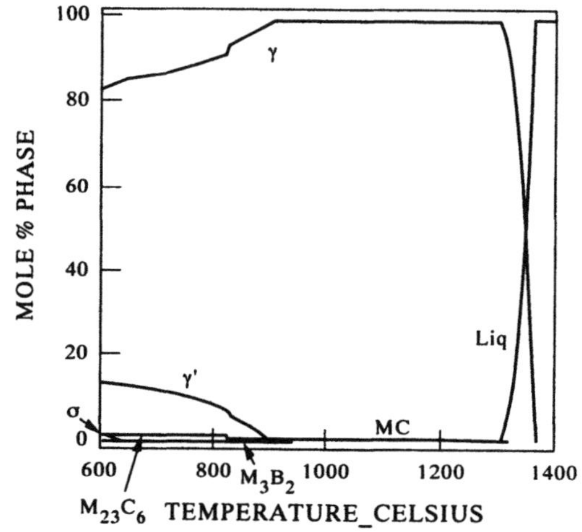

Figure 11.5. Calculated phase % vs temperature plot for a Nimonic 263 Ni-based superalloy with stable η-phase suspended.

11.2.2 Rapid-solidification processing

A simple method for predicting limits for glass-forming ability (GFA) in metallic alloys was proposed by Saunders and Miodownik (1983). They utilised the T_0 criterion to predict the limit to the glass-forming range (GFR) of a number of binary and ternary systems. The T_0 criterion follows the premise that, if cooling conditions

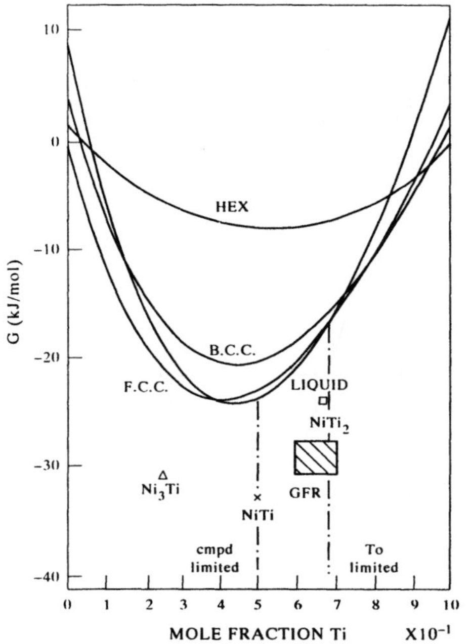

Figure 11.6. Calculated G/x curves for Ni–Ti at T_g (760 K). Experimental glass-forming range (GFR) from Polk *et al.* (1978).

are sufficiently fast to preclude solute redistribution in the liquid, the GFR would be limited by the composition region where the liquid is more stable than other competing single phases at the glass transition temperature (T_g). It is straightforward to find this limit by calculating the G/x diagram for the system of interest at T_g.

CALPHAD calculations were made for a number of binary systems including Au–Si, (Hf,Ti,Zr)–Be and Ni–Ti (Saunders and Miodownik 1983) and a series of ternaries Hf–Ti–Be, Hf–Zr–Be and Hf–Ti–Be (Saunders *et al.* 1985). Figure 11.6 shows such a calculation for Ni–Ti. The results were encouraging in that they predicted with reasonable accuracy the limit to glass formation when the terminal solid solutions were considered. However, there was limited success when taking into account compound phases. To this end the approach was extended to include the kinetics of transformation more explicitly. Remarkably good results were then obtained for a wide variety of binary and ternary system and these are reported in Section 11.3.4.

11.2.3 Solid-state amorphisation

The use of G/x diagrams has found more extensive use in the prediction of GFA by solid-state amorphisation reaction (SSAR) (Johnson 1986, Schwarz 1988). SSAR

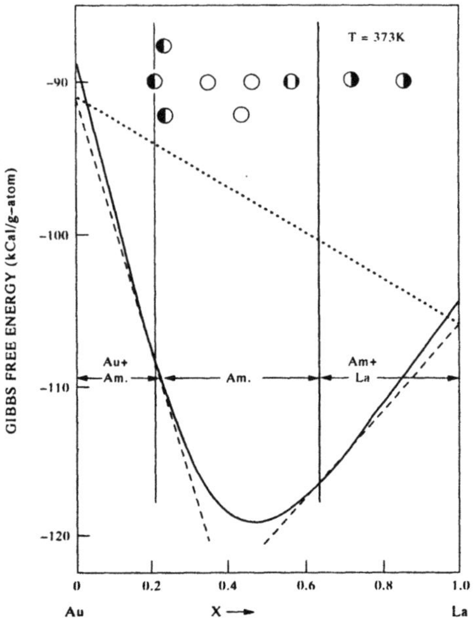

Figure 11.7. G/x curves for Au–La (from Schwarz and Johnson 1983). Open circles indicate fully amorphous structure, partly filled circles indicate partly amorphous structure.

can be achieved in a number of ways, for example by isothermal inter-diffusion reactions between thin films of elements (Schwarz and Johnson 1983) and mechanical alloying of elemental powders (Koch *et al.* 1983). The process relies on the fast diffusion of one of the elements into the other and the kinetics being sufficiently slow to prevent the formation of the intermetallic compounds. The thermodynamic stability of the amorphous phase is then the controlling factor in whether or not an alloy will form a glass.

The early work of Schwarz and Johnson (1983) used a prediction of the underlying thermodynamics of the Au–La system to explain the relative stability of the liquid/amorphous phase in their elemental layered composites (Fig. 11.7). However, they utilised the method proposed by Miedema (1976) for thermodynamic stability of the liquid/amorphous phase. There are clear limitations to the Miedema approach; firstly it is not guaranteed to produce the correct phase diagram and therefore phase competition is at best only approximated, and secondly, the thermodynamics of the terminal solid solutions are chosen quite arbitrarily.

There are a number of studies where the CALPHAD approach to GFA in amorphous systems has been used. Bormann and co-workers (1988, 1990, 1993) and Saunders and Miodownik (1986) have all used G/x diagrams to help explain the formation of amorphous phases in SSAR. This work has shown that as well as

References are listed on pp. 458–461.

predicting the phase diagram, it is possible to demonstrate on a more quantitative basis the relative underlying stability of the various phases. While Saunders and Miodownik (1986) extrapolated the properties of the high-temperature liquid to the SSAR temperature, Bormann *et al.* (1988, 1990) considered the possibility of further stabilisation due to the liquid-to-glass transition and this has significant advantages in helping to understand the thermodynamic properties of the amorphous state.

The use of the CALPHAD route also pointed to a potential reaction sequence that could explain the initial stages of the amorphisation reaction (Saunders and Miodownik 1986). It is clear from the G/x diagrams of a number of systems that a preceding step in the formation of an amorphous phase could be the supersaturation of an element such as Zr by fast-moving atoms such as Co and Ni (Fig. 11.8). This would then lead to a collapse of the crystalline structure to a more stable amorphous phase beyond a critical Co or Ni concentration. Some findings of the supersaturation of Zr by Co in layered structures have been reported (Samwer *et al.* 1988, Pampus *et al.* 1987), which would be consistent with the proposed mechanism, and this was again suggested by Gfeller *et al.* (1988). However, it seems that the main mechanism for formation of the amorphous phase is by reaction at the interface (Johnson 1986).

A further method of producing amorphous phases is by a strain-driven solid-state reaction (Blatter and von Allmen 1985, 1988, Blatter *et al.* 1987, Gfeller *et al.* 1988). It appears that solid solutions of some transition metal-(Ti,Nb) binary systems, which are only stable at high temperatures, can be made amorphous. This is done by first quenching an alloy to retain the high-temperature solid solution. The alloy is then annealed at low temperatures where the amorphous phase appears transiently during the decomposition of the metastable crystalline phase. The effect was explained by the stabilisation of the liquid phase due to the liquid→glass

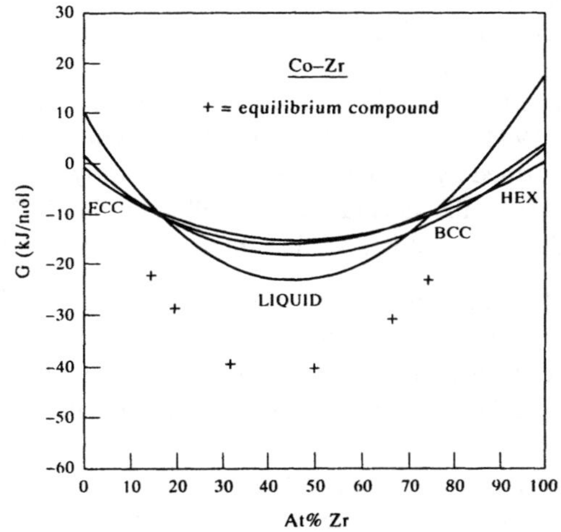

Figure 11.8. Calculated G/x curves for Co–Zr at 200°C (Saunders and Miodownik 1986).

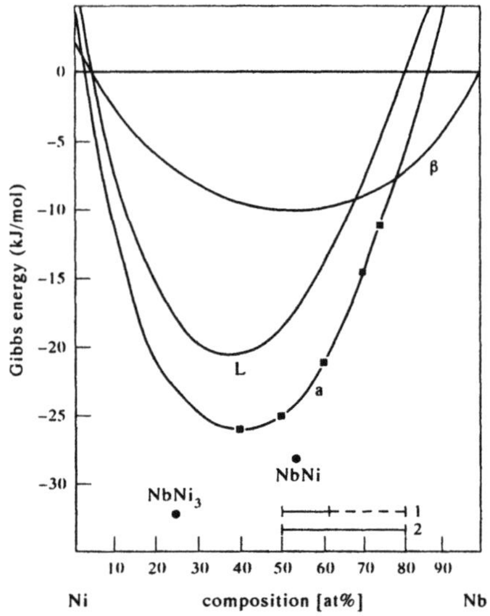

Figure 11.9. Calculated G/x curves for Nb–Ni at 580°C (Gfeller *et al.* 1988). L, β and a represent, respectively, the undercooled liquid, the b.c.c. solid solution and the amorphous phase. (■) are results from enthalpy of crystallisation experiments. Horizontal bars represent amorphous phase by (1) interdiffusion reaction and (2) by laser-quenching.

transition and supported using both calculated G/x curves and experimental measurements of crystallisation enthalpies (Fig. 11.9).

11.2.4 Vapour deposition

The approaches described previously assume that reactions will be suppressed, without giving any specific mechanism, and then rationalise the behaviour of the process using calculated metastable equilibria. A more innovative approach was taken by Saunders (1984) and Saunders and Miodownik (1985, 1987) for the prediction of phases formed by vapour co-deposition of alloys. It was postulated that the formation of phases on the substrate is controlled by the diffusional breakdown of fully intermixed depositing atoms so that three kinetic regimes are observed:

(i) At low substrate temperatures the surface mobility is insufficient for the necessary diffusional movement to transform to multi-phase mixtures. The final structure of the deposited film is, therefore, constrained to be single phase.
(ii) With increasing substrate temperature the decomposition to non-equilibrium multi-phase mixtures is observed.
(iii) When the substrate temperature is high enough, the atomic mobility at the surface is sufficient to allow the necessary atomic re-arrangements for the formation of equilibrium phases.

By using a simple kinetic model it was possible to predict the distance an atom could move on the surface, $\bar{d}$, and show that, when $\bar{d} < 5$ nm, a co-deposited alloy would be constrained to be single-phase. It was further hypothesised that, as depositing atoms lose their kinetic energy within a few atomic vibrations, any transformation involving an intermediate high-temperature phase, whether liquid or solid, was unlikely. As nucleation and growth processes are controlled by the substrate temperature, the phase formed in the first kinetic regime should be the most thermodynamically stable phase at the temperature of the substrate. G/x diagrams for a number of alloys which had been vapour co-deposited were constructed and used to validate the hypothesis. Figures 11.10–11.12 show a few such diagrams with the structures observed during deposition superimposed.

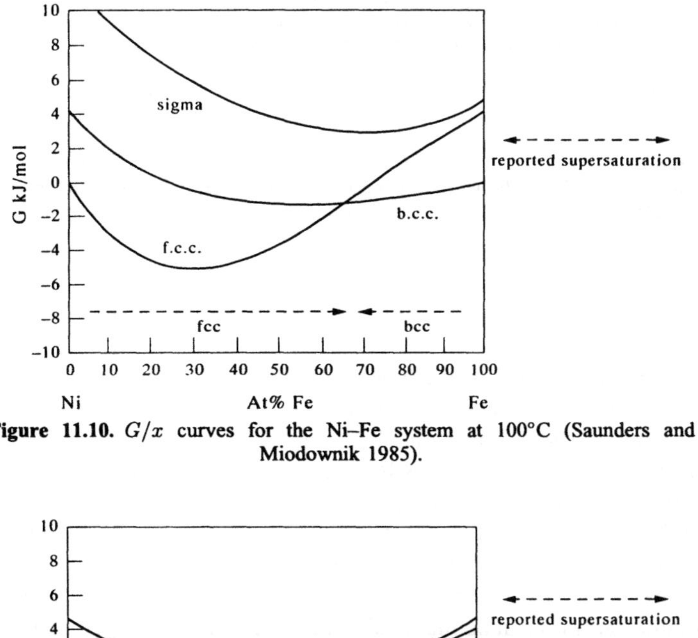

Figure 11.10. G/x curves for the Ni–Fe system at 100°C (Saunders and Miodownik 1985).

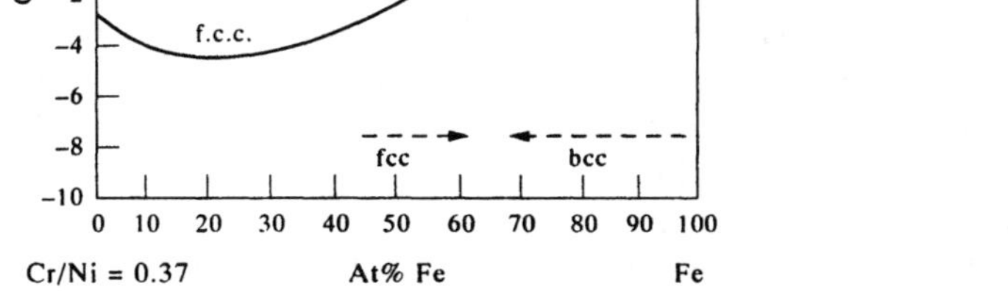

Figure 11.11. G/x curves for the section Cr/Ni = 0.37 of the Fe–Ni–Cr system at 100°C (Saunders and Miodownik 1985).

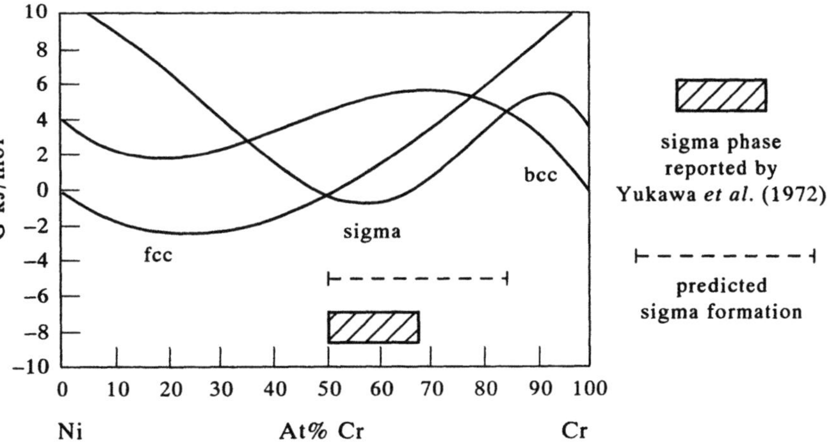

Figure 11.12. G/x curves for the Ni–Cr system at 25°C (Saunders and Miodownik 1985). Hashed box indicates σ phase observed in vapour-deposited alloys (Yukawa *et al.* 1972) and dashed line predicted region of formation.

11.3. THE DIRECT COUPLING OF THERMODYNAMICS AND KINETICS

Historically, the main area of this type of modelling has been in steels and for constructing transformation diagrams (Kirkaldy *et al.* 1978, Bhadeshia 1982, Kirkaldy and Venugopolan 1983, Enomoto 1992, Lee and Bhadeshia 1992). The models have considered simple formalisms for describing the Gibbs energy of the solution phases (Kirkaldy and Baganis 1978, Hasiguchi *et al.* 1984, Enomoto and Aaronson 1985, Sugden and Bhadeshia 1989) which have usually been austenite and ferrite, and in the dilute solution range. The models have been shown to work well and have been extensively used for low to medium alloyed steels. However, although quite complex carbides can be included within the formalisms (Kroupa and Kirkaldy 1993) they are generally limited in their application to only small additions of substitutional elements.

The use of more generally applicable concentrated solution models has, until recently, been somewhat neglected. This has been mainly due to the increase in computational time which is necessary to calculate equilibria with the more complex models and, to a somewhat lesser extent, the lack of good assessments using more generally applicable concentrated solution models. In recent years the Thermo-Calc software programme (Sundman 1991) has been used to calculate the necessary thermodynamic input for kinetic modelling and included in the DICTRA software programme for simulation of diffusional reactions in multi-component

References are listed on pp. 458–461.

alloys (Ågren 1992). This has led to a series of publications for a variety of material types (Liu 1994, Engström *et al.* 1994, Du and Ågren 1994) and has demonstrated the advantage of producing a more generalised approach to phase transformations.

The direct coupling of thermodynamics and kinetics for predicting phase formation in rapidly solidified (RS) alloys has proved a fruitful area of research. The use of explicitly derived thermodynamic input parameters in standard kinetic formalisms has shown that the entropy of fusion is one of the controlling features in the liquid→solid transformation at these high cooling rates. This has led to excellent predictions for the glass-forming ability (GFA) in metallic alloys (Saunders and Miodownik 1986, 1988) and the prediction of solubility extension in RS Al-alloy alloys as a function of cooling rate (Saunders and Tsakiropoulos 1988, Pan *et al.* 1989, Saunders and Miodownik 1991). More complex work which considers diffusional effects during growth has been undertaken by Jönsson and Ågren (1988) and Jönsson (1991).

At more conventional cooling rates the kinetic model proposed by Gulliver (1922) and developed further by Scheil (1942) has been used with excellent success for certain types of alloys. This formalism assumes that back diffusion in the solid during solidification can be considered small enough to be neglected and this process can be simulated by a simple modification of an equilibrium calculation. Predictions for such solidification in Al-alloys can almost completely match extensive and detailed experimental results (Backerud *et al.* 1990). Further, the use of calculated fraction solid vs temperature curves and associated heat evolution leads to superior simulations for the freezing behaviour of an LM25 alloy during sand-casting (Spittle *et al.* 1995).

More complex formalisms have been used with excellent success, particularly for steels where fast diffusion of C in the solid occurs. To account for this Yamada and Matsumiya (1992), Mettinen *et al.* (1992) and Matsumiya *et al.* (1993) have all included the effect of back diffusion in their modelling.

The next four sections will discuss some of the above work in detail. The first two sections concern solid state transformations with section (11.3.1) concentrating on steels while section (11.3.2) looks in more detail at the DICTRA programme. The last two sections concern liquid→solid transformations. Section (11.3.3) deals with conventional solidification while section (11.3.4) deals with rapid solidification.

11.3.1 Phase transformations in steels

This section will begin by looking at how thermodynamic and kinetic modelling has been combined to understand time–temperature–transformation diagrams in steels. The work, for the most part, is semi-empirical in nature, which is forced upon the topic area by difficulties associated with the diffusional transformations, particularly where nucleation aspects have to be considered. The approaches have considered how best to predict the time/temperature conditions for austenite to

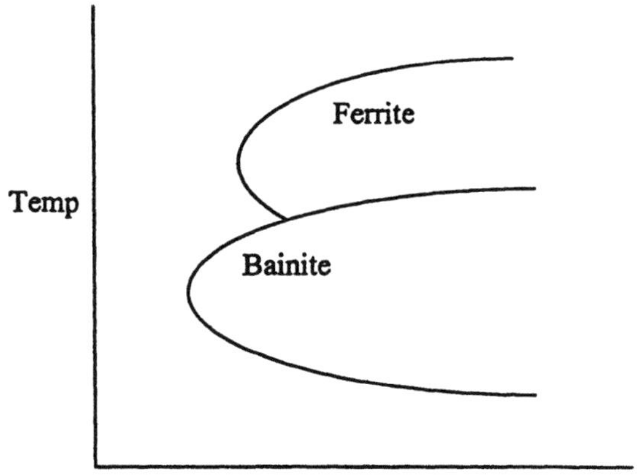

Figure 11.13. Schematic TTT diagram for a low-alloy steel showing ferrite and bainite formation.

transform to various products such as ferrite, pearlite, bainite and martensite. Figure 11.13 shows a simple schematic isothermal transformation diagram for a low-alloy steel and is characterised by the appearance of two 'noses' and a bay. The 'noses', or 'C' curves, correspond to (1) the diffusional transformation of austenite to ferrite and/or pearlite and (2) displacive reactions concerned with the formation of Widmanstätten ferrite and/or bainite. The bay corresponds to an area where there is insufficient undercooling necessary to nucleate the displacive transformations while there is insufficient diffusive movements to allow ferrite or pearlite to form.

***11.3.1.1 The prediction of transformation diagrams after Kirkaldy** et al.* (1978). A model for the calculation of ferrite and pearlite was first presented by Kirkaldy *et al.* (1978) based on Zener–Hillert type expressions (Zener 1946, Hillert 1957). In this first effort, no attempt was made to differentiate between the diffusive and displacive transformations and a overall 'C' curve was produced of the type shown schematically in Fig. 11.14. Kirkaldy *et al.* (1978) used the formalism below where the general formula for the time (τ) to transform x fraction of austenite at a temperature T is given by

$$\tau(x,T) = \frac{1}{\alpha(G)\,\mathrm{d}\Delta T^q} \int_0^x \frac{\mathrm{d}x}{x^{2(1-x)/3}(1-x)^{2x/3}} \qquad (11.1)$$

where $\alpha = \beta 2^{(G-1)/2}$, β is an empirical coefficient, G is the ASTMS grain size, ΔT is the undercooling below the temperature where austenite is unstable with respect to ferrite (the Ae_3 temperature) and q is an exponent dependent on the effective diffusion mechanism.

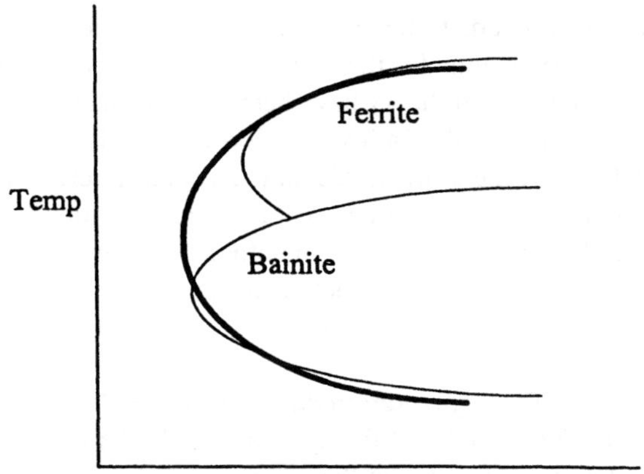

Figure 11.14. Schematic TTT diagram for a low-alloy steels showing ferrite and bainite formation with 'overall' C curve behaviour assumed by Kirkaldy *et al.* (1978).

They also assumed that there was an effective diffusion coefficient (D_{eff}) which was dependent on a series resistance relationship (R) involving the alloying elements such that

$$R_{\text{eff}} \propto \frac{1}{D_{\text{eff}}} \propto \mathrm{e}^{(Q_{\text{eff}}/RT)} \sum_{j=1}^{m} \alpha_j C_j \tag{11.2}$$

where α_j is an constant for each element, j, C_j is the concentration of the element, j, and Q_{eff} is an effective activation energy for diffusion. The modified formula is given below

$$\tau_{\text{TTT}} = \frac{(\Delta T)^{-3}}{2^{(G/8)}} \mathrm{e}^{(Q_{\text{eff}}/RT)} \sum_{j=1}^{m} \alpha_j C_j. \tag{11.3}$$

The critical input parameters are then (1) the grain size, which should be known for each case, (2) the Ae_3 temperature which is calculated from thermodynamics, (3) the effective diffusion activation energy, Q_{eff}, and (4) the empirical constants α_j for each element. Q_{eff} and α_j were determined by empirically fitting curves derived using Eq. (11.12) to experimentally observed TTT curves, and the final formula for calculating τ was given as

$$\tau_{0.1\%} = \frac{\exp^{2000/RT}}{2^{G/8}(Ae_3 - T)^3}(60\text{x}\%\text{C} + 90\text{x}\%\text{Si} + 160\text{x}\%\text{Cr} + 200\text{x}\%\text{Mo}). \tag{11.4}$$

The above derivation means that, once the composition and grain size are provided, only the Ae_3 temperature needs to be calculated. This was done using dilute solution theory via Taylor expansions for the activity coefficients after Wagner (1951). The limitations to dilute solution theory have been discussed earlier (Chapter 5) and mean that, at best, the total level of alloy addition cannot rise much beyond a few per cent. However, for the case of the hardenable steels of interest to Kirkaldy *et al.* (1978) this limitation does not present any real problems, as total levels of C do not rise above 0.7wt% and the total addition of the other alloying elements does not rise above 2.75wt%. Some comparisons between calculated and observed TTT diagrams are shown in Figs. 11.15((a)–(d)).

11.3.1.2 The prediction of transformation diagrams after Bhadeshia (1982). Later work by Bhadeshia (1982) noted that the approach of Kirkaldy *et al.* (1978) could not predict the appearance of the bay in the experimentally observed TTT diagrams of many steels, and he proposed that the onset of transformation was governed by nucleation. He considered that the time period before the onset of a detectable amount of isothermal transformation, τ, could be reasonably defined as the incubation period, τ_s, necessary to establish a steady-state nucleation rate. The following expression for τ_s was then utilised

$$\tau_s \propto \frac{T}{(G_m^v)^p D} \tag{11.5}$$

where D is an effective diffusion coefficient relating to boundary or volume diffusion depending on the coherency state of the nucleus, G_m^v is the maximum volume free-energy change accompanying the formation of a nucleus in a large amount of matrix phase and p is an exponent dependent on the nature of the nucleus. Taking a simple expression for the diffusion coefficient,

$$D \propto \exp(S/R)\exp(-Q/RT) \tag{11.6}$$

where Q and S are respectively the activation enthalpy and entropy for diffusion, assuming $\tau = \tau_s$ and multiplying G_m^v by the molar volume of ferrite, the following expression was obtained

$$\log_e\left[\frac{(G_m)^p \tau}{T}\right] = \frac{Q}{RT} + C_1 \tag{11.7}$$

where Q is the activation enthalpy for diffusion and C_1 is a constant. The critical input parameters are then: (1) G_m which is calculated from thermodynamics, (2) the diffusion coefficient, Q, (3) the exponent p and (4) the constant C_1.

Q, p and C_1 were then determined by empirically fitting curves, derived using Eq. (11.7), to experimentally observed TTT curves. Unfortunately, although good correlation coefficients were obtained during the fitting process, the results proved insufficiently accurate to predict further TTT diagrams. However, it was observed

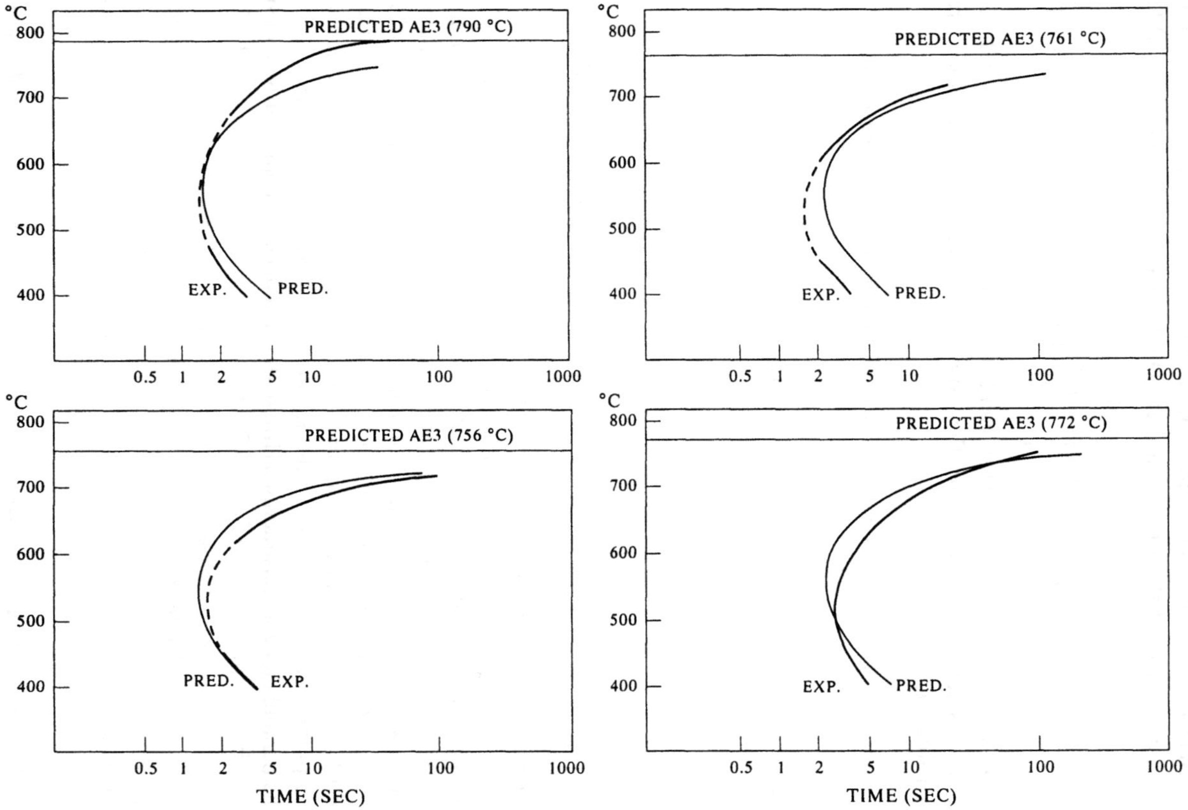

Figure 11.15. Comparison of predicted and experimental TTT start curves for steel types (a) 4135, (b) 5140, (c) 4050 and (d) 4135 from Kirkaldy *et al.* (1978).

that the graph of $\log_e(G_m\tau/T)$ against $1/RT$ exhibited a systematic curvature instead of being linear, as expected from the form of Eq. (11.7). Bhadeshia (1982) noted that this would be consistent with a temperature dependence of the activation enthalpy and entropy for diffusion and further manipulated Eqs (11.5) and (11.6) to derive

$$\log_e\left[\frac{(G_m)^p\tau}{T^z}\right] = \frac{Q'}{RT} + C' \tag{11.8}$$

where z, Q' and C' are constants. This yielded much improved results and the following values for the constants and the exponent p were obtained which are shown in Table 11.1 below.

Table 11.1. Parameters obtained by maximising the correlation coefficient R_1 using Eq. (11.8)

Constant $Q' \times 10^{-6}$ J mol^{-1}	Constant $-C' \times 10^{-2}$	Correlation coefficient R_1	Exponent p	Constant z	C curve
0.2432	1.35	0.97428	5	20	Shear
0.6031	1.905	0.91165	4	20	Diffusional

Bhadeshia (1982) associated the two sets of parameters with (1) diffusional transformations such as the formation of ferrite and (2) shear-related transformations such as bainite or Widmanstätten ferrite. Some additional conditions were made concerning the upper temperature limit to the shear C curve and the M_s temperature was calculated by a thermodynamic method (Bhadeshia 1981). The calculated TTT diagrams were then compared with experiment for a wide number of alloys (Figs. 11.16(a)–(d)) where there was good agreement, particularly with the more medium-alloyed steels where total levels of substitutional elements could be as high as 5wt%.

Bhadeshia (1982) utilised dilute solution thermodynamics to obtain the critical values of G_m for his calculations and also used the parallel tangent construction method to predict the composition of the ferrite which would nucleate from austenite. The parallel tangent construction (Hillert 1953) defines the composition which would have the maximum driving force to form α from γ and is shown graphically in Fig. 11.17, where G_m is the Gibbs energy to precipitate α with the maximum driving force and G_e is the Gibbs energy to form the equilibrium α and γ phases with compositions ${}^e x_C^\alpha$ and ${}^e x_C^\gamma$ respectively. It should be noted that Bhadeshia (1982) also assumed that no partitioning of the substitutional elements was possible, i.e., the transformation proceeded under conditions of para-equilibrium.

11.3.1.3 The prediction of transformation diagrams after Kirkaldy and Venugopolan (1984). Kirkaldy and Venugopolan (1984) refined the work of Kirkaldy *et al.* (1978) by considering that there should be C curves associated with

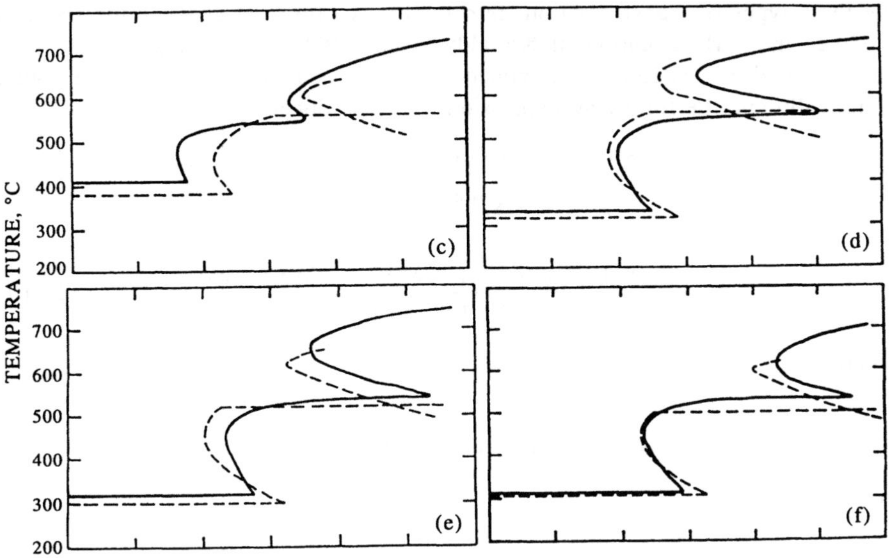

Figure 11.16. Comparison between calculated (– – –) and experimental (— —) TTT diagrams for steel types (a) En 36, (b) En 17, (c) En 24 and (d) En 23 from Bhadeshia (1982).

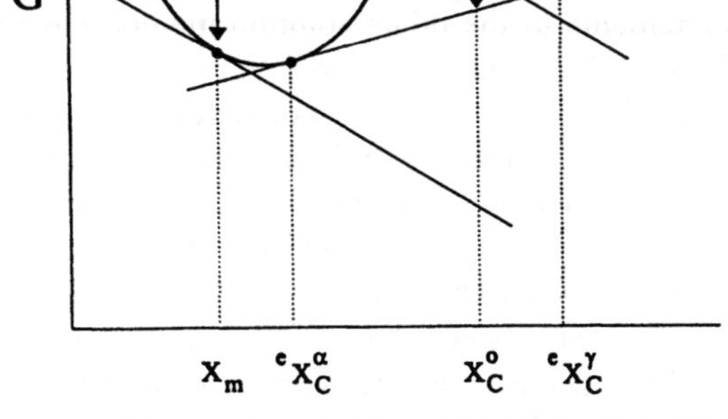

Figure 11.17. Parallel tangent construction used to derive the maximum driving force (G_m) to form α from a γ alloy of composition x_C^o.

the three types of transformation, ferritic (F), pearlitic (P) and bainitic (B) and modified the early equation of Kirkaldy *et al.* (1978), Eq. (11.4), so that three curves could be calculated. Exponents for the equations where again found empirically and the following expressions were produced:

$$\tau_F = \frac{59.6\%\text{Mn} + 1.45\%\text{Ni} + 67.7\%\text{Cr} + 24.4\%\text{Mo}}{2^{(G-1)/2}(\Delta T)^3 \exp(-23500/RT)}.I \qquad (11.9)$$

$$\tau_P = \frac{1.79 + 5.42(\%\text{Cr} + \%\text{Mo} + 4\%\text{Mo} \times \%\text{Ni})}{2^{(G-1)/2}(\Delta T)^3 D}.I \qquad (11.10)$$

where

$$\frac{1}{D} = \frac{1}{\exp(-27,500/RT)} + \frac{0.01\%\text{Cr} + 0.52\%\text{Mo}}{\exp(-37500/RT)} \qquad (11.11)$$

and

$$I = \int_0^x \frac{dx}{x^{2(1-x)/3}(1-x)^{2x/3}}. \qquad (11.12)$$

The bainite reaction was treated in a slightly different way and an expression for τ_B given as

$$\tau_B = \frac{(2.34 + 10.1\%\text{C} + 3.8\%\text{Cr} + 19\%\text{Mo}) \cdot 10^{-4}}{2^{(G-1)/2}(\Delta T)^2 \exp(-27500/RT)}.I' \qquad (11.13)$$

where I' is an empirical modification of I such that $I' > I$ to account for the sluggish bainite termination. It is then possible to account for the observed transformational effects; typical agreements are shown in Figs. 11.18(a)–(b). A more rigorous thermodynamic basis for the underlying thermodynamics was produced by Hashiguchi *et al.* (1984) and this was later incorporated by Kroupa and Kirkaldy (1993). Further refinements to the dilute solution method have been proposed by Sugden and Bhadeshia (1989) and Enomoto (1985).

Lee and Bhadeshia (1992) have recently compared the models of Kirkaldy and Venugopolan (1984) and Bhadeshia (1982) with experimental TTT diagrams of various low- to medium-alloyed steels. They found that low-alloy steels were best fitted using the Kirkaldy and Venugopolan model, while medium-alloy steels worked best with the Bhadeshia model. They further attempted to refine both models but, although some improvement was possible, they still performed best under different alloy regimes and the inherent limitations in both models persisted. The main limitation of the Kirkaldy and Venugopolan model was the underestimation of the transformation time around the 'bay' region and the bainite finish time, while for

References are listed on pp. 458–461.

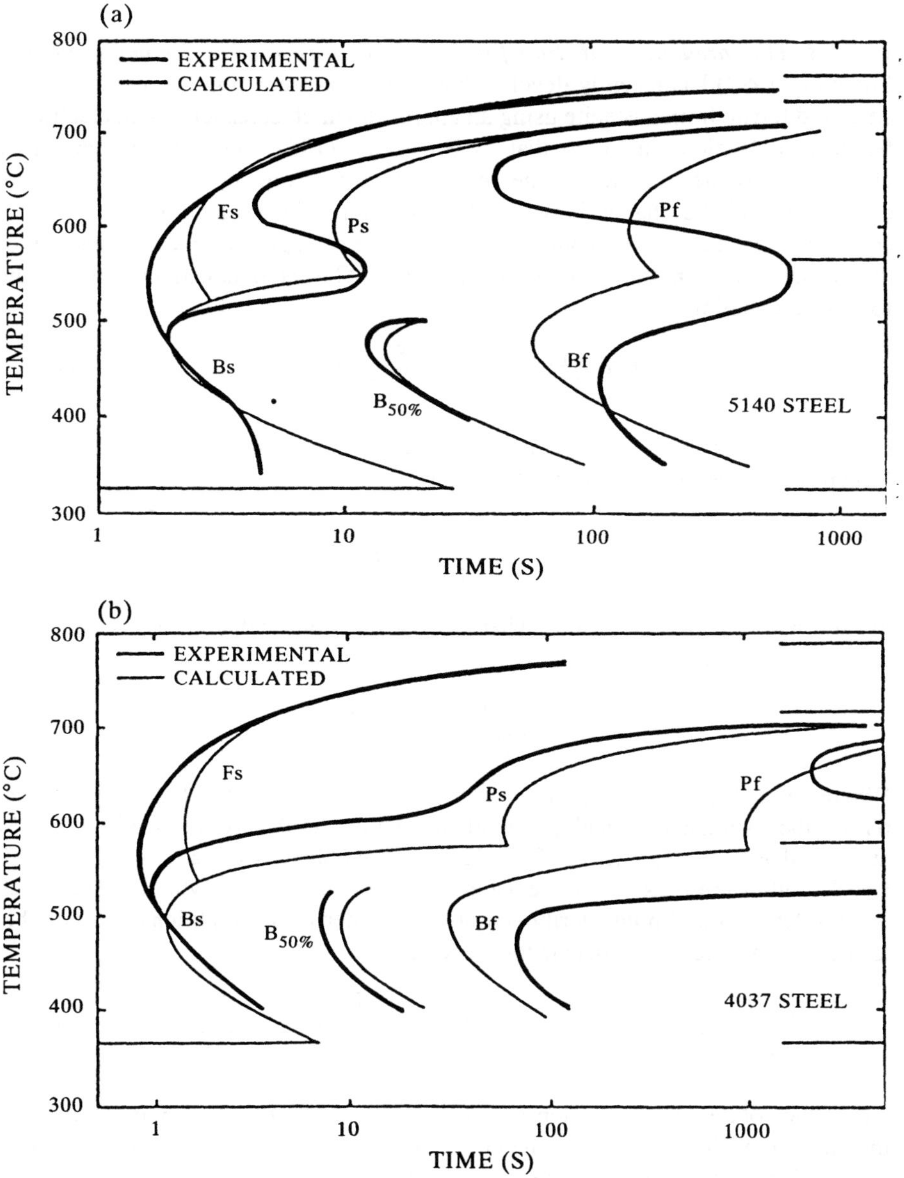

Figure 11.18. Comparison between predicted and experimental TTT diagrams for steel types (a) 5140 and (b) 8630 from Kirkaldy and Venugopolan (1983).

the Bhadeshia model it was only possible to calculate the transformation start time.

11.3.1.4 The prediction of transformation diagrams after Enomoto (1992). Enomoto (1992) has recently developed a model for predicting the growth of pro-eutectoid ferrite from austenite using an approach which considers the nucleation of ferrite in a fashion similar to that of Bhadeshia (1982). In addition, capillarity effects due to the volume change in transforming from austenite to ferrite are considered and the change between the condition of para-equilibrium and local equilibrium is taken into account. In the presence of capillarity the composition of the nucleus, x_i^N, in the ternary system Fe–M–C was obtained by solving the following equations:

$$\frac{\mu_{\mathrm{Fe}}^{\gamma}-\mu_{\mathrm{Fe}}^{N}}{\bar{V}_{\mathrm{Fe}}^{\alpha}}=\frac{\mu_{\mathrm{M}}^{\gamma}-\mu_{\mathrm{M}}^{N}}{\bar{V}_{\mathrm{M}}^{\alpha}}=\frac{\mu_{\mathrm{C}}^{\gamma}-\mu_{\mathrm{C}}^{N}}{\bar{V}_{\mathrm{C}}^{\alpha}}. \tag{11.14}$$

Once the composition of the critical nucleus is found, the driving force for its formation can be obtained using the equation:

$$G_v=\frac{1}{V_\alpha}\sum_{i=1}^{3}(\mu_i^{\gamma}-\mu_i^{N})x_i^N. \tag{11.15}$$

For the case of para-equilibrium, where M is a substitutional atom, x_i^N is calculated from the equation

$$\frac{\mu_s^{\gamma}-\mu_s^{N}}{\bar{V}_s^{\alpha}}=\frac{\mu_{\mathrm{C}}^{\gamma}-\mu_{\mathrm{C}}^{N}}{\bar{V}_{\mathrm{C}}^{\alpha}} \tag{11.16}$$

Where $\mu_s=(\mu_{\mathrm{Fe}}-\theta\mu_{\mathrm{M}})/(1+\theta)$ and $\bar{V}_s=(\bar{V}_{\mathrm{Fe}}+\theta\bar{V}_{\mathrm{M}})/(1+\theta)$ are, respectively, the average chemical potential and partial molar volumes of Fe and M atoms and $\theta=x_{\mathrm{M}}^{\mathrm{o}}/x_{\mathrm{Fe}}^{\mathrm{o}}$, where $x_{\mathrm{Fe}}^{\mathrm{o}}$ and $x_{\mathrm{M}}^{\mathrm{o}}$ are the bulk concentrations of Fe and M. Local equilibrium is assumed to exist at the interface of the growth of ferrite which is considered to be 'momentarily stationary'. Under these conditions it is possible to derive a position for a planar interface as

$$\frac{1-\Omega}{\Omega}\frac{\xi}{1-\xi}=1-\frac{8}{\pi^2}\sum_{n=0}^{\infty}\frac{1}{(2n+1)^2}\exp\left[-\frac{(2n+1)^2\pi^2\tau}{4(1-\xi)^2}\right] \tag{11.17}$$

where the solute supersaturation $\Omega=(x^{\gamma}-x^{\mathrm{o}})/(x^{\gamma}-x^{\alpha})$, $\xi=S/d$ (where S is the position of the interface at time t), $2d$ = the grain size and $\tau=Dt/d^2$, where D is the solute diffusivity. For a spherical particle Eq. (11.17) becomes

$$\frac{1-\Omega}{\Omega}\frac{\xi^3}{1-\xi^3}=1-\frac{6}{\pi^2}\sum_{n=1}^{\infty}\frac{1}{n^2}\exp\left[-\frac{n^2\pi^2\tau}{\xi^2}\right]. \tag{11.18}$$

References are listed on pp. 458–461.

A critical part of the calculations is to calculate the tie-line at the interface corresponding to local equilibrium, and Enomoto (1992) used the 'central atoms' model to predict the thermodynamic properties of α and γ. Some assumptions were made concerning the growth mode and the calculation of this tie-line is dependent on whether growth occurred under the following alternative conditions:

(1) Local equilibrium, i.e., both the substitutional elements and carbon partition between austenite and ferrite. This was called the PLE mode and was taken to occur at high temperatures.
(2) Para-equilibrium, where only carbon partitions between austenite and ferrite and the concentration of the substitutional elements Fe and Mn does not vary. This was called the PARA mode and occurred at low temperatures.
(3) An intermediate mode between PLE and PARA modes was also taken to occur. In this case only carbon partitions between austenite and ferrite, but there is also a narrow diffusion spike of Fe and Mn at the vicinity of the interface. This was called the NPLE mode.

The calculation of the composition of austenite and ferrite at the interface is further modified by the diffusivity of the various elements and, using the above approach, Enomoto (1992) was able to calculate TTT diagrams for a number of Fe–Mn–C alloys and compare these with experiment (Figs. 11.19(a) and (b)). Results were reasonable and the effect of assumptions concerning planar and spherical growth on results were also discussed. More recently papers by Tanaka *et al.* (1995a–c), Bourne *et al.* (1994) and Akbay *et al.* (1993) have sought to combine thermodynamics and kinetics in a similar fashion, mainly in the area of growth or dissolution of proeutectoid ferrite.

11.3.2 The DICTRA programme

Probably the most closely coupled thermodynamic and kinetic approach to phase transformations has been undertaken in the development of the DICTRA programme (Ågren 1992). The philosophy has been to make a software programme which can access Thermo-Calc and utilise a full range of thermodynamic models. It is therefore more generalised and can be applied to a variety of material types. For example, it is possible to deal with such diverse problems as the dissolution of cementite in Fe–Cr–C alloys (Liu *et al.* 1991), the nitriding and nitrocarburizing of Fe (Du and Ågren 1994), the carburizing of Ni-based alloys (Engström *et al.* 1994), the growth of σ in Ni-based superalloys (Jönsson *et al.* 1993) and solidification (Ågren 1995).

The DICTRA programme is based on a numerical solution of multi-component diffusion equations assuming that thermodynamic equilibrium is locally maintained at phase interfaces. Essentially the programme is broken down into four modules which involve: (1) the solution of the diffusion equations, (2) the calculation of

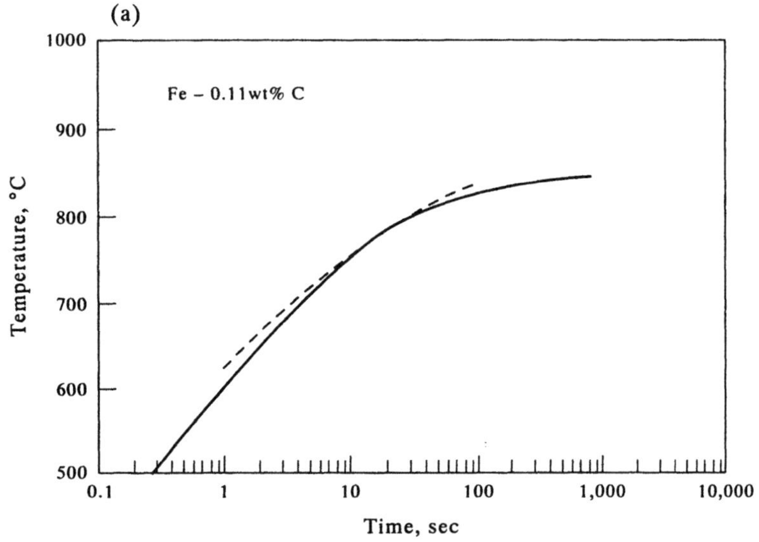

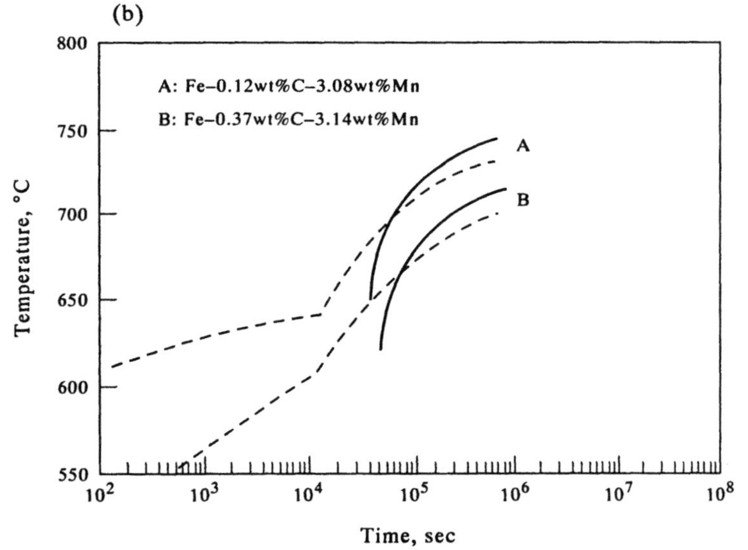

Figure 11.19. Comparison between predicted (——) and experimental (– – –) TTT curves for various steels (Enomoto 1992).

thermodynamic equilibrium (Thermo-Calc is called a sub-routine) (3) the solution of flux-balance equations and (4) the displacement of phase interface positions and adjustment of grid points.

A sample calculation for growth of a second phase in a matrix phase (i.e., the growth of ferrite in austenite) would take the following procedure. The flux balance equation at the moving phase interface is formulated as

$$v^{\alpha} c_i^{\alpha} - v^{\gamma} c_i^{\gamma} = J_i^{\alpha} - J_i^{\gamma} \tag{11.19}$$

where c_i^{α} and c_i^{γ} are the concentrations of component i in the α and γ phases at the phase interface, v^{α} and v^{β} are interface migration rates and J_i^{α} and J_i^{γ} correspond to the diffusional flux of component i in the α and γ phases either side of the phase interface. The latter can be obtained from Fick's law where

$$J_i^{\alpha} = -D_i^{\alpha} \frac{\partial c_i^{\alpha}}{\partial z} \quad \text{and} \quad J_i^{\gamma} = -D_i^{\gamma} \frac{\partial c_i^{\gamma}}{\partial z}. \tag{11.20}$$

A frame of reference for the phase interface is then taken, either 'number-fixed' with respect to the number of moles of components, i, (Ågren 1992) where

$$\sum_{i=1}^{n} J_i^{\alpha} = \sum_{i=1}^{n} J_i^{\gamma} = 0 \tag{11.21}$$

or 'volume-fixed' which is defined in such a way that there is no net flow of volume (Andersson *et al.* 1990)

$$\sum_{i=1}^{n} J_i^{\alpha} V_i^{\alpha} = \sum_{i=1}^{n} J_i^{\gamma} V_i^{\gamma} = 0. \tag{11.22}$$

This leaves $(n-1)$ independent flux balance equations. The rate v can be found straightforwardly for a binary system as the composition of α and γ are fixed at any temperature. In a ternary system there is a further degree of freedom as the number of thermodynamically possible tie-lines between the α and γ phases is infinitely large. However, each tie-line may be specified uniquely by the chemical potential of one of the three components and thus there are only two unknowns and two equations to solve. The above approach can be generalised for multi-component systems and forms the platform for the DICTRA software package.

Examples of DICTRA usage for steels can be found in papers by Liu *et al.* (1991) and Ågren (1992). Liu *et al.* (1991) studied the dissolution of cementite in Fe-2.06Cr-3.91C (at%) alloy and compared this to simulations using DICTRA. They were able to simulate both the rate of dissolution (Fig. 11.20) and the composition profiles of Cr in the sample as a function of time (Fig. 11.21) with a high degree of accuracy. Ågren (1992) reported on the construction of a CCT diagram for grain boundary ferrite in a Fe-0.2C (wt%) alloy previously studied by Jönsson (1991) (Fig. 11.22).

The advantage in using a software package such as DICTRA lies in its general applicability. However, as the solution of the diffusion equations is performed

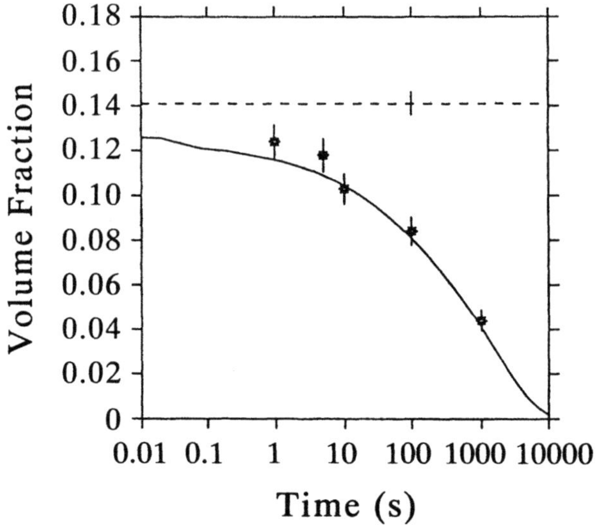

Figure 11.20. Comparison between calculated (— —) and experimental (★) volume fractions of cementite in a Fe–2.06Cr–3.91C (at%) alloy showing dissolution of cementite as a function of time from Liu *et al.* (1991).

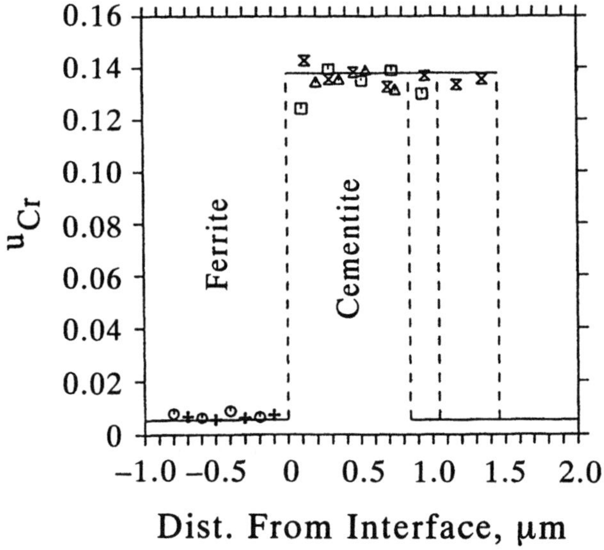

Figure 11.21. Comparison between calculated (— —) and measured (symbols) Cr distribution in ferrite and cemtitie after annealing at 735°C in a Fe–2.06Cr–3.91C (at%) alloy from Liu *et al.* (1991).

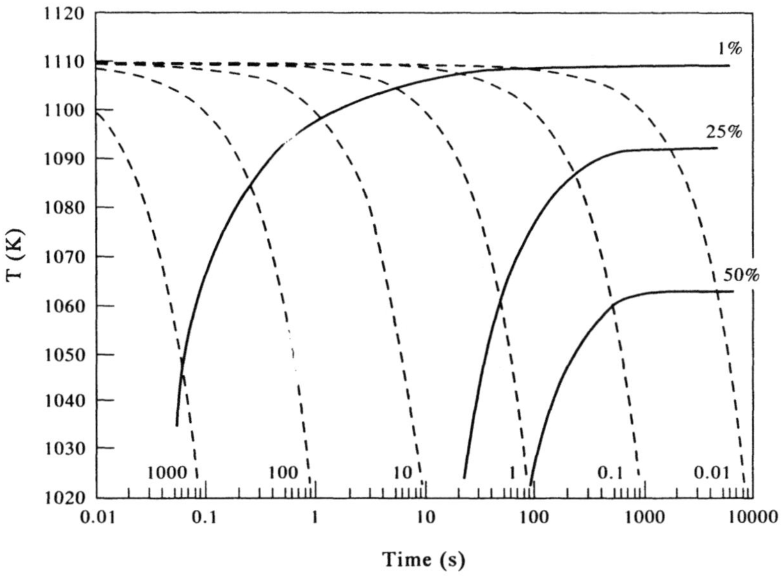

Figure 11.22. Calculated CCT diagram for grain-boundary ferrite in a Fe–0.2C (wt%) alloy from Ågren (1992).

rigorously, and complex thermodynamic modelling is used, there is a time penalty. This means that, although one can consider quite complex alloys and precipitation reactions involving a number of different phases, the computation time can become high if the more complex type of commercial alloys are considered. However, it should be stressed that problems associated with computational power will be overcome as computer speeds increase and the development of such packages will then present the user with a substantially more powerful platform on which phase transformations can be simulated.

11.3.2.1 Diffusion couple problems. Using the same governing principles as described in the previous section, the DICTRA package is also able to simulate situations which, broadly speaking, lie in the realm of diffusion couples; for example, coatings on Ni-based alloys (Engström and Ågren 1996, Engström *et al.* 1997), carburization (Engström *et al.* 1994) and nitriding (Du and Ågren 1994). Instead of starting with a simple alloy, the finite element grid is separated into two parts, one with a composition of the substrate and the other with the composition of the coating. Resulting simulations have been very encouraging.

Engström *et al.* (1997) looked at diffusion couples in Ni–Cr–Al alloys and found that extremely complex diffusion paths were possible, see for example (Figs. 11.23(a) and (b)). These seem strange, but they are all consistent with the observed

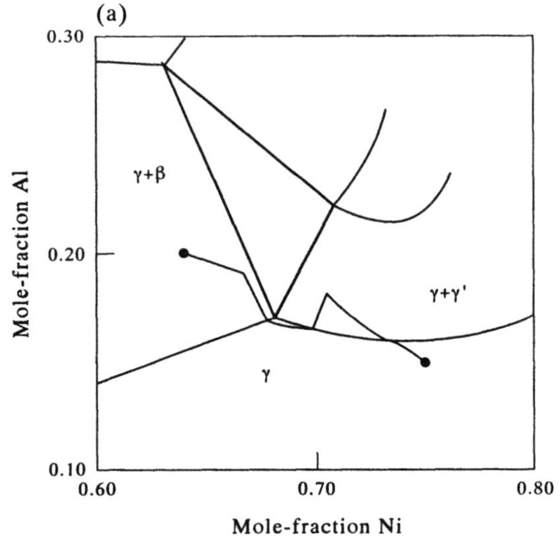

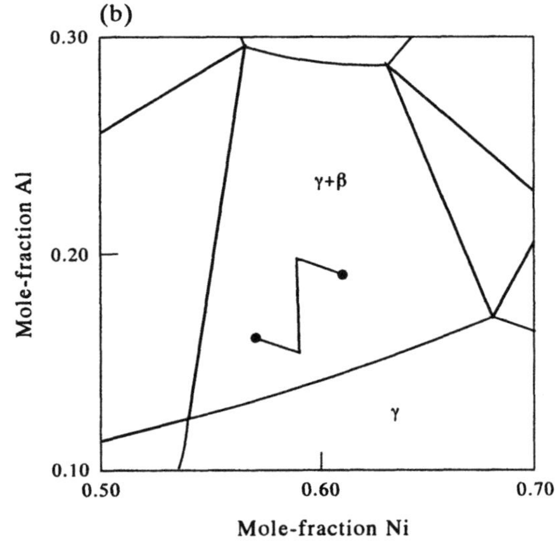

Figure 11.23. Calculated diffusion paths for (a) $\gamma/\gamma+\beta$ and (b) $\gamma+\beta/\gamma+\beta$ diffusion couples in the Ni–Cr–Al system after 100 h at 1200°C (Engström *et al.* 1997).

behaviour of such alloys and can be understood using the ideas proposed by Morral *et al.* (1996) concerning local equilibrium and the formation of three distinct boundary types. Such studies were applied to the more complex situation of a NiCrAl coating on a Ni–19Cr–1.5Al–14Co–4.3Mo–3Ti superalloy. Figure 11.24 shows the

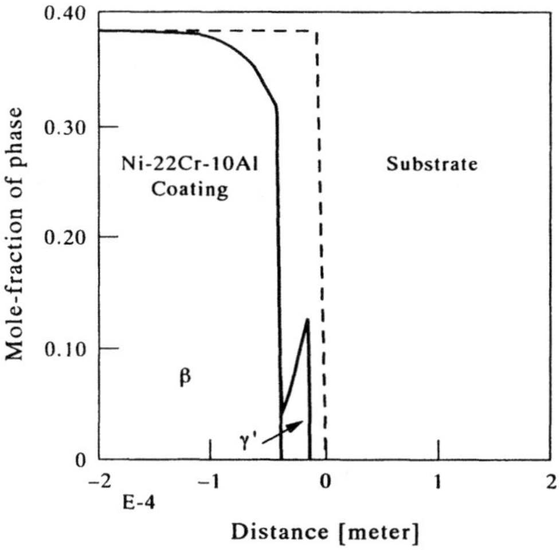

Figure 11.24. Calculated mole fractions phases across the interface of a NiCrAl-coated superalloy after 10 h at 1100°C. Dotted line shows the initial fraction of β in the coating.

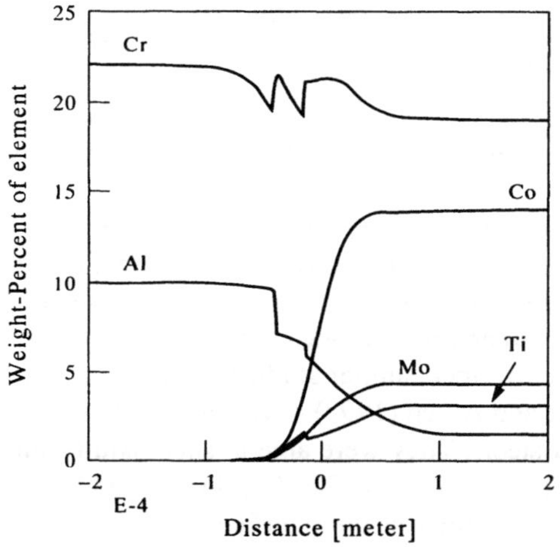

Figure 11.25. Calculated composition profiles (wt%) for various elements across the interface of a NiCrAl-coated superalloy after 10 h at 1100°C.

predicted behaviour of the coating after 10 hrs at 1100°C while Fig. 11.25 shows calculated composition profiles for the same conditions. It is interesting to note the 'spiky' appearance of the Cr plot in Fig. 11.25 around the original interface. This is a real effect and can only be understood when it is realised that, although the

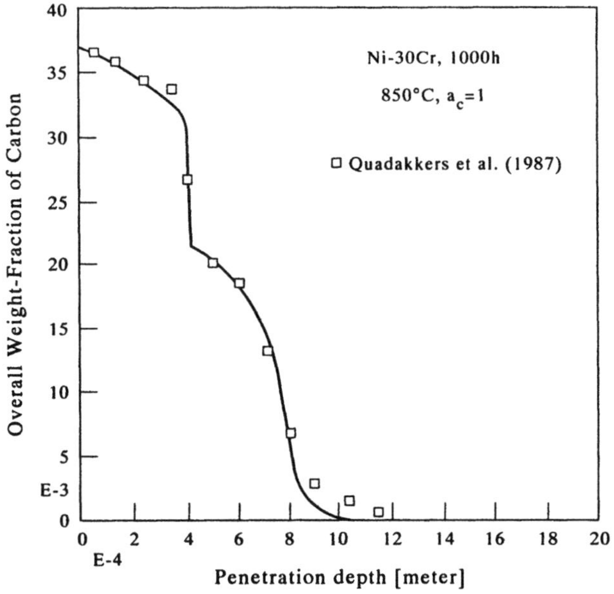

Figure 11.26. Comparison between calculated (— —) and experimental (□) total C concentration (wt fraction) vs penetration depth for a Ni–30Cr (wt%) alloy after carbonization at 850°C for 1000 h (Engström *et al.* 1994).

concentration of Cr can change in an abrupt fashion, its activity changes smoothly through this region.

Carburization and nitriding have proved to be a promising field of application for DICTRA. Engström *et al.* (1994) looked at carburization of Ni–Cr and Ni–Cr–Fe alloys and were able to accurately predict both the penetration of C into the alloys of interest (Fig. 11.26) and the carbides formed during the process (Fig. 11.27). Du and Ågren (1994) performed similar studies for nitriding and carbo-nitriding of Fe. Recently, Helander and Ågren (1997) have simulated multi-component diffusion in joints of dissimilar steels. They were able to successfully simulate features such as C activity and its concentration profile (Fig. 11.28) and the formation of high levels of $M_{23}C_6$ and M_7C_3 carbides at the interface (Fig. 11.29).

11.3.3 Conventional solidification

In almost all practical cases, solidification during more traditional processing routes such as investment casting, strip casting, etc., occurs in a non-equilibrium fashion, giving rise to 'cored' castings with large variations in chemistry across the sample. This leads to a number of problems. Some of these are associated with *metastable* low-melting-point eutectics which can hinder solution heat treatments, which

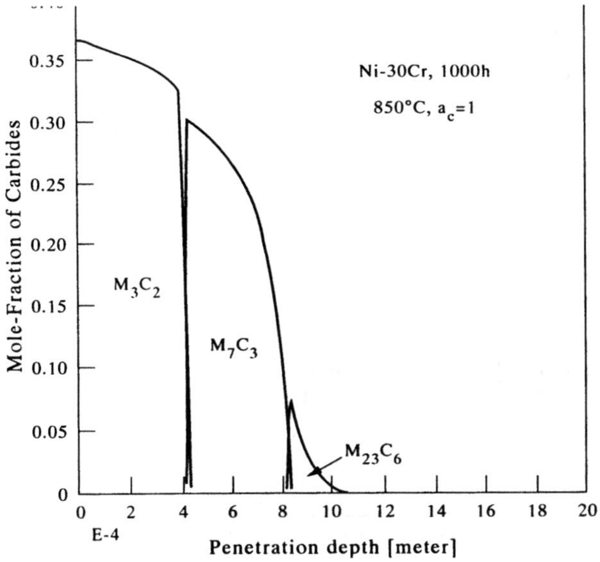

Figure 11.27. Calculated mole fraction of precipitated carbides vs penetration depth for a Ni–30Cr (wt%) alloy after carbonization at 850°C for 1000 h (Engström *et al.* (1994).

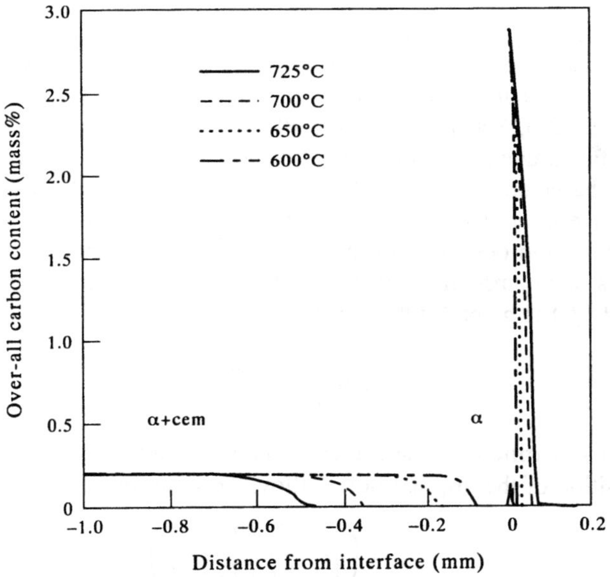

Figure 11.28. Calculated concentration profile for C in a diffusion couple between two dissimilar steels at various temperatures. (From Helander and Ågren 1997.)

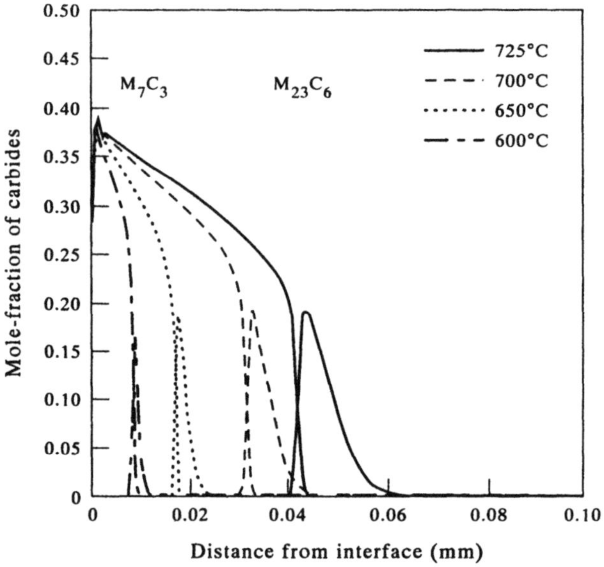

Figure 11.29. Calculated mole fraction of carbides in a diffusion couple between two dissimilar steels at various temperatures. (From Helander and Ågren 1997.)

themselves are designed to homogenise the 'cored' structure. Others are associated with the formation of secondary phases which are detrimental to the mechanical properties of the casting. Early classical treatments of non-equilibrium solidification by Gulliver (1922), Scheil (1942) and Pfann (1952) have led to an equation for predicting the amount of transformation in a binary alloy as a function of temperature based on knowledge of the liquid/solid elemental partition coefficients. It is instructive to briefly derive what has become known as the 'Scheil equation'.

For solidification described by the lever rule and assuming linear liquidus and solidus lines, the composition of the solid (C_s) as a function of the fraction solid transformed (f_s) is given by the equation

$$C_s = \frac{kC_0}{f_s(k-1)+1} \tag{11.23}$$

where k is the partition coefficient and is constant during solidification, and C_0 is the composition of the original liquid alloy. This can be re-arranged to give

$$f_s = \left(\frac{1}{1-k}\right)\left(\frac{T_L - T}{T_f - T}\right) \tag{11.24}$$

where T_L and T_f are the equilibrium liquidus and solidus temperatures. A complementary limiting case to equilibrium solidification is to assume that solute diffusion in the solid phase is small enough to be considered negligible, and that diffusion in the liquid is extremely fast, fast enough to assume that diffusion is complete. In this case Eq. (11.23) can be rewritten as

$$C_s = kC_0(1 - f_s)^{k-1} \tag{11.25}$$

and Eq. (11.24) as

$$f_s = 1 - \left(\frac{T_f - T}{T_f - T_L}\right). \tag{11.26}$$

The treatment above is the traditional derivation of the Scheil equation. However, it is not possible to derive this equation, using the same mathematical method, if the partition coefficient, k, is dependent on temperature and/or composition. The Scheil equation is applicable only to dendritic solidification and cannot, therefore, be applied to eutectic-type alloys such Al–Si-based casting alloys, or even for alloys which may be mainly dendritic in nature but contain some final eutectic product. Further, it cannot be used to predict the formation of intermetallics during solidification.

Using a CALPHAD route all of the above disadvantages can be overcome. The process that physically occurs during 'Scheil' solidification can be envisaged as follows (Fig. 11.30). A liquid of composition C_0 is cooled to a small amount below its liquidus to T_1. It precipitates out solid with a composition C_1^S and the liquid changes its composition to C_1^L. However, on further cooling to T_2 the initial solid cannot change its composition owing to lack of back-diffusion and it is effectively

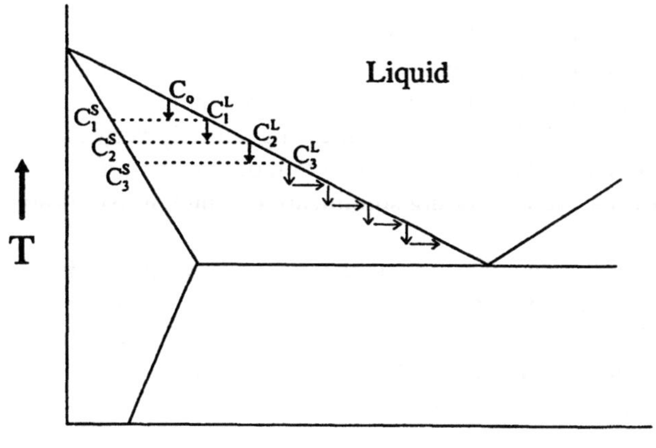

Figure 11.30. Schematic representation of solidification occurring under 'Scheil' conditions.

'isolated'. A local equilibrium is then set up where the liquid of composition C_1^L transforms to a liquid of composition C_2^L and a solid with composition C_2^S which is precipitated onto the original solid with composition C_1^S. This process occurs again on cooling to T_3 where the liquid of composition C_2^L transforms to a liquid of composition C_3^L, and a solid with composition C_3^S grows on the existing solid. This process occurs continuously during cooling and, when $k < 1$, leads to the solid phase becoming lean in solute in the centre of the dendrite and the liquid becoming more and more enriched in solute as solidification proceeds. Eventually, the composition of the liquid will reach the eutectic composition and final solidification will occur via this reaction.

Any appearance of secondary phases can be easily taken into account in this approach with the assumption that no back-diffusion is involved. Therefore all transformations can be handled, including the final eutectic solidification. This approach is based on a series of isothermal steps but, as the temperature step size becomes small, it provides results which are almost completely equivalent to those which would be obtained from continuous cooling.

The procedure described above is simple to model in a computer programme and has a number of significant advantages: (1) The 'Scheil equation' is only applicable to binary alloys and is not easily derived with multiple k values, which would be necessary for a multi-component alloy. A calculation as described above can be applied to an alloy with any number of elements. (2) The partition coefficients need not be constant, which is a prerequisite of the 'Scheil equation'. (3) The 'Scheil equation' cannot take into account other phases which may form during such a solidification process. This is handled straightforwardly by the above calculation route.

11.3.3.1 Using the Scheil solidification model. Although it is realised that some back-diffusion will occur, results on the calculation of solidification using the Scheil simulation have proved to be successful in a number of cases. For Al-alloys it appears to be particularly successful, allowing not only very accurate predictions for fraction solid transformed (f_s) as a function of temperature but also for predicting the phases which appear during solidification, which in Al-alloys, is a complex phenomenon.

Saunders (1997a) made extensive comparisons between predicted solidification behaviour and the experimental work of Backerud *et al.* (1986) who examined nearly 40 Al-alloys, encompassing all types ranging from the 1000 series through the various casting alloys to 7000 series alloys such as AA7075. Their work provides 'fraction solid (f_s)' vs temperature curves for all alloys and detailed information on the phases formed during solidification. All of the alloys studied by Backerud *et al.* (1986) were modelled and results predicted for both f_S vs temperature and phase formation. Figure 11.31 shows comparisons for some of these which encompass most of the major types of Al-alloys, ranging from the

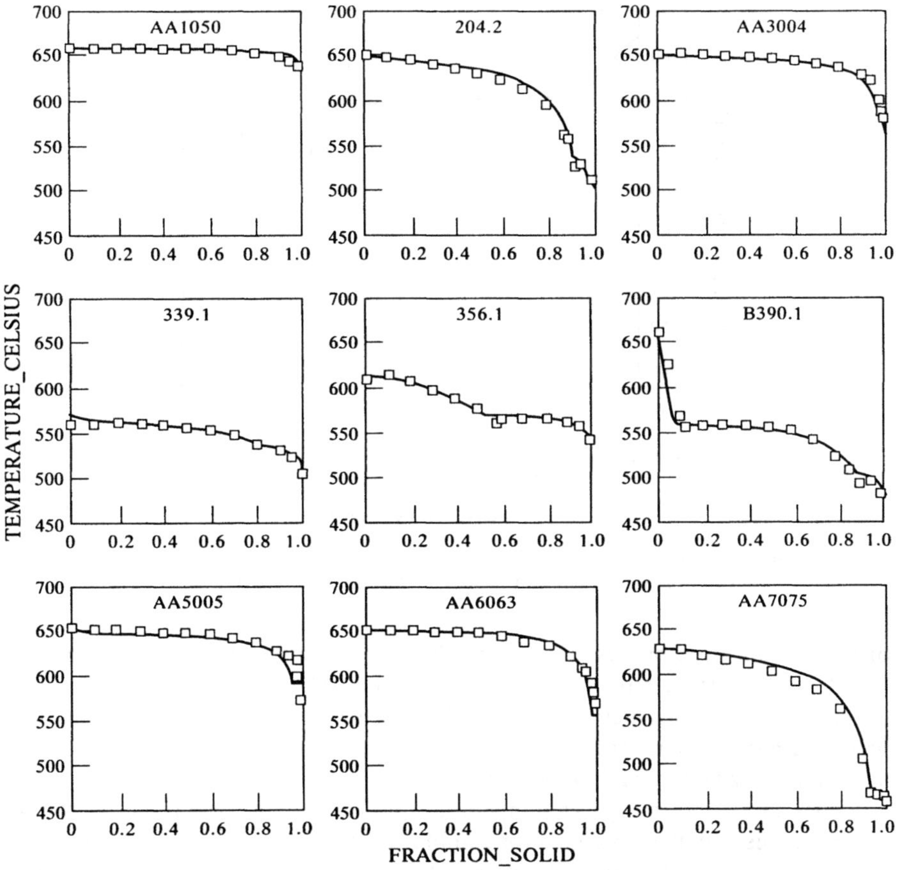

Figure 11.31. Fraction solid vs temperature plots for various Al-alloys calculated under 'Scheil' conditions (Saunders 1997a) with experimental results (□) of Backerud *et al.* (1986) shown for comparison.

commercially pure grade AA1050, through the casting alloys such as the hypo-eutectic type 356/LM25, a complex eutectic type 339.1 and a hyper-eutectic type B390.1, to the wrought types such as AA3004, AA6063 and AA7075. The level of accuracy achieved for these alloys is typical of that attained overall in the comparison with Bakerud *et al.* (1986), and excellent results were also obtained for the phases formed during solidification.

Another case where the 'Scheil' simulation may work reasonably well is in eutectic cast irons (Harding and Saunders 1997). While it is well known that C diffusion in the solid is rapid and that the assumption of no diffusion during solidification is incorrect, cast irons are unique in that graphite is involved for most

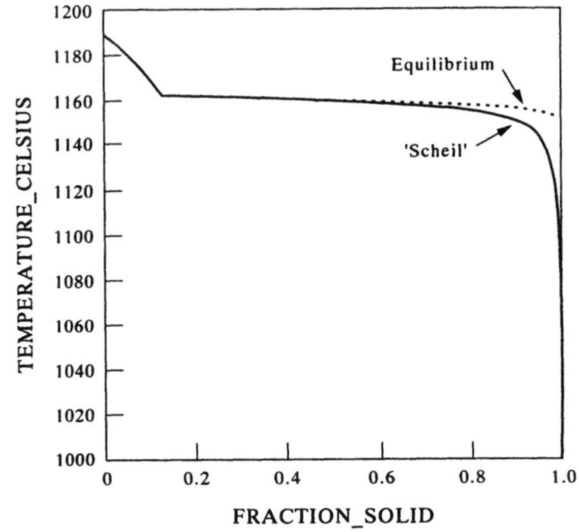

Figure 11.32. Predicted equilibrium and 'Scheil' solidification of an LGI cast iron (Harding and Saunders 1997, Saunders 1997b).

of the solidification process. Therefore the concentration of C in austenite at the liquid–solid interface remains fairly constant and little diffusion of C will actually occur. Some questions may still be raised about the degree to which Si diffuses, but predicted composition profiles in austenite appear consistent with those observed experimentally (Jolley and Gilbert 1987).

Figure 11.32 shows a f_s vs temperature plot for a lamellar grey cast iron and compares this with that predicted for equilibrium solidification. While the behaviour in both cases is similar up to about 70–80% of the transformation, the two curves deviate sharply above this point as the liquid becomes increasingly enriched with elements such as P, Mo and Cr. Final solidification ($f_s > 0.97$) involves both cementite and Fe_3P, both of which can be observed in interdendritic regions of cast irons (Elliott 1988). Although the P level is only 0.05wt%, its formation is consistent with experimental observations of interdendritic phosphide (Janowak and Gundlack 1982). Figure 11.33 shows the solute profile in the liquid and the behaviour of all elements is consistent with experimental observation, particularly the sharp increase of Mo in the final interdendritic liquid. Both the Cr and Mn decrease at the very end of transformation which is associated with the formation of Fe_3C in this last part of solidification.

Simulations for Ni-based superalloys have been carried out by Saunders (1995, 1996b), Chen *et al.* (1994) and Boettinger *et al.* (1995). Saunders (1995) used a straightforward Scheil simulation to predict chemical partitioning and the formation of interdendritic phases in a modified 'single crystal' U720 alloy containing

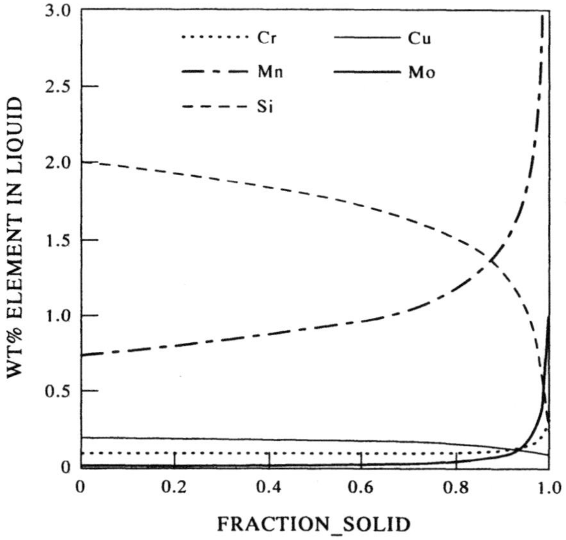

Figure 11.33. Predicted elemental partitioning to the liquid during 'Scheil' solidification of a lamellar grey cast iron (Harding and Saunders 1997, Saunders 1997b).

no B or C. The simulation predicted that the interdendritic regions would contain both γ' and Ni_3Ti, and experimental results (Small 1993) indeed confirmed the presence of both phases, showing that the Scheil simulation was able to produce good results for liquid/solid partitioning.

Recent attempts at solidification modelling have been made for René N4 (Chen *et al.* 1994) and IN718 (Boettinger *et al.* 1995) using a database collected from early Kaufman work. The work on IN718 by Boettinger *et al.* (1995) is in reasonable agreement with the known behaviour of this alloy, and the form of their 'Scheil' plot is similar to that predicted by Saunders (1996b), who also included B and C in the calculation (Fig. 11.34). However, the results for René N4 suggested that a 'Scheil' simulation for this alloy produced a freezing range greater than observed in practice. To account for this apparent discrepancy a new type of solidification mechanism was proposed and a modified 'Scheil' plot was produced. However, it is noted that a straightforward 'Scheil' plot using a recently assessed Ni-database (Saunders 1996b) gives a similar freezing range to the *modified* plot of Chen *et al.* (1994). Clearly there is little need to resort to new, complex solidification mechanisms in this particular case.

11.3.3.2 Modifying the Scheil solidification model. The 'Scheil equation' can be modified to allow some back diffusion into the solid during a small isothermal increment of solidification. The composition of the liquid is then modified to a new

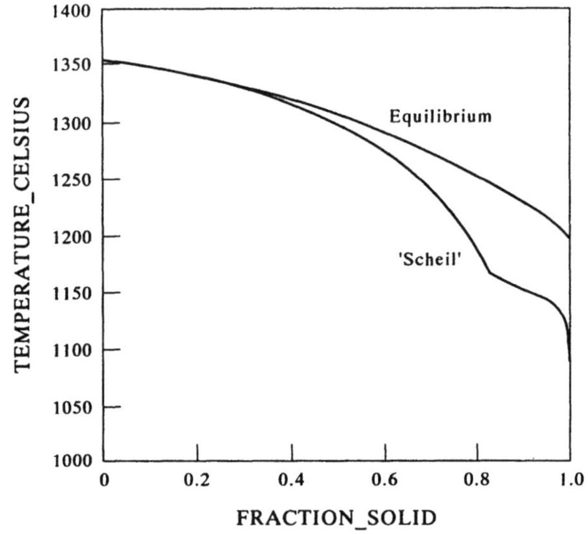

Figure 11.34. Predicted equilibrium and 'Scheil' solidification of an IN718 superalloy.

value and this liquid then proceeds to the next isothermal step. Flux balance needs to be maintained but, if k is kept constant, it is possible to derive a modification to Eqs (11.21) and (11.22) (Brody and Flemings 1966) such that, for parabolic growth,

$$C_s = kC_0\left[1 - (1 - 2\alpha k)f_s\right]^{(k-1)/(1-2\alpha k)} \tag{11.27}$$

and

$$f_s = \left(\frac{1}{1 - 2\alpha k}\right)\left[1 - \left(\frac{T_f - T}{T_f - T_L}\right)^{(1-2\alpha k)/(k-1)}\right] \tag{11.28}$$

where α is a function which takes into account solute back-diffusion given by

$$\alpha = \frac{4D_s t_f}{\lambda^2} \tag{11.29}$$

where D_S is the solute diffusivity, t_f is the local solidification time and λ is the primary dendrite arm spacing. A number of different expressions have been developed for α, notably by Clyne and Kurz (1981) and recently by Ohnaka (1986) and Chen *et al.* (1991).

By ensuring that correct flux balance is maintained, it is quite possible to modify the liquid composition during a simulation and this has been successfully implemented for Al–Li-based alloys by Chen *et al.* (1991) and by Yamada *et al.* (1991), Yamada and Matsumiya (1992) and Matsumiya *et al.* (1993) for steels.

Chen *et al.* (1991) studied the solidification of two Al–Li–Cu alloys in an attempt to predict f_s vs temperature behaviour and the phases formed during solidification. They modified the composition of the liquid in a fashion very similar to Brody and Flemings (1966) but solved the diffusion equations within their programme. Their results produced good predictions for the formation of the secondary phases for both alloys and for f_s of primary phase in one alloy (Al–9at%Li–1at%Cu). However, for a Al–9.9at%Li–7.5at%Cu alloy their predicted f_s for the primary phase was significantly different from that observed experimentally.

Yamada *et al.* (1991) and Yamada and Matsumiya (1992) tried a similar modification but with the composition of the liquid being modified according to Clyne and Kurz (1981). In this case the concentration of the solute, i, in the liquid changes by dC_i given by the equation

$$(1 - f_s)dC_i + 2\Omega_i k_i f_s dC_i = C_i(1 - k_i)df_s \tag{11.30}$$

where k_i is the partition coefficient of solute i between the liquid and solid, df_s is the increment of solid transformed and Ω_i is given by

$$\Omega_i = \alpha_i\left[1 - \exp\left(\frac{-1}{\alpha_i}\right)\right] - \frac{1}{2}\exp\left(\frac{-1}{2}\alpha_i\right) \tag{11.31}$$

where α_i is given by Eq. (11.29). Yamada *et al.* (1991) analysed a simple Fe–Ni–Cr alloy to observe the effect of the diffusion parameter on the f_s vs temperature plot and this is shown in Fig. 11.35. In this case it is clear that the diffusion of Cr and Ni

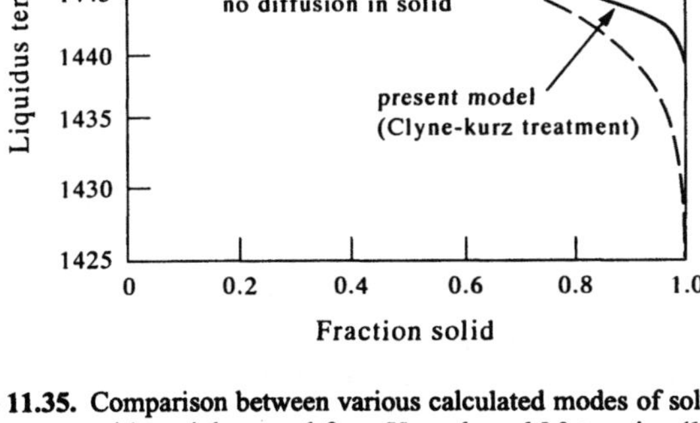

Figure 11.35. Comparison between various calculated modes of solidification in an austenitic stainless steel from Yamada and Matsumiya (1992).

in δ-ferrite is significant and cannot be ignored. They then proceeded to look at the inclusions which were likely to form during solidification, and the work was further extended by Yamada and Matsumiya (1992).

11.3.3.3 More explicit methods of accounting for back-diffusion. It is also possible to take into account the effects of back-diffusion by more explicitly solving the diffusion equations during the solidification process. This requires that the thermodynamic and kinetic parts of the transformation are solved simultaneously. One of the strengths of the DICTRA programme is that, in addition to handling solid-state transformations, it can be applied to solidification. This has been done for stainless steels (DICTRA 1996) and Ni-based superalloys (Saunders 1997b). Figure 11.36 shows solidification of a 'single-crystal' Ni-based superalloy under three conditions: (1) equilibrium, (2) 'Scheil' and (3) taking into account back-diffusion in the solid.

Meittinen *et al.* (1992) and Matsumiya *et al.* (1993) have also attempted to explicitly solve the diffusion equations using a finite difference approach for the diffusion of solute in the solid phase. Meittinen *et al.* (1992) used different approaches dependent on whether the solute was fast-moving or slow-moving and treated solidification involving δ-ferrite and austenite in different ways. This may give reasonable answers for steels but the programme then loses general applicability to other material types.

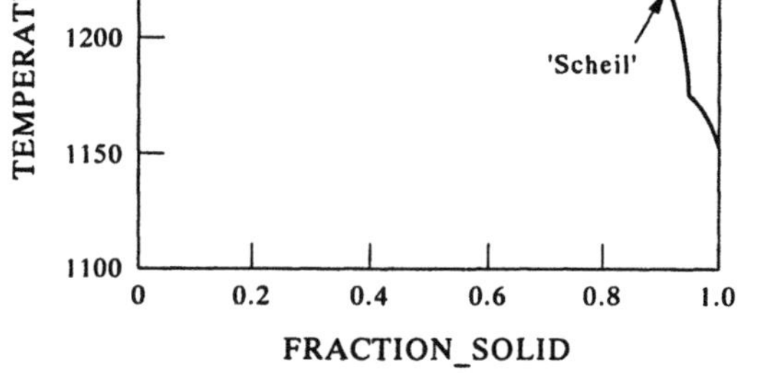

Figure 11.36. Comparison of equilibrium and 'Scheil' solidification for an 'ideal' IN100 superalloy with DICTRA simulation for back-diffusion (Saunders 1997b).

11.3.4 Rapid solidification

Rapid solidification (RS) requires different solutions concerning kinetic questions than to conventional solidification. While solidification at conventional cooling rates mainly concerns itself with diffusion in the solid, and usually takes diffusion in the liquid to be complete, the time scales involved in rapid solidification are so small that diffusion in the solid state can effectively be ignored and the main questions concerning diffusivity lie within the liquid. Also, a number of phenomena such as glass formation and extension of solid solubility limits have been treated by modelling nucleation effects as well as those involving growth.

One of the earliest attempts to explicitly combine thermodynamics and kinetics in rapid solidification was by Saunders *et al.* (1985). They examined the equations derived by Davies (1976) and Uhlmann (1972) for predicting TTT diagrams. These were based on Johnson–Mehl–Avrami kinetics for predicting glass formation during rapid solidification where the ruling equation could be given as

$$t \approx \frac{9.3\eta}{kT}\left(\frac{[K\exp(G^*/kT)]}{[1-\exp(-G_m/RT)]^3}\right)^{1/4} \tag{11.32}$$

where t is the time taken to transform 10^{-6} volume fraction of crystalline phase (the limiting case for glass formation), η the viscosity of the liquid, K is an alloy-dependent constant, G^* the activation energy for nucleation and G_m the Gibbs energy of the liquid→crystal transformation.

The three most important factors in the equation are the viscosity and the thermodynamic parameters G^* and G_m. The viscosity can be approximated between the liquidus temperature, T_{liq}, and the liquid→glass transition temperature, T_g, by a Doolittle expression involving the relative free volume (Ramachandrarao *et al.* 1977) while G^* can be calculated using the relationship

$$G^* = \frac{16\pi}{3}\left(\frac{\sigma_m^3}{G_m^2}\right) \tag{11.33}$$

where σ_m is the liquid/crystal interfacial energy which can in turn be related to the molar heat of fusion, H_m^f, such that

$$\sigma_m = A \cdot H_m^f \tag{11.34}$$

where A is 0.3–0.45 (Turnbull 1950). G_m and H_m^f are calculated directly from thermodynamic modelling.

Equation (11.32) is largely controlled by η and G^*. η is strongly controlled by the value of T_g and alloys with high values of T_{liq}/T_g are generally better glass formers than those where T_{liq}/T_g is low. This is in accord with the finding that the best glass-forming alloys occur at low-melting-point eutectics. The value of G^* is also a powerful factor and most previous treatments had assumed it was constant for all types of liquid→solid transformations. Saunders *et al.* (1985), however, were able to show that G^* could vary markedly both between various alloy systems and between various competing phases in the same alloy system. This was first demonstrated in the Ti–Be system (Saunders *et al.* 1985) where the metastable compound, TiBe, controls the ability of Ti–Be alloys to form glasses. This was successfully modelled (Fig. 11.37(a) and (b)) and the method was further applied to the Au–Si and Ag–Si systems (Kambli *et al.* 1985) where GFA was also controlled by phases not observed in the stable phase diagram (Figs. 11.38 (a) and (b)).

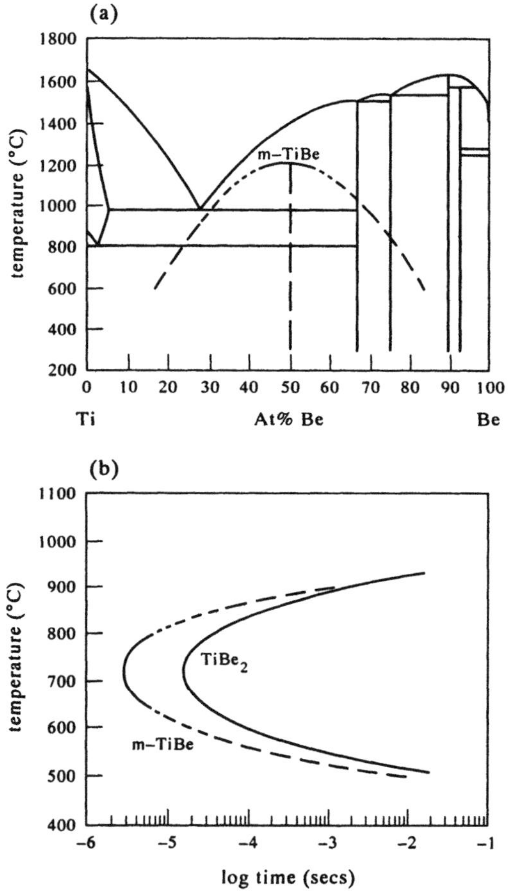

Figure 11.37. (a) Calculated phase diagram for Ti–Be showing liquidus for the metastable compound TiBe and (b) calculated TTT curves for the precipitation of TiBe and $TiBe_2$ from a liquid Ti–40Be (at%) alloy (Saunders *et al.* 1985).

References are listed on pp. 458–461.

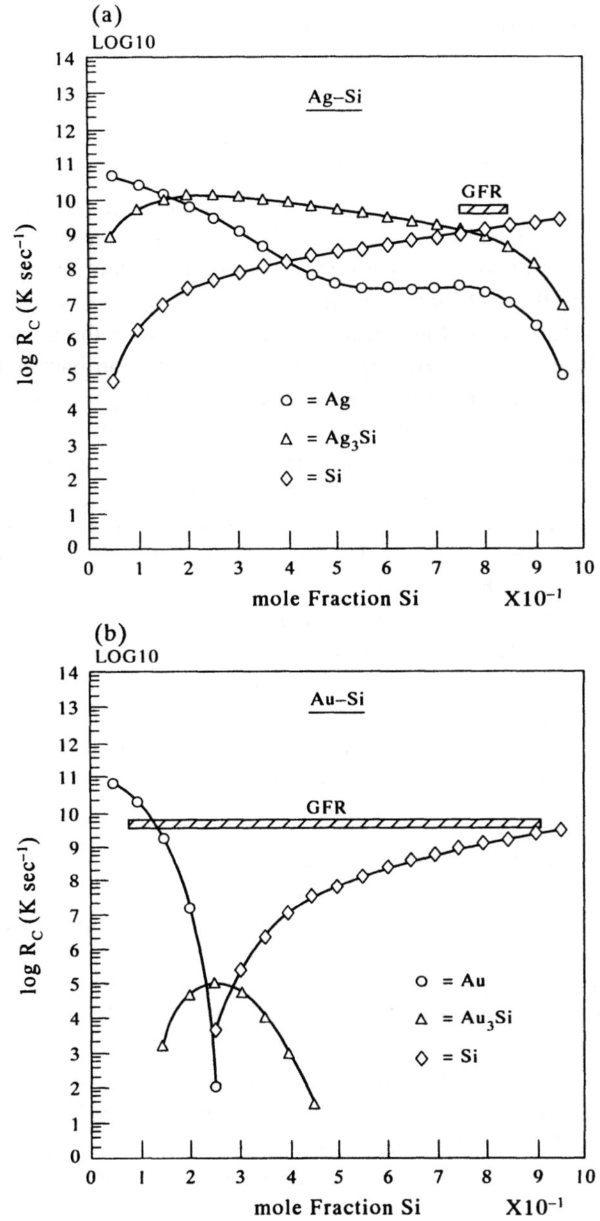

Figure 11.38. Calculated critical cooling rates for suppression of various phases in (a) Ag–Si and (b) Au–Si from Kambli *et al.* (1985). Horizontal bars indicate experimental glass-forming range by laser quenching.

Such effects are largely controlled by the entropy of fusion, S^f_m. A simple derivation of Eqs (11.29) and (11.30) shows that

$$\frac{G^*}{kT} = \frac{S^f_m}{\Delta T_r^2 T_r} \cdot K' \tag{11.35}$$

where $\Delta T_r = (T_{\text{liq}} - T)$, $T_r = T/T_{\text{liq}}$ and K' is a geometric constant based on the shape of the nucleus. Therefore, phases which have a high entropy of fusion will be more difficult to form than those with low entropies of fusion. This can perhaps be best understood by the fact that the work needed to form the interface relates back to the entropy of fusion. As the entropy of fusion increases the work necessary to create the interface will become larger, hence the activation barrier to nucleation increases. The approach culminated in a paper where the glass-forming ability of nearly 20 binary and ternary systems was predicted with excellent results (Saunders and Miodownik 1988). Figure 11.39(a)–(d) show results from some of the binary systems and Fig. 11.40 for the Ti–Zr–B ternary systems. A great strength of the combined thermodynamic and kinetic model is that once an alloy system has been thermodynamically characterised, the only additional input parameter for the software programme is the value of T_g.

A nucleation approach was also applied to predicting the extension of solid solubility obtained in Al-alloys by rapid solidification. In this case the competition for forming either the supersaturated Al solid solution or the equilibrium compound was taken to be nucleation controlled (Saunders and Tsakiropoulos 1988). The model was first applied to high-pressure gas-atomised (HPGA) Al–Cr and Al–Zr alloys and was immediately able to predict a number of important general effects, all of which are consistent with experimental observation.

(1) The necessary undercooling for the formation of compounds is far larger than those for the supersaturated Al phase, undercooling being defined as $(T_{\text{liq}} - T_{\text{crit,d}})$, where $T_{\text{crit,d}}$ is the critical nucleation temperature of the phase in question.
(2) The formation of supersaturated Al is favoured by decreasing powder size/increasing cooling rate and the composition at which the intermetallic phases become rate limiting is shifted to higher concentrations of solute with decreasing powder size/increasing cooling rate.
(3) As had been found in the glass-forming systems, the predicted undercoolings for compounds could vary markedly both between various types alloy systems and between various competing phases in the same alloy system.

Predominant nucleation maps could then be constructed (Figs. 11.41(a) and (b)) which show how the Al solution phase is extended as a function of composition and powder size. Pan *et al.* (1989) improved the model and the approach was further applied by Saunders and Miodownik (1991) to a wide range of Al binary and

References are listed on pp. 458–461.

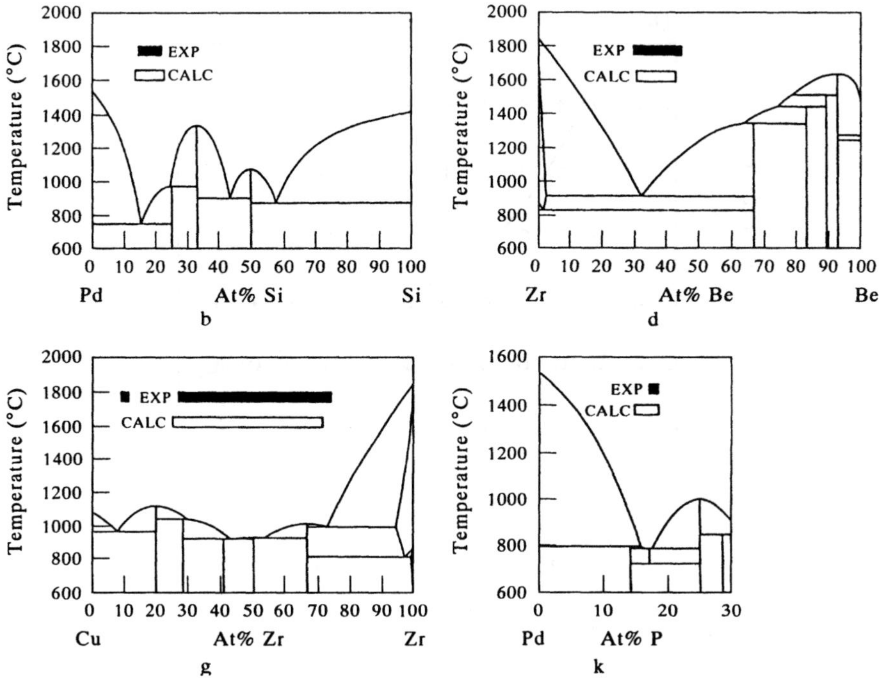

Figure 11.39. Comparison between calculated (□) glass-forming ranges at ~10^7 K sec^{-1} and observed (■) glass-forming ranges at melt spin cooling rates for (a) Pd–Si, (b) Zr–Be, (c) Cu–Zr and (d) Pd–P (Saunders and Miodownik 1988).

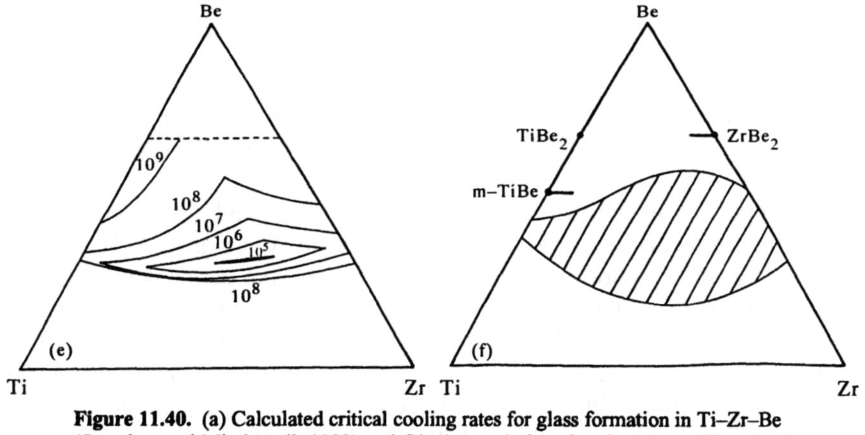

Figure 11.40. (a) Calculated critical cooling rates for glass formation in Ti–Zr–Be (Saunders and Miodownik 1988) and (b) observed glass-forming range (Saunders *et al.* 1985).

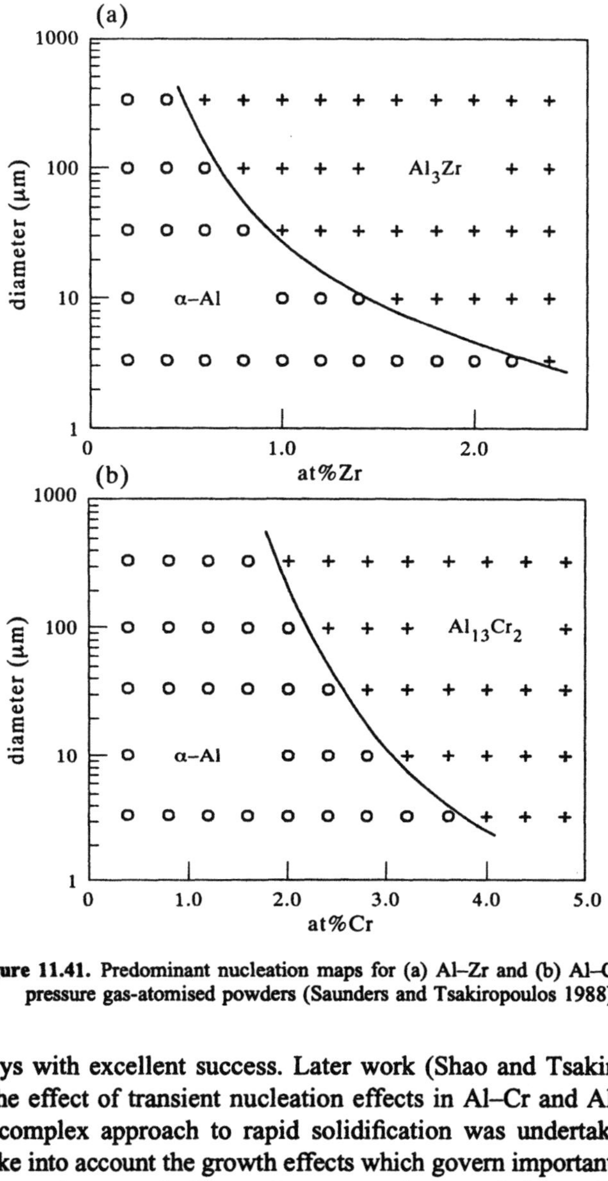

Figure 11.41. Predominant nucleation maps for (a) Al–Zr and (b) Al–Cr high-pressure gas-atomised powders (Saunders and Tsakiropoulos 1988).

ternary alloys with excellent success. Later work (Shao and Tsakiropoulos 1994) examined the effect of transient nucleation effects in Al–Cr and Al–Ti alloys.

A more complex approach to rapid solidification was undertaken by Jonsson (1990) to take into account the growth effects which govern important features of the final RS alloy such as morphology, microsegregation, coupled growth structures, etc. Firstly, the heat-flow characteristics of the RS technique were modelled to define the temperature conditions felt by the undercooled liquid. A nucleation expression similar to Eq. (11.33) was then used, taking into account more specifically the effects of heterogeneous nucleation. Growth was then modelled considering solute diffusivity in the liquid and 'solute drag' effects (Ågren 1989). The solute-drag model considers the deviation from local equilibrium condition such that

$$G_i^1 = G_i^s + \left(\Delta G_f + \Delta G_\sigma + \Delta\mu_i^{sd}\right) \tag{11.36}$$

where G_i^1 and G_i^s are the ordinary chemical potentials of component i in the liquid and solid next to the phase interface, ΔG_f and ΔG_σ are the deviations from local equilibrium due to the finite interface kinetics and the Gibbs–Thompson effect respectively and $\Delta\mu_i^{sd}$ is the deviation due to the solute drag effect of component i. ΔG_f is given by

$$\Delta G_f = -RT\ln\left(1 - \frac{v}{v_0}\right) \tag{11.37}$$

where v is the interface velocity, v_0 is a kinetic parameter of the order of the speed of sound and ΔG_σ is given by

$$\Delta G_\sigma = \sigma\left(\frac{1}{\rho_1} + \frac{1}{\rho_2}\right)V_m \tag{11.38}$$

where σ is the surface tension and ρ_1 and ρ_2 are the principal radii of the interface curvature. To determine $\Delta\mu_i^{sd}$ further limiting conditions have to be made, such as the assumption that growth is steady state and that there is proportionality between the diffusive flux at the liquid side of the interface and the difference in diffusion potential acting over the interface. The following equation can be derived for a binary system A–B (Ågren 1989):

$$v = \frac{-L}{x^{L/s} - x^{s/L}}\left(\Delta\mu_A^{sd} - \Delta\mu_B^{sd}\right) \tag{11.39}$$

where $x^{L/s}$ and $x^{s/L}$ are the interfacial concentrations of B in the liquid and the solid phases respectively. L can be estimated from the equation

$$L = \frac{D_{ti}}{\delta RT}\left(1 - x^{L/s}\right)x^{L/s} \tag{11.40}$$

where D_{ti} is the trans-interphase diffusivity and δ is the width of the interphase. There are further equations which interrelate the various parameters and these can then be used to define a system of non-linear equations which can be solved and allow the calculation of the crystal growth rate. From this it is possible to account for a transition between partitionless massive growth and partitioned growth, using a Cu–13%Ag alloy as an example.

The two growth regimes are associated with correspondingly distinct rates of latent heat evolution and, at some critical heat extraction rate, an interesting oscillation between the two growth types is predicted to occur. At sufficiently high undercoolings, i.e., at high heat extraction rates, nucleation occurs below the temperature where partitionless growth is favoured. Growth is then rapid, which gives rise to sufficient recalescence to bring its temperature back above the temperature where partitioned growth occurs. The rate of latent heat evolution then slows down and the heat extraction rate is sufficient so that the temperature begins to drop again and at some point falls below the critical temperature for partitionless transformation. The cycle can then begin again and a banded microstructure is predicted to occur. Such an effect was found to be consistent with microstructures observed experimentally by Boettinger *et al.* (1984).

REFERENCES

Ågren, J. (1989) *Acta Met.*, **37**, 181.
Ågren, J. (1992) *ISIJ International*, **32**, 291.
Ågren, J. (1995), unpublished research.
Akbay, T., Reed, R. and Enomoto, M. (1993) in *Computer Aided Innovation of New Materials: Vol. 1*, eds Doyama, M *et al.* (Elsevier Science, Amsterdam), p. 771.
Andersson, J.-O., Höglund, L., Jönsson, B. and Ågren, J. (1990) in *Fundamentals and Applications of Ternary Diffusion*, ed. Purdy, G. R. (Pergamon Press, New York), p. 153.
Ansara, I. (1979) *Int. Met. Reviews*, **22**, 1.
Backerud, L., Krol, E. and Tamminen, J. (1986) *Solidification Characteristics of Aluminium Alloys: Vols. 1 and 2* (Tangen Trykk A/S, Oslo).
Betteridge, W. and Heslop, J. (1974) in *The NIMONIC Alloys and Other Ni-Based High Temperature Alloys: 2nd Edition* (Edward Arnold Ltd, London).
Bhadeshia, H. K. D. H. (1981) *Met. Sci.*, **15**, 175.
Bhadeshia, H. K. D. H. (1982) *Met. Sci.*, **16**, 159.
Blatter, A. and von Allmen, M. (1985) *Phys. Rev. Lett.*, **54**, 2103.
Blatter, A. and von Allmen, M. (1988) *Mat. Sci. Eng.*, **97**, 93.
Blatter, A., von Allmen, M. and Baltzer, N. (1987) *J. Appl. Phys.*, **62**, 276.
Boettinger, W. J., Shechtman, D., Schaefer, R. J. and Biancaniello, F. S. (1984) *Met. Trans. A*, **15**, 55.
Boettinger, W. J., Kattner, U. R., Coriell, S. R., Chang, Y. A. and Mueller, B. A. (1995) in *Modeling of Casting, Welding and Advanced Solidification Processes, VII*, eds Cross, M. and Campbell, J. (TMS, Warrendale. PA), p. 649.
Bormann, R. (1993) *Mat. Sci. Eng.*, **A179/A180**, 31.
Bormann, R. and Büsch, R. (1990) *J. Non-Crystal. Solids*, **117–118**, 539.
Bormann, R., Gärtner, F. and Haider, F. (1988) *Mat. Sci. Eng.*, **97**, 79.
Bourne, J. P., Atkinson, C. and Reed, R. C. (1994) *Met. Mater. Trans. A*, **24A**, 2683.
Brody, H. D. and Flemings, M. C. (1966) *Trans. Met. Soc. AIME*, **236**, 615.
Chen, S.-L., Oldfield, W., Chang, Y. A. and Thomas, M. K. (1994) *Met. Mater. Trans. A*, **25A**, 1525.
Chen, S.-W., Chuang, Y.-Y., Chang, Y. A. and Chu, M. G. (1991) *Met. Trans. A*, **22A**, 2837.
Clyne, T. W. and Kurz, W. (1981) *Met. Trans. A*, **12A**, 965.
Davies, H. A. (1976) *Phys. Chem. Glasses*, **17**, 159.

DICTRA (1996) in *DICTRA Examples Manual: Ver. 18* (Div. Physical Metallurgy, Dept. Mater. Sci. Eng., Royal Institute of Technology, S-100 44 Stockholm, Sweden, February), Examples b4a-b4c.
Du, H. and Ågren, J. (1994) Trita-Mac 0570, Report, Royal Institute of Technology, S-100 44 Stockholm, Sweden.
Elliott, R. (1988) in *Cast Iron Technology* (Butterworths, London).
Engström, A. and Ågren, J. (1996) to be published in *Defect and Diffusion Forum*.
Engström, A., Höglund, L. and Ågren, J. (1994) *Met. Mater. Trans. A*, **25A,** 1127.
Engström, A., Morral, J. E. and Ågren, J. (1997) *Acta Mater.*, **45,** 1189.
Enomoto, M. and Aaronson, H. I. (1985) *CALPHAD*, **9,** 43.
Enomoto, M. (1992) *ISIJ International*, **32,** 297.
Gfeller, J., Blatter, A. and Kambli, U. (1988) *J. Less Common Metals*, **145,** 105.
Gulliver, G. M. (1922) *Metallic Alloys* (Griffen, London).
Gustafson, P. (1985) *Scand. J. Metallurgy*, **14,** 259.
Harding, R. and Saunders, N. (1997) to be published in *Trans. American Foundrymen's Society*.
Hashiguchi, K., Kirkaldy, J. S., Fukuzumi, T. and Pavaskar, V. (1984) *CALPHAD*, **8,** 173.
Helander, T. and Ågren, J. (1997) *Met. Mater. Trans. A*, **28A,** 303.
Hillert, M. (1953) *Acta Met.*, **1,** 764.
Hillert, M. (1957) *Jerkontorets Ann.*, **141,** 758.
Janowak, J. F. and Gundlach, R. B. (1982) *Trans. American Foundrymen's Society*, **90,** 847.
Johnson, W. L. (1986) *Progr. Mater. Sci.*, **30,** 81.
Jolley, G. and Gilbert, G. N. J. (1987) *The British Foundryman*, March vol., 79.
Jönsson, B. (1990) "Thermodynamics and kinetics of rapid solidification", Ph.D. Thesis, The Royal Institute of Technology, Stockholm, Sweden.
Jönsson, B. (1991) *Met. Trans. A*, **22A,** 2475.
Jönsson, B. and Ågren, J. (1988) *J. Less Common Metals*, **145,** 153.
Jönsson, B., Ågren, J. and Saunders, N. (1993) unpublished research.
Kambli, U., Von Allmen, M., Saunders, N. and Miodownik, A. P. (1985) *Appl. Phys. A*, **36,** 189.
Kirkaldy, J. S. and Baganis, E. A. (1978b) *Met. Trans.*, **9,** 495.
Kirkaldy, J. S. and Venugopolan, D. (1984) in *Phase Transformations in Ferrous Alloys*, eds Marder, A. R. and Goldstein, J. I. (AIME, Warrendale), p. 125.
Kirkaldy, J. S., Thomson, B. A. and Baganis, E. A. (1978) in *Hardenability Concepts with Applications to Steel*, eds Kirkaldy, J. S. and Doane, D. V. (AIME, Warrendale), p. 82.
Koch, C. C., Cavin, O. B., McKamey, C. G. and Scarbrough, J. O. (1983) *Appl. Phys. Lett.*, **43,** 1017.
Kroupa, A. and Kirkaldy, J. S. (1993) *J. Phase Equilibria*, **14,** 150.
Lee, J.-L. and Bhadeshia, H. K. D. H. (1992) *China Steel Report*, **6,** 19.
Liu, Z.-K. (1994) in *Solid→Solid Phase Transformations*, eds Johnson, W. C., Howe, J. M., Laughlin, D. E. and Soffa, W. A. (TMS, Warrendale), p. 39.
Liu, Z.-K., Höglund, L., Jönsson, B. and Ågren, J. (1991), *Met. Trans. A*, **22A,** 1745.
Matsumiya, T., Yamada, W. and Koseki, T. (1993) in *Computer Aided Innovation of New Materials II*, eds Doyama, M. *et al.* (Elsevier Science, Amsterdam), p. 1723.
Miettinen, J. (1992) *Met. Trans. A*, **23A,** 1155.
Miedema, A. R (1976) *Phillips Tech. Rev.*, **36,** 217.

Morral, J. E., Jin, C., Engström, A. and Ågren, J. (1996) *Scripta Metall.*, **34,** 12.
Oehring, M. and Bormann, R. (1991) *Mat. Sci. Eng. A*, **A134,** 1330.
Ohnaka, I. (1986) *Trans. Iron Steel Inst. Japan*, **26,** 1045.
Pampus, K., Bøttiger, J., Torp, B., Schröder, H. and Samwer, K. (1987) *Phys. Rev. B*, **35,** 7010.
Pan, L.-M., Saunders, N. and Tsakiropoulos, P. (1989) **5,** 609.
Pfann, W. G. (1942) *Trans. AIME*, **194,** 70.
Polk, D. E., Calka, A. and Giessen, B. C. (1978) *Acta Met.*, **26,** 1097.
Ramachandrarao, P., Cantor, B. and Cahn, R. W. (1977) *J. Mat. Sci.*, **12,** 2488.
Samwer, K., Schroder, H. and Pampus, K. (1988) *Mat. Sci. Eng.*, **97,** 63.
Saunders, N. (1984) Ph.D. Thesis, University of Surrey, Guildford, UK.
Saunders, N. (1995) *Phil. Trans. A*, **351,** 543.
Saunders, N. (1996a) in *Titanium '95: Science and Technology*, eds Blenkinsop, P. *et al.* (Inst. Materials, London), p. 2167.
Saunders, N. (1996b) in *Superalloys 1996*, eds Kissinger, R. D. *et al.* (TMS, Warrendale, PA), p. 101.
Saunders, N. (1997a) in *Light Metals 1997*, ed. Huglen, R. (TMS, Warrendale, PA), p. 911.
Saunders, N. (1997b) in "Solidification Processing 1997" eds Beech, J. and Jones, H. (University of Sheffield, Sheffield, UK), p. 362.
Saunders, N. and Miodownik, A. P. (1983) *Ber. Bunsenges. Phys. Chem.*, **87,** 830.
Saunders, N. and Miodownik, A. P. (1985) *CALPHAD*, **9,** 283.
Saunders, N. and Miodownik, A. P. (1986) *J. Mater. Res.*, **1,** 38.
Saunders, N. and Miodownik, A. P. (1987) *J. Mater. Sci.*, **22,** 629.
Saunders, N. and Miodownik, A. P. (1988) *Mater. Sci. Tech.*, **4,** 768.
Saunders, N. and Miodownik, A. P. (1991) in *High Temperature–Low Density Powder Metallurgy*, eds Frazier, W. E., Koczak, M. J. and Lee, P. W. (ASM, Materials Park, OH), p. 3.
Saunders, N. and Tsakiropoulos, P. (1988) *Mat. Sci. Tech.*, **4,** 157.
Saunders, N., Miodownik, A. P. and Tanner, L. E. (1985) in *Rapidly Quenched Metals*, eds Steeb, S. and Warlimont, H. (Elsevier Science, Amsterdam, 1985), p. 191.
Scheil, E. (1942) *Z. Metallkde.*, **34,** 70.
Schwarz, R. B. (1988) *Mat. Sci. Eng.*, **97,** 71.
Schwarz, R. B. and Johnson, W. L. (1983) *Phys. Rev. Lett.*, **51,** 415.
Shao, G. and Tsakiropoulos, P. (1994) *Acta Met. Mater.*, **42,** 2937.
Small, C. (1993) private communication. Rolls-Royce plc, Derby, UK.
Spittle, J. A., Brown, S. G. R. and Al Meshhedani, M. (1995) presented at the 9th Int. Conf. on Numerical Methods in Thermal Problems, Atlanta, USA, July 1995.
Sugden, A. A. B. and Bhadeshia, H. K. D. B. (1989) *Mat. Sci. Tech.*, **5,** 977.
Sundman, B. (1991) in *User Aspects of Phase Diagrams*, ed. Hayes, F. H. (Inst. Metals, London), p. 130.
Tanaka, T., Aaronson, H. I. and Enomoto, M. (1995a) *Met. Mater. Trans. A*, **26A,** 535.
Tanaka, T., Aaronson, H. I. and Enomoto, M. (1995b) *Met. Mater. Trans. A*, **26A,** 547.
Tanaka, T., Aaronson, H. I. and Enomoto, M. (1995c) *Met. Mater. Trans. A*, **26A,** 560.
Turnbull, D. (1950) *J. Appl. Phys.*, **21,** 1022.
Uhlmann, D. R. (1972) *J. Non-Cryst. Sol.*, **7,** 337.
Wagner, C. (1951) in *Thermodynamics of Alloys* (Addison Wesley, Reading, Mass), p. 51.

Yamada, W. and Matsumiya, T. (1992) Nippon Steel Tech. Rpt. No. 52, January 1992, p. 31.
Yamada, W., Matsumiya, T. and Sundman, B. (1991) in *Computer Aided Innovation of New Materials*, eds Doyama, M., *et al.* (Elsevier Science, Amsterdam), p. 587.
Yukawa, N., Hida, M., Imura, T., Kawamura, K. and Mizuno, Y. (1972) *Met. Trans.*, **3**, 887.
Zener, C. (1946) *Trans. AIME*, **167**, 550.

Chapter 12

Future Developments

As summarised at the end of Chapter 2, CALPHAD has come of age. Complex phase equilibria calculations can be performed as a routine operation and, as a subject area, it is now at a watershed. In terms of thermodynamic modelling, many current ideas are based on concepts which are actually quite old, e.g. the sub-lattice model, the quasi-chemical model and the cluster variation method. Although there have been clear advances in programming, the current Gibbs energy minimisation packages utilise code whose main principles remain firmly based on historical mathematical concepts. Major advances in these areas are therefore unlikely to be achieved by modifications to existing models and software. It is more likely that CALPHAD will be invigorated in the new millennium by an expansion of CALPHAD methods into new fields and the discovery of vital new concepts.

As far as new applications are concerned, the existence of validated multi-component databases means that their output can, effectively, be considered as equivalent to adding further experimental data into larger projects. This will lead to application in wider areas of materials production and processing. Input into the field of materials design will also expand. For example, recent work by Olson (1997) shows how a traditional CALPHAD approach can be integrated into a modern systems approach, incorporating a variety of numerical treatments for materials properties, in order to design better ultra-high strength steels.

However, it is clear that, if the CALPHAD methodology is to be applied to materials with new combinations of components and where sound validated databases do not exist, its predictive capability must be extended. Otherwise, its future application will always have a significant, and potentially unacceptable, time lag associated with building up the necessary binary and ternary assessments. This emphasises the need to incorporate more results from First Principles Calculations into CALPHAD software programmes. The difficulties in achieving this should not be underestimated. However, if one looks back at the history of the subject area it is unlikely that the early researchers would have predicted that phase equilibria in highly complex multi-component materials could be estimated so accurately (and so quickly) as is currently possible. The advantages would be not only a better estimation of unknown thermodynamic parameters but, also, the associated generation of many other physical properties which are required in more comprehensive approaches to the design of new materials. This is a challenging project but the rewards will be proportionately great.

At present electron energy codes relate to relatively small arrays of atoms with fixed atomic placement in crystallographically distinct configurations. At the other end of the structural spectrum, better methods must be found which can look at random atomic placements on both crystallographically distinct *and* indistinct arrays. Amongst the major stumbling blocks to incorporating ab-initio methods into CALPHAD software is the current lack of any reliable treatment of the liquid phase, and the inability to properly account for the effect of temperature on the Gibbs energy. If such features cannot be included in the future, ab-initio approaches will remain of interest only as a source of data for ground state configurations and will not be capable of being directly used for high temperature phase equilibria.

Fundamentally based ordering models which consider short-range ordering phenomena are rarely incorporated in CALPHAD software, and this is another area which the authors would like to see expanded. However, before this can be done in a general fashion, it is clear that ordering models need to be made more flexible, while retaining a sound physical basis, and that vibrational entropy as well as complex compositional effects must be included.

In order to achieve these goals, the speed of calculation must be substantially increased. However, one has only to look at the huge leaps which have occurred in computing power over the last 10–15 years to appreciate that new theoretical foundations should be put in place now, in order to take advantage of any new advances in computing which will undoubtedly occur in the future. Codes based on principles of molecular dynamics have not been explored in any depth by the CALPHAD community; so perhaps it is also time to see what this subject area can offer.

In terms of more conventional modelling it is surprising to find that there is still incompatibility between databases that handle complex metallic materials and slags. Sound databases exist for both separately but as yet there is no major integrated database that can handle the very important area of overlap. In terms of a total materials modelling capability it is clearly of benefit to have one self-consistent database rather than a series of separate ones to span a greater range of applications.

The trend to directly incorporate codes for Gibbs energy minimisation into software packages which deal with kinetic and/or process modelling is certain to be a growth area. The DICTRA programme (Ågren 1992) marked one of the first attempts to incorporate a fully functional, generalised minimisation code into a software package which solved a series of complex diffusional equations. The advantages of such treatments are only just being appreciated by industry and the concept of extending such calculations still further to include the mobility of charge carriers in semiconductors and opto-electronic materials (Anderson 1997) is an exciting possibility. However, progress is being held back by the lack of an extensive database for such mobilities. The direct inclusion of CALPHAD software into codes that deal with process modelling has begun and Thermo-Calc and

References are listed on p. 465

ChemSage have developed a common interface with the direct aim of enabling such applications (Eriksson *et al.* 1994). Recently, the PMLFKT code, developed for the Lukas programmes (Lukas *et al.* 1982) and extended by Kattner *et al.* (1996), has been directly coupled with the complex finite element package ProCAST (Samonds and Waite 1993) to predict heat evolution and solidification paths during casting (Bannerjee *et al.* 1997).

The forthcoming years will offer a great deal of excitement if the CALPHAD method is extended and infused with new ideas and directions. For our part, we would like to see efforts continue which place CALPHAD on the soundest possible physical basis so that the semi-empirical nature of the subject area can be reduced. This, alongside the exciting new areas of application which are continually appearing, promises to provide the necessary scientific stimulus to keep alive the long pioneering tradition of the early workers in this field.

REFERENCES

Ågren, J. (1992) *ISIJ International*, **32**, 291.

Anderson, T. (1997) private communication at 1997 Ringberg III Workshop, Schloss Ringberg, Bavaria, Nov. 30–Dec. 5.

Bannerjee, D. K., Kattner, U. and Boettinger, W. (1997) *Solidification Processing 1997*, eds Beech, J. and Jones, H. (University of Sheffield, UK), p. 354.

Eriksson, G., Sippola, H. and Sundman, B. (1994) *A Proposal for a General Thermodynamic Software Interface*, Proceedings of The Colloquium on Process Simulation, ed. Jokilaakso, H. (Helsinki University of Technology, Finland), p. 67.

Kattner, U. R., Boettinger, W. J. and Coriell, S. R. (1996) *Z. Metallkde.*, **87**, 522.

Lukas, H. L., Weiss, J. and Henig, E.-Th. (1982) *CALPHAD*, **6**, 229.

Olson, G. (1997) *Science*, **277**, 1237.

Samonds, M. and Waite, D. (1993) *Mathematical Modelling for Materials Processing*, eds Cross, M. *et al.* (Oxford University Press, Oxford), p. 283.

Author index

Subject index

ALLOY INDEX

SYSTEMS INDEX

Binary Systems

9780080421292